Digital Signal Processing: Principles, Algorithms and System Design

Digital Signal Processing: Principles, Algorithms and System Design

Winser E. Alexander
Cranos M. Williams
North Carolina State University
NC, USA

AMSTERDAM • BOSTON • HEIDELBERG • LONDON • NEW YORK
OXFORD • PARIS • SAN DIEGO • SAN FRANCISCO • SINGAPORE
SYDNEY • TOKYO
Academic Press is an imprint of Elsevier

ACADEMIC
PRESS

Academic Press is an imprint of Elsevier
125 London Wall, London EC2Y 5AS, United Kingdom
525 B Street, Suite 1800, San Diego, CA 92101-4495, United States
50 Hampshire Street, 5th Floor, Cambridge, MA 02139, United States
The Boulevard, Langford Lane, Kidlington, Oxford OX5 1GB, United Kingdom

Library of Congress Cataloging-in-Publication Data
A catalog record for this book is available from the Library of Congress

British Library Cataloguing-in-Publication Data
A catalogue record for this book is available from the British Library

ISBN: 978-0-12-804547-3

For information on all Academic Press publications
visit our website at https://www.elsevier.com

Working together
to grow libraries in
developing countries

www.elsevier.com • www.bookaid.org

Publisher: Todd Green
Acquisition Editor: Steve Merken
Editorial Project Manager: Peter Jardim
Production Project Manager: Mohana Natarajan
Designer: Victoria Pearson

Typeset by VTeX

Contents

List of Tables

List of Figures

About the Authors

Winser E. Alexander is a Professor of Electrical and Computer Engineering at North Carolina State University. He served as Interim Provost and Vice-Chancellor for Academic Affairs (2011–2013) and as Interim Dean of the College of Engineering at North Carolina Agricultural and Technical (A&T) State University (2009–2011) while on leave of absence from North Carolina State University. He received the BS Degree in Electrical Engineering from North Carolina A&T State University and both the MS Degree in Engineering and the PhD in Electrical Engineering from the University of New Mexico. He served as an officer in the US Air Force (highest rank of captain), he was a Member of Technical Staff at Sandia Laboratories, Albuquerque, New Mexico, and he was previously Chair of the Department of Electrical Engineering at North Carolina A&T State University. His research interests include digital signal processing (DSP), genomic signal processing, parallel algorithms, and special purpose multiprocessor architectures for DSP. He has taught courses in DSP, DSP architecture, and fundamentals of logic systems design at North Carolina State University since 1982.

Dr. Alexander is a senior life member of IEEE, a member of Sigma Xi, and he is registered as a professional engineer in North Carolina.

Cranos M. Williams is an Associate Professor of Electrical and Computer Engineering at North Carolina State University. He received the BS Degree in Electrical Engineering from North Carolina A&T State University and both the MS and PhD degrees in Electrical Engineering from North Carolina State University. His research focuses on developing computational and analytical solutions for modeling and understanding the combinatorial interactions of biomolecular, physiological, and structural processes that impact plant growth, development, and adaptation. His research involves the development of methodologies familiar to other areas of electrical and computer engineering (e.g., computational intelligence, system identification, uncertainty propagation, experimental design, nonlinear systems analysis, control, and signal processing) to predict the impact that internal (genetic) and external (environmental) perturbations have on overall plant response (e.g., biomass, cellulose content, and plant cell wall strength). He has taught courses in analytical foundations of electrical and computer engineering, DSP, and analysis of nonlinear complex systems at North Carolina State University since 2008.

Dr. Williams is a member of IEEE.

Preface

Digital signal processing (DSP) has been applied to a very wide range of applications. This includes voice processing, image processing, digital communications, the transfer of data over the internet, image and data compression, etc. Engineers who develop DSP applications today, and in the future, will need to address many implementation issues including mapping algorithms to computational structures, computational efficiency, power dissipation, the effects of finite precision arithmetic, throughput, and hardware implementation. It is not practical to cover all of these in a single text. However, this text emphasizes the practical implementation of DSP algorithms as well as the fundamental theories and analytical procedures that form the basis for modern DSP applications.

This text provides an introduction to the principals of digital signal processing along with a balanced analytical and practical treatment of algorithms and applications for digital signal processing. It is intended to serve as a suitable text for a one semester junior or senior level undergraduate course. It is also intended for use in a following one semester first-year graduate level course in digital signal processing. It may also be used as a reference by professionals involved in the design of embedded computer systems, application specific integrated circuits or special purpose computer systems for digital signal processing, multimedia, communications, or image processing.

The following list gives an overview of the contents of the chapters of the text.

- Chapter 1 discusses the general concepts related to the practical application of digital signal processing algorithms and discrete time systems. The introductory concepts presented include the description of elementary time domain operations, sampling, filtering, and frequency representation of discrete time signals. It also includes an introduction to Matlab and gives examples for generating and plotting some fundamental discrete time sequences using Matlab.
- Chapter 2 discusses the fundamental concepts related to the use of discrete time signals and systems for digital signal processing applications. This includes topics such as the representation, classification, and transformation of discrete time signals, sampling continuous time signals to obtain discrete time signals, the discrete time impulse response, discrete time convolution, the representation of discrete time signals and systems using difference equations, Z Transforms, the properties of Z Transforms, system transfer functions, and poles and zeros for a system transfer function. Additional topics include regions of convergence for Z Transforms and stability analysis of system transfer functions. It covers inverse Z Transforms including Z Transforms with complex and multiple poles as well as the reconstruction of a discrete time signal to obtain the corresponding continuous time signal.
- Chapter 3 covers the frequency representation and frequency domain analysis of discrete time signals and systems. This includes the frequency domain representa-

tion of discrete time systems using the Discrete Fourier Transform and the Discrete Cosine Transform, frequency response estimates using pole–zero plots, convolution using the Discrete Fourier Transform, and frequency domain sampling. It covers applications such as filtering of discrete time sequences using the Discrete Fourier Transform, filtering of long data sequences, and algorithms for the Fast Fourier Transform.

- Chapter 4 covers several approaches for the design of discrete time systems using finite impulse response and infinite impulse response digital filters as examples. Generally, the concepts presented for the design of digital filters can be used to develop algorithms for the modeling or simulation of other discrete time signals and systems. It covers the design of FIR digital filters using the window based design approach and the frequency sampling approach. It covers the design of IIR digital filters using the impulse invariant design approach, the bilinear transformation, analog filter prototypes, and frequency transformations.

- Chapter 5 covers the development of computational structures for the implementation of discrete time systems. It presents several computational structures for the implementation of FIR and IIR digital systems as examples. This includes computational structures for the direct, transpose direct, cascade, lattice, and parallel forms for digital filters. It discusses partitioning the filters in second and fourth order sections, and simulation of the systems to verify performance and accuracy. It also presents the state space representation for discrete time systems.

- Chapter 6 covers methods used to represent numbers and the impact of the use of finite precision arithmetic for the implementation of discrete time systems. It discusses the representation of numbers using the IEEE floating point representation, computational errors due to rounding, and the multiplication of numbers that are represented using floating point. It covers the analytical basis for the two's complement representation of numbers and computational procedures for numbers represented using two's complement numbers. It covers the scaling of the coefficients for discrete time systems for given word sizes and for a restriction to avoid overflow during computations. It also presents a concept for statistical analysis of rounding errors due to word length effects.

- Chapter 7 covers fundamental concepts for changing the sample rate of discrete time signals. This includes multirate digital signal processing, frequency based interpolation using the Discrete Time Fourier Transform, polyphase digital filters, filter banks, and multilevel decomposition of digital filters.

- Chapter 8 covers concepts for the design of digital signal processing systems including the development of computational structures for the efficient implementation of discrete time systems. The concepts presented are intended to lead to computationally efficient algorithms for the implementation of discrete time systems on general purpose computer systems, on special purpose digital systems, or application specific digital systems.

- Chapter 9 presents concepts that can be applied to the design of discrete time systems to implement digital signal processing applications in hardware. It presents examples for the implementation of discrete time systems based upon the computational structures developed in Chap. 8. The concept presented involves the development of a Matlab model to verify the correctness of the algorithm using floating point computations, the development of Matlab models using the arithmetic to be used in the hardware design, and a model using a hardware description language, such as Verilog, to capture the design.

- Chapter 10 presents fundamental concepts for two-dimensional signal processing and gives an introduction to multidimensional signal processing. It covers some of the concepts for one-dimensional discrete time systems that can be readily extended to two-dimensional digital signal processing applications. It covers concepts related to the regions of support and stability analysis of two-dimensional discrete time systems, two-dimensional convolution, the two-dimensional Discrete Cosine Transform, and the filtering of large two-dimensional discrete time sequences using overlap–add and overlap–save procedures.

Introduction to Digital Signal Processing

1.1 INTRODUCTION

Advances in digital circuit and systems technology have had a dramatic impact on modern society related to the use of computer technology for many applications that affect our daily lives. These advances have enabled corresponding advances in digital signal processing (DSP) which have led to the use of DSP for many applications such as digital noise filtering, frequency analysis of signals, speech recognition and compression, noise cancellation and analysis of biomedical signals, image enhancement, and many other applications related to communications, television, data storage and retrieval, information processing, etc. [1].

A signal can be considered to be something that conveys information [2]. For example, a signal can convey information about the state or behavior of a physical system or a physical phenomena, or it can be used to transmit information across a communication media. Signals can be used for the purpose of communicating information between humans, between humans and machines, or between two or more machines. The information in a signal is represented as variations in the patterns for some quantity that can be manipulated, stored, or transmitted by a physical process [3]. For example, a speech signal can be represented as a function of time, and an image can be represented as a function of two spatial variables. The speech signal can be considered to be a one-dimensional signal because it has one independent variable, which is time. The image can be considered to be a two-dimensional signal because it has two independent variables such as width and height. It is common to use the convention of expressing the independent variable for one-dimensional signals as time, although the actual independent variable may not be time. This convention will generally be used in this text.

The independent variable(s) for a signal may be continuous or discrete. A signal is considered to be a continuous time signal if it is defined over a continuum of the independent variable. A signal is considered to be discrete time if the independent variable only has discrete values. The values of a discrete time signal are often quantized, for many practical applications, to obtain numbers that can be represented for use in a digital circuit or system. A quantized, discrete time, signal is considered to be a digital signal. Thus, if both the independent and dependent variables are only defined at discrete values, then the signal is considered to be a digital signal. Digital signals can be represented as a sequence of finite precision numbers.

Digital Signal Processing. DOI: 10.1016/B978-0-12-804547-3.00001-2

Signals play an important role in many activities in our daily lives. Signals such as speech, music, video, etc., are routinely encountered. A signal is a function of an independent variable such as time, distance, position, temperature, and pressure. For example, the speech and music we hear are signals represented by the air pressure at a point in space as a function of time. The ear converts the signal into a form that the brain can interpret. The video signal in a television consist of a sequence of images called frames and each frame can be considered to be an image. The video signal is a function of three variables: two spatial coordinates and time.

The independent variables such as time, distance, temperature, etc., for many of the signals we interact with daily, can be considered to be continuous. Signals with continuous independent variables are considered to be continuous time signals. Advances in computer and digital systems technology have made it practical to sample and quantize many of these signals and process them using digital circuits and systems for practical applications. The processing of signals using computers and other digital systems is called digital signal processing. Digital signal processing involves the sampling, quantization and processing of these signals for many applications including communications, voice processing, image processing, digital communications, the transfer of data over the internet, and various kinds of data compression.

Many applications that involve continuous time signals are implemented using digital signal processing. The continuous time signals are quantized and coded in digital format to be processed by digital circuits and systems. The output from these digital systems is then either stored for later use of converted to continuous time signals to meet the requirements of the application. There are many reasons why digital signal processing has become a cost effective approach to implement many applications including speech processing, video processing and transmission, transmission of signals over communications media, and data retrieval and storage. Some of these reasons follow [4]:

1. A programmable digital system provides the flexibility to configure a system for different applications. The processing algorithm can be modified by changing the system parameters or by changing the order of the operations through the use of software. Reconfiguring a continuous time system often means redesigning the system and changing or modifying its components.
2. Tolerances in continuous time or analog system components make it difficult for a designer to control the accuracy of the output signal. On the other hand, the accuracy of the output signal for a digital system is predictable and controllable by the type of arithmetic used and the number of bits used in the computations.
3. Digital signals can be stored in digital computers, on disks or other storage media, without the loss of fidelity beyond that introduced by acquiring the signal through some process such as converting a continuous time signal to a digital signal. Storage media for continuous time signals are prone to the loss of signal accuracy over time and/or to the addition of noise due to surroundings.
4. Digital implementation permits the easy sharing of a given processor among a number of signals by timesharing. Several digital signals can be combined, as

one, using multiplexing. The multiplexed signal can then be processed by a single processor as needed for a particular application. The corresponding individual outputs can then be separated from the output of the digital system with the results being the same as if the signals were processed by different systems. This permits the use of a single high speed digital system to process several different digital signals with relatively low sampling frequencies.

5. Digital signal processing can be used to easily process very low frequency signals such as seismic signals. Continuous time processing of these signals would require very large components such as large capacitors and/or large inductors.

6. The implementation cost of digital systems is often very low due to the manufacture of a large number of microprocessors or microchips with a single design. This has made it very cost effective to implement digital systems that can take advantage of being manufactured in large quantities.

7. Encryption can be used to provide security with digital signals. This is important for internet security as well as security for wireless communications and the protection of personal data.

There are some disadvantages associated with digital signal processing:

1. A digital signal processing system, for a particular application, is often more complicated than a corresponding analog signal processing system.

2. The upper frequency that can be represented for digital systems is determined by the sampling frequency. Thus, continuous time systems are still used for many high frequency applications.

3. Digital systems use active circuits that consume power. Analog systems can be designed that use passive circuits which can result is the design of a system that consumes less power than a corresponding digital system.

Discrete time signal processing is used in many applications considered to be in the category of information technology. Information technology includes such diverse subjects as speech processing, image processing, multimedia applications, computational engineering, visualization of data, database management, teleconferencing, remote operation of robots, autonomous vehicles, computer networks, simulation and modeling of physical systems, etc. Information technology, which is largely based upon the use of digital signal processing concepts, is essential for solving critical national problems in areas such as fundamental science and engineering, environment, health care, and government operations.

1.2 DETERMINISTIC AND RANDOM SIGNALS

A deterministic signal is a function of one or more independent variables such as time, distance, position, temperature, and pressure. It can be uniquely determined by a well-defined process such as a mathematical expression of one or more independent variables, or by table look up. For example,

$$s(t) = 3 \sin(2.1\pi t + 0.3198) \tag{1.1}$$

is a deterministic signal with independent variable t.

There are some important signals that cannot be uniquely represented by these methods, and therefore, they are not deterministic signals. Generally, speech is not considered to be a deterministic signal because it cannot be described functionally by a mathematical expression. However, a recorded segment of speech can be represented, to a high degree of accuracy, as the sum of several sinusoids of different amplitudes and frequencies such as [4]

$$s(t) = \sum_{k=1}^{N} A_k(t) \sin[2\pi F_k(t)t + \theta_k(t)] \tag{1.2}$$

where $A_k(t)$ represents the amplitude of sinusoid k at time t, $F_k(t)$ represents the frequency of sinusoid k at time t, and $\theta_k(t)$ represents the phase of sinusoid k at time t.

A signal that is determined in a random way and cannot be predicted ahead of time is a random signal. Statistical approaches are often used to analyze random signals. This text primarily covers deterministic signals.

1.3 MATHEMATICAL REPRESENTATION OF SIGNALS

The following is an example of a continuous time signal with one independent variable:

$$x(t) = 2t^2 + 3t + 5. \tag{1.3}$$

A new function can be formed by replacing each instance of the independent variable by a given function. For example, each t in Eq. (1.3) can be replaced by $2t$ to obtain

$$\begin{aligned} g(t) &= x(2t) = 2(2t)^2 + 3(2t) + 5 \\ &= 8t^2 + 6t + 5. \end{aligned} \tag{1.4}$$

This procedure can be used to easily derive many signals of interest from a few basic signals.

A signal involving more than one independent variable is often of interest. An image is a good example of this type of signal. For example, an image segment can be represented as

$$g(x, y) = \sum_{k_1=0}^{K_1} \sum_{k_2=0}^{K_2} A_{k_1,k_2}(x, y) \sin\left[2\pi F_{k_1,k_2}(x, y) + \theta_{k_1,k_2}(x, y)\right]. \tag{1.5}$$

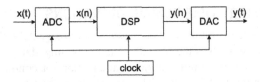

FIGURE 1.1

Conceptual block diagram of a typical system

The image in Eq. (1.5) is represented as a finite sum of two-dimensional sinusoidal signals, in this case $A_{k_1,k_2}(x, y)$ represents the magnitude of the two-dimensional sinusoidal signal, $F_{k_1,k_2}(x, y)$ represents the frequency of the two-dimensional sinusoidal signal, and $\theta_{k_1,k_2}(x, y)$ represents the phase of the two-dimensional sinusoidal signal. This is an example of a two-dimensional signal because it has two independent variables, x and y. A signal with several independent variables would be called a multidimensional signal.

A discrete time signal $x(n)$ is a function of an independent variable that only has discrete values. It is important to note that a discrete time signal is not defined at instants between two successive samples. Thus, the discrete time signal $x(n)$ is not defined at $n = 0.5$, $n = 1.5$, $n = 2.5$, etc.

1.4 TYPICAL SIGNAL PROCESSING OPERATIONS

Most of the signals encountered in science and engineering are continuous time signals or functions of one or more continuous independent variables. We are typically interested in the case where

1. The continuous time signal is converted to a digital signal (analog to digital converter – ADC)
2. The digital signal is processed with a digital system (such as a digital signal processor – DSP) to meet the requirements of some practical application
3. The digital output signal is converted to a continuous time output signal (digital-to-analog converter – DAC).

Many applications involve both continuous and discrete time signals. In a typical application, a sensor or detector is used to obtain a continuous time signal representing speech, temperature, light intensity, etc. An ADC is then used to convert the continuous time signal to a digital signal. A digital system is then used to process the signal and finally, a DAC is used to convert the output digital signal to a continuous time signal for use in the application. A conceptual representation of such a system is given in Fig. 1.1.

1.4.1 ELEMENTARY TIME DOMAIN OPERATIONS

The three most basic operations for processing digital signals in the domain of the independent variable, such as the time domain, are scaling, delay, and addition. The scaling operation involves amplification or attenuation for continuous time signals and multiplication for digital signals. This operation can be represented as

$$y(t) = \alpha x(t) \tag{1.6}$$

for a continuous time system and by

$$y(n) = \alpha x(n) \tag{1.7}$$

for a discrete time system where n is the sample number.

The delay operation generates a signal that is a delayed replica of the original signal. This can be represented by

$$y(t) = x(t - t_0) \tag{1.8}$$

for a continuous time signal where the signal is delayed by the amount t_0. The corresponding representation for the discrete time system is

$$y(n) = x(n - m) \tag{1.9}$$

where m and n are integers and the signal is delayed by m samples.

Many applications involve two or more signals to generate a new signal. For example,

$$y(t) = x_1(t) + x_2(t) + x_3(t) \tag{1.10}$$

is a signal generated by the addition of three continuous time signals. Similarly,

$$y(n) = x_1(n) + x_2(n) + x_3(n) \tag{1.11}$$

is a signal generated by the addition of three discrete time signals.

1.4.2 FILTERING

The filter operation is one of the most widely used signal processing operations. It involves passing a certain range of frequencies in the signal to the output and blocking the others. The range of frequencies allowed to pass is called the pass band and the range of frequencies blocked is called the stop band. For example, an ideal low pass filter passes all frequencies below a specified cutoff frequency and blocks all frequencies above the specified cutoff frequency. An ideal high pass filter passes all frequencies above a specified cutoff frequency and blocks all frequencies below the specified cutoff frequency. An ideal band pass filter passes all frequencies between two specified cutoff frequencies and blocks all other frequencies. An ideal band stop filter blocks all frequencies between two specified cutoff frequencies and passes all other frequencies.

A signal may get corrupted unintentionally by an interfering signal, or noise may be added to it by some natural phenomena. In such cases, the desired signal can be recovered by filtering if the range of frequencies for the interfering signal or noise is different from those occurring in the original signal. In practical cases, there is usually some overlap and the original signal can only be partially recovered.

1.5 SIGNAL PROCESSING WITH MATLAB

Matlab, which is a product of The Mathworks, Inc., is a high level language and interactive environment that facilitates the exploration and visualization of ideas across disciplines including signal processing, image processing, communications, control systems, and computational finance. Matlab can often be used to solve technical computing problems faster than with the use of traditional programming languages, such as C, C++, and Fortran. Matlab provides a DSP System Toolbox that implements the common algorithms used for the design, analysis, and implementation of linear, shift invariant, discrete time systems. It also provides an extensive help utility that includes documentation on the use of Matlab, help documentation on the use of Matlab functions, and many examples of the use of Matlab to solve problems. Many of the examples, presented in this text, will take advantage of the Matlab environment to facilitate the discussion on concepts for the design, analysis, and implementation of linear, shift invariant, discrete time systems.

Matlab is easy to learn and it can easily be used to develop a simulation for a new DSP algorithm. Matlab has a very useful "help" utility. Help can be obtained for a particular function by using the help utility at the Matlab prompt. For example, help on the Matlab function to generate a sequence of ones can be obtained by using the following command in the Matlab command window:

```
help ones
```

The response is

```
ones    Ones array.
    ones(N) is an N-by-N matrix of ones.

    ones(M,N) or ones([M,N]) is an M-by-N matrix of ones.

    ones(M,N,P,...) or ones([M N P ...]) is an M-by-N-by
    -P-by-... array of ones.

    ones(SIZE(A)) is the same size as A and all ones.

    ones with no arguments is the scalar 1.

    ones(..., CLASSNAME) is an array of ones of class
    specified by the string CLASSNAME.
```

```
ones(..., 'like', Y) is an array of ones with the
same data type, sparsity, and complexity (real or
complex) as the numeric variable Y.

Note: The size inputs M, N, and P... should be
nonnegative integers.
Negative integers are treated as 0.

Example:
    x = ones(2,3,'int8');

See also eye, zeros.

Overloaded methods:
    distributed/ones
    codistributor2dbc/ones
    codistributor1d/ones
    codistributed/ones
    gpuArray/ones

Reference page in Help browser
    doc ones
```

Many other Matlab functions will be used during this course including the following functions:

1. rand
2. filter
3. conv
4. fir1
5. butter
6. cheby1
7. cheby2
8. ellip
9. residuez
10. zplane

Information on the use of these functions can be obtained by using the Matlab help utility.

The indices for arrays and sequences within Matlab must be positive, finite integers. Therefore, it is a good idea to generate the corresponding time sequence to use with plots instead of using the indices for the sequence for the plotted values on the x axis. The "stem" function has been provided within Matlab to plot discrete time sequences. It plots the data sequence Y as stems from the X axis terminated with circles for the data values.

The Matlab functions "ones" can be used to generate a discrete time step function and the Matlab function "zeros" can be used in addition to the "ones" function to generate a delayed step function, an impulse function or a delayed impulse function. The following examples illustrate the use of Matlab to generate discrete time sequences.

EXAMPLE 1.1
(Step Functions)

Problem:
Use the appropriate Matlab "ones" functions and "zero" functions to generate the following sequence. Use the "stem" function to plot the sequence

$$x(n) = \begin{cases} 0 & \text{for} \quad 0 \leq n \leq 4, \\ 1 & \text{for} \quad 5 \leq n \leq 19, \\ 0 & \text{for} \quad 20 \leq n \leq 29, \\ -1 & \text{for} \quad 30 \leq n \leq 44, \\ 0 & \text{for} \quad 45 \leq n \leq 49. \end{cases} \tag{1.12}$$

Solution:
The following Matlab script can be used to generate and plot this sequence.

Matlab Script 1.1.

```
% Matlab Script for Example 1.1
tn=0:49;
xn=[zeros(1,5),ones(1,15),zeros(1,10),-ones(1,15),zeros(1,5)];
figure(1);
stem(tn,xn);
xlabel('Sample Numbers');
ylabel('Magnitude');
axis([(min(tn)-0.5) (max(tn)+0.5) (min(xn)-0.1) (max(xn)+0.1)]);
grid on;
```

End of the Script

 Fig. 1.2 gives the resulting plot.
End of the Example

EXAMPLE 1.2
(A Cosine Sequence)

Problem:
Generate 60 samples of the cosine signal in Eq. (1.13) using a normalized sampling interval of 1. Use the Matlab "stem" function to plot the sequence.

$$x(n) = 3.0\cos(0.221\pi n - 0.65\pi); \tag{1.13}$$

Solution:
The following Matlab script can be used to generate and plot the sequence.

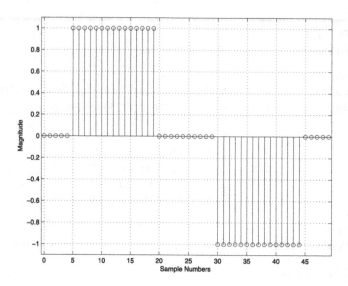

FIGURE 1.2

Stem plot for sequence in Example 1.1

Matlab Script 1.2.

```
% Matlab Script for Example 1.2
tn=0:59;
xn=3*cos(0.221*pi*tn-0.65*pi);
H = gcf;
figure(H+1)
stem(tn,xn);
xlabel('Sample Numbers');
ylabel('Magnitude');
axis([(min(tn)-0.5) (max(tn)+0.5) (min(xn)-0.3) (max(xn)+0.3)]);
grid on;
```

End of the Script

   Fig. 1.3 gives the resulting plot.

End of the Example

EXAMPLE 1.3
(A Delayed Impulse)

Problem:

Assume that a sequence is a digital impulse delayed by 25 samples. Plot 35 samples of this sequence using the Matlab "stem" function.

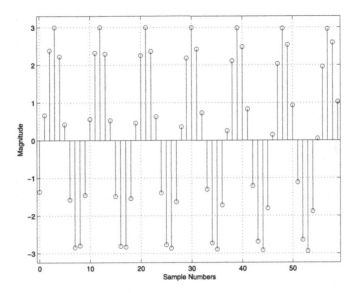

FIGURE 1.3

Stem plot of cosine sequence for Example 1.2

Solution:

The following Matlab script can be used to generate and plot this sequence.

Matlab Script 1.3.

```
% Matlab Script for Example 1.3
tn=0:34;
xn=[zeros(1,25), ones(1,1), zeros(1,9)];
H = gcf;
figure(H+1)
stem(tn,xn);
axis([(min(tn)-0.5) (max(tn)+0.5) (min(xn)-0.1) (max(xn)+0.1)]);
xlabel('Sample Numbers');
ylabel('Magnitude');
grid on;
```

End of the Script

Fig. 1.4 gives the resulting plot.

End of the Example

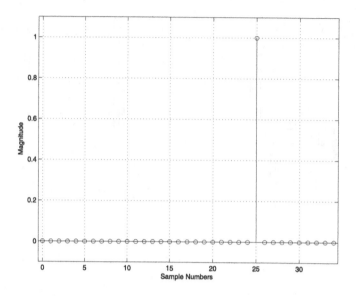

FIGURE 1.4

Stem plot of a delayed impulse for Example 1.3

EXAMPLE 1.4
(Sinusoidal Signal)

Problem:
Matlab script 1.4 generates and plots a section of a sinusoidal signal. The sample interval is small enough that the Matlab "plot" function essentially represents a continuous time function. Derive a formula for the corresponding continuous time sinusoidal signal.

Matlab Script 1.4. (Sinusoidal Signal)

```
% Matlab Script for Example 1.4
dt = 1/450;
tt =   -1: dt: 1;
Fo = 3.75;
xx = 5*sin(2*pi*(Fo*tt - 0.65*pi));
H = gcf;
figure(H+1);
plot(tt, xx),
xstrt = min(tt)-0.1;
xend = max(tt) + 0.1;
ybot = min(xx) - 0.1*abs(min(xx));
ytop = max(xx) + 0.1*abs(max(xx));
axis([xstrt xend ybot ytop]);
```

```
xlabel('Time (sec)')
ylabel('Magnitude');
grid on;
```

End of the Script

Solution:
Fig. 1.5 gives the output plot from the Matlab script. The following parameters can be determined from the Matlab script:

1. The continuous time frequency $\Omega = 2(3.75)\pi = 7.5\pi$.
2. The sampling interval $T = \frac{1}{450} = 0.0022$.
3. The normalized discrete time radial frequency $\omega = \frac{2(3.75)\pi}{450} = \frac{7.5\pi}{450} = 0.0524$.
4. The phase angle $\theta = 2(-0.65)\pi = -1.30\pi = -4.0841$.

Since $\omega < \pi$, there is no aliasing of the signal. Thus, the sinusoidal signal given in the Matlab script is

$$xx(n) = 5\sin(0.0524n - 4.0841). \tag{1.14}$$

Since the sinusoidal signal was sampled at 450 samples per second, the corresponding continuous time signal

$$xx(t) = 5\sin(7.5\pi t - 4.0841). \tag{1.15}$$

End of the Example

EXAMPLE 1.5
(Complex Exponential Signals)

Problem:
Define $s(t)$ as

$$s(t) = 10e^{j\frac{3\pi}{5}}e^{j13\pi t}. \tag{1.16}$$

Use Matlab to make a plot of $s_r(t) = \Re e\{s(t)\}$. Pick a range of values that will include exactly four periods of the signal. Use a total of 200 samples for the plot.

Solution:
The requirement is to plot $s_r(t) = \Re e\{s(t)\}$ over exactly four periods, so the period of $s_r(t)$ should be determined first:

$$2\pi f = 13\pi. \tag{1.17}$$

Thus,

$$T = \frac{1}{f} = \frac{2}{13}. \tag{1.18}$$

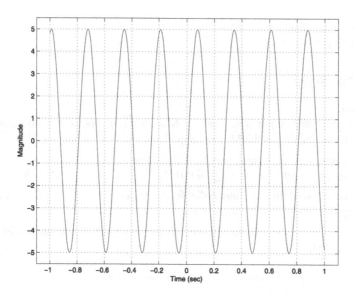

FIGURE 1.5

Plot of section of a sinusoidal signal for Example 1.4

Since we know the time interval and the function, we can plot it using Matlab. The following Matlab script will accomplish this:

Matlab Script 1.5. (Complex Exponential)

```
% Matlab Script for Example 1.5
th=0.6*pi;
per=2/13;
dt=4*per/200;      %time interval to plot the function.
tt=(0:199)*dt;
si=10*real(exp(j*(13*pi*tt+th)));
H = gcf;
figure(H+1);
plot(tt,si);
% title('Plot of the Real Part of a Complex Exponential');
xlabel('Time(Seconds)');
ylabel('Magnitude');
axis([0.0 0.6154 -10.1 10.1]);
grid on;
```

End of the Script

Fig. 1.6 gives the resulting figure.

End of the Example

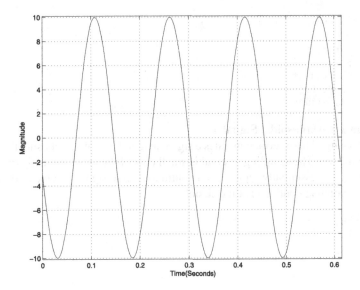

FIGURE 1.6

Plot of the real part of a complex exponential in Example 1.5

1.6 PROBLEMS FOR CHAPTER 1

Problem 1.1. (Complex Numbers)
Simplify the following complex-valued expressions and give the answers in both Cartesian form $((x + jy))$ and polar form $((re^{j\theta}))$:

1. $7e^{j\frac{2\pi}{3}} + 3e^{-j(\frac{3\pi}{4})}$
2. $(\sqrt{7} + j3)^5$
3. $(\sqrt{11} - j7)^{-3}$
4. $(\sqrt{6} + j9)^{\frac{1}{3}}$
5. $\Im m\{je^{-j\frac{\pi}{7}} + e^{j\frac{2\pi}{5}}\}$

1.7 MATLAB PROBLEMS FOR CHAPTER 1

Problem 1.2. (Sinusoidal Signals)
The following Matlab script generates a signal and makes a plot of it. The sampling interval is small enough that the plot using the Matlab "plot" function essentially represents a continuous time signal. Derive a formula for the corresponding continuous time signal.

```
dt = 1/200;
tt = -1: dt :1;
Fo = 3;
```

```
xx = 20*cos(2*pi*Fo*(tt - 0.35));
%
plot(tt, xx),
grid on;
title('SECTION of a SINUSOID');
xlabel('TIME (sec)');
ylabel('Magnitude');
```

Problem 1.3. (Sinusoidal Signals)
The following Matlab script generates a signal and makes a plot of it using the Matlab
"plot" function. The sampling interval is small enough that the plot using the Matlab
"plot" function essentially represents a continuous time signal. Derive a formula for
the corresponding continuous time signal.

```
dt = 1/400;
tt = -1: dt :1;
Fo = 4.5;
xx = 20*sin(2*pi*Fo*(tt - 0.65));
plot(tt, xx);
grid on;
title('SECTION of a SINUSOID');
xlabel('TIME (sec)');
ylabel('Magnitude');
```

Problem 1.4. (Complex Exponential Signals)
Define $s(t)$ as

$$s(t) = 10e^{j\frac{3\pi}{4}}e^{j23\pi t}. \tag{1.19}$$

1. Use the Matlab "plot" function to make a plot of the real part of $s(t)$, $r(t) = \Re\{s(t)\}$. Use a sampling interval $T = 0.05$. Pick a range of values that will include exactly four periods of the signal.
2. Use Matlab to make a plot of the imaginary part of $s(t)$, $q(t) = \Im\{s(t)\}$. Again, use a sampling interval $T = 0.05$. Pick a range of values that will include exactly four periods of the signal.

Problem 1.5. (Representation of Sinusoidal Signals)
Define $x(t)$ as

$$x(t) = 5\cos\left(\omega_0 t + \frac{3\pi}{5}\right) + 7\cos\left(\omega_0 t - \frac{2\pi}{3}\right) + 4\cos\left(\omega_0 t + \frac{\pi}{4}\right) \tag{1.20}$$

Express $x(t)$ in the form $x(t) = A\cos(\omega_0 t + \phi)$ by finding the numerical values of
A and ϕ.

Problem 1.6. (Generating Sequences)
Use the appropriate Matlab functions to generate the following sequences. Use the
"stem" function to plot each of these. Turn in your Matlab script and the resulting
plots.

1.

$$x(n) = \begin{cases} 0 & \text{for} & 0 \le n \le 19, \\ 1 & \text{for} & 20 \le n \le 39, \\ 0 & \text{for} & 40 \le n \le 49, \\ -1 & \text{for} & 50 \le n \le 69, \\ 0 & \text{for} & 70 \le n \le 79. \end{cases} \qquad (1.21)$$

2. Plot 40 values of

$$x(n) = 5.0 * \sin(0.37\pi n - 0.47\pi). \qquad (1.22)$$

Problem 1.7. (Generating Sequences)
Use the appropriate Matlab functions to generate the following sequences. Use the "stem" function to plot each of these. Turn in your Matlab script and the resulting plots.

1.

$$x(n) = \begin{cases} 0 & \text{for} & 0 \le n \le 9, \\ -1 & \text{for} & 10 \le n \le 24, \\ 0 & \text{for} & 25 \le n \le 34, \\ 1 & \text{for} & 35 \le n \le 49, \\ 0 & \text{for} & 50 \le n \le 59. \end{cases} \qquad (1.23)$$

2. Generate 60 values of

$$x(n) = 3.0 \cos(0.421\pi n - 0.65\pi). \qquad (1.24)$$

3. Assume that a sequence is a digital impulse delayed by 25 samples. Plot 35 samples of this sequence.

4. Assume that a sequence is a step function delayed by 20 samples. Plot 40 samples of this sequence.

Fundamental DSP Concepts

2.1 INTRODUCTION

This chapter covers several important fundamental concepts that are important to the study of discrete time signals and systems. It is generally acceptable to consider time to be the independent variable during the discussion of discrete time signals and systems with one independent variable. However, the concepts developed, using this approach, can be generalized to accommodate other independent variables such as length, altitude, etc. Discrete time signals are frequently obtained by sampling a continuous time signal at regular intervals. The sampling of a speech signal at regular intervals to obtain 8000 individual samples for each second of speech is an example of a discrete time signal. The samples occur at intervals of 125 μs and the magnitude can have a continuous range of values. The samples are often quantized using an analog to digital (A/D) converter for practical applications. The magnitudes of the samples can only have a finite number of possible values after the samples have been quantized. The number of possible values depends upon the number of bits used for the quantization. The resulting sampled and quantized signal is called a digital signal.

This chapter also covers some details related to the characteristics of discrete time sinusoidal signals. The sinusoidal signal is an important elementary signal that can be used as a basic building block to construct more complex signals. Fundamental building block signals, such as the sinusoidal signal, can be considered to be basis functions that can be used to represent more complicated signals. There are other elementary signals that are important to signal processing, and this chapter covers some of these signals as well.

This chapter considers the characterization of discrete time systems with emphasis on the class of linear, time invariant (LTI) systems. The term linear, shift invariant (LSI) is a more general term because it represents a system that does not change its characteristics in response to shifts in the independent variable. Thus, an LTI system is a special case of an LSI system with time as the independent variable. A study of LTI systems is important because a large number of important practical systems are either LTI systems or they can be approximated or represented as LTI systems. In addition, a large number of mathematical techniques can be applied to model and analyze LTI systems [4]. This chapter covers many of these mathematical techniques.

Digital Signal Processing. DOI: 10.1016/B978-0-12-804547-3.00002-4

Table 2.1 A tabular representation of a discrete time signal

n	0	1	2	3	4	5	6
$x(n)$	−3	−1	−2	5	0	4	−1

2.2 REPRESENTING DISCRETE TIME SIGNALS

It is convenient to represent the values of the independent variable, for a discrete time signal, in terms of integer multiples of the sampling interval. For example, the sampling interval for a speech signal sampled at 8000 samples per second would be $T = 125$ μs. If the first sample occurs at time $t = 0$, the second at $t = T$, the third at $2T$, etc., then the time of each sample can be determined from its index n. The first sample occurs at $n = 0$, the second sample at $n = T$, the third sample at $n = 2T$, etc. Thus, it is convenient to represent the value of the independent variable using the sampling interval, T, along with the index for the sample, n. For example, if a continuous time signal $x(t)$ was sampled using a sampling interval T, then the individual samples can be represented as $x(nT)$. This notation is typically shortened to the form $x(n)$ when the sampling is performed at regular intervals. The independent variable is therefore represented as the variable n using this convention. This representation implies a normalization of the sampling interval to $T = 1$ units. The impact of this normalization will be discussed later in this chapter.

There are several ways to represent a discrete time signal. Some of these ways are given below:

1. A table as given in Table 2.1 and shown in Fig. 2.1,
2. A functional representation as given in Eq. (2.1) and also shown in Fig. 2.2,

$$x(n) = \begin{cases} 0 & \text{for } n < 0, \\ (0.6)^n & \text{for } 0 \le n \le 10, \\ 0 & \text{for } n > 10. \end{cases} \tag{2.1}$$

3. A sequence representation as given in Eq. (2.2) and shown in Fig. 2.3,

$$x(n) = \left\{ -2, -1, 1, \; 5, \; 3, -1, -3 \atop \uparrow \right\} \tag{2.2}$$

The up-arrow (↑) in Eq. (2.2) indicates the sample for $n = 0$.

2.3 ELEMENTARY DISCRETE TIME SIGNALS

There are several elementary discrete time signals that can be conveniently used to represent more complicated discrete time signals. The unit impulse, the unit step, and the unit ramp are among these. In addition, many of the discrete time signals

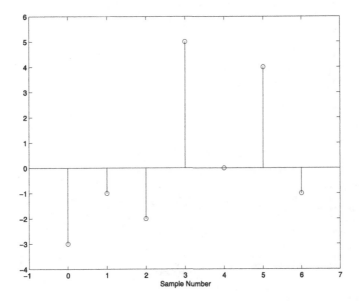

FIGURE 2.1

Example of a discrete time signal as represented by Table 2.1

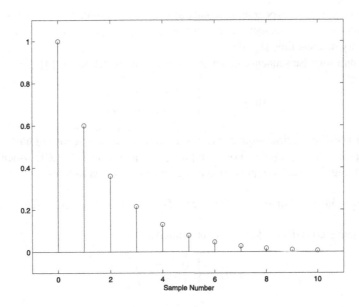

FIGURE 2.2

Example of a discrete time signal represented by a functional representation

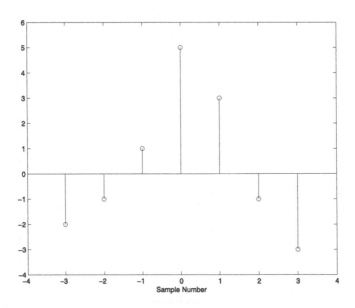

FIGURE 2.3

Example of a discrete time signal represented by a sequence representation

of interest either are exponential signals or can be represented as a combination of exponential signals. This section presents definitions and examples of the use of these elementary discrete time signals.

The unit impulse sequence is denoted as $\delta(n)$ and is defined as [4]

$$\delta(n - m) = \begin{cases} 1 & \text{for} \quad n = m, \\ 0 & \text{for} \quad n \neq m. \end{cases} \qquad (2.3)$$

An arbitrary discrete time sequence can be represented as a sequence of unit impulses with appropriate magnitudes. For example, the sequence in Table 2.1, which is also shown in Fig. 2.1, can be represented using unit impulses as follows:

$$s_1(n) = -3\delta(n) - \delta(n - 1) - 2\delta(n - 2) + 5\delta(n - 3) + 4\delta(n - 5) - \delta(n - 6). \quad (2.4)$$

The unit step is denoted $u(n)$ and is defined as

$$u(n - m) = \begin{cases} 0 & \text{for} \quad n < m, \\ 1 & \text{for} \quad n \geq m. \end{cases} \qquad (2.5)$$

Note that the unit step sequence has a value of 1.0 if the argument $(n - m)$ is zero or positive and it has a value of zero if the argument is negative. Example 2.1 illustrates the utilization of the unit step sequence to represent a discrete time signal.

EXAMPLE 2.1
(Unit Step Sequence)

Problem:
A discrete time sequence is defined by Eq. (2.6),

$$s_2(n) = \begin{cases} 0 & \text{for } n < 2, \\ 3.0 & \text{for } 2 \le n \le 10, \\ 0 & \text{for } n > 10. \end{cases} \qquad (2.6)$$

Show how to represent this sequence using unit step sequences.
Solution:
The sequence, $s_2(n)$, as given in Eq. (2.6) can be represented as follows using unit step sequences:

$$s_2(n) = 3.0u(n-2) - 3.0u(n-11). \qquad (2.7)$$

Alternatively, $s_2(n)$ can be represented as

$$s_2(n) = 3.0u(n-2)u(-n+10). \qquad (2.8)$$

Note that

$$u(n-2) = \begin{cases} 0 & \text{for } n < 2, \\ 1 & \text{for } n \ge 2 \end{cases} \qquad (2.9)$$

and

$$u(-n+10) = \begin{cases} 1 & \text{for } n \le 10, \\ 0 & \text{for } n \ge 11. \end{cases} \qquad (2.10)$$

Thus, the product $3u(n-2)u(-n+10)$ gives the correct values for the sequence.
Fig. 2.4 gives a stem plot of the sequence $s_2(n)$.
End of the Example

The unit ramp can be denoted $u_r(n)$ and is defined as

$$u_r(n-m) = \begin{cases} 0 & \text{for } n < m, \\ (n-m) & \text{for } n \ge m. \end{cases} \qquad (2.11)$$

Example 2.2 illustrates the use of the unit ramp to represent a discrete time sequence.

EXAMPLE 2.2
(Unit Ramp Sequence)

Consider the discrete time signal $s_3(n)$ which has been represented in segments using unit ramp sequences,

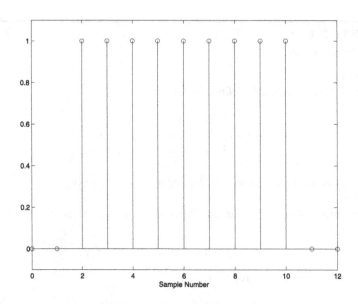

FIGURE 2.4

Stem plot of a sequence using unit step sequences

$$s_3(n) = 2.0u_r(n) - 2.0u_r(n-12) - 2.0u_r(n-24) + 2.0u_r(n-36), \quad (2.12)$$

$$u_r(n) = \begin{cases} 0 & \forall n < 0, \\ n & \forall n \geq 0, \end{cases} \quad (2.13)$$

$$u_r(n-12) = \begin{cases} 0 & \forall n < 12, \\ n-12 & \forall n \geq 12, \end{cases} \quad (2.14)$$

$$u_r(n-24) = \begin{cases} 0 & \forall n < 24, \\ n-24 & \forall n \geq 24, \end{cases} \quad (2.15)$$

$$u_r(n-36) = \begin{cases} 0 & \forall n < 36, \\ n-36 & \forall n \geq 36. \end{cases} \quad (2.16)$$

The individual contributions to $s_3(n)$ can be added to obtain

$$s_3(n) = \begin{cases} 0 & \forall & n < 0, \\ 2(n) & \forall & 0 \leq n \leq 12, \\ 24 & \forall & 12 \leq n \leq 24, \\ 24 - 2(n-24) & \forall & 24 \leq n \leq 36, \\ 0 & \forall & n \geq 36. \end{cases} \quad (2.17)$$

Fig. 2.5 gives a stem plot of $s_3(n)$.

End of the Example

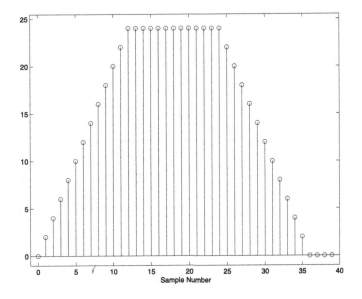

FIGURE 2.5

Stem plot of sequence using the unit ramp

The exponential sequence can be defined as

$$c^n u(n) = \begin{cases} 0 & \text{for} \quad n < 0, \\ c^n & \text{for} \quad n \geq 0. \end{cases} \tag{2.18}$$

Exponential signals, which can be either real or complex, are very important for digital signal processing applications. Consider the case where the constant c has a complex value such that

$$c = \alpha + j\beta. \tag{2.19}$$

The constant, c, can also be represented in the form

$$c = re^{j\theta} \tag{2.20}$$

where

$$r = \sqrt{\alpha^2 + \beta^2},$$

$$\theta = \tan^{-1}\left(\frac{\beta}{\alpha}\right). \tag{2.21}$$

It follows that if c is complex, as given in Eq. (2.19), then a sequence

$$x(n) = c^n \tag{2.22}$$

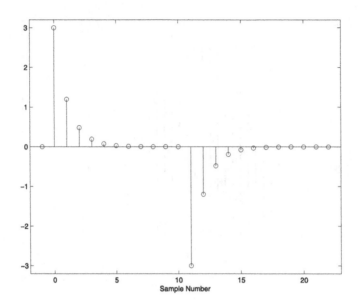

FIGURE 2.6

Stem plot of sequence using exponential signals

can be written as

$$x(n) = r^n (e^{j\theta})^n = r^n e^{jn\theta} = r^n \left[\cos(n\theta) + j \sin(n\theta) \right]. \qquad (2.23)$$

Example 2.3 gives an exponential sequence with the parameter c being real.

EXAMPLE 2.3
(Real Exponential Sequence)

Consider the example where $c = 0.4$, which is real, and

$$s_4(n) = 3(0.4)^n u(n)u(-n + 10) - 3(0.4)^{n-10} u(n - 11)u(-n + 22). \quad (2.24)$$

This example uses the unit step to start and end the different segments of the signal. The unit step sequence $u(n)$ is a right sided sequence that starts at $n = 0$ and ends at $n = \infty$. The unit step sequence $u(-n + 10)$ is a left sided sequence that starts at $n = -\infty$ and ends at $n = 10$. The product of the two step sequences starts at $n = 0$ and ends at $n = 10$. In a similar way, the product $u(n - 11)u(-n + 22)$ starts at $n = 11$ and ends at $n = 22$. Fig. 2.6 gives a stem plot of $s_4(n)$.
End of the Example

Example 2.4 gives an exponential sequence with the parameter a being complex.

EXAMPLE 2.4
(Complex Exponential Sequence)
Consider a discrete time sequence

$$
\begin{aligned}
s_5(n) \;=\;& 3\left[e^{(-0.4+0.7j)n} + e^{(-0.4-0.7j)n}\right]u(n)u(-n+10) \\
&- 3e^{(-0.4+0.7j)(n-10)}u(n-11)u(-n+22) \\
&+ 3e^{(-0.4+0.7j)(n-10)}u(n-11)u(-n+22).
\end{aligned} \tag{2.25}
$$

This sequence has two complex constants

$$
\begin{aligned}
c_1 &= e^{(-0.4+0.7j)}, \\
c_2 &= e^{(-0.4-0.7j)}, \\
c_1 &= c_2^*.
\end{aligned} \tag{2.26}
$$

Define

$$
r_1(n) = 3\left[e^{(-0.4+0.7j)n} + e^{(-0.4-0.7j)n}\right]u(n)u(-n+10) \tag{2.27}
$$

and

$$
\begin{aligned}
r_2(n) \;=\;& -3e^{(-0.4+0.7j)(n-10)}u(n-11)u(-n+22) \\
&+ 3e^{(-0.4+0.7j)(n-10)}u(n-11)u(-n+22). \tag{2.28}
\end{aligned}
$$

It follows that

$$
s_5(n) = r_1(n) + r_2(n). \tag{2.29}
$$

Note that $r_1(n)$ is real because $e^{(-0.4+0.7j)n}$ and $e^{(-0.4-0.7j)n}$ are complex conjugates. Thus, $r_1(n)$ begins at $n=0$ and ends at $n=10$ as determined by the unit step sequences $u(n)$, which is right sided, and $u(-n+10)$, which is left sided. Also note that $r_2(n)$ is real because the terms $3e^{(-0.4+0.7j)(n-10)}$ and $3e^{(-0.4+0.7j)(n-10)}$ are complex conjugates. The subsequence, $r_2(n)$, begins at $n=11$ and ends at $n=22$ as determined by the unit step sequences $u(n-11)$, which is right sided, and $u(-n+22)$, which is left sided. It follows that $s_5(n)$, which is the sum of $r_1(n)$ and $r_2(n)$, is real and extends from $n=0$ to $n=22$. Fig. 2.7 gives a stem plot of $s_5(n)$ for this example.
End of the Example

2.4 CLASSIFICATION OF DISCRETE TIME SIGNALS

It is desirable to classify signals and systems according to their general properties for analysis as well as for system design. It is important to stress the point that for a

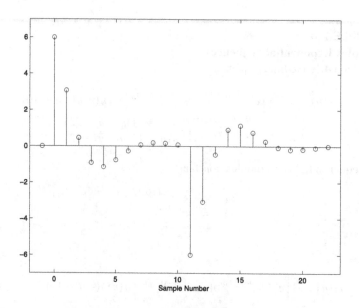

FIGURE 2.7

Stem plot of sequence using complex exponential signals

system to possess a given property, the property must hold for every possible input and output combination. If the property holds for some inputs and/or outputs but not for others, then the system does not possess that property. Similarly, in order for a class of signals to have a given property, each signal in the class must possess that property.

2.4.1 EVEN AND ODD SIGNALS

A signal is even if

$$x(-n) = x(n). \tag{2.30}$$

Even signals are symmetric with respect to the origin ($n = 0$). The signal

$$s(n) = 2\cos(0.279n) + \cos(0.813n) \quad \forall\, -20 \le n \le 20 \tag{2.31}$$

is an example of an even signal. This signal is shown in Fig. 2.8.

A signal is odd if

$$x(-n) = -x(-n). \tag{2.32}$$

An odd signal is antisymmetric with respect to the origin. The signal

$$s(n) = 2\sin(0.381n) + \sin(0.792n) \quad \forall\, -20 \le n \le 20 \tag{2.33}$$

is an example of an odd signal. This signal is shown in Fig. 2.9.

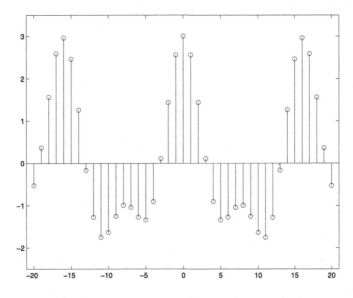

FIGURE 2.8

Stem plot of an even discrete time signal

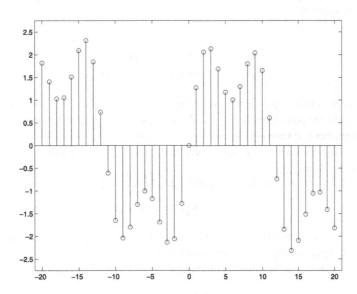

FIGURE 2.9

Stem plot of an odd discrete time signal

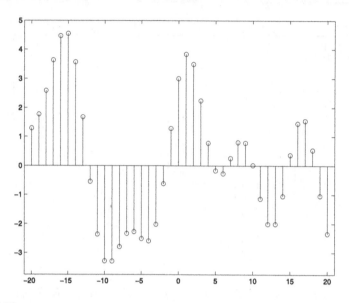

FIGURE 2.10

Stem plot of a discrete time signal that is neither even nor odd

An arbitrary signal, $x(n)$, can be separated into its even and odd parts using the following equations:

$$
\begin{aligned}
x(n) &= x_e(n) + x_o(n), \\
x_e(n) &= 0.5\,[x(n) + x(-n)], \\
x_o(n) &= 0.5\,[x(n) - x(-n)].
\end{aligned}
\tag{2.34}
$$

This concept can be demonstrated by adding the even signal in Eq. (2.30), and shown in Fig. 2.8, to the odd signal in Eq. (2.33), and shown in Fig. 2.9, to obtain the signal which is neither even nor odd as shown in Fig. 2.10.

Fig. 2.11 shows $x(-n)$ which is obtained by performing a left–right flip of the signal in Fig. 2.10.

Fig. 2.12 shows the results of computing the even part of $x(n)$ using

$$
x_e(n) = 0.5\,[x(n) + x(-n)]
\tag{2.35}
$$

compared with the even signal in Fig. 2.8. The signals are the same (except for possible rounding errors during the computations).

Fig. 2.13 shows the results of computing the odd part of $x(n)$ using

$$
x_o(n) = 0.5\,[x(n) - x(-n)]
\tag{2.36}
$$

compared with the odd signal in Fig. 2.9. The signals are the same (except for possible rounding errors during the computations).

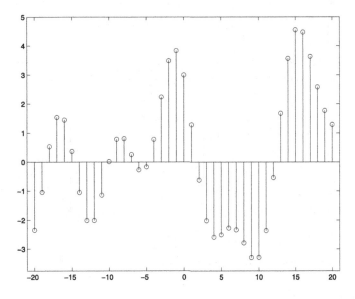

FIGURE 2.11

Stem plot of the discrete time signal in Fig. 2.10 after flipping it left to right to form $x(-n)$

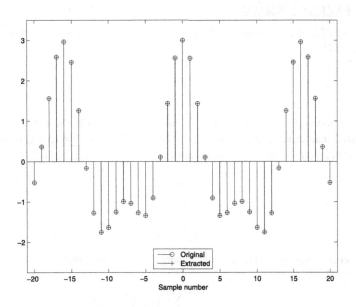

FIGURE 2.12

Stem plot of the comparison of the original even signal with the extracted even signal

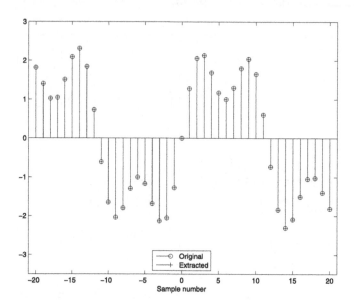

FIGURE 2.13

Stem plot of the comparison of the original odd signal with the extracted odd signal

2.4.2 ENERGY SIGNALS

The energy of a discrete time signal, that is defined over the range $N_1 \leq n \leq N_2$, is defined as [4]

$$E_x = \sum_{n=N_1}^{N_2} |x(n)|^2.$$ (2.37)

This definition applies to both real and complex sequences because Eq. (2.37) uses the squared magnitude of $x(n)$ in the computation. If $x(n)$ is finite and both N_1 and N_2 are finite, then the energy of the signal is finite. However, if either

$$N_1 \rightarrow -\infty$$ (2.38)

or

$$N_2 \rightarrow \infty$$ (2.39)

then the energy may not be finite. If the energy is finite such that

$$E_x = \lim_{\substack{N_1 \to -\infty \\ N_2 \to \infty}} \sum_{n=N_1}^{N_2} |x(n)|^2 < \infty$$ (2.40)

then $x(n)$ is considered to be an energy signal.

2.4.3 POWER SIGNALS

The average power of a discrete time signal that is defined over the range $N_1 \leq n \leq N_2$ is defined as

$$P_x = \frac{1}{N_2 - N_1 + 1} \sum_{n=N_1}^{N_2} |x(n)|^2 . \tag{2.41}$$

Generally, if $N_1 \to -\infty$ and/or $N_2 \to \infty$, then

$$P_x = \lim_{\substack{N_1 \to -\infty \\ N_2 \to \infty}} \frac{1}{N_2 - N_1 + 1} \sum_{n=N_1}^{N_2} |x(n)|^2 . \tag{2.42}$$

Note that the average power of a discrete time signal with infinite energy may be finite. Thus, while the energy of a discrete time signal, as defined in Eq. (2.37) may be infinite, the average power as defined in Eq. (2.42) may be finite. The definition of the average power applies to both real and complex signals because Eq. (2.42) uses the squared magnitude of $x(n)$ in the computation.

2.4.4 PERIODIC AND APERIODIC SIGNALS

A discrete time signal is periodic, with period N, if and only if [4]

$$x(n + N) = x(n) \quad \forall -\infty \leq n \leq \infty. \tag{2.43}$$

The smallest value of N for which Eq. (2.43) holds is called the fundamental period. The signal is aperiodic if there is no value of N which satisfies Eq. (2.43).

Observe that discrete time sinusoidal signals of the form

$$x(n) = A \sin(2\pi f_0 n) \tag{2.44}$$

are periodic if f_0 is a rational number that can be expressed as

$$f_0 = \frac{k}{N} \tag{2.45}$$

where both k and N are integers. If f_0 as given in Eq. (2.44) is not a rational number, then the corresponding sinusoidal signal is aperiodic.

The power of a periodic signal, with fundamental period N, can be computed as

$$P = \frac{1}{N} \sum_{n=0}^{N-1} |x(n)|^2 , \tag{2.46}$$

provided $x(n)$ is finite over the period $0 \leq n \leq N - 1$. Alternatively, the energy of a periodic signal with infinite extent, is infinite because it has finite power over each period and its extent is infinite. Consequently, periodic signals are power signals [4].

2.4.5 STATIC VERSUS DYNAMIC SYSTEMS

A discrete time system is called static if it is memoryless and depends only on the current input. Any system that uses memory is said to be a dynamic system. In other words, a dynamic system has memory and therefore can respond to past inputs and possible past outputs as well. The systems described by the following input–output relationships are static:

$$
\begin{aligned}
y_1(n) &= 0.6x(n), \\
y_2(n) &= 2x(n) + 4x^2(n).
\end{aligned}
\tag{2.47}
$$

Note that both $y_1(n)$ and $y_2(n)$ only depend upon the current input $x(n)$. Thus, the system does not need to use memory to store values for later use. However, $y_2(n)$ is nonlinear.

On the other hand, the systems described by the following input–output relationships are dynamic:

$$
\begin{aligned}
y_1(n) &= 0.5x(n) + x(n-1) + 0.5x(n-2), \\
y_2(n) &= \sum_{k=0}^{n} x(n), \\
y_3(n) &= 0.78x(n) - 0.49y_3(n-1).
\end{aligned}
\tag{2.48}
$$

The first two of these systems use past values of the input. Therefore, there is a need for memory to store input values for later use. The third system uses a past value of the output. Note that any system that uses previous output values to compute the current output is a dynamic system that needs memory to store one or more output values for later use.

2.4.6 TIME INVARIANT VERSUS TIME VARIANT SYSTEMS

If the output is $y(n)$ for a relaxed, time invariant, or shift invariant system for the input $x(n)$, then the output is $y(n-m)$ for the shifted input $x(n-m)$. Furthermore, the difference equation and the system transfer function for a shift invariant system have constant coefficients. In general, if a relaxed, shift invariant system can be described using the difference equation

$$
y(n) = \sum_{k=0}^{K_1} b(k)x(n-k) - \sum_{k=1}^{K_2} a(k)y(n-k)
\tag{2.49}
$$

then all of the $a(k)$ and $b(k)$ are constants and do not depend on n. This means that if the sequence $x(n)$ is delayed by m samples to obtain $x(n-m)$, then the corresponding delayed output $y(n - m)$ can be computed as

$$y(n - m) = \sum_{k=0}^{K_1} b(k)x(n - m - k) - \sum_{k=1}^{K_2} a(k)y(n - m - k). \qquad (2.50)$$

Alternatively, a system whose coefficients depend on the independent variable n is a time variant or shift variant system. Consequently, the delayed output $y(n - m)$ corresponding to the delayed input sequence $x(n - m)$ cannot be computed using Eq. (2.50).

2.5 TRANSFORMATION OF SIGNALS

This section considers concepts to transform a given discrete time sequence to another discrete time sequence. These concepts can be used to transform a standard sequence to a sequence of interest.

2.5.1 TIME SCALING

Time scaling is a basic transformation of the time (or other independent variable) axis for a sequence. Time scaling of sequences can be performed by scaling the independent variable, time. For example, let $x(n)$ be a standard discrete time signal. The independent variable, n, can be scaled by multiplying all instances of n by a scale factor a to obtain $g(n) = x(an)$. Then $g(n)$ is an expansion of $x(n)$ if $a < 1$ and it is a compression of $x(n)$ if $a > 1$. Keep in mind that $g(n)$ is only defined where an is an integer. Example 2.5 illustrates the time scaling concept.

EXAMPLE 2.5
(Time Scaling)

Consider the sinusoidal sequence given by

$$
\begin{aligned}
f_0 &= 25.0, \\
f_s &= 25 f_0, \\
T &= \frac{1}{f_s}, \\
x(n) &= 4.0 \sin(2\pi f_0 n T) \ \forall \ 0 \le n \le 50.
\end{aligned}
\qquad (2.51)
$$

The sequence $x(n)$ is shown in Fig. 2.14.

Consider the case where the independent variable, n, is scaled by a factor $a = 0.5$ ($a < 1.0$) such that the new independent variable is $n_2 = 0.5n$. Note that the possible values of n_2 occur at the following values of the original independent

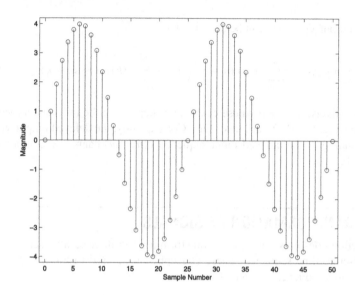

FIGURE 2.14

Stem plot of a sinusoidal sequence

variable n:

$$n = [0.0, 0.5, 1.0, 1.5, \ldots, 49.5, 50]. \qquad (2.52)$$

The equivalent values of the new independent variable n_2 are given by

$$n_2 = [0.0, 1.0, 2.0, 3.0, \ldots, 100]. \qquad (2.53)$$

However, the sequence $x(n)$ is undefined at the odd values of n_2 which occur at non-integer values of the original independent variable n. Thus, the scaled sequence, $x(n_2)$, is undefined at the values

$$n_2 = [1.0, 3.0, \ldots, 99]. \qquad (2.54)$$

It is typical to assign a value of 0 to these undefined values of the scaled sequence. The sequence $x_2(n)$, with the undefined values assigned a value of 0, is shown in Fig. 2.15.

Consider another scale factor $a = 1.5$ such that $n_3 = 1.5n$. Note that the possible values of n_3 occur at the following values of the original independent variable n:

$$n = [0.0, 1.5, 3.0, 4.5, \ldots, 48, 49.5, \ldots, 73.5, 75]. \qquad (2.55)$$

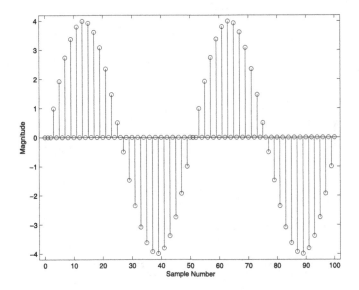

FIGURE 2.15

Stem plot of a sinusoidal sequence with independent variable scaled by a factor of $a = 0.5$

However, the sequence $x(n)$ is undefined at the values of n_3 that occur at non-integer values of the original independent variable n. Additionally, the original sequence is undefined outside the range $0 \leq n \leq 50$. Thus, the scaled sequence, $x_3(n)$, will also not be defined for values of n_3 outside the range of $0 \leq n_3 \leq 33$. The equivalent values of the new independent variable n_3 are given by

$$n_3 = [0.0, 1.0, 2.0, 3.0, \ldots, 33]. \tag{2.56}$$

It is typical to assign a value of 0 to the values of n_3 that map into the undefined values of the scaled sequence. The resulting sequence, $x_3(n)$, with the undefined values assigned a value of 0, is shown in Fig. 2.16
End of the Example

2.5.2 TIME SHIFT

A shift by a value of m in the independent variable for a discrete time sequence, $x(n)$, can be represented by replacing the independent variable n by $n - m$ for all instances of n to obtain $s(n) = x(n - m)$, which will be a shifted version of $x(n)$. If m is positive, then $x(n)$ will be delayed by m samples. If m is negative, then $x(n)$ will be advanced by m samples. For example, if

$$x(n) = A \sin(\omega n) \tag{2.57}$$

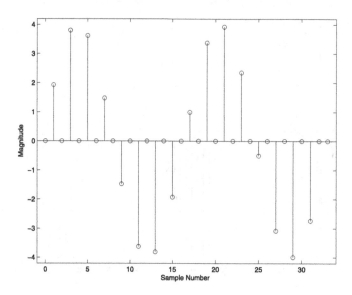

FIGURE 2.16

Stem plot of a sinusoidal sequence with independent variable scaled by a factor of $a = 1.5$

then a new sequence, formed by shifting $x(n)$ by m samples, can be represented as

$$s(n) = x(n - m) = A \sin [\omega(n - m)]. \qquad (2.58)$$

Example 2.6 illustrates this concept.

EXAMPLE 2.6
(Time Shift)

The time shift property can be illustrated by using the sinusoidal sequence used in Example 2.5 as shown again in Fig. 2.17. Fig. 2.18 shows this sequence shifted by a value of $m = 6$ which results in a shift to the right (delay) by 6 samples. Fig. 2.19 shows this sequence shifted by a value of $m = -6$ which results in a shift to the left (advance) by 6 samples.
End of the Example

2.5.3 TIME REVERSAL

Time reversal is another basic transformation of the independent variable axis. A discrete time signal can be reversed in time (or other independent variable) by changing the sign of the independent variable for all instances. The modification of the range of the values of the modified independent variable should be considered as well. For

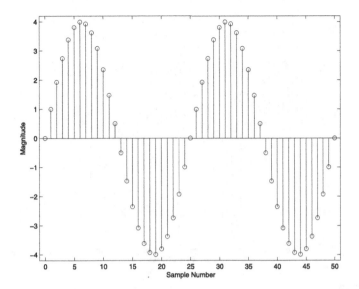

FIGURE 2.17

Stem plot of a sinusoidal sequence

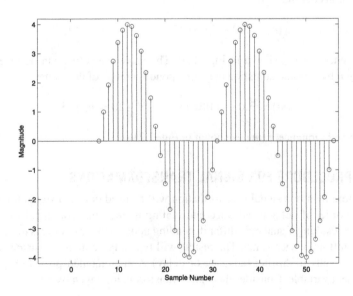

FIGURE 2.18

Stem plot of a sinusoidal sequence with independent variable shifted by a factor of $m = 6$

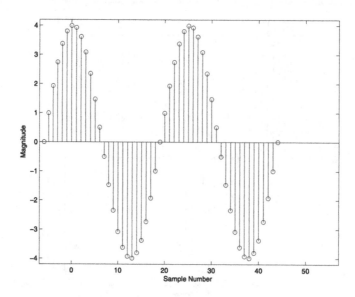

FIGURE 2.19

Stem plot of a sinusoidal sequence with independent variable shifted by a factor of $m = -6$

example, consider the signal

$$x(n) = 3.0 - 3.0(0.75)^n \quad \forall -5 \leq n \leq 20. \tag{2.59}$$

The sequence $x(n)$ is shown in Fig. 2.20. The sequence, $x(n)$, can be reversed by replacing n by $-m$ and adjusting the corresponding range of the values of m to obtain

$$r(m) = 3.0 - 3.0(0.75)^{-m} \quad \forall -20 \leq m \leq 5. \tag{2.60}$$

The reversed sequence, $r(m)$, is shown in Fig. 2.21.

2.5.4 PROCEDURE FOR SIGNAL TRANSFORMATIONS

Often, more than one signal transformation will be used during a signal transformation. The order of delay or advance and scaling is important for signal transformations. The resulting signal may differ depending upon whether scaling is implemented prior to shifting or vice versa. The results will typically be more consistent with the expected results if shifting (advance or delay) is implemented prior to scaling the independent variable. Consider the signal transformation given by

$$y(n) = x(\alpha n + \beta). \tag{2.61}$$

A systematic procedure for implementing such a signal transformations is as follows:

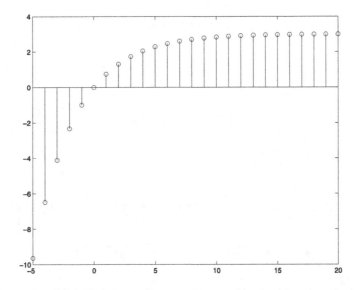

FIGURE 2.20

Stem plot of a sample sequence

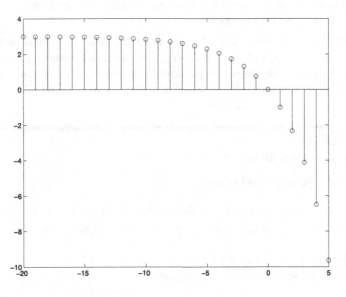

FIGURE 2.21

Stem plot of the reversed sample sequence

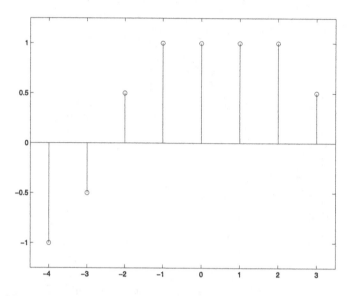

FIGURE 2.22

Original discrete time signal for Example 2.7

1. First, delay or advance the signal in accordance with the value of β.
2. Perform scaling of the signal in accordance with the magnitude of α.
3. If $\alpha < 0$, perform time reversal.

Signal transformations are used to represent physical phenomena such as time shift in sonar signals, speeding up or slowing down an audio sequence, and for general analysis of signals and systems. In particular, convolution and correlation will later be discussed as examples that use signal transformations. Example 2.7 gives some examples of signal transformations.

EXAMPLE 2.7
(Signal Transformations)

A discrete time signal $y(n)$ is given by

$$x(n) = -\delta(n+4) - 0.5\delta(n+3) + 0.5\delta(n+2) + \delta(n+1)$$
$$+ \delta(n) + \delta(n-1) + \delta(n-2) + 0.5\delta(n-3). \qquad (2.62)$$

This signal is shown in Fig. 2.22.
    Determine each of the following signals:

1.

$$y_1(n) = x(n-4). \qquad (2.63)$$

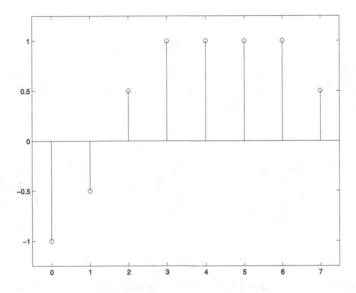

FIGURE 2.23

Transformation $y_1(n) = x(n-4)$ for Example 2.7

Solution: The original sequence is shifted to the right by 4 samples for this case. The stem plot $y_1(n) = x(n-4)$ is given in Fig. 2.23.

2.

$$y_2(n) = y(3-n). \tag{2.64}$$

Solution: In this case, the sequence is shifted 3 samples to the left and then time reversed. The stem plot $y_2(n) = x(3-n)$ is given in Fig. 2.24.

3.

$$y_3(n) = x(3n). \tag{2.65}$$

Solution: In this case, the original sequence is compressed by a factor of 3 in the time axis. Only every third sample is used in the output sequence since samples can only occur on integer values of the independent variable. The stem plot $y_3(n) = x(3n)$ is given in Fig. 2.25.

4.

$$y_4(n) = x(3n+1). \tag{2.66}$$

Solution: In this case, the original sequence is shifted to the left by one sample and then it is compressed by a factor of 3 in the time axis. Only every third sample is used again since this is a discrete time sequence. The stem plot $y_4(n) = x(3n+1)$ is given in Fig. 2.26.

5.

$$y_5(n) = x(n)\, u(2 - n). \tag{2.67}$$

Solution: Note that $u(2 - n) = 0 \ \forall\, n > 2$. Thus $y_5(n)$ is the same as $x(n)$ except that the sample at $n = 3$ has been removed. The stem plot $y_5(n) = x(n)\, u(2 - n)$ is given in Fig. 2.27.

6.

$$y_6(n) = x(n - 2)\delta(n - 2). \tag{2.68}$$

Solution: In this case, the original sequence is first shifted to the right by two samples. Then the solution is just a single impulse sequence with magnitude equal to the sample located at $n = 2$ for the shifted sequence. The stem plot $y_6(n) = x(n - 2)\delta(n - 2)$ is given in Fig. 2.28.

7.

$$y_7(n) = \frac{1}{2}x(n) + \frac{1}{2}(-1)^n x(n). \tag{2.69}$$

Solution: This transformation results in setting every other sample equal to zero. The signal $y_7(n) = \frac{1}{2}x(n) + \frac{1}{2}(-1)^n x(n)$ is given in Fig. 2.29.

8.

$$y_8(n) = x\left(\frac{n}{2}\right). \tag{2.70}$$

Solution: This transformation results in the expansion of the independent variable axis by a factor of 2. The signal $y_8(n) = x\left(\frac{n}{2}\right)$ is given in Fig. 2.30.

End of the Example

2.6 OPERATIONS ON SEQUENCES

It is often desirable to perform an operation on an input sequence $x(n)$ using a set of prescribed rules to obtain an output sequence. The basic operations for this course include

1. Modulation, $y(n) = h(n)x(n)$, as shown in Fig. 2.31.
2. Addition, $y(n) = h(n) + x(n)$, as shown in Fig. 2.32.
3. Multiplication by a constant, $y(n) = ax(n)$, as shown in Fig. 2.33.
4. Delay, $y(n) = x(n - 1)$, as shown in Fig. 2.34.

2.7 SAMPLING CONTINUOUS TIME SIGNALS

Often, a discrete time sequence is generated by uniformly sampling a continuous time signal. The continuous time signal can be uniquely reconstructed from these samples

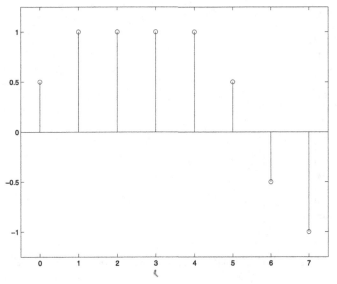

FIGURE 2.24

Transformation $y_2(n) = x(3 - n)$ for Example 2.7

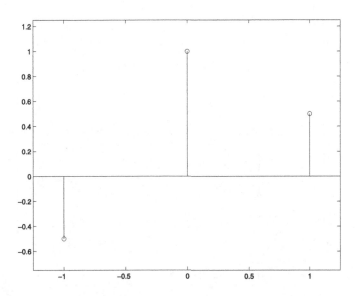

FIGURE 2.25

Transformation $y_3(n) = x(3n)$ for Example 2.7

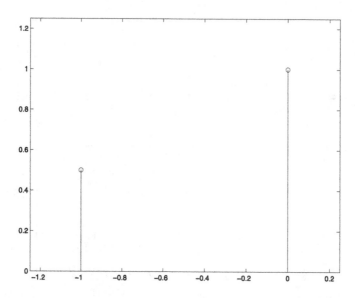

FIGURE 2.26

Transformation $y_4(n) = x(3n + 1)$ for Example 2.7

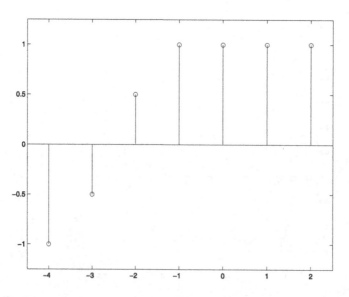

FIGURE 2.27

Transformation $y_5(n) = x(n) \, u(2 - n)$ for Example 2.7

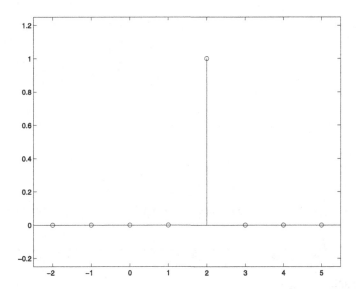

FIGURE 2.28

Transformation $y_6(n) = x(n-2)\,\delta(n-2)$ for Example 2.7

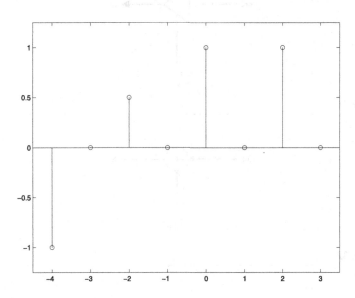

FIGURE 2.29

Transformation $y_7(n) = \frac{1}{2}x(n) + \frac{1}{2}(-1)^n x(n)$ for Example 2.7

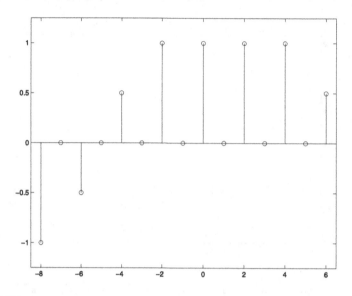

FIGURE 2.30

Transformation $y_8(n) = x\left(\frac{n}{2}\right)$ for Example 2.7

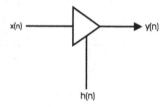

FIGURE 2.31

Modulation

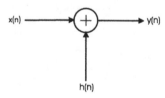

FIGURE 2.32

Addition

if the sampling frequency is greater than 2 times the highest frequency in the signal according to the Sampling Theorem.

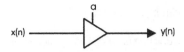

FIGURE 2.33

Multiplication

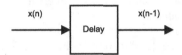

FIGURE 2.34

Delay

Theorem 2.1 (Sampling Theorem [4]). *If the highest frequency contained in an analog signal $x_a(t)$ is $F_{max} = B$ and the signal is sampled at a rate of $F_s > 2F_{max} \equiv 2B$, then $x_a(t)$ can be exactly recovered from its sample values using the interpolation function*

$$g(t) = \frac{\sin(2\pi Bt)}{2\pi Bt}. \tag{2.71}$$

Thus $x_a(t)$ may be expressed as

$$x_a(t) = \sum_{n=-\infty}^{\infty} x_a\left(\frac{n}{F_s}\right) g\left(t - \frac{n}{F_s}\right) \tag{2.72}$$

where $x_a\left(\frac{n}{F_s}\right) = x_a(nT) \equiv x(n)$ are the samples of $x_a(t)$.

If the sampling frequency is not high enough, then aliasing occurs. Example 2.8 illustrates the concept of aliasing through the sampling of continuous time sinusoidal signals.

EXAMPLE 2.8
(Aliasing)

Assume that the signal

$$x_1(t) = \cos(700\pi t)u(t) \tag{2.73}$$

is sampled with a sampling interval of

$$T = 0.001. \tag{2.74}$$

The resulting sequence can be represented as

$$x_1(n) = \cos[700\pi(0.001n)] = \cos(0.7\pi n)u(n). \tag{2.75}$$

Now consider another signal given by

$$x_2(t) = \cos(2700\pi t)u(t). \tag{2.76}$$

If $x_2(t)$ is sampled with the same sampling interval, then the resulting sampled sequence can be represented as

$$x_2(n) = \cos[2700\pi(0.001n)] = \cos(2.7\pi n)u(n). \tag{2.77}$$

However,

$$
\begin{aligned}
\cos(2.7\pi n) &= \cos(2.0\pi n + 0.7\pi n)u(n) \\
&= \cos(2.0\pi n)\cos(0.7\pi n) - \sin(2.0\pi n)\sin(0.7\pi n) \quad (2.78)
\end{aligned}
$$

where the trigonometric identity

$$\cos(a+b) = \cos(a)\cos(b) - \sin(a)\sin(b) \tag{2.79}$$

has been used. Then, since

$$
\begin{aligned}
\cos(2.0\pi n) &= 1.0 \ \forall \, 0 \le n \le \infty, \\
\sin(2.0\pi n) &= 0.0 \ \forall \, 0 \le n \le \infty, \quad (2.80)
\end{aligned}
$$

it follows that

$$\cos(2.7\pi n)u(n) = \cos(0.7\pi n)u(n) \ \forall \, 0 \le n \le \infty. \tag{2.81}$$

Thus, the sequences resulting from sampling $x_1(t)$ and from sampling $x_2(t)$ are identical. Fig. 2.35 shows the continuous time signal $x_1(t)$ with a solid line, the continuous time signal $x_2(t)$ with a dashed line, the samples for $x_1(n)$ with the circles o, and the samples for $x_2(n)$ with the pluses $+$.

This figure illustrates that it would not be possible to distinguish the samples from sampling $x_1(t)$ with a sampling rate of $T = 0.001$ from the samples of $x_2(t)$ with a sampling rate of $T = 0.001$ since the samples would be identical.

Actually, since the cosine sequence is even, all of the cosine signals with frequencies

$$f_0 = 700\pi t \pm 2\pi \left(\frac{mt}{0.001}\right) \ \forall \ 0 \le m < \infty \tag{2.82}$$

have the same sequence when sampled with a sampling interval of $T = 0.001$.
End of the Example

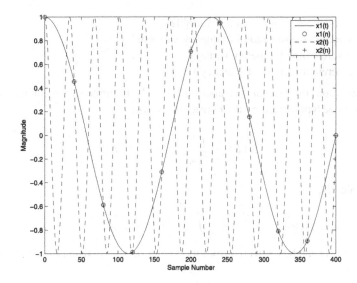

FIGURE 2.35

Signals $x_1(t)$, $x_2(t)$, and their samples on the same plot

Aliasing occurs for the higher frequency sinusoids with the result that the sample sequences for the aliased, higher frequency sinusoids are identical to the sample sequence for the lower frequency sinusoid with the signal frequency that is less than 1/2 the sampling frequency.

Digital signal processing is used for many applications that involve continuous time signals. These continuous time signals are typically sampled and quantized to form discrete time signals that are represented using a finite number of bits. The process of sampling a continuous time signal to obtain a discrete time signal is called analog-to-digital (A/D) conversion. The process of constructing the band limited, continuous time signal from a sequence of samples is called digital-to-analog (D/A) conversion. Both processes are very important for many signal processing applications. The conversion of a continuous time signal to a discrete time signal normally involves periodically sampling the continuous time signal with a sampling interval of T. The samples are also typically quantized to a finite word size.

If the continuous time signal is $x(t)$, then the sampled sequence can be represented as [4]

$$x(n) = x(t)|_{t=nT} \quad \forall -\infty \leq n \leq \infty. \tag{2.83}$$

The sampling interval, T, must be large enough such that the sampling process does not cause any loss or distortion of the spectral information due to aliasing.

The following notations will be used for continuous time and discrete time signals in the following development:

1. $F \rightarrow$ continuous time frequency variable
2. $\Omega \rightarrow$ continuous time radial frequency variable
3. $f \rightarrow$ discrete time frequency variable
4. $\omega \rightarrow$ discrete time radial frequency variable

If the continuous time signal is an aperiodic signal with finite energy, then its spectral content can be represented by the Fourier transform

$$X(F) = \int_{-\infty}^{\infty} x(t)e^{-2\pi jFt}dt, \qquad (2.84)$$

or equivalently,

$$X(\Omega) = \int_{-\infty}^{\infty} x(t)e^{-j\Omega t}dt. \qquad (2.85)$$

The signal $x(t)$ can be recovered from its Fourier transform using the inverse Fourier transform

$$x(t) = \int_{-\infty}^{\infty} X(F)e^{j2\pi Ft}dF, \qquad (2.86)$$

or equivalently,

$$x(t) = \frac{1}{2\pi} \int_{-\infty}^{\infty} X(\Omega)e^{j\Omega t}d\Omega. \qquad (2.87)$$

Note that all frequency components for the infinite range $-\infty \leq F \leq \infty$ must be used to reconstruct the signal if the signal is not band limited.

The spectrum of a discrete time signal $x(n)$, obtained by sampling a continuous time signal $x(t)$, is periodic with period $\frac{2\pi}{T}$ where T is the sampling interval. This spectrum is given by [4]

$$X(\omega) = \sum_{n=-\infty}^{\infty} x(n)e^{-j\omega n}, \qquad (2.88)$$

or equivalently,

$$X(f) = \sum_{n=-\infty}^{\infty} x(n)e^{-j2\pi fn}. \qquad (2.89)$$

The relationship between the independent variables, t, for the continuous time signal and n, for the discrete time signal, is given by

$$t = nT = \frac{n}{F_s} \qquad (2.90)$$

where T is the sample interval and F_s is the sample frequency.

The discrete time sequence can be recovered from its spectrum with the use of the inverse transform

$$x(n) = \int_{-1/2}^{1/2} X(f)e^{j2\pi fn} df, \qquad (2.91)$$

or equivalently,

$$x(n) = \frac{1}{2\pi} \int_{-\pi}^{\pi} X(\omega)e^{j\omega n} d\omega. \qquad (2.92)$$

Eq. (2.90) shows the relationship between the continuous time variable t and the discrete time variable T. This relationship can be used to develop a relationship between the continuous time frequency variable, F, and the discrete time frequency variable, f. Consider the inverse transform of the continuous time signal as given by

$$x(t) = \frac{1}{2\pi} \int_{-\infty}^{\infty} X(\Omega)e^{j\Omega t} d\Omega. \qquad (2.93)$$

The relationship in Eq. (2.90) can be used to substitute nT for t to obtain

$$x(n) = x(nT) = x(t)|_{t=nT} = \frac{1}{2\pi} \int_{-\infty}^{\infty} X(\Omega)e^{j\Omega nT} d\Omega. \qquad (2.94)$$

The inverse Fourier transform for the discrete time sequence, $x(n)$, is given by

$$x(n) = \frac{1}{2\pi} \int_{-\pi}^{\pi} X(\omega)e^{j\omega n} d\omega. \qquad (2.95)$$

It follows that

$$x(n) = \frac{1}{2\pi} \int_{-\pi}^{\pi} X(\omega)e^{j\omega n} d\omega = \frac{1}{2\pi} \int_{-\infty}^{\infty} X(\Omega)e^{j\Omega nT} d\Omega. \qquad (2.96)$$

The integration range for the integral, on the right side of Eq. (2.96), is infinite. However, the frequency space for a discrete time system is periodic with period $\frac{2\pi}{T}$. The integration can be broken up into separate integrals over the ranges

$$(k-1)\frac{\pi}{T} \le \omega \le (k+1)\frac{\pi}{T} \quad \forall\, -\infty \le k \le \infty. \qquad (2.97)$$

It follows that

$$\frac{1}{2\pi} \int_{-\pi}^{\pi} X(\omega)e^{j\omega n} d\omega = \frac{1}{2\pi} \sum_{k=-\infty}^{\infty} \int_{(k-1)\frac{\pi}{T}}^{(k+1)\frac{\pi}{T}} X\left(\Omega - \frac{2\pi k}{T}\right) e^{j\Omega nT} d\Omega. \qquad (2.98)$$

The discrete time frequency, f, is the normalized frequency for the discrete time signal, and the relationship between the normalized discrete time frequency and the continuous time frequency is given by

$$\omega = \Omega T, \qquad (2.99)$$

or

$$\Omega = \frac{\omega}{T}. \qquad (2.100)$$

If the Fourier Transform of $x(t)$ exists, then the order of summation and integration in Eq. (2.98) can be changed. Thus, if the Fourier Transform of $x(t)$ exists, then

$$\frac{1}{2\pi} \int_{-\pi}^{\pi} X(\omega) e^{j\omega n} d\omega = \frac{1}{T} \sum_{k=-\infty}^{\infty} \frac{1}{2\pi} \int_{(k-1)\pi}^{(k+1)\pi} X\left(\frac{\omega - 2\pi k}{T}\right) e^{j\omega n} d\omega. \qquad (2.101)$$

Since $X(\omega)$ is periodic with period 2π, it follows that

$$\int_{(k-1)\pi}^{(k+1)\pi} X\left(\frac{\omega - 2\pi k}{T}\right) e^{j\omega n} d\omega = \int_{-\pi}^{\pi} X(\omega) e^{j\omega n} d\omega. \qquad (2.102)$$

Therefore,

$$\frac{1}{T} \sum_{k=-\infty}^{\infty} \frac{1}{2\pi} \int_{-\pi}^{\pi} X(\omega - 2\pi k) e^{j\omega n} d\omega = \frac{1}{2\pi} \int_{-\pi}^{\pi} X(\omega) e^{j\omega n} d\omega \qquad (2.103)$$

and

$$\frac{1}{2\pi} \int_{-\pi}^{\pi} X(\omega) e^{j\omega n} d\omega = \frac{1}{2\pi} \left[\frac{1}{T} \sum_{k=-\infty}^{\infty} \int_{-\pi}^{\pi} X(\omega - 2\pi k) \right] e^{j\omega n} d\omega. \qquad (2.104)$$

Thus, the relationship between the discrete time and continuous time frequency representations is given by

$$X(\omega) = \frac{1}{T} \sum_{k=-\infty}^{\infty} X(\omega - 2\pi k), \qquad (2.105)$$

or

$$X(\omega) = \frac{1}{T} \sum_{k=-\infty}^{\infty} X\left(\Omega - \frac{2\pi k}{T}\right). \qquad (2.106)$$

It follows that the discrete time frequency representation is obtained by scaling the continuous time frequency representation by the factor $\frac{1}{T}$ and replicating it at intervals of $\frac{2\pi k}{T}$ for $-\infty \leq k \leq \infty$.

2.8 DISCRETE TIME CONVOLUTION

Convolution is a mathematical operation involving two functions to produce a third function that is typically considered to be a modification of one of the original functions. It can be used to compute the zero state response of a linear, shift invariant, discrete time system. The zero state refers to the state of a system when all of the initial conditions are equal to zero. Thus, a discrete time convolution can be used to compute the output sequence $y(n)$, for a linear, shift invariant, discrete time system with an impulse response $h(n)$, an input sequence $x(n)$, and with all initial conditions equal to zero.

The convolution summation, or superposition summation, is defined as [4]

$$y(n) = \sum_{k=-\infty}^{\infty} h(k)x(n-k) \tag{2.107}$$

where $h(k)$ is the discrete time system impulse response and $x(n-k)$ is the input sequence shifted by k time intervals. Note that either $h(n)$ or $x(n)$ can be shifted to form the convolution summation (2.107). Thus, the convolution of $h(n)$ and $x(n)$ can also be represented as

$$y(n) = \sum_{k=-\infty}^{\infty} h(n-k)x(k). \tag{2.108}$$

Eigenfunctions have the property that if an eigenfunction is an input to the system, then the output will be the same eigenfunction multiplied by a real or complex constant that does not depend on the independent variable. The input $x(n) = C_1 a^n$, where C_1 is either a real or complex constant, is an eigenfunction for a linear, shift invariant, discrete time system as an example. If an input, $C_1 a^n$, is applied to a linear, shift invariant, discrete time system at $n = 0$, then this input can be represented as the eigenfunction

$$x(n) = C_1 a^n u(n). \tag{2.109}$$

Assume that the impulse response of the system in question can be represented by

$$h(n) = C_2 b^n u(n), \quad 0 \le n \le \infty. \tag{2.110}$$

Note that a and b can be either real or complex numbers. The discrete time system impulse response, $h(n)$, shifted as shown in Eq. (2.107), is given by

$$h(n-k) = C_2 b^{n-k} u(n-k). \tag{2.111}$$

Thus, the output sequence, $y(n)$, can be computed using the convolution summation as follows:

$$y(n) = \sum_{k=-\infty}^{\infty} \left[C_1 a^k u(k) \right] \left[C_2 b^{n-k} u(n-k) \right]. \tag{2.112}$$

This can be rewritten as

$$y(n) = \sum_{k=-\infty}^{\infty} \left[C_1 C_2 a^k b^{n-k} \right] u(k)u(n-k). \tag{2.113}$$

Note that

$$u(k)u(n-k) = \begin{cases} 1 & \text{if } 0 \le k \le n, \\ 0 & \text{otherwise.} \end{cases} \tag{2.114}$$

The terms that are not functions of k can be factored out of the summation to obtain

$$y(n) = C_1 C_2 b^n \sum_{k=0}^{n} a^k b^{-k}, \tag{2.115}$$

$$y(n) = C_1 C_2 b^n \sum_{k=0}^{n} \left(\frac{a}{b}\right)^k. \tag{2.116}$$

Note that the above equation for $y(n)$ is a geometric series. Thus, if $a \ne b$, then the geometric series can be written in closed form to obtain

$$y(n) = C_1 C_2 b^n \left[\frac{1 - (\frac{a}{b})^{k+1}}{1 - \frac{a}{b}} \right]. \tag{2.117}$$

If $a = b$, then the result is given by

$$y(n) = C_1 C_2 b^n \sum_{k=0}^{n} 1 = C_1 C_2 b^n (n+1) \tag{2.118}$$

where

$$\sum_{n=0}^{n} 1 = n + 1. \tag{2.119}$$

Since a and b can be real or complex, the above solution can be used for many different problems. If the system is linear and shift invariant, then the principle of superposition can be used to form the output for a combination of different inputs represented as linear combinations of eigenfunctions. Example 2.9 gives an example for the partitioning of an input sequence into subsequences so that Eqs. (2.117) and (2.118) can be used to obtain the solution using the principal of superposition.

EXAMPLE 2.9
(Convolution)

Consider an input sequence, $x(n)$, given by

$$x(n) = \left[5(0.4)^n + 3\cos(0.3n + 0.42) \right] u(n). \tag{2.120}$$

The cosine term can be written as

$$\cos(0.3n + 0.42) = 0.5\left[e^{j(0.3n+0.42)} + e^{-j(0.3n+0.42)}\right]$$
$$= 0.5e^{j0.42}(e^{j0.3})^n + 0.5e^{-j0.42}(e^{-j0.3})^n. \quad (2.121)$$

It follows that

$$x(n) = \left[5(0.4)^n + 1.5e^{j0.42}(e^{j0.3})^n + 1.5e^{-j0.42}(e^{-j0.3})^n\right]u(n). \quad (2.122)$$

Thus, if the following assignments are made:

$$\begin{aligned} C_1 &= 5, \\ a_1 &= 0.4, \\ C_2 &= 1.5e^{j0.42}, \\ a_2 &= e^{j0.3}, \\ C_3 &= 1.5e^{-j0.42}, \\ a_3 &= e^{-j0.3}, \end{aligned} \quad (2.123)$$

then

$$x(n) = C_1 a_1^n u(n) + C_2 a_2^n u(n) + C_3 a_3^n u(n). \quad (2.124)$$

Eq. (2.117) can be used to determine the output for each term of $x(n)$. The principal of superposition can then be used to determine the output as the sum of the outputs for each individual term.
End of the Example

Example 2.10 gives an example of the use of the convolution summation to compute the step response for a discrete time system.

EXAMPLE 2.10
(System Step Response)

Problem:
The impulse response for a discrete time system is given by

$$h(n) = 2.0(0.7)^n u(n). \quad (2.125)$$

Use the convolution summation to compute the step response for this discrete time system.
Compute the convolution summation

$$y(n) = \sum_{k=-\infty}^{\infty} h(k)x(n-k) = \sum_{k=-\infty}^{\infty} h(n-k)x(k) \quad (2.126)$$

for the unit step as the input sequence,

$$x(n) = u(n). \tag{2.127}$$

The output $y(n)$ will be the step response for the system with impulse response $h(n)$ because the input $x(n)$ is a unit step function.

Solution:

The required input is the unit step function $u(n)$. Thus,

$$y(n) = \sum_{k=-\infty}^{\infty} h(n-k)u(n), \tag{2.128}$$

$$h(n-k) = 2.0(0.7)^{n-k}u(n-k). \tag{2.129}$$

Thus,

$$y(n) = \sum_{k=-\infty}^{\infty} 2.0(0.7)^{n-k}u(n-k)u(k). \tag{2.130}$$

Note that

$$u(n-k)u(k) = \begin{cases} 0 & \text{if } k < 0, \\ 1 & \text{if } 0 \le k \le n, \\ 0 & \text{if } k > n. \end{cases} \tag{2.131}$$

Thus

$$y(n) = \sum_{k=0}^{n} 2.0(0.7)^n(0.7)^{-k} = 2.0(0.7)^n \sum_{k=0}^{n}(0.7)^{-k}. \tag{2.132}$$

The summation is a geometric series. It follows that

$$y(n) = 2(0.7)^n \left[\frac{1 - (\frac{1}{0.7})^{n+1}}{1 - (\frac{1}{0.7})} \right] u(n), \tag{2.133}$$

$$y(n) = 2 \left[\frac{(\frac{1}{0.7}) - (0.7)^n}{\frac{1}{0.7} - 1} \right] u(n), \tag{2.134}$$

$$y(n) = 4.667[1.4286 - (0.7)^n]u(n). \tag{2.135}$$

End of the Example

2.9 OUTPUT RESPONSE OF A DISCRETE TIME SYSTEM TO A GIVEN INPUT SEQUENCE

The discrete time impulse response for a linear, shift invariant discrete time system can be used with the convolution summation to compute the output of the system to an arbitrary discrete time input. If the discrete time impulse $\delta(n)$ if applied to a linear, shift invariant, discrete time system, then its output will be the impulse response $h(n)$. The computation of the output response of a digital system for a given input sequence $x(n)$ in terms of the *convolution summation* is given by

$$y(n) = \sum_{k=-\infty}^{\infty} h(k)x(n-k) = \sum_{k=-\infty}^{\infty} h(n-k)x(n) \qquad (2.136)$$

where $h(k)$ is the impulse response for a discrete time impulse input applied at $k = 0$. The convolution summation uses the linear shift property for discrete time systems. If an input $x(n)$ is applied at time k, then the corresponding output will be $h(k)x(n-k)$. The convolution summation formula uses this result to compute the output for a given input sequence as a sum of shifted discrete time impulses $\delta(n-k)$ with magnitude $x(n)$.

The effects of initial conditions on the system are often ignored. When all of the initial conditions of a system are equal to zero, then the system is said to be *relaxed*. However, in many practical applications, initial conditions as well as the input sequence must also be considered to determine the output of a discrete time system. If the system is linear and shift invariant, the principle of superposition can be used to compute the output as a linear combination of the output due to the initial state (initial conditions) and the output due to the input sequence.

Discrete time systems are often characterized in terms of the nature of their impulse responses. If the impulse response is of finite duration, then the system is called a *finite impulse response* (FIR) system. If the system has an infinite impulse response, then the system is called an *infinite impulse response* (IIR) system. The following is an impulse response for a FIR system:

$$h(n) = \left\{ 0.125, 0.25, \ \ 0.5, \ \ 0.25, 0.125 \atop \uparrow \right\} \qquad (2.137)$$

where the symbol ($\uparrow$) identifies the sample at $n = 0$.

Infinite impulse response systems are usually recursive. The definition of a recursive system is given by [4]

> *Definition:* A system whose output $y(n)$ at time n depends on any number of past output values $y(n-1)$, $y(n-2)$, ..., is called a recursive system.

The following is an impulse response for an IIR system:

$$h(n) = \left[(0.76)^n + (-0.59)^n \right] \ \ \forall \ 0 \le n \le \infty. \qquad (2.138)$$

2.10 DIFFERENCE EQUATION REPRESENTATION

A difference equation is an alternative approach to representing linear, shift invariant, discrete time systems. It is appropriate to represent a discrete time system by a difference equation when the discrete time impulse response for the system can be represented using only a few samples. Consider the discrete time system with impulse response

$$h(n) = s(n)u(n)u(N - n).$$ (2.139)

Note that

$$u(n) = \begin{cases} 0 & \forall n < 0, \\ 1 & \forall n \geq 0, \end{cases}$$ (2.140)

$$u(N - n) = \begin{cases} 0 & \forall n > N, \\ 1 & \forall n \leq N. \end{cases}$$

It follows that

$$u(n)u(N - n) = \begin{cases} 0 & \forall n < 0, \\ 1 & \forall 0 \leq n \leq N, \\ 0 & \forall n > N. \end{cases}$$ (2.141)

Also note that

$$\delta(n - k) = \begin{cases} 0 & \forall n \neq k, \\ 1 & \forall n = k. \end{cases}$$ (2.142)

It follows that the impulse response given in Eq. (2.139) can also be written as

$$h(n) = \sum_{k=0}^{N} s(k)\delta(n - k).$$ (2.143)

The convolution summation can be used to find the output, $y(n)$, of a discrete time system for a given input, $x(n)$, as

$$y(n) = \sum_{k=-\infty}^{\infty} h(k)x(n - k).$$ (2.144)

The output for the discrete time system with impulse response given by Eq. (2.143) for an arbitrary input sequence $x(n)$ can be obtained by substituting for $h(k)$ in Eq. (2.144),

$$y(n) = \sum_{k=-\infty}^{\infty} \left(\sum_{m=0}^{N} s(k)\delta(k - m) \right) x(n - k).$$ (2.145)

This equation can be easily evaluated to obtain

$$y(n) = s(0)x(n) + s(1)x(n - 1) + s(2)x(n - 2) \cdots s(N)x(n - N).$$ (2.146)

Eq. (2.146) is the difference equation representation of the system with the impulse response given in Eq. (2.143).

Table 2.2 Impulse response of discrete time system for Example 2.12

n	-3	-2	-1	0	1	2	3
$h(n)$	0.55	0.0	0.75	1.0	0.75	0.0	0.55

EXAMPLE 2.11
(FIR Discrete Time System Difference Equation)

Let $h(n)$ be given by

$$h(n) = 1.0\delta(n) + 0.5\delta(n-1) + 0.25\delta(n-2). \tag{2.147}$$

The corresponding difference equation for $y(n)$ is given by

$$y(n) = x(n) + 0.5x(n-1) + 0.25x(n-2). \tag{2.148}$$

Now consider that the input $x(n)$ to the corresponding discrete time system is given by

$$x(n) = (0.6)^n u(n). \tag{2.149}$$

It follows that

$$
\begin{aligned}
y(0) &= (0.6)^0 = 1.0, \\
y(1) &= (0.6) + 0.5(1.0) = 1.1, \\
y(2) &= (0.6)^2 + 0.5(0.6) + 0.25(1.0) = 0.91, \\
y(3) &= (0.6)^3 + 0.5(0.6)^2 + 0.25(0.6) = 0.5420, \\
&\vdots \\
y(n) &= (0.6)^n + 0.5(0.6)^{n-1} + 0.25(0.6)^{n-2}, \\
y(n) &= (0.6)^{n-2}\left[(0.6)^2 + 0.5(0.6) + 0.25\right], \\
y(n) &= (0.6)^{n-2}(0.91), \quad 3 \le n \le \infty, \\
y(n) &= (0.6)^n\left[\frac{0.91}{(0.6^2}\right] = 2.5278(0.6)^n, \quad 3 \le n \le \infty. \tag{2.150}
\end{aligned}
$$

End of the Example

EXAMPLE 2.12
(Difference Equation from the Impulse Response)

The impulse response $h(n)$ for a discrete time system is given in Table 2.2. All of the values of $h(n)$ that are not in the table are equal to zero.

1. Write $h(n)$ using unit impulse functions
   Solution:

$$h(n) = 0.55\delta(n+3) + 0.75\delta(n+1) + \delta(n) + 0.75\delta(n-1) + 0.55\delta(n-3).$$

$$(2.151)$$

2. Write $h(n-k)$ using unit impulse functions.
   Solution:

$$\begin{aligned} h(n-k) &= 0.55\delta(n-k+3) + 0.75\delta(n-k+1) + \delta(n-k) \\ &+ 0.75\delta(n-k-1) + 0.55\delta(n-k-3). \end{aligned} \qquad (2.152)$$

3. Write a difference equation showing the relationship between the input $x(n)$ and the output $y(n)$.
   Solution: Eq. (2.152) can be used along with the convolution summation to write

$$\begin{aligned} y(n) &= \sum_{k=-\infty}^{\infty} x(k)h(n-k), \\ y(n) &= x(k)\left[0.55\delta(n-k+3) + 0.75\delta(n-k+1) + \delta(n-k)\right] \\ &+ x(k)\left[0.75\delta(n-k-1) + 0.55\delta(n-k-3)\right]. \end{aligned} \qquad (2.153)$$

The required difference equation can be obtained by evaluating the sum for arbitrary values of n,

$$y(n) = 0.55x(n+3) + 0.75x(n+1) + x(n) + 0.75x(n-1) + 0.55x(n-3).$$

$$(2.154)$$

End of the Example

2.11 NORMALIZED FREQUENCY REPRESENTATION

Generally, the discrete time samples resulting from sampling a continuous time signal can be represented by replacing the variable t by nT for the case where t is the independent variable for the continuous time system and n is the independent variable for the discrete time system. The parameter T is the sampling interval. A normalized frequency representation for the discrete time signal can then be obtained by replacing the continuous time radial frequency Ω by the discrete time frequency $\omega = \Omega T$.

Consider the signal

$$x(t) = Ae^{j\Omega t}u(t). \qquad (2.155)$$

The corresponding discrete time signal, after sampling, can be represented as

$$x(n) = Ae^{j\Omega nT}u(n) = Ae^{\omega n}u(n). \qquad (2.156)$$

Thus, the normalized discrete time frequency can be obtained by multiplying the continuous time frequency Ω by the sampling interval T.

The Sampling Theorem states that the largest frequency that can be represented in the sampled sequence is given by

$$\Omega_N < 2\pi f_N = \left(\frac{1.0}{2.0}\right)\left(\frac{2.0\pi}{T_{max}}\right) = \frac{\pi}{T_{max}}. \qquad (2.157)$$

It follows that the largest possible sampling interval to avoid aliasing is

$$T_{max} = \frac{\pi}{\Omega_N}. \qquad (2.158)$$

Thus, the normalized discrete time Nyquist frequency is always

$$\omega_N = \Omega_N T_{max} = \Omega_N \left(\frac{\pi}{\Omega_N}\right) = \pi. \qquad (2.159)$$

The appropriate fundamental radial frequency interval of concern for discrete time normalized frequencies is the range

$$-\pi \leq \omega \leq \pi. \qquad (2.160)$$

2.12 PERIODICITY FOR DISCRETE TIME SEQUENCES

The sample sequence obtained from sampling a periodic, continuous time signal is not necessarily periodic. The ratio between the sampling frequency and the fundamental frequency of the original, periodic, continuous time signal must be a rational number in order for the corresponding sampled sequence to be periodic [2]. This can be shown by considering the continuous time sinusoidal signal

$$x(t) = A\cos(\Omega t + \phi). \qquad (2.161)$$

The corresponding discrete time signal, after sampling $x(t)$, can be represented by

$$x(nT) = A\cos(\Omega nT + \phi). \qquad (2.162)$$

The requirement for periodicity can be stated as

$$x(nT) = A\cos(\Omega nT + \Omega NT + \phi) = A\cos(\Omega nT + \phi) \qquad (2.163)$$

where N is either a positive or negative integer. The following relationship must be true in order for $x(nT)$ to be periodic:

$$\Omega N T = 2\pi k, \qquad (2.164)$$

or

$$\Omega T = \frac{2\pi k}{N} \qquad (2.165)$$

where k is also an integer. If f is the frequency of the original, continuous time, signal and the sampling frequency $F_s = \frac{1}{T}$, then the requirement for periodicity is that

$$\Omega T = \frac{2\pi f}{F_s} = \frac{2\pi k}{N}, \qquad (2.166)$$

or

$$\frac{f}{F_s} = \frac{k}{N}. \qquad (2.167)$$

This means that the ratio of the sampling frequency and the fundamental frequency of the original, periodic signal must be a rational number. Thus, if a sampled, continuous time signal is periodic, the period, N, can be determined by finding the smallest integer values of k and N for which

$$\frac{f}{F_s} = \frac{k}{N}. \qquad (2.168)$$

Note that if normalized frequencies are used, where $F_s = \frac{1}{T} = 1$, then the requirement for periodicity becomes

$$\omega_0 = \frac{2\pi k}{N}, \qquad (2.169)$$

or

$$N = \frac{2\pi k}{\omega_0} \qquad (2.170)$$

where ω_0 is the normalized, radial frequency of the original, periodic, continuous time signal. Example 2.13 illustrates this point.

EXAMPLE 2.13
(Periodicity)

Problem:
Determine whether or not each of the following discrete time signals is periodic. If the signal is periodic, determine its fundamental period.

(a)

$$x(n) = \sin\left(\frac{8\pi}{7}n + 1\right).\tag{2.171}$$

Solution:
The requirement for the discrete time signal to be periodic requires that

$$N = \frac{2\pi k}{\omega_0} = \frac{(2\pi k)(7)}{8\pi} = \frac{7k}{4}.\tag{2.172}$$

The smallest integer value for N can be obtained if k is set to $k = 4$. Then, $x(n)$ is periodic with $N = 7$ samples.

(b)

$$x(n) = \cos\left(\frac{n}{5} - \pi\right).\tag{2.173}$$

Solution:
The normalized, original, continuous time, fundamental frequency is

$$\omega_0 = \frac{1}{5}.\tag{2.174}$$

The requirement for the discrete time signal to be periodic requires that

$$N = \frac{2\pi k}{\omega_0} = 2(5)\pi k = 10\pi k.\tag{2.175}$$

Since π is not a rational number, $x(n)$ is not periodic.

(c)

$$x(n) = \cos\left(\frac{\pi n}{2}\right)\cos\left(\frac{\pi n}{4}\right).\tag{2.176}$$

Solution:
The requirement for the discrete time signal to be periodic requires that

$$N_1 = \frac{2\pi k_1}{\omega_1} = \frac{(2\pi k_1)(2)}{\pi} = 4k_1,\tag{2.177}$$

$$N_2 = \frac{2\pi k_2}{\omega_2} = \frac{(2\pi k_2)(4)}{\pi} = 8k_2.\tag{2.178}$$

Choosing $k_1 = 2$ and $k_2 = 1$ results in $N_1 = N_2 = 8$, and $x(n)$ is periodic with period equal to 8 samples.

End of the Example

2.13 INTERCONNECTION OF SYSTEMS

The properties of discrete time systems can be used to determine the overall input–output relationship when two or more discrete time systems are interconnected.

EXAMPLE 2.14
(Serial Connection of Systems)

This example explores the relationships for the serial interconnection of two linear, shift invariant, discrete time systems.

1. Consider a system S with input $x(n)$ and output $y(n)$. This system is obtained through a series interconnection of a system S_1 followed by a system S_2. The input–output relationships for S_1 and S_2 are

$$(S_1) \quad y_1(n) = 2x_1(n) + 4x_1(n-1),$$
$$(S_2) \quad y_2(n) = x_2(n-2) + \frac{1}{2}x_2(n-3) \qquad (2.179)$$

where $x_1(n)$ and $x_2(n)$ denote input signals.

(a) Determine the input–output relationship of the system S.
Solution: The input to S_2 is the output from S_1 with the series interconnection and the output $y_2(n)$ is also the output $y(n)$. The input $x_1(n)$ is the input $x(n)$. Thus

$$y(n) = y_1(n-2) + \frac{1}{2}y_1(n-3), \qquad (2.180)$$

$$y_1(n-2) = 2x(n-2) + 4x(n-3), \qquad (2.181)$$

$$y_1(n-3) = 2x(n-3) + 4x(n-4). \qquad (2.182)$$

It follows that

$$y(n) = 2x(n-2) + 4x(n-3) + \frac{1}{2}[2x(n-3) + 4x(n-4)],$$
$$y(n) = 2x(n-2) + 5x(n-3) + 2x(n-4). \qquad (2.183)$$

(b) Does the input–output relationship of the system S change if the order in which S_1 and S_2 are connected in series is reversed (i.e., if S_1 follows S_2)?
Solution: Both systems are linear and time invariant. Therefore, the input–output relationship should be the same if the systems are reversed. If the systems are reversed, then the input to S_1 is $y_2(n)$ when the input $x_2(n)$ is $x(n)$. The output $y_1(n)$ is the output $y(n)$. Thus,

$$y(n) = 2y_2(n) + 4y_2(n-1),$$
$$y_2(n) = x(n-2) + \frac{1}{2}x(n-3), \qquad (2.184)$$

$$y_2(n-1) = x(n-3) + \frac{1}{2}x(n-4). \qquad (2.185)$$

Thus,

$$y(n) = 2\left[x(n-2) + \frac{1}{2}x(n-3)\right] + 4\left[x(n-3) + \frac{1}{2}x(n-4)\right],$$
(2.186)

$$y(n) = 2x(n-2) + 5x(n-3) + 2x(n-4)$$ (2.187)

which is the same input–output relationship.

End of the Example

2.14 DISCRETE TIME CORRELATION

Correlation is a mathematical abstraction to represent the relationship of two signals or of a signal with itself. One application is to determine the best shift of one signal relative to another to provide the best match. Assume that there are two signals $x_1(n)$ and $x_2(n)$ such that

$$x_1(n) = C_1 a^n u(n), \quad 0 \le n \le \infty$$ (2.188)

and

$$x_2(n) = C_2 b^n u(n), \quad 0 \le n \le \infty.$$ (2.189)

Note that a and b can be real or complex numbers.

The correlation sequence is defined as

$$
\begin{aligned}
y(m) &= \sum_{n=-\infty}^{\infty} x_1(n)x_2(n-m) \\
&= \sum_{n=-\infty}^{\infty} x_1(n+m)x_2(n).
\end{aligned}
$$
(2.190)

Note that either $x_1(n)$ or $x_2(n)$ can be shifted and the shift can be either positive or negative. The sequence $x_2(n)$ will be shifted in this example. It follows that

$$x_2(n-m) = C_2 b^{n-m} u(n-m).$$ (2.191)

Thus, the corresponding discrete time correlation is given by

$$y(m) = \sum_{n=-\infty}^{\infty} \left[C_1 a^n u(n)\right]\left[C_2 b^{n-m} u(n-m)\right].$$ (2.192)

This can be rewritten as

$$y(m) = \sum_{n=-\infty}^{\infty} \left[C_1 C_2 a^n b^{n-m} \right] \left[u(n)u(n-m) \right]. \tag{2.193}$$

Note that

$$u(n)u(n-m) = \begin{cases} 0 & \text{if } n < 0, \\ 1 & \text{if } m \le n \le \infty, \ m > 0, \\ 1 & \text{if } n > 0, \ m < 0, \\ 0 & \text{otherwise.} \end{cases} \tag{2.194}$$

Consider the case for $m > 0$ first. The terms or constants that are not functions of n can be factored out of the summation to obtain

$$y(m) = C_1 C_2 b^{-m} \sum_{n=m}^{\infty} a^n b^n, \tag{2.195}$$

$$y(m) = C_1 C_2 b^{-m} \sum_{n=m}^{\infty} (ab)^n. \tag{2.196}$$

Note that the above equation for $y(m)$ is a geometric series. Also, note that

$$y(m) = C_1 C_2 b^{-m} \sum_{n=m}^{\infty} (ab)^n = C_1 C_2 b^{-m} \sum_{n=0}^{\infty} (ab)^n - C_1 C_2 b^{-m} \sum_{n=0}^{m-1} (ab)^n. \tag{2.197}$$

Thus, if $a \neq b$, the result can be written as

$$y(m) = C_1 C_2 b^{-m} \lim_{N \to \infty} \left[\frac{1 - (ab)^{N+1}}{1 - (ab)} \right] - C_1 C_2 b^{-m} \left[\frac{1 - (ab)^m}{1 - (ab)} \right]. \tag{2.198}$$

If $|ab| < 1$, then

$$\lim_{N \to \infty} \left[\frac{1 - (ab)^{N+1}}{1 - (ab)} \right] = \frac{1}{1 - (ab)} \tag{2.199}$$

and

$$y(m) = C_1 C_2 \left[\frac{a^m}{1 - (ab)} \right]. \tag{2.200}$$

If $|ab| \ge 1$, then the correlation sequence is undefined.

If $m < 0$, then

$$u(n)u(n-m) = \begin{cases} 1 & \text{if } n \ge 0, \\ 0 & \text{otherwise.} \end{cases} \tag{2.201}$$

In that case

$$y(m) = C_1 C_2 b^{-m} \sum_{n=0}^{\infty} (ab)^n. \tag{2.202}$$

If $|ab| < 1$, then

$$y(m) = C_1 C_2 \left[\frac{b^{-m}}{1 - (ab)} \right]. \tag{2.203}$$

If $|ab| \geq 1$, then the correlation sequence is undefined.

The cross-correlation sequence has many practical applications. For example, it can be used to estimate the phase shift of a system by applying a known signal to the system and computing the cross-correlation between the input and the output of the system. The peak in the cross-correlation sequence determines the lag between the signals for them to have the maximum correlation. This concept is illustrated in Example 2.15.

EXAMPLE 2.15
(Cross-Correlation)

Consider the two signals

$$x_1 = \begin{cases} 0 & \text{for} \quad 0 \leq n \leq 4, \\ 1 & \text{for} \quad 5 \leq n \leq 15, \\ 0 & \text{for} \quad 16 \leq n \leq 20, \end{cases} \tag{2.204}$$

$$x_2 = \begin{cases} 1 & \text{for} \quad 0 \leq n \leq 10, \\ 0 & \text{for} \quad 11 \leq n \leq 20. \end{cases} \tag{2.205}$$

Alternatively, the signals can be written in terms of unit step sequences as follows:

$$x_1(n) = u(n - 5) - u(n - 16), \tag{2.206}$$

$$x_2(n) = u(n) - u(n - 11). \tag{2.207}$$

Matlab can be used to define the sequences as follows:

```
x1 = [zeros(1,5) ones(1,11) zeros(1,5)];
x2 = [ones(1,11) zeros(1,10)];
```

Fig. 2.36 gives the stem plot of $x_1(n)$, and Fig. 2.37 gives the stem plot of $x_2(n)$.

The Matlab "xcorr" cross-correlation function can be used to compute the cross-correlation of the two sequences using the following Matlab script:

```
y = xcorr(x1, x2);
m1 = length(y);
m2 = fix(0.5*(m1-1));
t = -m2:m2
```

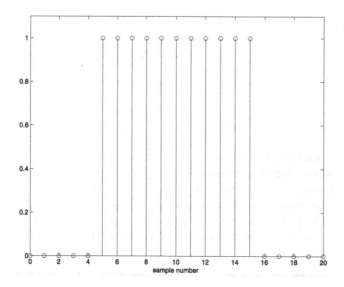

FIGURE 2.36

Input 1 for cross-correlation example

```
clf
figure(1)
stem(t, y);
```

Fig. 2.38 gives the cross-correlation sequence for the two signals. Note that the peak value of the cross-correlation sequence occurs where $n = 5$. This means that the best correlation between the two signals would be obtained if signal $x_2(n)$ is delayed by 5 samples. This is obvious from the plots.
End of the Example

2.15 THE Z TRANSFORM

The Z Transform provides a convenient way to analyze linear, shift invariant, discrete time systems. The Z Transform plays a role for discrete time signals and systems that is similar to the role played by the Laplace Transform for continuous time signals and systems. Thus, many of the properties of the Z Transform correspond to related properties of the Laplace Transform. This section introduces the Z Transform and its properties and discusses how to use the Z Transform for the analysis and characterization of discrete time signals and systems.

The Z Transform for a discrete time sequence, $x(n)$, is defined as

$$X(z) = \sum_{n=-\infty}^{\infty} x(n)z^{-n}. \qquad (2.208)$$

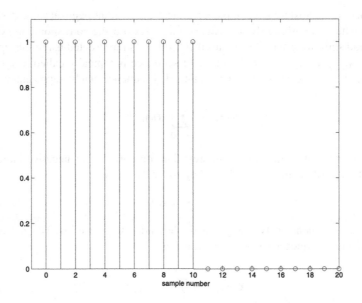

FIGURE 2.37

Input 2 for cross-correlation example

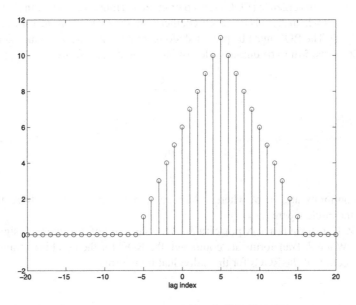

FIGURE 2.38

Block diagram of the overall system

Sequences of interest for many typical practical applications begin at some designated sample which can arbitrarily be assigned the index 0 if the corresponding system is linear and shift invariant. This means that sequences represented using this approach do not exist prior to $n = 0$. This type of sequence is typically called a right sided sequence [2]. The Z Transform for a right sided sequence can be represented as

$$X(z) = \sum_{n=0}^{\infty} x(n)z^{-n}. \tag{2.209}$$

There are some sequences of interest that are nonzero for negative indices and zero for all positive indices such that they cover the range

$$-\infty \leq n < 0. \tag{2.210}$$

This type of sequence is typically called a left sided sequence [2]. The Z Transform for a left sided sequence can be represented as

$$X(z) = \sum_{n=-\infty}^{-1} x(n)z^{-n}. \tag{2.211}$$

2.15.1 REGION OF CONVERGENCE

The region of convergence (ROC) for a particular Z Transform, $X(z)$, includes those values of z for which $X(z)$ is finite. Often, there is more than one ROC for a particular Z Transform. The ROC must be provided along with the Z Transform in order for the inverse Z Transform to be uniquely determined. The Z Transform

$$F(z) = \frac{Cz}{z - p} \tag{2.212}$$

has two ROCs:

1. $|z| < |p|$,
2. $|z| > |p|$. $\qquad (2.213)$

The discontinuity at $z = p$, where p may be complex, prohibits the ROC from including the circle where $|z| = |p|$.

The Z Transform is a very powerful tool for the study of discrete time signals and systems. When Z Transforms are combined, the ROC for the combined transform is the intersection of the ROCs for the individual transforms.

2.16 PROPERTIES OF THE Z TRANSFORM

Some useful properties of the Z Transform will be presented in this section.

2.16.1 LINEARITY

The linearity property is an important property for the analysis and synthesis of discrete time systems. This property permits the simple analysis of complex systems by combining the analysis of individual parts of the system using the principal of superposition.

Property 2.1. (Linearity)
If individual sequences, $x_k(n)$, have Z Transforms such that

$$x_k(n) \Leftrightarrow X_k(z) \ \forall \ 1 \le k \le K, \tag{2.214}$$

then

$$\sum_{k=1}^{K} a_k x_k(n) \Leftrightarrow \sum_{k=1}^{K} a_k X_k(z). \tag{2.215}$$

End of the Property

   This is the well known principle of superposition as it relates to Z Transforms. This property can be generalized to an arbitrary number of discrete time signals or systems. Example 2.16 verifies this property for the simple case of adding two sequences.

EXAMPLE 2.16
(Linearity Property)

Assume that two sequences and their corresponding Z Transforms are represented as follows

$$x_1(n)u(n) \Leftrightarrow X_1(z),$$
$$x_2(n)u(n) \Leftrightarrow X_2(z). \tag{2.216}$$

A new sequence can be formed as the sum of $x_1(n)$ and $x_2(n)$ such that

$$x_3(n) = c_1 x_1(n)u(n) + c_2 x_2(n)u(n). \tag{2.217}$$

Then the Z Transform of $x_3(n)$ is given by

$$X_3(z) = \sum_{n=0}^{\infty} [c_1 x_1(n) + c_2 x_2(n)]\, z^{-n}. \tag{2.218}$$

Since neither c_1 nor c_2 depend on n, $X_3(n)$ can be written in the form

$$X_3(z) = c_1 \sum_{n=0}^{\infty} x_1(n)z^{-n} + c_2 \sum_{n=0}^{n} x_2(n)z^{-n}. \tag{2.219}$$

However, by definition,

$$X_1(z) = \sum_{n=0}^{\infty} x_1(n)z^{-n} \tag{2.220}$$

and

$$X_2(z) = \sum_{n=0}^{\infty} x_2(n)z^{-n}. \tag{2.221}$$

It follows that

$$X_3(z) = c_1 X_1(z) + c_2 X_2(z). \tag{2.222}$$

End of the Example

The result in Example 2.16 is consistent with the linearity property. Note that the results for Example 2.16 are also based upon the fact that the constants c_1 and c_2 do not depend on the independent variable n.

2.16.2 TIME SHIFT

Linear, shift invariant, discrete time system are invariant regarding shifts in their independent variables. This property permits the initial value of the independent variable to be arbitrarily set to zero ($n = 0$) during the analysis of such systems. This can often simplify the analysis.

Property 2.2. (Time Shift)
Consider a linear, shift invariant, discrete time sequence with Z transform such that

$$x(n) \Leftrightarrow X(z). \tag{2.223}$$

If the sequence, $x(n)$, is shifted by k sample intervals to obtain $x(n-k)$, then the Z Transform of the shifted sequence, $x(n-k)$, is given as follows:

$$x(n-k) \Leftrightarrow z^{-k} X(z). \tag{2.224}$$

The ROC for the shifted sequence is the same as that for $X(z)$ except possibly at $z = 0$ for $k > 0$ and except possibly at $z = \infty$ for $k < 0$.
End of the Property

Example 2.6 verifies this property for the simple case where $x(n) = a^n u(n)$.

EXAMPLE 2.17
(Time Shift Property)

Consider the sequence defined by

$$x_1(n) = a^n \ \forall \ 0 \le n \le \infty. \tag{2.225}$$

The corresponding Z Transform is given by

$$X_1(z) = \sum_{n=0}^{\infty} a^n z^{-n}. \tag{2.226}$$

Now consider the sequence $x_2(n)$ that is formed by delaying $x_1(n)$ by k samples such that

$$x_2(n) = a^{n-k} \ \forall \ k \le n \le \infty. \tag{2.227}$$

The Z Transform for $x_2(n)$ is given by

$$X_2(z) = \sum_{n=k}^{\infty} a^{n-k} z^{-n}. \tag{2.228}$$

Let

$$m = n - k \tag{2.229}$$

which implies that

$$
\begin{aligned}
n &= m + k, \\
n = k &\rightarrow m = 0, \\
n = \infty &\rightarrow m = \infty.
\end{aligned}
\tag{2.230}
$$

If $m + k$ is substituted for n in Eq. (2.228) for $X_2(z)$, then the result is

$$X_2(z) = \sum_{m=0}^{\infty} a^m z^{-m-k} = z^{-k} \sum_{m=0}^{\infty} a^m z^{-m} = z^{-k} X_1(z). \tag{2.231}$$

End of the Example

Thus, we see that the time shift property holds for Example 2.17.

2.16.3 THE SCALING PROPERTY OF THE Z TRANSFORM

The scaling property of the Z Transform provides a procedure for obtaining the Z Transform of many useful sequences by using the scaling property with a simple Z Transform such as the Z Transform for the unit step. The scaling property can be represented as follows.

Property 2.3. (Z Transform Scaling)
If

$$x(n) \Leftrightarrow X(z), \quad ROC \quad r_1 < |z| < r_2, \tag{2.232}$$

then

$$c^n x(n) \Leftarrow X(c^{-1}z), \quad ROC \quad |c|r_1 < |z| < |c|r_2 \tag{2.233}$$

for any constant c, real or complex.
End of the Property

Example 2.18 illustrates the scaling property of the Z Transform domain through the use of a simple example.

EXAMPLE 2.18
(Scaling Property)

The Z Transform for the unit step function $u(n)$ is given by

$$U(z) = \sum_{n=0}^{\infty} z^{-n}, \quad ROC \quad 1 < |z| < \infty, \tag{2.234}$$

$$U(z) = \frac{1}{1 - z^{-1}}, \quad |z| > 1. \tag{2.235}$$

Define a new sequence, $x(n)$, by scaling the unit step function:

$$x(n) = c^n u(n), \quad |c| < 1. \tag{2.236}$$

The corresponding Z Transform of the sequence $x(n)$ is given by

$$X(z) = \sum_{n=0}^{\infty} c^n z^{-n} = \sum_{n=0}^{\infty} (c^{-1}z)^{-1}, \quad ROC \quad |c| < |z| < \infty.$$

$$X(z) = U(c^{-1}z) = \frac{1}{1 - (c^{-1}z^{-1})},$$

$$X(z) = \frac{1}{1 - cz^{-1}}, \quad ROC \quad |c| < |z| < \infty. \tag{2.237}$$

End of the Example

Thus, we see that the Z Transform scaling property holds for Example 2.18.

2.16.4 TIME REVERSAL

The time reversal property shows the relationship between a sequence and its transform with the sequence reversed in time and the corresponding Z Transform for the reversed sequence [4].

Property 2.4. (Time Reversal)
If

$$x(n) \Leftrightarrow X(z), \quad ROC \;\; r_1 < |z| < r_2, \tag{2.238}$$

then

$$x(-n) \Leftrightarrow X(z^{-1}), \quad ROC \;\; \frac{1}{r_2} < |z| < \frac{1}{r_1}. \tag{2.239}$$

End of the Property

Example 2.19 illustrates the time reversal property of the Z Transform.

EXAMPLE 2.19
(Time Reversal Property)

Consider the sequence

$$x_1(n) = a^n \; \forall \; 0 \le n \le \infty. \tag{2.240}$$

The Z Transform of $x_1(n)$ is given by

$$X_1(z) = \sum_{n=0}^{\infty} a^n z^{-n}, \quad ROC \; |z| > |a|. \tag{2.241}$$

Now consider $x_2(n)$ that is formed by reversing $x_1(n)$ such that

$$x_2(n) = x_1(-n) = a^{-n} \; \forall \; -\infty \le n \le 0. \tag{2.242}$$

The Z Transform of $x_2(n)$ is given by

$$X_2(n) = \sum_{n=-\infty}^{0} a^{-n} z^{-n}. \tag{2.243}$$

Let $k = -n$ in the equation for $X_2(z)$ to obtain

$$X_2(z) \;\; = \;\; \sum_{k=\infty}^{0} a^k z^k = \sum_{k=0}^{\infty} a^k (z^{-1})^{-k},$$

$$X_2(z) \;\; = \;\; X_1(z^{-1}), \quad ROC \; |z^{-1}| > |a| \text{ or } |z| < |a|. \tag{2.244}$$

End of the Example

Thus, the time reversal property holds for Example 2.19.

2.16.5 DIFFERENTIATION IN THE Z TRANSFORM DOMAIN

The differentiation of a Z Transform in the Z Transform domain can often be used to determine the Z Transform for sequences that are not easily represented as a geometric series.

Property 2.5. (Differentiation of the Z Transform)
If

$$x(n) \Leftrightarrow X(z) \tag{2.245}$$

then

$$nx(n) \Leftrightarrow -z\frac{dX(z)}{dz}. \tag{2.246}$$

End of the Property

Example 2.20 illustrates the property of differentiation of the Z Transform.

EXAMPLE 2.20
(Differentiation of the Z Transform)

Problem:
Show that the Z Transform resulting from the expansion of a geometric series is consistent with the Differentiation of the Z Transform property, where

$$X(z) = \sum_{n=0}^{\infty} a^n z^{-n}. \tag{2.247}$$

Solution:
The Z Transform of the expansion of a geometric series with parameter a is given by

$$X(z) = \sum_{n=0}^{\infty} a^n z^{-n} = \lim_{N\to\infty} \frac{1 - (az^{-1})^{N+1}}{1 - az^{-1}}. \tag{2.248}$$

If $|a| < 1$, then $X(z)$ can be written in closed form as

$$X(z) = \frac{1}{1 - az^{-1}} = \frac{z}{z - a}. \tag{2.249}$$

The differentiation of $X(z)$ in the Z plane is given by

$$\begin{aligned}
\frac{dX(z)}{dz} &= \frac{1}{z-a} - \frac{z}{(z-a)^2} \\
&= \frac{z-a}{(z-a)^2} - \frac{z}{(z-a)^2} \\
&= \frac{-a}{(z-a)^2}. \tag{2.250}
\end{aligned}$$

It follows that

$$-z\frac{dX(z)}{dz} = \frac{-z(-a)}{(z-a)^2} = \frac{az}{(z-a)^2}. \tag{2.251}$$

The derivative of $H(z)$ can also be determined from its geometric expansion as follows:

$$\frac{dX(z)}{dz} = \frac{d\left[\sum_{n=0}^{\infty} a^n z^{-n}\right]}{dz}$$

$$= \sum_{n=0}^{\infty} -na^n z^{-n-1}. \tag{2.252}$$

Thus,

$$-z\frac{dX(z)}{dz} = -z\left[\sum_{n=0}^{\infty} -na^n z^{-n-1}\right] = \sum_{n=0}^{\infty} na^n z^{-n}. \tag{2.253}$$

This result is the Z Transform of the discrete time sequence

$$g(n) = na^n \quad \forall\, 0 \le n \le \infty. \tag{2.254}$$

It follows that

$$G(z) = -z\frac{dX(z)}{dz} \Leftrightarrow nx(n), \tag{2.255}$$

$$G(z) = \frac{az}{(z-a)^2}. \tag{2.256}$$

This results verifies the differentiation property for $X(z)$.
End of the Example

2.16.6 CONVOLUTION OF TWO SEQUENCES

The convolution property of the Z Transform makes it convenient to obtain the Z Transform for the convolution of two sequences as the product of their respective Z Transforms.

Property 2.6. (Convolution using the Z Transform)
If two sequences $x_1(n)$ and $x_2(n)$ and their corresponding Z Transforms are given by

$$x_1(n) \Leftrightarrow X_1(z) \tag{2.257}$$

and

$$x_2(n) \Leftrightarrow X_2(z), \tag{2.258}$$

then the Z Transform of the convolution of the two sequences $x_1(n)$ and $x_2(n)$ is the product of their corresponding Z transforms. Thus,

$$y(n) = x_1(n) * x_2(n) \Leftrightarrow X_1(z)X_2(z). \tag{2.259}$$

End of the Property

2.16.7 CORRELATION OF TWO SEQUENCES

Property 2.7. (Correlation Property)
If two sequences $x_1(n)$ and $x_2(n)$ and their corresponding Z Transforms are given by

$$x_1(n) \Leftrightarrow X_1(z) \tag{2.260}$$

and

$$x_2(n) \Leftrightarrow X_2(z), \tag{2.261}$$

then the Z Transform of the correlation of the two sequences $x_1(n)$ and $x_2(n)$ is the product of $X_1(z)$ and $X_2(z^{-1})$ or the product of $X_1(z^{-1})$ and $X_2(z)$. Thus,

$$r_{x_1,x_2}(m) = \sum_{n=-\infty}^{\infty} x_1(n)x_2(n-m) \Leftrightarrow X_1(z)X_2(z^{-1}) \quad \text{or}$$

$$r_{x_1,x_2}(m) = \sum_{n=-\infty}^{\infty} x_1(n)x_2(n-m) \Leftrightarrow X_1(z^{-1})X_2(z). \tag{2.262}$$

End of the Property

2.16.8 THE INITIAL VALUE THEOREM

Property 2.8. (Initial Value)
If $x(n)$ is causal, then

$$x(0) = \lim_{z \to \infty} X(z). \tag{2.263}$$

End of the Property

2.16.9 THE FINAL VALUE THEOREM

Property 2.9. (Final Value)
If $x(n)$ is causal, then

$$\lim_{n \to \infty} x(n) = \lim_{z \to 1}(1 - z^{-1})X(z). \tag{2.264}$$

End of the Property

2.17 SYSTEM TRANSFER FUNCTION

A causal, linear, shift invariant, discrete time system may be characterized by a general linear constant coefficient difference equation which is a linear combination of current and past inputs and past inputs as follows:

$$y(n) = \sum_{k=0}^{L} b(k)x(n-k) - \sum_{k=1}^{L} a(k)y(n-k). \qquad (2.265)$$

The Z Transform of this difference equation is a rational system function that can also be used to characterize the system. The shift property for the Z Transform can be used to obtain

$$Y(z) = \sum_{k=0}^{L} b(k)z^{-k}X(z) - \sum_{k=1}^{L} a(k)z^{-k}Y(z). \qquad (2.266)$$

This can be simplified to obtain the relationship

$$Y(z) = \frac{\displaystyle\sum_{k=0}^{L} b(k)z^{-k}}{1.0 + \displaystyle\sum_{k=1}^{L} a(k)z^{-k}} X(z). \qquad (2.267)$$

The system transfer function is defined as the ratio of the Z Transforms of the output and the input. Thus, the system transfer function for the system corresponding to the difference equation in Eq. (2.265) is given by

$$H(z) = \frac{Y(z)}{X(z)} = \frac{\displaystyle\sum_{k=0}^{L} b(k)z^{-k}}{1.0 + \displaystyle\sum_{k=1}^{L} a(k)z^{-k}}. \qquad (2.268)$$

A linear, shift invariant, discrete time system can also be characterized by its impulse response. The impulse response can be determined by applying a discrete time impulse, $\delta(n)$, to its input. The corresponding output will be the system's impulse response, $h(n)$. The impulse response for a causal system is only defined for $n \geq 0$ when the impulse is applied at $n = 0$. The system transfer function for a linear, shift invariant, discrete time system is the Z Transform of its impulse response. Thus, the system transfer function for a causal, linear, shift invariant discrete time system is given by

$$H(z) = \sum_{n=0}^{\infty} h(n)z^{-n}. \qquad (2.269)$$

Discrete time systems are typically categorized in terms of the nature of their impulse responses. If the impulse response is of finite duration, the system is called a *finite impulse response* (FIR) system. If the system has an infinite impulse response, then the system is called an *infinite impulse response* (IIR) system.

Example 2.21 gives an example of an FIR discrete time system.

EXAMPLE 2.21
(Finite Impulse Response)

The impulse response for a discrete time system is given by

$$h(n) = \left\{ -0.0105,\ 0.0269, 0.4836, 0.4836, 0.0269, -0.0105 \right\} \qquad (2.270)$$
$$\uparrow$$

where the symbol ↑ identifies the sample at $n = 0$. The Z Transform of this impulse response is the system transfer function for the corresponding FIR discrete time system and is given by

$$\begin{aligned} H(z) &= -0.0105 + 0.0269z^{-1} + 0.4836z^{-2} + 0.4836z^{-3} \\ &\quad + 0.0269z^{-4} - 0.0105z^{-5}. \end{aligned} \qquad (2.271)$$

End of the Example

Infinite impulse response systems are usually recursive. The definition of a recursive system is given by [5]

> *Definition:* A system whose output $y(n)$ at time n depends on any number of past output values $y(n-1)$, $y(n-2)$, ..., is called a recursive system.

Example 2.22 gives an example of an IIR.

EXAMPLE 2.22
(Infinite Impulse Response)

The following is an impulse response for a discrete time system

$$h(n) = \begin{cases} (0.5)^n & \text{for } n \geq 0, \\ 0 & \text{for } n < 0. \end{cases} \qquad (2.272)$$

The Z Transform of this impulse response is the transfer function of the system and is given by

$$H(z) = \sum_{n=0}^{\infty} (0.5)^n z^{-n}. \tag{2.273}$$

The Z Transform, $H(z)$, can be represented in closed form provided that

$$\lim_{N \to \infty} (0.5z^{-1})^{N+1} = 0, \quad |z| > 0.5. \tag{2.274}$$

If the condition in Eq. (2.274) is met, then

$$H(z) = \lim_{N \to \infty} \frac{1 - (0.5z^{-1})^{N+1}}{1 - 0.5z^{-1}} = \frac{1}{1 - 0.5z^{-1}}. \tag{2.275}$$

$H(z)$ can be simplified to obtain the system transfer function for the corresponding IIR discrete time system as

$$H(z) = \frac{z}{z - 0.5}. \tag{2.276}$$

End of the Example

Note that in this case $H(z)$ is expressed in closed form as a ratio of two polynomials in z. In contrast, the transfer function for the FIR system has only a single polynomial in z, rather than a ratio of polynomials. In general, the closed form expression of the transfer function for an IIR system will be the ratio of two polynomials in z. On the other hand, the closed form expression of the transfer function for a FIR system will be a single polynomial in z.

Example 2.23 presents an example involving the determination of the impulse response of an FIR system from a difference equation representation of the system.

EXAMPLE 2.23
(FIR System Impulse Response from Difference Equation)
Problem:
Determine the difference equation for the FIR system as given by the following difference equation:

$$\begin{aligned} y(n) &= 0.007x(n) + 0.164x(n-1) + 0.4x(n-2) \\ &+ 0.164x(n-3) + 0.007x(n-4). \end{aligned} \tag{2.277}$$

Solution: The impulse response can be determined as the output, $y(n)$, with the input, $x(n)$, being a discrete time impulse:

$$x(n) = \delta(n) = \begin{cases} 1 & \text{for } n = 0, \\ 0 & \text{otherwise.} \end{cases} \tag{2.278}$$

Table 2.3 Impulse response of FIR system in Example 2.23

n	0	1	2	3	4	5	6	...
$h(n)$	0.007	0.164	0.4	0.164	0.007	0	0	...

The values of $h(n)$ can be determined by assigning values of n to the difference equation in Eq. (2.277). The corresponding values of $h(n)$ are given in Table 2.3.
End of the Example

Observe that the values of $h(n)$ are the same as the coefficients for the difference equation. This is always the case for the impulse response of an FIR discrete time system.

The impulse response of an IIR system can be obtained by applying an impulse to its input. Thus, the impulse response is the output of the system for the following input:

$$x(n) = \begin{cases} 1 & \text{for } n = 0, \\ 0 & \text{otherwise.} \end{cases} \qquad (2.279)$$

The Z Transform of the discrete time impulse is therefore

$$X(z) = \sum_{n=0}^{\infty} x(n)z^{-n} = 1.0. \qquad (2.280)$$

The corresponding Z Transform of the output can be determined as

$$Y(z) = H(z)X(z) = H(z). \qquad (2.281)$$

Thus, the impulse response for a linear, shift invariant, discrete time system is the inverse Z Transform of its system transfer function, $H(z)$. This applies for both FIR and IIR discrete time systems.

EXAMPLE 2.24
(System Impulse Response)

Problem:
The system function for a discrete time system is given by

$$H(z) = \frac{0.0653z^2 + 0.1306z + 0.0653}{(z - 0.81)(z + 0.375)}. \qquad (2.282)$$

Determine the impulse response for the system $h(n)$, $0 \leq n \leq \infty$.
Solution:
The impulse response for a discrete time system is the response of the system to a unit impulse. Thus,

$$x(n) = \delta(n). \qquad (2.283)$$

It follows that

$$X(z) = 1 \tag{2.284}$$

and

$$Y(z) = H(z)X(z) = H(z). \tag{2.285}$$

Thus, the impulse response for a discrete time system can be obtained as the inverse Z Transform of the system transfer function, $H(z)$. The inverse Z Transform of $H(z)$ can be obtained as follows:

$$\frac{H(z)}{z} = \frac{0.0653z^2 + 0.1306z + 0.0653}{z(z - 0.81)(z + 0.375)}$$

$$= \frac{A}{z - (-0.375)} + \frac{B}{z - 0.81} + \frac{C}{z}. \tag{2.286}$$

Unknown A, B, and C can be determined as follows:

$$A = \left[\frac{H(z)}{z}\right](z - 0.375)\Big|_{z=-0.375}$$

$$= \frac{0.0653(-0.375)^2 + 0.1306(-0.375) + 0.0653}{(-0.375)(-0.375 - 0.81)},$$

$$A = 0.0574,$$

$$B = \left[\frac{H(z)}{z}\right](z - 0.81)\Big|_{z=0.81} = \frac{0.0653(0.81)^2 + 0.1306(0.81) + 0.0653}{(0.81)(0.81 + 0.375)},$$

$$B = 0.2228,$$

$$C = \left[\frac{H(z)}{z}\right](z)\Big|_{z=0} = \frac{0.0653(0)^2 + 0.1306(0) + 0.0653}{(0.0 - 0.81)(0.0 + 0.375)},$$

$$C = -0.2149. \tag{2.287}$$

Thus, $H(z)$ can be written as

$$H(z) = \frac{0.0574z}{z - (-0.375)} + \frac{0.2228z}{z - 0.81} - 0.2149. \tag{2.288}$$

The inverse Z Transform of $H(z)$ can be determined term by term to obtain $h(n)$. Thus,

$$h(n) = 0.0574(-0.375)^n u(n) + 0.2228(0.81)^n u(n) - 0.2149\delta(n). \tag{2.289}$$

End of the Example

2.18 POLES AND ZEROS

The category of Z Transforms that can be represented as a ratio of two polynomials is an important category for discrete time systems. If the Z Transform, $X(z)$, is a rational function, then $X(z)$ can be written in the form [4]

$$X(z) = \frac{B(z)}{A(z)} \qquad (2.290)$$

where the numerator polynomial, $B(z)$, can be written in the form

$$B(z) = \sum_{k=0}^{M} b(k)z^{-k} \qquad (2.291)$$

and the denominator polynomial, $A(z)$, can be written in the form

$$A(z) = \sum_{k=0}^{N} a(k)z^{-k}. \qquad (2.292)$$

It follows that if a Z Transform is a rational function, then it can be written in the form

$$X(z) = \frac{\displaystyle\sum_{k=0}^{M} b(k)z^{-k}}{\displaystyle\sum_{k=0}^{N} a(k)a^{-k}}. \qquad (2.293)$$

The poles of $X(z)$ are defined as the values of z for which $X(z) = \infty$. The zeros of $X(z)$ are defined as the values of z for which $X(z) = 0$.

The terms $b(0)z^{-M}$ and $a(0)z^{-N}$ can be factored out of the numerator and denominator of $X(z)$, respectively, to obtain

$$X(z) = \left[\frac{b(0)z^{-M}}{a(0)z^{-N}}\right]\left[\frac{z^M + \frac{b(1)}{b(0)}z^{M-1} + \cdots + \frac{b(M)}{b(0)}}{z^N + \frac{a(0)}{a(0)}z^{N-1} + \cdots + \frac{a(N)}{a(0)}}\right]. \qquad (2.294)$$

The polynomials $B(z)$ and $A(z)$ can then be expressed in factored form to represent $X(z)$ in the form

$$X(z) = \left[\frac{b(0)z^{-M}}{a(0)z^{-N}}\right]\left[\frac{(z - q_1)(z - q_2)\cdots(z - q_M)}{(z - p_1)(z - p_2)\cdots(z - p_N)}\right]. \qquad (2.295)$$

There are three possible cases:

Case 1 If $M > N$, then $M - N$ zeros are located at $q_k \; \forall \; 1 \le k \le M - N$ and $M - N$ zeros are located at $z = 0$. There are N poles located at $p_k \; \forall \; 1 \le k \le N$ and $M - N$ poles are located at $z = \infty$.

Case 2 If $M = N$, then there are M zeros located at q_k $\forall$ $1 \le k \le M$ and there are N poles p_k $\forall$ $1 \le k \le N$.

Case 3 If $N > M$ then there are $N - M$ zeros located at q_k $\forall$ $1 \le k \le M$ and there are $N - M$ zeros at $z = \infty$. There are N poles located at p_k $\forall$ $1 \le k \le N - M$ and there are $N - M$ poles located at $z = 0$.

Thus, if all of the poles and zeros of $X(z)$ located at both $z = 0$ and $z = \infty$ are counted, then $X(z)$ has the same number of poles and zeros. Example 2.25 illustrates this concept.

EXAMPLE 2.25
(Poles and Zeros)

Problem:
Determine the poles and zeros for the following system transfer function:

$$X(z) = \frac{0.0439 + 0.1317z^{-1} + 0.1317z^{-2} + 0.0439z^{-3}}{1.000 - 0.5000z^{-1} - 0.2599z^{-2} + 0.2813z^{-3} - 0.1702z^{-4}}. \quad (2.296)$$

Solution: $M = 3$ and $N = 4$ for this system transfer function. Thus, $X(z)$ can be written in the form

$$X(z) = \frac{0.0439z^{-3}\left[\frac{0.0439}{0.0439}z^3 + \frac{0.1317}{0.0439}z^2 + \frac{0.1317}{0.0439}z + \frac{0.0439}{0.0439}\right]}{z^{-4}\left[z^4 - 0.5000z^3 - 0.2599z^2 + 0.2813z - 0.1702\right]}, \quad (2.297)$$

$$X(z) = \frac{0.0439z\left[z^3 + 3.00z^2 + 3.00z + 1.00\right]}{\left[z^4 - 0.5000z^3 - 0.2599z^2 + 0.2813z - 0.1702\right]}. \quad (2.298)$$

We can factor the polynomials in the numerator and the denominator to obtain

$$X(z) = \frac{0.0439z(z + 1)^3}{(z + 0.75)(z - 0.75)(z - 0.25 - 0.49j)(z - 0.25 + 0.49j)}. \quad (2.299)$$

It follows directly that the poles are located at

$$
\begin{aligned}
p_1 &= -0.75, \\
p_2 &= 0.75, \\
p_3 &= 0.25 + 0.49j, \\
p_4 &= 0.25 - 0.49j.
\end{aligned} \quad (2.300)
$$

The zeros are given by

$$
\begin{aligned}
q_1 = q_2 = q_3 &= -1.00, \\
q_4 &= 0.0.
\end{aligned} \quad (2.301)
$$

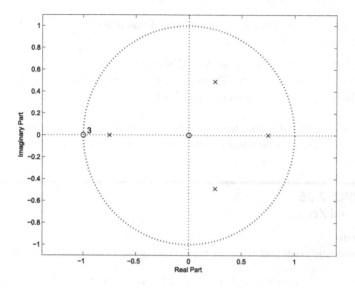

FIGURE 2.39

Pole–zero plot for Example 2.25

There are 4 poles and 4 zeros for $X(z)$ which verifies that the number of poles and the number zeros are the same for $X(z)$. Fig. 2.39 gives a pole–zero plot in the complex Z plane of the poles and zeros of $X(z)$.
End of the Example

2.19 THE DIFFERENCE EQUATION AND THE TRANSFER FUNCTION

The general form for the system transfer function of a causal, linear, shift invariant discrete time system is given by

$$H(z) = \frac{\displaystyle\sum_{k=0}^{L} b(k)z^{-k}}{1.0 + \displaystyle\sum_{k=1}^{L} a(k)z^{-k}}. \tag{2.302}$$

Fig. 2.40 gives a block diagram of the discrete time system. The system transfer function, $H(z)$, can also be written in terms of the ratio of the Z Transforms of the

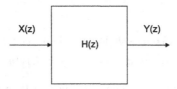

FIGURE 2.40

Block diagram for a discrete time system

output and the input as

$$H(z) = \frac{Y(z)}{X(z)}. \tag{2.303}$$

It follows that

$$Y(z)\left[1.0 + \sum_{k=1}^{L} a(k)z^{-k}\right] = X(z)\left[\sum_{k=0}^{L} b(k)z^{-k}\right]. \tag{2.304}$$

The time shift property of the Z Transform can be used to obtain the difference equation. Thus, if $x(n-k) = 0 \ \forall \, n < k$, the time shift property indicates that

$$z^{-k}X(z) = \sum_{n=0}^{\infty} x(n-k)z^{-n}. \tag{2.305}$$

Similarly, if $y(n-k) = 0 \ \forall \, n < k$, the time shift property indicates that

$$z^{-k}Y(z) = \sum_{n=0}^{\infty} y(n-k)z^{-n}. \tag{2.306}$$

This result can be used to modify Eq. (2.304) to obtain

$$\sum_{n=0}^{\infty}\left[y(n) + a(1)y(n-1) + \cdots + a(L)y(n-L)\right]z^{-n}$$

$$= \sum_{n=0}^{\infty}\left[b(0)x(n) + b(1)y(n-1) + \cdots + b(L)x(n-L)\right]z^{-n}. \tag{2.307}$$

The difference equation for the system can be obtained by equating powers of z

$$\begin{aligned} y(n) \ &+ \ a(1)y(n-1) + \cdots + a(L)y(n-L) \\ &= \ b(0)x(n) + b(1)x(n-1) \\ &\quad + \cdots + b(L)x(n-L), \end{aligned} \tag{2.308}$$

or

$$y(n) = b(0)x(n) + b(1)y(n-1) + \cdots + b(L)x(n-L) \\ - a(1)y(n-1) - \cdots - a(L)y(n-L). \tag{2.309}$$

It follows that the causal, linear shift invariant discrete time system can be represented in either of the following standard forms:

$$H(z) = \frac{\displaystyle\sum_{k=0}^{L} b(k)z^{-k}}{1.0 + \displaystyle\sum_{k=1}^{L} a(k)z^{-k}},$$

$$y(n) = \sum_{k=0}^{L} b(k)x(n-k) - \sum_{k=1}^{L} a(k)y(n-k). \tag{2.310}$$

Note the relationship between the coefficients $b(k)$ and $a(k)$ in the system transfer function and the corresponding coefficients in the difference equation. This relationship makes it easy to write the difference equation for a given system transfer function by inspection and vice versa. This relationship can be used to write the difference equation for a given system transfer function by inspection as follows:

1. Represent the system transfer function in standard form

$$H(z) = \frac{\displaystyle\sum_{k=0}^{L} b(k)z^{-k}}{1.0 + \displaystyle\sum_{k=1}^{L} a(k)z^{-k}}. \tag{2.311}$$

2. Form the equation for the weighted inputs as

$$r(n) = \sum_{k=0}^{L} b(k)x(n-k). \tag{2.312}$$

3. Form the equation for the weighted outputs as

$$s(n) = -\sum_{k=1}^{L} a(k)y(n-k). \tag{2.313}$$

4. Form the equation for the output as the sum of $r(n)$ and $s(n)$ as

$$y(n) = r(n) + s(n) = \sum_{k=0}^{L} b(k)x(n-k) - \sum_{k=1}^{L} a(k)y(n-k). \tag{2.314}$$

Example 2.26 illustrates this concept.

EXAMPLE 2.26
(Difference Equation from the System Transfer Function)

Problem:
Determine the difference equation corresponding to the system transfer function given by

$$H(z) = \frac{0.3118 - 0.9355z^{-1} + 0.9355z^{-2} - 0.3118z^{-3}}{1.0000 - 0.8682z^{-1} + 0.5344z^{-2} - 0.0921z^{-3}}. \tag{2.315}$$

Solution:
Designate the input sequence as $x(n)$ and the output sequence as $y(n)$. Note that $a(0) = 1$. Therefore, the system transfer function is in the standard form. The part of difference equation related to the weighted sum of inputs can be designated as $r(n)$. It follows that

$$r(n) = 0.3118x(n) - 0.9355x(n-1) + 0.9355x(n-2) - 0.3118x(n-3). \tag{2.316}$$

The part of the difference equation related to the weighted sum of past values of outputs can be obtained by forming the weights as the negated values of the corresponding denominator coefficients of the system transfer function. Designate this part of the difference equation as $s(n)$. It follows that

$$s(n) = 0.8682y(n-1) - 0.5344y(n-2) + 0.0921y(n-3). \tag{2.317}$$

Form the difference equation as the sum of $r(n)$ and $s(n)$:

$$\begin{aligned} y(n) &= r(n) + s(n), \\ y(n) &= 0.3118x(n) - 0.9355x(n-1) + 0.9355x(n-2) - 0.3118x(n-3) \\ &\quad + 0.8682y(n-1) - 0.5344y(n-2) + 0.0921y(n-3). \end{aligned} \tag{2.318}$$

End of the Example

2.20 THE Z TRANSFORM OF A SEQUENCE

The Fourier Transform, or the Fourier Series for periodic sequences, is defined on the unit circle in the Z Transform domain where $|z| = 1$. The Fourier Transform or the Fourier series converges when the ROC for the corresponding Z Transform includes the unit circle. Thus, if a particular Z Transform has more than one ROC and one of the ROCs includes the unit circle, then the ROC that includes the unit circle should be chosen to compute the inverse of the Z Transform to ensure that the resulting sequence is bounded. The following example illustrates this concept.

EXAMPLE 2.27
(Z Transform of a Sequence)

Consider the sequence

$$x(n) = \begin{cases} a^n, & n \geq 0, \\ -b^n, & n \leq -1. \end{cases} \tag{2.319}$$

The definition of the Z Transform, as given in Eq. (2.208), can be used to determine the Z Transform for this sequence as

$$X(z) = \sum_{n=-\infty}^{-1} (-b^n)z^{-n} + \sum_{n=0}^{\infty} a^n z^{-n}. \tag{2.320}$$

A closed form expression for $X(z)$ can be obtained for those values of z which are in the ROC for $X(z)$ since it is a geometric series. Let

$$S_1(z) = \sum_{n=-\infty}^{-1} (-b^n)z^{-n} \tag{2.321}$$

and

$$S_2(z) = \sum_{n=0}^{\infty} a^n z^{-n}. \tag{2.322}$$

Then

$$S_1(z) = -\sum_{m=1}^{\infty} b^{-m} z^m = -\sum_{m=1}^{\infty} (b^{-1}z)^m. \tag{2.323}$$

A geometric series in the form

$$Y = \sum_{n=0}^{N} c^n \tag{2.324}$$

can be written in closed form as

$$Y = \frac{1 - c^{N+1}}{1 - c}. \tag{2.325}$$

This result can be used to obtain a closed form expression for $S_1(z)$ as

$$S_1(z) = -\left[\sum_{m=0}^{\infty}(b^{-1}z)^m - 1\right] \qquad (2.326)$$

where the term for $m = 0$ has been added and subtracted. Thus,

$$S_1(z) = -\lim_{M\to\infty}\left[\frac{1-(b^{-1}z)^{M+1}}{1-(b^{-1}z)} - 1\right]. \qquad (2.327)$$

If

$$\left|b^{-1}z\right| < 1 \Rightarrow (|z| < |b|) \qquad (2.328)$$

then the ROC includes the unit circle where $|z| = 1$ if $|b| > 1$. It follows that for the case where $|b| > 1$,

$$\lim_{M\Rightarrow\infty}(b^{-1}z)^{M+1} = 0 \qquad (2.329)$$

and

$$S_1(z) = -\left[\frac{1}{1-(b^{-1}z)} - 1\right], \qquad (2.330)$$

$$S_1(z) = -\left[\frac{1}{1-(b^{-1}z)} - 1\right], \qquad (2.331)$$

$$S_1(z) = -\left[\frac{1-1+(b^{-1}z)}{1-(b^{-1}z)}\right] = -\left[\frac{b^{-1}z}{1-(b^{-1}z)}\right], \qquad (2.332)$$

$$S_1(z) = -\left[\frac{1}{b}\right]\left[\frac{z}{1-(b^{-1}z)}\right]. \qquad (2.333)$$

This result can be simplified to obtain

$$S_1(z) = \frac{z}{z-b}. \qquad (2.334)$$

A closed form expression for $S_2(z)$ can be obtained in a similar way:

$$S_2(z) = \lim_{N\to\infty}\left[\frac{1-(az^{-1})^{N+1}}{1-(az^{-1})}\right]. \qquad (2.335)$$

If

$$\left|az^{-1}\right| < 1 \Rightarrow (|a| < |z|) \qquad (2.336)$$

then the unit circle is in the ROC if $|a| < 1$. It follows that for the case where $|a| < 1$

$$\lim_{N \to \infty} (az^{-1})^{N+1} = 0 \tag{2.337}$$

and

$$S_2(z) = \left[\frac{1}{1 - (az^{-1})} \right]. \tag{2.338}$$

This result can be simplified to obtain

$$S_2(z) = \frac{z}{z - a}. \tag{2.339}$$

The two partial results can be combined to obtain

$$X(z) = S_1(z) + S_2(z) = \frac{z}{z - a} + \frac{z}{z - b}, \tag{2.340}$$

$$X(z) = \frac{2z^2 - (a+b)z}{z^2 - (a+b)z + ab}. \tag{2.341}$$

This Z Transform has finite values for the region

$$|a| < |z| < |b| \ \forall \ |a| < |b|. \tag{2.342}$$

The ROC includes the unit circle and the sequence is bounded if

$$|a| < 1.0 \ \text{ and } \ |b| > 1.0. \tag{2.343}$$

On the other hand, if

$$|b| < |a| \tag{2.344}$$

then the wrong way has been chosen to partition $X(z)$ and the two sequences $S_1(z)$ and $S_2(z)$ have no common region of convergence. Fig. 2.41 shows the region of convergence for $S_1(z)$ with $b = 3$. The ROC is inside the circle with radius $r = b$. Fig. 2.42 shows the region of convergence for $a = 0.5$. The ROC is outside the circle with radius $r = 0.5$. Fig. 2.43 shows the region of convergence for $X(z)$ with $a = 0.5$ and $b = 3.0$. The ROC is the region between the two circles with radii $r = -0.5$ and $r = 3.0$.
End of the Example

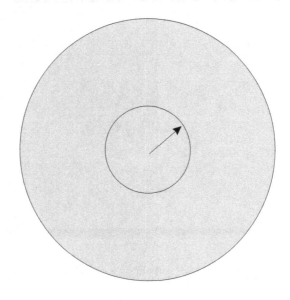

FIGURE 2.41

Region of convergence for $S_1(z)$ with $b = 3$

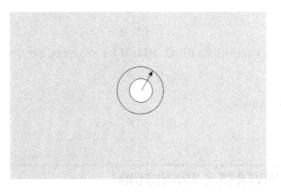

FIGURE 2.42

Region of convergence for $S_2(z)$ with $a = 0.5$

2.21 BOUNDED INPUT, BOUNDED OUTPUT STABILITY

Bounded input, bounded output (BIBO) stability is a form of stability often used for signal processing applications. The requirement for a linear, shift invariant, discrete time system to be BIBO stable is for the output to be bounded for every input to the system that is bounded. Assume that R is a real or complex number with finite

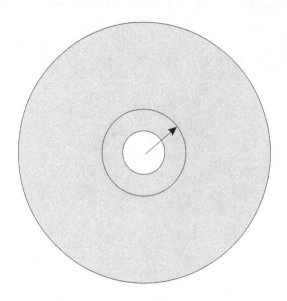

FIGURE 2.43

Region of convergence for $X(z)$ with $a = 0.5$, $b = 3.0$

magnitude such that

$$|R| < \infty. \tag{2.345}$$

A discrete time system will be BIBO stable if for all inputs, $x(n)$, such that

$$|x(n)| < |R| < \infty, \tag{2.346}$$

the output, $y(n)$, satisfies

$$|y(n)| < |R| < \infty. \tag{2.347}$$

2.22 THE INVERSE Z TRANSFORM

Many practical applications of the Z Transform involve the determination of a signal sequence for a given Z Transform. This process in called the inverse Z Transform.

2.22.1 PARTIAL FRACTION EXPANSION

Several illustrations have been presented that show that a Z Transform in the form

$$X(z) = C \sum_{n=0}^{\infty} a^n z^{-n}, \quad |a| < 1,$$

$$X(z) = \frac{C}{1 - az^{-1}} \tag{2.348}$$

is a geometric series with an inverse Z Transform given by

$$x(n) = Ca^n u(n) \tag{2.349}$$

which is a right sided sequence. If $|a| > 1$ then

$$x(n) = a^n u(-1 - n), \tag{2.350}$$

which is a left sided sequence. Thus, a convenient way to determine the inverse Z Transform for a Z Transform with distinct poles would be to expand the transform into the sum of terms like the term in Eq. (2.348). The inverse Z Transform could then be determined as the sum of the inverse Z Transforms of each of the individual parts using the principle of superposition, which holds for linear, shift invariant discrete time systems.

Consider a Z Transform with distinct poles such that

$$X(z) = \frac{b_0 z^3 + b_1 z^2 + b_2 z + b_3}{z^3 + a1 z^2 + a2 z + a3} = \frac{b_0 z^3 + b_1 z^2 + b_2 z + b_3}{(z - p_1)(z - p_2)(z - p_3)} \tag{2.351}$$

where p_1, p_2, and p_3 are the three distinct poles of $X(z)$. The goal is to expand $X(z)$ into terms that look like

$$X(z) = C_0 + \frac{C_1}{1 - p_1 z^{-1}} + \frac{C_2}{1 - p_2 z^{-1}} + \frac{C_3}{1 - p_3 z^{-1}}. \tag{2.352}$$

However, a straightforward partial fraction expansion of $X(z)$ would yield the form

$$X(z) = D_0 + \frac{D_1}{z - p_1} + \frac{D_1}{z - p_1} + \frac{D_1}{z - p_1}. \tag{2.353}$$

A partial fraction expansion, in the desired form, can be obtained by expanding $z^{-1} X(z)$ into partial fractions as follows:

$$\frac{X(z)}{z} = \frac{E_0}{z} + \frac{E_1}{z - p_1} + \frac{E_2}{z - p_2} + \frac{E_3}{z - p_3}. \tag{2.354}$$

The coefficient, E_0, can be determined using the equation

$$E_0 = \left. \frac{z X(z)}{z} \right|_{z=0} = X(z)|_{z=0}. \tag{2.355}$$

The coefficients, E_k, $\forall\, 1 \leq k \leq 3$, can be determined using the equation

$$E_k = \left. \frac{(z - p_k) X(z)}{z} \right|_{z=p_k}. \tag{2.356}$$

The following example illustrates how to use this approach to find a bounded sequence as the inverse for a given Z Transform with distinct poles.

EXAMPLE 2.28
(Inverse Z Transform)

Problem: Find a bounded sequence with the following Z Transform:

$$X(z) = \frac{1.3915z + 1.6670}{z^2 + 2.2329z + 1.0000}. \qquad (2.357)$$

Solution: The poles of $X(z)$ are given by

$$
\begin{aligned}
p_1 &= -0.6200, \\
p_2 &= -1.6129. \qquad (2.358)
\end{aligned}
$$

Thus, $z^{-1}X(z)$ can be written in the form

$$\frac{X(z)}{z} = \frac{1.3915z + 1.6670}{z(z + 0.6200)(z + 1.6129)} = \frac{C_0}{z} + \frac{C_1}{z + 1.6129} + \frac{C_2}{z + 0.6200}. \qquad (2.359)$$

The coefficient, C_0, can be computed as follows:

$$C_0 = \left.\frac{zX(z)}{z}\right|_{z=0} = \frac{1.3915(0) + 1.6670}{(0 + 0.62)(0 + 1.6129)} = 1.6670. \qquad (2.360)$$

The coefficients, C_1 and C_2, can be computed as follows:

$$
\begin{aligned}
C_1 &= \left.\frac{(z + 1.6129)(1.3915z + 1.6670)}{z(z + 0.6200)(z + 1.6129)}\right|_{z=-1.6129} \\
&= \frac{1.3915(-1.6129) + 1.6670}{(-1.6129)(-1.6129 + 0.6200)} = -0.3605, \\
C_2 &= \left.\frac{(z + 0.6200)(1.3915z + 1.6670)}{z(z + 0.6200)(z + 1.6129)}\right|_{z=-0.6200} \\
&= \frac{1.3915(-0.6200) + 1.6670}{(-0.6200)(-0.6200 + 1.6129)} = -1.3065. \qquad (2.361)
\end{aligned}
$$

It follows that

$$
\begin{aligned}
X(z) &= \frac{1.6670z}{z} - \frac{0.3605z}{z + 1.6129} - \frac{1.3065z}{z + 0.62} \\
&= 1.6670 - \frac{0.3605}{1.0 + 1.6129z^{-1}} - \frac{1.3065}{1.0 + 0.62z^{-1}}. \qquad (2.362)
\end{aligned}
$$

Let

$$X_0(z) = 1.6670, \qquad (2.363)$$

$$X_1(z) = \frac{-0.3605}{1.0 + 1.6129z^{-1}}, \qquad (2.364)$$

and

$$X_2(z) = \frac{-1.3605}{1.0 + 0.6200z^{-1}} \qquad (2.365)$$

such that

$$X(z) = X_0(z) + X_1(z) + X_2(z). \qquad (2.366)$$

It will be practical to find the appropriate bounded sequence for each term and then combine the sequences.

The inverse Z Transform of $X_0(z)$, which is a constant, can readily be determined as

$$x_0(n) = 1.6670,$$
$$x_0(n) = 0 \; \forall \, n \neq 0. \qquad (2.367)$$

The Z Transform, $X_1(z)$, has a pole with a magnitude greater than 1.0. Therefore, its inverse Z Transform must be a left sided sequence in order for it to be bounded. If a pole, p, has a magnitude greater than 1.0, and if $|z| = 1$, then $\left|p^{-1}z\right| < 1$, and

$$\lim_{N \to -\infty} \left(pz^{-1}\right)^{N+1} = 0. \qquad (2.368)$$

Thus, $X_1(z)$ can be written in the form

$$X_1(z) = 0.3605 \left[- \sum_{n=-\infty}^{-1} (-1.6129)^n z^{-n} \right]. \qquad (2.369)$$

And so, with the requirement that the ROC includes the unit circle, the inverse Z Transform for $X_1(z)$ can be readily determined as

$$x_1(n) = 0.3605(-1.6129)^n, \quad -\infty \leq n \leq -1. \qquad (2.370)$$

Since the magnitude of the pole for $X_2(z)$ is less than 1.0, the requirement that the ROC includes the unit circle leads to the result that the inverse Z Transform of $X_2(z)$ is a right sided sequence. Thus, with the requirement that $|z| = 1$, it follows that

$$\lim_{N \to \infty} (-0.6200z^{-1})^{N+1} = 0 \qquad (2.371)$$

and

$$X_2(z) = -1.3065 \sum_{n=0}^{\infty} (-0.6200)^n z^{-n}. \qquad (2.372)$$

The inverse Z Transform for $X_2(z)$ can easily be determined to be

$$x_2(n) = -1.3065(-0.6200)^n u(n). \qquad (2.373)$$

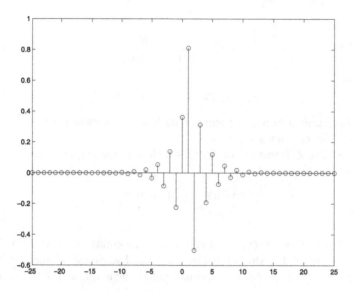

FIGURE 2.44

Stem plot of $x(n)$ in Example 2.28 for $n = -25$ to $n = 25$

The three inverse Z Transforms can be combined to obtain $x(n)$ as follows:

$$
\begin{aligned}
x(n) &= -0.3605(-1.6129)^n \quad \text{for } -\infty \leq n \leq -1, \\
x(0) &= 1.6670 - 1.3065 = 0.3605, \\
x(n) &= -1.3065(-0.6200)^n \quad \text{for } 1 \leq n \leq \infty.
\end{aligned}
\tag{2.374}
$$

Fig. 2.44 gives a stem plot of $x(n)$ for $n = -25$ to $n = 25$.
End of the Example

The impulse response of a linear, shift invariant, discrete time system by computing the inverse Z Transform of its system transfer function, $H(z)$. Example 2.29 illustrates this concept.

EXAMPLE 2.29
(Impulse Response of IIR Discrete Time System)

Problem:
The system transfer function for an IIR discrete time system is given by

$$
H(z) = \frac{2z^2 + 4z + 2}{z^2 + 0.9000z + 0.2000}
\tag{2.375}
$$

which can also be written as

$$H(z) = \frac{2(z+1)^2}{(z+0.5)(z+0.4)}. \tag{2.376}$$

Determine the impulse response for the corresponding discrete time system.
Solution:
If the input is an impulse, then the Z Transform of the input, $\{x(n)\}$, is given by

$$X(z) = \sum_{n=-\infty}^{\infty} \delta(n)z^{-n} = 1. \tag{2.377}$$

Thus

$$Y(z) = H(z)X(z) = H(z). \tag{2.378}$$

It follows that the impulse response of a linear, shift invariant, discrete time system can be obtained by computing the inverse Z Transform of the system transfer function, $H(z)$. The Z Transform, $z^{-1}H(z)$, can be expanded into partial fractions as follows.

$$\frac{H(z)}{z} = \frac{C_0}{z} + \frac{C_1}{z+0.5} + \frac{C_2}{z+0.4}. \tag{2.379}$$

The partial fraction coefficients can be computed as follows:

$$C_0 = \left.\frac{2z^2 + 4z + 2}{(z+0.5)(z+0.4)}\right|_{z=0} = \frac{2(0)+4(0)+2}{(0+0.5)(0+0.4)} = 10.0,$$

$$C_1 = \left.\frac{2z^2 + 4z + 2}{z(z+0.4)}\right|_{z=-0.5} = \frac{2(-0.5)^2+4(-0.5)+2}{(-0.5)(-0.5+0.4)} = 10.0, \tag{2.380}$$

$$C_2 = \left.\frac{2z^2 + 4z + 2}{z(z+0.5)}\right|_{z=-0.4} = \frac{2(-0.4)^2+4(-0.4)+2}{(-0.4)(-0.4+0.5)},$$

$$C_2 = -18.0 \tag{2.381}$$

Thus,

$$H(z) = 10 + \frac{10z}{z+0.5} - \frac{18z}{z+0.4} \tag{2.382}$$

which can also be written in the form

$$H(z) = 10 + 10\left(\frac{1.0}{1-(-0.5z^{-1})}\right) - 18\left(\frac{1.0}{1-(-0.4z^{-1})}\right). \tag{2.383}$$

The poles of $H(z)$ both have magnitudes less than 1.0. Thus, the inverse Z Transform of $H(z)$ is a right sided sequence and $H(z)$ can be written in the following form:

$$H(z) = 10 + 10\sum_{n=0}^{\infty}(-0.5)^n z^{-n} - 18\sum_{n=0}^{\infty}(-0.4)^n z^{-n}. \qquad (2.384)$$

The powers of z can be equated in the above equation to obtain

$$
\begin{aligned}
h(0) &= & 10 + 10 - 18 &= 2, \\
h(1) &= & 10(0.5) - 18(-0.4) &= 2.2, \\
h(2) &= & 10(-0.5)^2 - 18(-0.4)^2 &= -0.38, \\
h(n) &= & 10(-0.5)^n - 18(-0.4)^n, & 3 \le n \le \infty. \qquad (2.385)
\end{aligned}
$$

The conditions on the above equations are that

$$|-0.5z^{-1}| < 1, \quad |-0.4z^{-1}| < 1, \qquad (2.386)$$

or

$$|z| > 0.5, \quad |z| > 0.4. \qquad (2.387)$$

The more restrictive of these is that

$$|z| > 0.5. \qquad (2.388)$$

Therefore, the ROC for $H(z)$ is given by

$$|z| > 0.5. \qquad (2.389)$$

End of the Example

Example 2.30 involves the determination of the causal impulse response discrete time system from its system transfer function.

EXAMPLE 2.30
(Impulse Response of a Discrete Time System)

Problem:
The transfer function for a digital system is given by

$$H(z) = \frac{0.14(z + 1)^2}{(z - 0.5)(z + 0.4)} \qquad (2.390)$$

Determine the impulse response of the system assuming that the system is causal.

Solution:

The system transfer function, $H(z)$, can be expanded into partial fractions as follows:

$$\frac{H(z)}{z} = \frac{(0.14)(z+1)^2}{(z)(z-0.5)(z+0.4)},$$

$$\frac{H(z)}{z} = \frac{B_1}{z} + \frac{B_2}{z-0.5} + \frac{B_3}{z+0.4}. \tag{2.391}$$

It follows that

$$B_1 = \left.\frac{(0.14)(z+1)^2}{(z-0.5)(z+0.4)}\right|_{z=0} = \frac{(0.14)(1)^2}{(-0.5)(0.4)} = -0.7,$$

$$B_2 = \left.\frac{(0.14)(z+1)^2}{z(z+0.4)}\right|_{z=0.5} = \frac{(0.14)(0.5+1)^2}{(0.5)(0.5+0.4)} = 0.7,$$

$$B_3 = \left.\frac{(0.14)(z+1)^2}{z(z-0.5)}\right|_{z=-0.4} = \frac{(0.14)(-0.4+1)^2}{(-0.4)(-0.4-0.5)} = 0.14. \tag{2.392}$$

Thus,

$$H(z) = \frac{-0.7z}{z} + \frac{0.7z}{z-0.5} + \frac{0.14z}{z+0.4}, \tag{2.393}$$

$$H(z) = -0.7 + 0.7\sum_{n=0}^{\infty}(0.5)^n z^{-n} + 0.14\sum_{n=0}^{\infty}(-0.4)^n z^{-n}. \tag{2.394}$$

We recognize the above equation as a constant plus the sum of two geometric series. It follows directly that

$$h(n) = \begin{cases} -0.7 + 0.7 + 0.14 = 0.14 & \text{for } n = 0, \\ 0.7(0.5)^n + 0.14(-0.4)^n & \text{for } 1 \le n \le \infty. \end{cases} \tag{2.395}$$

End of the Example

2.22.2 SYSTEM TRANSFER FUNCTION FROM AN IMPULSE RESPONSE

Example 2.29 verified that the impulse response of a linear, shift invariant, discrete time system is the inverse Z Transform of its system transfer function. Example 2.31 involves the computation of the system transfer function, $H(z)$, for a discrete time system from its impulse response.

EXAMPLE 2.31
(System Transfer Function from an Impulse Response)
Problem:
The impulse response of a discrete time system is given by

$$h(n) = 5 - 5(0.3)^n \text{ for } 0 \leq n \leq \infty,$$
$$h(n) = 0 \text{ otherwise.} \tag{2.396}$$

Determine the system transfer function, $H(z)$, for the corresponding system as the Z Transform of $h(n)$. Represent $H(z)$ as a ratio of polynomials in z.
Solution:
The Z Transform of $h(n)$ is given by

$$H(z) = 5 \sum_{n=0}^{\infty} \left[1 - (0.3)^n \right] z^{-n}, \tag{2.397}$$

$$H(z) = 5 \left[\sum_{n=0}^{\infty} (z^{-1})^n - \sum_{n=0}^{\infty} (0.3z^{-1})^n \right]. \tag{2.398}$$

Observing that $H(z)$ is a sum of two geometric series, we can write

$$H(z) = 5 \left[\lim_{N \to \infty} \left(\frac{1 - (z^{-1})^{N+1}}{1 - z^{-1}} \right) - \lim_{N \to \infty} \left(\frac{1 - (0.3z^{-1})^{N+1}}{1 - 0.3z^{-1}} \right) \right]. \tag{2.399}$$

Since the requirement $|z^{-1}| > 1$ is more restrictive than the requirement that $|0.3z^{-1}| > 1$, the ROC for $H(z)$ is $|z| > 1.0$. Thus,

$$H(z) = 5 \left[\left(\frac{1}{1 - z^{-1}} \right) - \left(\frac{1}{1 - 0.3z^{-1}} \right) \right] \ \forall \ |z| > 1, \tag{2.400}$$

$$H(z) = \frac{3.5z}{z^2 - 1.3z + 0.3} \ \forall \ |z| > 1. \tag{2.401}$$

End of the Example

2.23 SYSTEMS WITH COMPLEX POLES

Consider the case where the transfer function of a discrete time system has complex poles. If the coefficients of both the numerator and denominator polynomials of $H(z)$ are real, then the poles and zeros of $H(z)$ are real or they occur in complex conjugate pairs. Partial fraction expansion can be used to expand the system transfer function,

$H(z)$, as

$$H(z) = \sum_{m=1}^{N_1} F_m(z) + \sum_{k=1}^{N_2} \frac{C_k z}{z - p_k} \tag{2.402}$$

where all of the C_k and p_k are real and

$$F_m(z) = \frac{D_{1m} z}{z - p_{1m}} + \frac{D_{2m} z}{z - p_{2m}}. \tag{2.403}$$

The constants D_{1m} and D_{2m} are complex conjugates and the poles p_{1m} and p_{2m} are complex conjugates.

The inverse Z transform of each $F_m(z)$ can be obtained in the form

$$f_m(n) = \left[D_{1m}(p_{1m})^n + D_{2m}(p_{2m})^n \right] u(n). \tag{2.404}$$

The complex poles, p_{1m} and p_{2m}, can be written in complex exponential form as

$$\begin{aligned} p_{1m} &= e^{\alpha + j\omega}, \\ p_{2m} &= e^{\alpha - j\omega}. \end{aligned} \tag{2.405}$$

It follows that the inverse Z transform of $F_m(z)$ can be written in the form

$$y_m(n) = \left[D_{1m} e^{(\alpha + j\omega)n} + D_{2m} e^{(\alpha - j\omega)n} \right]. \tag{2.406}$$

However,

$$\begin{aligned} e^{j\omega n} &= \cos(\omega n) + j \sin(\omega n), \\ e^{-j\omega n} &= \cos(\omega n) - j \sin(\omega n). \end{aligned} \tag{2.407}$$

Thus,

$$y_m(n) = \left[(D_{1m} + D_{2m}) e^{\alpha n} \cos(\omega n) + j (D_{1m} - D_{2m}) e^{\alpha n} \sin(\omega n) \right] u(n). \tag{2.408}$$

However,

$$\begin{aligned} D_{1m} + D_{2m} &= 2\Re e(D_{1m}), \\ D_{1m} - D_{2m} &= 2j\Im m(D_{1m}). \end{aligned} \tag{2.409}$$

Thus

$$y_m(n) = \left[2\Re e\,(D_{1m}) e^{\alpha n} \cos(\omega n) - 2\Im m\,(D_{1m}) e^{\alpha n} \sin(\omega n) \right] u(n). \tag{2.410}$$

The equation for $y_m(n)$ can further be simplified by using the trigonometric equation

$$\cos(\omega n + \phi) = \cos(\phi) \cos(\omega n) - \sin(\phi) \sin(\omega n). \tag{2.411}$$

The equation for $y_m(n)$ can also be written in the form

$$y_n(m) = Ae^{\alpha n}\cos(\omega n + \phi) \tag{2.412}$$

where

$$\phi = \tan^{-1}\left\{\frac{\Im m\,(D_{1m})}{\Re e\,(D_{1m})}\right\},$$

$$A = 2\sqrt{\Im m\,(D_{1m})^2 + \Re e\,(D_{1m})^2}. \tag{2.413}$$

Example 2.32 gives an example of the inverse Z Transform for a Z Transform with complex poles.

EXAMPLE 2.32
(Complex Poles)

Problem:
The system transfer function for a discrete time system is given by

$$H(z) = \frac{0.5276z^3 + 1.5829z^2 + 1.5829z + 0.5276}{z^3 + 1.76z^2 + 1.1829z + 0.2781}. \tag{2.414}$$

Determine the impulse response of the system assuming that the system is causal.
Solution:
If $x(n)$ is an impulse, then its Z transform is given by

$$X(z) = \sum_{n=-\infty}^{\infty}\delta(n) = 1. \tag{2.415}$$

Thus,

$$Y(z) = H(z)X(z) = H(z). \tag{2.416}$$

The relationship, $\frac{H(z)}{z}$, can be expanded into partial fractions to make it convenient to obtain the inverse Z Transform in the form of geometric series. Thus,

$$\frac{H(z)}{z} = \frac{0.5276z^3 + 1.5829z^2 + 1.5829z + 0.5276}{z(z^3 + 1.76z^2 + 1.1829z + 0.2781)}$$

$$= \frac{D_1}{z - p_1} + \frac{D_2}{z - p_2} + \frac{C_1}{z - p_3} + \frac{C_2}{z}. \tag{2.417}$$

It follows that

$$D_1 = -0.3221 + 0.1501j,$$
$$D_2 = -0.3221 - 0.1501j = D_1^*,$$

$$C_1 = -0.7254,$$
$$C_2 = 1.8972. \tag{2.418}$$

The complex conjugate pairs of the partial fractions can be combined as follows:

$$F_1(z) = \frac{D_1 z}{z - p_1} + \frac{D_2 z}{z - p_2}. \tag{2.419}$$

The inverse Z Transform of $F_1(z)$ will be the part of the impulse response corresponding to the complex pair of poles p_1 and p_2:

$$p_1 = e^{\alpha + j\omega} = e^{\alpha}(\cos(\omega) + j\sin(\omega)),$$

$$\tan(\omega) = \frac{\Im m(p_1)}{\Re e(p_1)} = \frac{0.3935}{-0.6252},$$

$$\omega = \tan^{-1}(\frac{0.3935}{-0.6252}) = 2.5799,$$

$$e^{\alpha}\cos(\omega) = -0.6252 \Rightarrow e^{\alpha} = \frac{-0.6252}{\cos(\omega)} = 0.7387,$$

$$\alpha = \log_e(0.7387) = -0.3029. \tag{2.420}$$

It follows that

$$f_1(n) = \left[2(-0.3221)e^{-0.3029n}\cos(2.5799n)\right]u(n)$$
$$- \left[2(0.1501)e^{-0.3029n}\sin(2.5799n)\right]u(n),$$

$$f_1(n) = \left[-0.6442e^{-0.3029n}\cos(2.5799n)\right]u(n)$$
$$- \left[0.3002e^{-0.3029n}\sin(2.5799n)\right]u(n). \tag{2.421}$$

The equation for $f_1(n)$ can be simplified further as

$$f_1(n) = Ae^{-0.3029n}\cos(2.5799n + \phi)u(n) \tag{2.422}$$

where

$$A = 2\sqrt{(0.1501)^2 + (-0.3221)^2} = 0.7107,$$

$$\phi = \tan^{-1}\left[\frac{0.1501}{-0.3221}\right] = 2.7055. \tag{2.423}$$

Thus,

$$h(n) = \left[1.8972\delta(n) - 0.7254 - 0.5097)^n\right]u(n) + f_1(n), \tag{2.424}$$

$$
\begin{aligned}
h(n) &= 1.8972\delta(n) - \left[0.7254(-0.5097)^n\right]u(n) \\
&\quad + 0.7107e^{-0.3029n}\cos(2.5799n + 2.7055)u(n).
\end{aligned}
\tag{2.425}
$$

End of the Example

2.24 SYSTEMS WITH MULTIPLE POLES

The system transfer function for a discrete time system may have multiple poles. However, this is not a common occurrence. Example 2.33 illustrates a procedure for obtaining the impulse response for a system with a double pole.

EXAMPLE 2.33
(Multiple Poles)

Problem:
The system transfer function for a discrete time system is given by

$$
H(z) = \frac{0.1584z^3 + 0.4572z^2 + 0.4572z + 0.1584}{z^3 + 0.2000z^2 - 0.2275z - 0.0612}.
\tag{2.426}
$$

Determine the impulse response of the system assuming that the system is causal.
Solution:
The impulse response of the system is the inverse Z Transform of the system transfer function since

$$
\begin{aligned}
Y(z) &= H(z)X(z), \\
X(z) &= \sum_{n=0}^{\infty}\delta(n) = 1.
\end{aligned}
\tag{2.427}
$$

Thus,

$$
y(n) = h(n).
\tag{2.428}
$$

The system transfer function, $H(z)$, can be written in the form

$$
H(z) = \frac{0.1584(z + 1)^3}{z^3 + 0.2000z^2 - 0.2275z - 0.0612}.
\tag{2.429}
$$

The poles of $H(z)$ are located at

$$
\begin{aligned}
p(1) &= 0.5000, \\
p(2) &= p(3) = -0.3500.
\end{aligned}
\tag{2.430}
$$

The partial fraction of $z^{-1}H(z)$ can be used to facilitate the representation of $H(z)$ in the form of geometric series:

$$\frac{H(z)}{z} = \frac{0.1584(z+1)^3}{z(z-0.5)(z+0.35)^2}$$

$$= \frac{C_0}{z} + \frac{C_1}{z-0.5} + \frac{C_2}{z+0.35} + \frac{C_3 z}{(z+0.35)^2}. \tag{2.431}$$

The partial fraction coefficients, C_0, C_1, and C_3, can be computed as follows:

$$C_0 = \left.\frac{zH(z)}{z}\right|_{z=0} = \left.\frac{0.1584(z+1)^3}{(z-0.5)(z+0.35)^2}\right|_{z=0}$$

$$= \frac{0.1584(0+1)^3}{(0-0.5)(0+0.35)^2} = -2.5861,$$

$$C_1 = \left.\frac{(z-0.5)H(z)}{z}\right|_{z=0.5} = \left.\frac{0.1584(z+1)^3}{z(z+0.35)^2}\right|_{z=0.5}$$

$$= \frac{0.1584(0.5+1)^3}{(0.5)(0.5+0.35)^2} = 1.4799,$$

$$C_3 = \left.\frac{(z+0.35)^2 H(z)}{z}\right|_{z=-0.35} = \left.\frac{0.1584(z+1)^3}{z^2(z-0.5)}\right|_{z=-0.35}$$

$$= \frac{0.1584(-0.35+1)^3}{(-0.35)^2(-0.35-0.5)} = -0.4178.$$

$$\tag{2.432}$$

The coefficient C_2 can be determined by solving the partial fraction expansion equation for C_2:

$$\frac{H(z)}{z} = \frac{C_0}{z} + \frac{C_1}{z-0.5} + \frac{C_2}{z+0.35} + \frac{C_3 z}{(z+0.35)^2}, \tag{2.433}$$

$$\frac{(z+0.35)^2 H(z)}{z} = \frac{C_0(z+0.35)^2}{z} + \frac{C_1(z+0.35)^2}{z-0.5} + C_2(z+0.35) + C_3 z, \tag{2.434}$$

$$\frac{(z+0.35)^2 H(z)}{z} = \frac{0.1584(z+1)^3}{z(z-0.5)}$$

$$= \frac{C_0(z+0.35)^2}{z} + \frac{C_1(z+0.35)^2}{z-0.5}$$

$$+ C_2(z+0.35) + C_3 z. \tag{2.435}$$

It follows that

$$\frac{0.1584(z+1)^3}{z(z-0.5)} = \frac{C_0(z+0.35)^2}{z} + \frac{C_1(z+0.35)^2}{z-0.5}$$
$$+ C_2(z+0.35) + C_3 z. \tag{2.436}$$

The coefficient C_2 can be determined from

$$\frac{d}{dz}\left[\frac{0.1584(z+1)^3}{z(z-0.5)}\right]\Bigg|_{z=-0.35}$$
$$= \frac{d}{dz}\left[\frac{C_0(z+0.35)^2}{z} + \frac{C_1(z+0.35)^2}{z-0.5}\right]\Bigg|_{z=-0.35}$$
$$+ \frac{d}{dz}\left[C_2(z+0.35) + C_3 z\right]\Bigg|_{z=-0.35}, \tag{2.437}$$
$$\frac{d}{dz}\left[\frac{0.1584(z+1)^3}{z(z-0.5)}\right]\Bigg|_{z=-0.35} = C_2 + C_3. \tag{2.438}$$

However,

$$\frac{d}{dz}\left[\frac{0.1584(z+1)^3}{z(z-0.5)}\right] = \frac{3(0.1584)(z+1)^2(z^2-0.5z) - (2z-0.5)(z+1)^3}{(z^2-0.5z)^2}.$$
$$\tag{2.439}$$

It follows that

$$C_2 + C_3 = \frac{3(0.1584)(z+1)^2(z^2-0.5z) - (2z-0.5)(z+1)^3}{(z^2-0.5z)^2}\Bigg|_{z=-0.35}$$
$$= \frac{3(0.1584)(-0.35+1)^2((-0.35)^2 - 0.5(-0.35))}{((-0.35)^2 - 0.5(-0.35))^2}$$
$$- \frac{(2(-0.35) - 0.5)(-0.35+1)^3}{((-0.35)^2 - 0.5(-0.35))^2},$$
$$C_2 + C_3 = 1.2647,$$
$$C_2 = 1.2646 - C_3 = 1.6824. \tag{2.440}$$

Thus,

$$H(z) = -2.5861 + \frac{1.4799z}{z-0.5} + \frac{1.6824z}{z+0.35} - \frac{0.4178z^2}{(z+0.35)^2}, \tag{2.441}$$

$$H(z) = -2.5861 + \frac{1.4799}{1.0000 - 0.5000z^{-1}} + \frac{1.6824}{1.0000 + 0.3500z^{-1}}$$
$$- \frac{0.4178}{(1.0 + 0.35z^{-1})^2}. \tag{2.442}$$

The corresponding inverse Z Transform is given by

$$
\begin{aligned}
h(n) \;=\; & -2.5861\delta(n) + \left[1.4799(0.5000)^n + 1.6824(-0.35)^n\right]u(n) \\
& - \left[0.4178n(-0.35)^n\right]u(n).
\end{aligned} \tag{2.443}
$$

End of the Example

2.25 NONZERO INITIAL CONDITIONS

The initial conditions for a discrete time system are often ignored during its analysis. When all of the initial conditions of a system are equal to zero, the system is designated to be *relaxed*. However, in many practical applications, initial conditions as well as the input sequence must be considered to determine the output of a discrete time system. If the system is linear and shift invariant, then the principle of superposition can be used and the output can be determined as a linear combination of the output due to the initial state (initial conditions) and the output due to the input sequence.

The one-sided Z Transform, which is appropriate for causal systems, can be used to obtain the response of pole–zero systems with nonzero initial conditions. It is defined for a causal discrete time signal $x(n)u(n)$ as follows:

$$
X(z) = \sum_{n=0}^{\infty} x(n)z^{-n}. \tag{2.444}
$$

The properties of the one-sided Z Transform are the same as those of the regular Z Transform except for the shifting property:

$$
x(n-k) \Leftrightarrow z^{-k}X(z) + \sum_{n=1}^{k} x(-n)z^{n-k}. \tag{2.445}
$$

Thus, the shifting property can be used to determine the Z Transforms of the following delayed signals:

$$
\begin{aligned}
x(n-1) \;&\Leftrightarrow\; z^{-1}X(z) + x(-1), \\
x(n-2) \;&\Leftrightarrow\; z^{-2}X(z) + x(-1)z^{-1} + x(-2), \\
x(n-3) \;&\Leftrightarrow\; z^{-3}X(z) + x(-1)z^{-3} + x(-2)z^{-2} + x(-3), \\
&\;\;\vdots \\
x(n-k) \;&\Leftrightarrow\; z^{-k}X(z) + x(-1)z^{-k+1} + x(-2)z^{-k+2} + \cdots + x(-k), \\
&\;\;\vdots
\end{aligned} \tag{2.446}
$$

Example 2.34 gives an example of the use of the shifting property for the Z Transform to obtain the response of a discrete time system with nonzero initial conditions.

EXAMPLE 2.34

(Nonzero Initial Conditions)

Problem:

Determine the unit step response of the system described by the difference equation

$$
\begin{aligned}
y(n) &= 0.2131x(n) + 0.4262x(n-1) + 0.2131x(n-2) \\
&\quad - 0.1y(n-1) + 0.2475y(n-2), \\
y(-1) &= 0.40, \quad y(-2) = 0.40.
\end{aligned}
\tag{2.447}
$$

Solution:

The shifting property for the one-sided Z Transform can be used to obtain

$$
\begin{aligned}
Y(z) &= 0.2131X(z) + 0.4262\left[z^{-1}X(z) + x(-1)\right] \\
&\quad + 0.2131\left[z^{-2}X(z) + z^{-1}x(-1) + x(-2)\right] \\
&\quad - 0.1\left[z^{-1}Y(z) + y(-1)\right] \\
&\quad + 0.2475\left[z^{-2}Y(z) + z^{-1}y(-1) + y(-2)\right].
\end{aligned}
\tag{2.448}
$$

Eq. (2.448) can be solved for $Y(z)$ to obtain

$$
\begin{aligned}
Y(z) &= \frac{\left[0.2131 + 0.4262z^{-1} + 0.2131z^{-2}\right]X(z)}{\left[1.0 + 0.1z^{-1} - 0.2475z^{-2}\right]} \\
&\quad + \frac{-0.1y(-1) + 0.2475\left[y(-1)z^{-1} + y(-2)\right]}{\left[1.0 + 0.1z^{-1} - 0.2475z^{-2}\right]}.
\end{aligned}
\tag{2.449}
$$

The Z Transform of the output, $Y(z)$, can be partitioned into the Z Transform of the response due to the input and the Z Transform of the response due to the initial conditions as follows:

$$
\begin{aligned}
Y(z) &= Y_1(z) + Y_2(z), \\
Y_1(z) &= \frac{\left[0.2131 + 0.4262z^{-1} + 0.2131z^{-2}\right]X(z)}{\left[1.0 + 0.1z^{-1} - 0.2475z^{-2}\right]}, \\
Y_2(z) &= \frac{-0.1y(-1) + 0.2475\left[y(-1)z^{-1} + y(-2)\right]}{\left[1.0 + 0.1z^{-1} - 0.2475z^{-2}\right]}.
\end{aligned}
\tag{2.450}
$$

The Z Transform of the input, $X(z)$, is needed to solve for $Y_1(z)$:

$$X(z) = \sum_{n=0}^{\infty} z^{-1} = \frac{1}{1 - z^{-1}}, \qquad |z| > 1. \qquad (2.451)$$

Thus,

$$Y_1(z) = \frac{[0.2131 + 0.4262z^{-1} + 0.2131z^{-2}]}{(1.0 + 0.1z^{-1} - 0.2475z^{-2})(1 - z^{-1})}, \qquad (2.452)$$

$$Y_1(z) = \frac{0.2131z^3 + 0.4262z^2 + 0.2131z}{z^3 - 0.9z^2 - 0.3475z + 0.2475}. \qquad (2.453)$$

The Z Transform, $\frac{Y_1(z)}{z}$, can be expanded into partial fractions to obtain

$$\begin{aligned}
\frac{Y_1(z)}{z} &= \frac{0.2131z^3 + 0.4262z^2 + 0.2131z}{z(z^3 - 0.9z^2 - 0.3475z + 0.2475)} \\
&= \frac{0.2131z^2 + 0.4262z + 0.2131}{z^3 - 0.9z^2 - 0.3475z + 0.2475} \\
&= \frac{C_1}{z - 1} + \frac{C_2}{z - 0.55} + \frac{C_3}{z + 0.45}. \qquad (2.454)
\end{aligned}$$

The partial fraction expansion coefficients are given by

$$C_1 = 1.0, \quad C_2 = 0.0278, \quad C_3 = -0.8146. \qquad (2.455)$$

Thus,

$$Y_1(z) = \frac{z}{z - 1} + \frac{0.0278z}{z + 0.55} - \frac{0.8146z}{z - 0.45}. \qquad (2.456)$$

The inverse Z Transform of $Y_1(z)$ is given by

$$y_1(n) = [1.0 + 0.0278(-0.55)^n - 0.8146(-0.45)^n] u(n). \qquad (2.457)$$

Similarly, the inverse transform of $Y_2(z)$ can be obtained by expanding $\frac{Y_2(z)}{z}$ into partial fractions:

$$\begin{aligned}
\frac{Y_2(z)}{z} &= \frac{[-0.1(0.4) + 0.2475(0.4)]z^2 + 0.2475(0.4)z}{z(z^2 + 0.1z - 0.2475)} \\
&= \frac{0.0509z + 0.0990}{z^2 + 0.1z - 0.2475} \\
&= \frac{C_3}{z - 0.55} + \frac{C_4}{z + 0.45}, \qquad (2.458)
\end{aligned}$$

$$C_3 = -0.071, \quad C_4 = 0.1219, \qquad (2.459)$$

$$Y_2(z) = -\frac{0.071z}{z+0.55} + \frac{0.1219z}{z-0.45}. \tag{2.460}$$

Thus,

$$y_2(n) = \left[-0.071(-0.55)^n + 0.1219(-0.45)^n\right]u(n). \tag{2.461}$$

The two partial solutions, $y_1(n)$ and $y_2(n)$, can be combined to obtain

$$\begin{aligned} y(n) &= \left[1.0 + 0.0278(-0.55)^n - 0.8146(-0.45)^n\right]u(n) \\ &+ \left[-0.071(-0.55)^n + 0.1219(-0.45)^n\right]u(n), \end{aligned} \tag{2.462}$$

$$y(n) = \left[1.0 - 0.0432(-0.55)^n - 0.6927(0.45)^n\right]u(n). \tag{2.463}$$

End of the Example

2.26 RECONSTRUCTION OF DISCRETE TIME SIGNALS

Assume that a normalized sampling interval ($T = 1$) has been used to sample a continuous function $x(t)$ to obtain the sequence $x(n)$. The Fourier transform of the sampled sequence is given by

$$X(\omega) = \sum_{n=-\infty}^{\infty} x(n)e^{-j\omega n}. \tag{2.464}$$

The original signal, $x(t)$, can be recovered from the sample sequence if aliasing did not occur during the sampling process. This can be accomplished by passing the sample sequence through the ideal low pass filter with frequency response given by

$$H(\omega) = T \text{ for } |\omega| \le \frac{\pi}{T}, \quad T = 1, \tag{2.465}$$

$$H(\omega) = 0 \text{ for } |\omega| > \frac{\pi}{T}. \tag{2.466}$$

In order to show that this is true, let $S(\omega)$ represent the Fourier transform of the output from the filter. Thus,

$$S(\omega) = H(\omega)X(\omega). \tag{2.467}$$

Note that $S(\omega) = 0$ for $|\omega| > \frac{\pi}{T}$. We can find the inverse Fourier transform of $S(\omega)$ to obtain $s(t)$ as follows:

$$s(t) = \frac{1}{2\pi}\int_{-\frac{\pi}{T}}^{\frac{\pi}{T}}\left[\sum_{n=-\infty}^{\infty}x(n)e^{-j\omega n}\right]e^{j\omega t}d\omega, \quad T = 1. \tag{2.468}$$

If aliasing did not occur during the sampling of $x(t)$, then $s(t) = x(t)$. Eq. (2.468) can be rewritten as follows:

$$s(t) = \frac{1}{2\pi} \sum_{n=-\infty}^{\infty} x(n) \int_{-\pi}^{\pi} e^{j\omega(t-n)} d\omega. \qquad (2.469)$$

If $t = n$ then

$$s(n) = \frac{1}{2\pi} x(n) \int_{-\pi}^{\pi} d\omega = x(n). \qquad (2.470)$$

Thus, $s(n) = x(n)$ at the sample points. On the other hand, if $t \neq n$ then

$$s(t) = \frac{1}{2\pi} \sum_{n=-\infty}^{\infty} x(n) \left[\frac{e^{j\omega(t-n)}}{j(t-n)} \right] \Bigg|_{-\pi}^{\pi}, \qquad (2.471)$$

$$s(t) = \frac{1}{2\pi} \sum_{n=-\infty}^{\infty} \left[\frac{x(n)}{\pi(t-n)} \right] \left[\frac{e^{j\pi(t-n)} - e^{-j\pi(t-n)}}{2j} \right], \qquad (2.472)$$

$$s(t) = \sum_{n=-\infty}^{\infty} \left[\frac{x(n) \sin[\pi(t-n)]}{\pi(t-n)} \right]. \qquad (2.473)$$

The mathematical expression, $\frac{\sin(u)}{u}$, is called the sinc function. However, the normalized form of the sinc function, as given by the form $\frac{\sin(\pi u)}{\pi u}$, is commonly used for the sinc(u) in digital signal processing and information theory [6]. Therefore, the reconstruction formula, using the normalized form of the sinc function, is given by

$$x(t) = \sum_{n=-\infty}^{\infty} x(n)\text{sinc}(t-n) \;\; \forall \, t \neq n, \qquad (2.474)$$

$$x(t) = x(n), \quad \forall \, t = n, \qquad (2.475)$$

when aliasing does not occur. The reconstructed signal can be obtained by multiplying each point in the original sample sequence, $x(n)$, by sinc$(t-n)$ and summing for each value of t. Note also that all of the samples must be obtained before the signal can be reconstructed. Thus, this ideal reconstruction formula is not a practical solution to the interpolation problem.

2.27 SAMPLE RATE CONVERSION BY INTERPOLATION

Many applications require the change in the sample rate as signals are transferred from one part of a system to another. This involves either interpolation or decimation. Interpolation involves the reconstruction of a sequence from its samples. This can be done without errors if the sequence is band limited. Common practical methods to perform interpolation include:

1. Linear point connector,
2. Sample and hold,
3. Frequency domain interpolation, and
4. Zero padding followed by filtering.

Common methods to perform decimation include:

1. Sample-and-hold,
2. Filtering and downsampling.

We will consider various methods for interpolation and decimation in the following sections of this chapter.

2.27.1 SAMPLE-AND-HOLD

One approach to obtaining a practical reconstruction of $x(t)$ is to use the first order hold or sample-and-hold. The sample-and-hold is also very convenient to implement in either hardware or software realizations of digital signal processing systems. The sample-and-hold approach involves inserting new samples between samples that have the same value as the most recent sample. We define the sample-and-hold as follows:

$$s(t) = x(n-1) \text{ for } (n-1)T \le t < nT .\qquad(2.476)$$

Let us consider the significance of using this function to reconstruct $x(t)$ by evaluating its frequency response in comparison to the frequency response of the ideal reconstruction filter. We can obtain the frequency response by computing the Fourier Transform of the impulse response of the sample-and-hold. The impulse response for the sample-and-hold is given by

$$h(t) = \begin{cases} 0 & \text{for} \quad -\infty \le t < 0, \\ 1 & \text{for} \quad 0 \le t < T, \\ 0 & \text{for} \quad T \le t \le \infty. \end{cases}\qquad(2.477)$$

It follows that

$$H_S(\omega) = \int_0^T e^{-j\omega t} dt,\qquad(2.478)$$

$$H_S(\omega) = \frac{1 - e^{-j\omega T}}{j\omega}.\qquad(2.479)$$

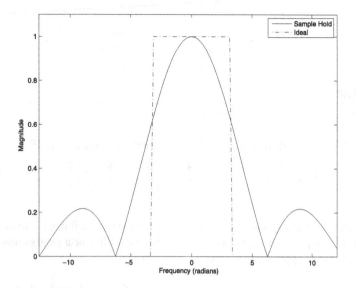

FIGURE 2.45

Comparison of the sample-and-hold and the ideal low pass filter

We can simplify $H_S(w)$ to obtain

$$H_S(\omega) = Te^{-j\omega T/2}\left[\frac{\sin(\omega T/2)}{\omega T/2}\right], \tag{2.480}$$

$$H_S(\omega) = Te^{-j\omega T/2}\text{sinc}(\omega T/2). \tag{2.481}$$

In Fig. 2.45, we compare the ideal low pass reconstruction filter with the magnitude response of the sample-and-hold reconstruction filter.

Note the phase term $e^{-j\omega T/2}$ for $H_S(\omega)$. The phase is given by

$$\phi(\omega) = \begin{cases} -\omega T/2 \pm 2k\pi & \text{when sinc}(\omega T/2) \geq 0, \\ -\omega T/2 \pm (2k+1)\pi & \text{when sinc}(\omega T/2) < 0, \\ \qquad\qquad -\infty \leq k \leq \infty. \end{cases} \tag{2.482}$$

Note also that the sample-and-hold algorithm yields very significant frequency content outside the desired region $-\pi/T \leq \omega \leq \pi/T$. This problem is alleviated by using a practical low pass filter after applying the sample-and-hold.

Fig. 2.46 gives a conceptual block diagram of a reconstruction system using a sample-and-hold and a post analog filter. The rejection frequency for the low pass filter should be $\omega_r = \frac{\pi}{T}$ to remove the unwanted part of the signal resulting from the use of the sample-and-hold.

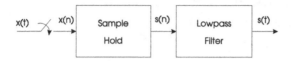

FIGURE 2.46

Reconstruction using a sample-and-hold and a low pass filter

Table 2.4 Tabulation of a sample sequence for Example 2.35

n	0	1	2	3	4	5	6	7	8	9
$x(n)$	0.0	0.5	1.0	1.5	2.0	2.0	1.5	1.0	0.5	0.0

The sample-and-hold can also be used to interpolate or decimate a sequence using non-integer sampling ratios. However, an antialiasing filter or a post sampling filter will be needed as appropriate to obtain a good result.

- An antialiasing low pass filter is needed prior to the sample-and-hold for decimation.
- A low pass filter is needed after the sample-and-hold for interpolation.

The output from the sample-and-hold is the most recent input. One or more input samples can be skipped if the sequence is to be decimated. The new samples do not have to be coincident with the previous samples.

2.27.2 INTERPOLATION BY A NON-INTEGER RATIO

Example 2.41 illustrates the use of the sample-and-hold for interpolation by a non-integer ratio.

EXAMPLE 2.35
(Non-Integer Ratio Interpolation)

Consider the sample sequence given in Table 2.4 and also shown in Fig. 2.47.

The sequence will be interpolated with a sampling interval of 0.35 times the original sampling interval for this example. Table 2.5 gives a tabulation of the output sequence after using the sample-and-hold, and Fig. 2.48 gives the stem plot of the output sequence after using the sample-and-hold operation with a sampling interval equal to 0.35 times the original sampling interval.

Fig. 2.49 gives the stem plot of the original sequence and the output sequence after using the sample-and-hold operation with a sampling interval equal to 0.35 times the original sampling interval on the same plot. Fig. 2.50 gives the stem plot of the original sequence and the output sequence after adjusting the output

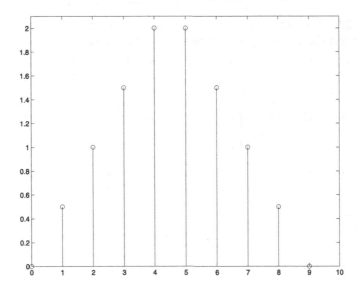

FIGURE 2.47

Stem plot of the sample sequence for Example 2.35

to compensate for the delay in the sample-and-hold operator (1/2 of the original sample interval) on the same plot.

The output can be filtered with a low pass filter to remove the frequencies in the output that are outside the range of the frequency content of the original sequence. An FIR low pass filter with a stop pass band frequency of 0.9 times the inverse of the interpolation ratio, which is a normalized radial frequency of $0.9(0.35)\pi = 0.315\pi$, was chosen for this Example.

The following Matlab script illustrates the use of Matlab to design the filter using the Matlab *fir1* function.

Matlab Script 2.1.

```
% FIR Filter Design
wc = 0.9*0.35;
order = 20;
b = fir1(order, wc);
y = conv(b, s);
```

End of the Script

Fig. 2.51 gives the stem plots of the original sequence and the filtered output sequence after compensating for the delays in the sample-and-hold and the filter on the same plot. Fig. 2.51 shows that the sample-and-hold followed by filtering gives

Table 2.5 Tabulation of a output sample sequence for Example 2.35 after interpolation using a sample-and-hold operation

n_2	0	1	2	3	4	5	6	7	8	9
$0.35n_2$	0.00	0.35	0.70	1.05	1.40	1.75	2.10	2.45	2.80	3.15
$\lfloor 0.35n_2 \rfloor$	0	0	0	1	1	1	2	2	2	3
$y(n)$	0.0	0.0	0.0	0.5	0.5	0.5	1.0	1.0	1.0	1.5
n_2	10	11	12	13	14	15	16	17	18	19
$0.35n_2$	3.50	3.85	4.20	4.55	4.90	5.25	5.60	5.95	6.30	6.65
$\lfloor 0.35n_2 \rfloor$	3	3	4	4	4	5	5	5	6	6
$y(n)$	1.5	1.5	2.0	2.0	2.0	2.0	2.0	2.0	1.5	1.5
n_2	20	21	22	23	24	25	26	27	28	
$0.35n_2$	7.00	7.35	7.70	8.05	8.40	8.75	9.10	9.45	9.80	
$\lfloor 0.35n_2 \rfloor$	7	7	7	8	8	8	9	9	9	
$y(n)$	1.0	1.0	1.0	0.5	0.5	0.5	0.0	0.0	0.0	

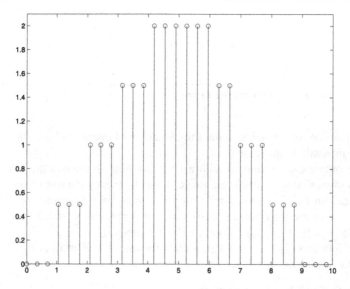

FIGURE 2.48

Stem plot of the output sample sequence for Example 2.35 after using the sample-and-hold operation

a satisfactory result for this example. This approach will generally provide good results for interpolation ratios of approximately 3 or higher.

2.27.3 LINEAR POINT CONNECTOR

An alternative practical approach to interpolation is to use a linear point connector. This approach involves inserting samples on a straight line drawn between adjacent

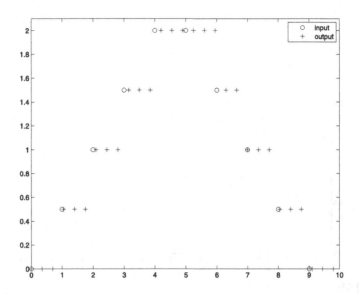

FIGURE 2.49

Stem plot of the input and output sample sequences for Example 2.35 on the same plot

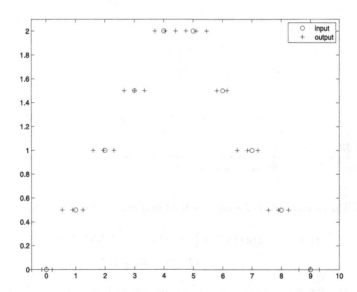

FIGURE 2.50

Stem plot of the input and output sample sequences for Example 2.35 after compensating for the delay in the sample-and-hold operator

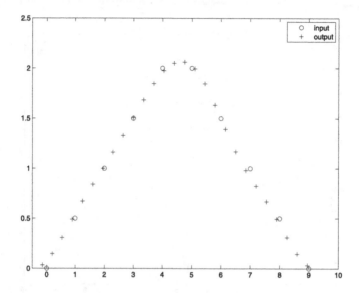

FIGURE 2.51

Stem plot of the input and filtered output sample sequences for Example 2.35 after compensating for the delays in the sample-and-hold operator and the filter

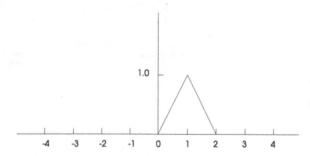

FIGURE 2.52

Impulse response for the linear point connector

samples. The reconstruction formula for the linear point connector is given by

$$s(t) = x(n)\left[\frac{t - nT}{T}\right] + x(n-1)\left[\frac{(n+1)T - t}{T}\right] \qquad (2.483)$$

$$\text{for } nT \le t < (n+1)T$$

where $x(n)$ is the input sequence and $s(t)$ is the interpolated sequence. Note that the output cannot be computed for the interval $nT \le t < (n+1)T$ until the sample at $(n+1)T$ is available to be used for reconstruction. Thus, the impulse response is delayed as shown in Fig. 2.52.

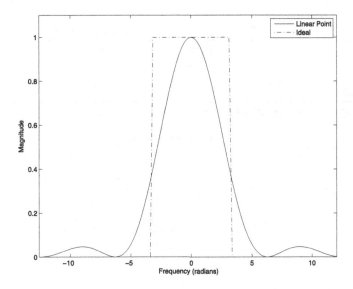

FIGURE 2.53

Comparison of the linear point connector and the ideal low pass filter

An equation for the impulse response of the linear point connector can be written as follows:

$$h_L(t) = \begin{cases} \frac{t}{T} & \text{for } 0 \le t < T, \\ \frac{2T-t}{T} & \text{for } T \le t < 2T, \\ 0 & \text{otherwise.} \end{cases} \tag{2.484}$$

The Fourier transform of $h_L(t)$ can be obtained as follows:

$$H_L(\omega) = \int_0^T \frac{t}{T} e^{-j\omega t} dt + \int_T^{2T} \frac{(2T-t)}{T} e^{-j\omega t} dt, \tag{2.485}$$

$$H_L(\omega) = T e^{-j\omega T} \left[\frac{\sin(\omega T/2)}{\omega T/2} \right]^2, \tag{2.486}$$

$$H_L(\omega) = T e^{-j\omega T} \left[\text{sinc}(\omega T/2) \right]^2. \tag{2.487}$$

Fig. 2.53 provides a comparison between the magnitude of the frequency response of the linear point connector and the ideal low pass reconstruction filter. Note the phase term $e^{-j\omega T}$ in $H_L(\omega)$. This results in a delay of one sample interval as previously discussed. Thus, the phase is given by

$$\phi(\omega) = -\omega T \pm 2\pi k, \quad -\infty \le k \le \infty. \tag{2.488}$$

The performance of the linear point connector can also be improved by using a post analog filter. Fig. 2.54 gives a conceptual diagram of the use of a linear point connector and a post analog filter to perform signal reconstruction.

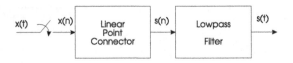

FIGURE 2.54

Reconstruction using a linear point connector and a low pass filter

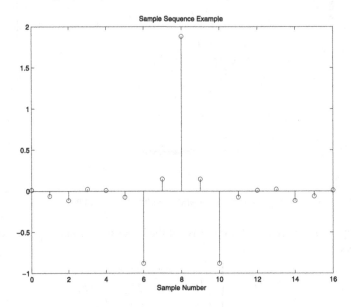

FIGURE 2.55

Sample sequence for Example 2.36

2.27.4 FREQUENCY DOMAIN INTERPOLATION

Interpolation can be performed in the time domain by adding zeros to the DFT of a sampled sequence and then computing the inverse DFT. Adding zeros to the DFT is equivalent to decreasing the sampling interval which results in interpolation when the inverse DFT is obtained. Example 2.36 illustrates this procedure.

EXAMPLE 2.36
(Frequency Domain Interpolation)

Consider the sample sequence given in Fig. 2.55. The magnitude and phase responses for the DFT of this sequence are given in Fig. 2.56. The length of the sequence is 17 samples. An interpolation ratio of 3 to 1 can be obtained by adding 34 zeros to the DFT between the last frequency sample and the normalized fre-

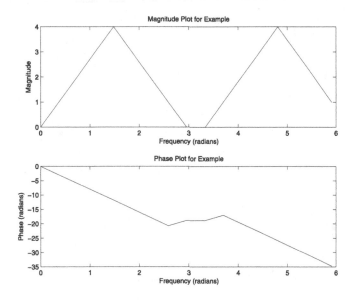

FIGURE 2.56

Magnitude and phase responses for the sample sequence in Example 2.36

quency π. The resulting DFT will be 51 samples long. The frequency sample at the normalized frequency, π, should have zero magnitude. Nonzero content at normalized frequency π indicates that aliasing will occur with the implementation of this procedure. The following Matlab script can be used to add the zeros for this example.

Matlab Script 2.2.

```
% Frequency Domain Interpolation
xfft = fft(x);
n1 = length(xfft);
n2 = fix(0.5*n1) + 1;
hf2 = [xfft(1, 1:n2) zeros(1, 34) xfft(1, (n2+1):n1)];
```

End of the Script

Fig. 2.57 gives the magnitude and phase responses for the modified sequence. The interpolated sequence can then be obtained by computing the inverse DFT of the modified sequence. The magnitude also needs to be adjusted by the interpolation ratio since the inverse DFT uses the inverse of the number of samples in the sequence. Without this adjustment, the magnitude of the resulting inverse DFT will be attenuated by the interpolation ratio. The following Matlab script can be used to obtain the interpolated sequence for an interpolation ration of 3 to 1.

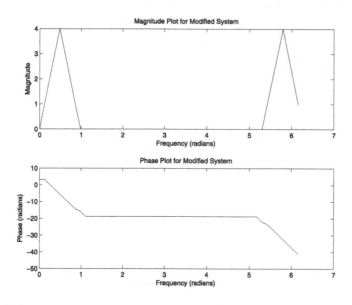

FIGURE 2.57

Magnitude and phase responses for the modified sample sequence in Example 2.36

Matlab Script 2.3.

```
ratio = 3;
y = ratio*real(ifft(hf2));
```

End of the Script

  Fig. 2.58 gives the stem plot of the sequence after interpolation.
End of the Example

2.27.5 ZERO PADDING AND FILTERING

A practical approach to the reconstruction of $x(t)$ from the sample sequence $x(n)$
is to pad the sequence $x(n)$ with zeros and then filter the result with an ideal filter
whose cutoff frequency is equal to the original Nyquist frequency. This approach is
based upon the assumption that $x(t)$ is band limited and the sampling frequency is
high enough to avoid aliasing. The sampled sequence, $x(n)$, has a periodic Fourier
transform with a period of $2\pi/T$ in the frequency space. If the sequence $x(n)$ is
padded to form the sequence $s(n)$ by adding m zeros between each sample, then the
Fourier transform of $s(n)$ is equal to the Fourier transform of the sequence $x(n)$ in
the range $-\pi/T \leq \omega < \pi/T$ and has images of this transform in the range $\pi/T <
|\omega| \leq m\pi/T$. The padded sequence, $s(n)$, now has $m+1$ times as many samples over
the same time interval as the original sequence. An interpolated version of $x(n)$ can

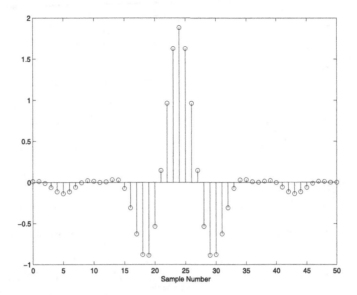

FIGURE 2.58

Sample sequence for Example 2.36 after interpolation

be obtained by filtering $s(n)$ using an ideal low pass filter with cut off frequency equal to π/T to remove the added images in the range $\pi/T < |\omega| \leq m\pi/T$. Although an ideal filter is desired, very good results can be obtained by using a practical low pass filter.

EXAMPLE 2.37
(Zero Padding and Filtering)

Consider the sample sequence in Fig. 2.59. This sequence can be interpolated to a sequence three times as long by adding two zeros between each sample and then filtering with a low pass filter with normalized cutoff frequency of $\omega_c = \frac{\pi}{3}$. The zeros between samples can be added with the use of the following Matlab script.

Matlab Script 2.4.

```
% Zero Padding and Filtering
n1 = length(x);
n2 = 3*n1;
s = zeros(1,n2);
s(1, 1:3:n2) = x;
```

End of the Script

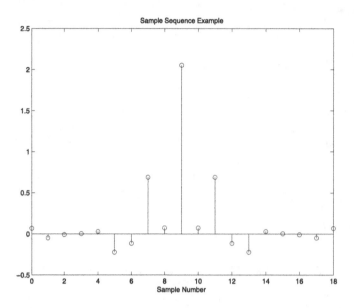

FIGURE 2.59

Stem plot of sample sequence for Example 2.37

Fig. 2.60 gives a stem plot of the sequence after it has been padded with zeros. The Matlab function *fir1* can be used to design an FIR filter to be used to filter the zero padded sequence, and the Matlab function *conv* can be used to filter the sequence $s(n)$ as shown in the following Matlab script.

Matlab Script 2.5.

```
% Lowpass Filer Design
wc = 1/3;
order = 32;
b = fir1(order, wc);
y1 = conv(b, s)
```

End of the Script

The linear phase filter designed by the *fir1* Matlab function has a delay of

$$m = 0.5 * (n3 - 1) \tag{2.489}$$

where $n3$ is the number of filter coefficients in the filter. Thus, finally, the interpolated and shifted sequence can be extracted the output $y1(n)$ to obtain the interpolated output $y2(n)$ as shown in the following Matlab script.

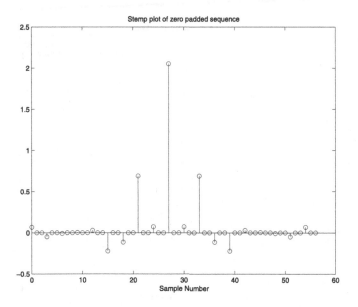

Stemp plot of zero padded sequence

FIGURE 2.60

Sample sequence after padding with zeros

Matlab Script 2.6.

```
%  Shifted Sequence
n3 = length(b);
m = fix(0.5*(n3-1));        % Cannot shift 1/2 sample
mp = n2 + m - 1;
y2 = y1(1, m:mp);
```

End of the Script

Fig. 2.61 gives the stem plot of the interpolated sequence.

Matlab has a function called *interp* that can be used to perform interpolation using the zero padding approach. Example 2.38 illustrates the use of the function *interp* to interpolate a sequence.

EXAMPLE 2.38
(Interpolation using Matlab interp)

Fig. 2.62 gives a simple signal composed of samples from a band limited pulse with a total of 256 points. Fig. 2.63 gives the output from the MATLAB function *interp* when it was used to increase the sampling rate by a factor of 3 to obtain 768 samples. The MATLAB function *interp* uses a symmetric filter to allow the

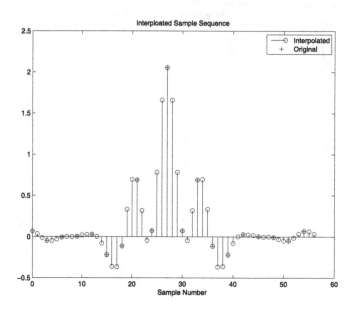

FIGURE 2.61

Interpolated output sequence for Example 2.37

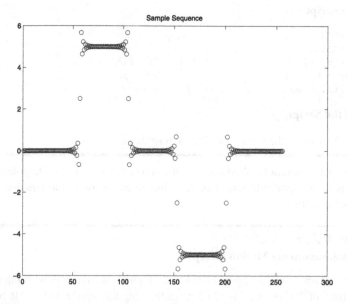

FIGURE 2.62

A band limited signal with 256 samples

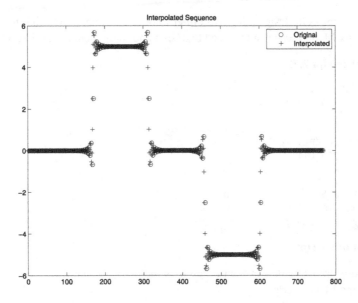

FIGURE 2.63

Interpolated signal upsampled by a factor of 3

original data to pass through unchanged and interpolates between so that the mean square error between them and their ideal values is minimized. The quality of the reconstruction for this approach depends upon the quality of the low pass filter and the type of input. We obtained excellent results for this example.
End of the Example

2.28 SAMPLE RATE CONVERSION BY DECIMATION

There is often a need to reduce the sampling rate of a sample sequence to match a system requirement. A sample sequence can be decimated by an integer factor by first filtering the sequence to avoid aliasing and then deleting samples as required to obtain the desired sampling rate. Example 2.39 illustrates this procedure.

EXAMPLE 2.39
(Decimation)

The same sample sequence used in Example 2.37, as also shown in Fig. 2.59, will be used for this example. The desire is to decimate the signal by a factor of 3.

The first step is to design a low pass filter to be used as an antialiasing filter. The following Matlab script can be used to design and implement the desired filter.

Matlab Script 2.7.

```
x = sampdata;
n2 = length(x);
wc = 1/3;
order = 32;
b = fir1(order, wc);
n3 = length(b);
n4 = n3-1;
y1 = conv(b, x);
m = fix(0.5*n4);      % Cannot shift 1/2 sample
mp = n2 + m - 1;
y2 = y1(1, m:mp);      % extract the output sequence
```

End of the Script

Every third sample can then be used as the output sequence as shown in the following Matlab script.

Matlab Script 2.8.

```
n5 = length(y2);
y3 = y2(1, 1:3:n5);
```

End of the Script

Fig. 2.64 gives a stem plot of the decimated sequence.

Matlab has a function called *decimate* that can be used to implement this approach. The following Matlab can be used to implement the decimation procedure using the Matlab *decimate* function to achieve similar results.

Matlab Script 2.9.

```
rate = 3;
y4 = decimate(x, rate);
```

End of the Script

Fig. 2.65 gives a stem plot of the sample sequence after decimation by a factor of 3 using the Matlab *decimate* function.

2.29 NON-INTEGER RATIO INTERPOLATION OR DECIMATION

Interpolation or decimation can be performed by a non-integer ratio by first using interpolation and then decimation. The final interpolation (decimation) ratio is the ratio

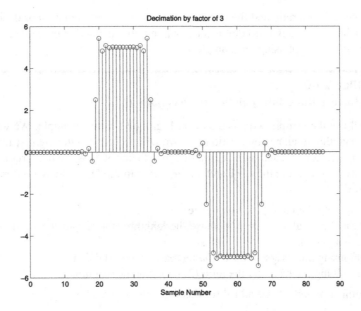

FIGURE 2.64

Sample sequence decimated by a factor of 3

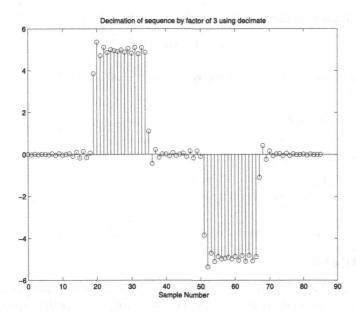

FIGURE 2.65

Sample sequence decimated by a factor of 3 using the *decimate* Matlab function

of the interpolation ratio and the decimation ratio. Thus, interpolation or decimation can be achieved for any ratio that is a rational number using this procedure. We can illustrate this concept using an example.

EXAMPLE 2.40

(Non-Integer Ratio Interpolation or Decimation)

We will use the sample sequence used in Fig. 2.62 for this example. We want to interpolate this sequence by a ratio of $\frac{2}{3}$. We will accomplish this by first interpolating by a factor of 2 and then decimating by a factor of 3. We will interpolate the sequence by obtaining the DFT and adding zeros to the DFT. We will interpolate the sequence by

1. Computing the DFT of the sequence,
2. Adding zeros at $\omega = \pi$ to increase the length of the sequence to 2 times the original length,
3. Computing the inverse DFT of the increased length DFT, and
4. Multiplying the result by the ratio (2) to adjust the magnitude.

The number of zeros to be added for this example is given by

$$
\begin{aligned}
n_1 &= 256, \\
n_2 &= (2-1)256 = 256.
\end{aligned}
\tag{2.490}
$$

The final length of the interpolated sequence will be

$$
k = 256 + 256 = 512.
\tag{2.491}
$$

We can obtain the interpolated sequence using the inverse DFT as shown in the following Matlab script.

Matlab Script 2.10.

```
ratio1 = 2;
x = sampdata;
xdft = fft(x);
n1 = length(x);
n2 = n1*(ratio1 - 1); % number of zeros to add
n3 = fix(0.5*n1);
x2dft = [xdft(1, 1:n3) zeros(1, n2) xdft((n3+1):end)];
x2 = real(ifft(x2dft)); % Obtain the interpolated sequence
```

End of the Script

We now need to filter the interpolated sequence to avoid aliasing when we decimate to obtain the final sequence. The cutoff frequency for the required low pass filter is the smaller of

- The cutoff frequency needed for the interpolation to remove the higher frequency components introduces by upsampling, etc., and
- The cutoff frequency needed to avoid aliasing before downsampling.

The normalized cutoff frequency for the filter required for the interpolation is

$$\omega_{c1} = \frac{\pi}{2}. \tag{2.492}$$

The normalized cutoff frequency for the filter required for the decimation is

$$\omega_{c2} = \frac{\pi}{3}. \tag{2.493}$$

Therefore, we should use $\omega_c = \omega_{c2}$ which is the smaller of the two. The following Matlab script can be used to design the filter and filter the interpolated sequence.

Matlab Script 2.11.

```
ratio1 = 2;  % Interpolate
ratio2 = 3;  % Decimate
wc = min(1/ratio1, 1/ratio2);
m1 = 32;
b = fir1(m1, wc);
x3 = conv(b, x2);
```

End of the Script

We can then decimate by a factor of 3 to obtain the output sequence using the following Matlab script.

Matlab Script 2.12.

```
m2 = fix(0.5*(length(b)-1));
m3 = fix(n1*ratio1) + m2;
y = x3(1, m2:ratio2:m3);
```

End of the Script

Fig. 2.66 gives a stem plot of the interpolated sequence (ratio = $\frac{2}{3}$).

We can illustrate the use of the sample-and-hold for decimation by a non-integer ratio with an example.

EXAMPLE 2.41
(Non-Integer Ratio Decimation)

Consider the sample sequence given in Table 2.4. We want to decimate this sequence by a ratio of 1.15. We must first filter the sequence with a low pass filter

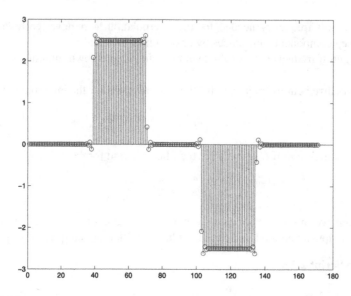

FIGURE 2.66

Stem plot for interpolated sequence for Example 2.40

Table 2.6 Tabulation of a sample sequence for Example 2.41

n	0	1	2	3	4
$x(n)$	0.0296	0.4829	1.0135	1.4839	1.9878
n	5	6	7	8	9
$x(n)$	1.9878	1.4839	1.0135	0.4829	0.0296

to avoid aliasing. We choose $\omega_s = \frac{\pi}{1.15} = 0.8696\pi$ for the stop band edge for the filter and $\omega_c = 0.7826\pi$ for the low pass filter. Table 2.6 gives a tabulation of the sequence after filtering with a low pass antialiasing filter. We want to decimate the sequence with a sampling interval of 1.15 times the original sampling interval. Table 2.7 gives a tabulation of the output sequence after using the sample-and-hold. Fig. 2.67 gives a stem plot of the sequence after filtering using the antialiasing low pass filter. Fig. 2.68 gives the stem plot of the output sequence after using the sample-and-hold operation with a sampling interval equal to 1.15 times the original sampling interval. Fig. 2.69 gives the stem plot of the original sequence and the output sequence after using the sample-and-hold operation with a sampling interval equal to 1.15 times the original sampling interval on the same plot. Fig. 2.70 gives the stem plot of the original sequence and the output sequence after adjusting the output to compensate for the delay in the sample-and-hold operator (1/2 of the original sample interval) on the same plot.

Table 2.7 Tabulation of a output sample sequence for Example 2.41 after decimation using a sample-and-hold operation

n_2	0	1	2	3	4
$1.15n_2$	0.00	1.15	2.30	3.45	4.60
$\lfloor 1.15n_2 \rfloor$	0	1	2	3	4
$y(n)$	0.0296	0.4829	1.0135	1.4839	1.9878
n_2	5	6	7	8	
$1.15n_2$	5.75	6.90	8.05	9.20	
$\lfloor 1.15n_2 \rfloor$	5	6	8	9	
$y(n)$	1.9878	1.4839	0.4829	0.0296	

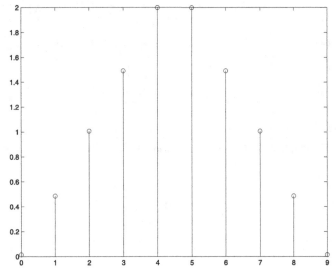

FIGURE 2.67

Stem plot of the sample sequence for Example 2.41 after filtering using the antialiasing low pass filter

2.30 PROBLEMS FOR CHAPTER 2

Problem 2.1. (Complex Numbers)
Simplify the following complex-valued expressions and give the answers in both Cartesian form $((x + jy))$ and polar form $((re^{j\theta}))$:

1. $7e^{j\frac{2\pi}{3}} + 3e^{-j\frac{3\pi}{4}}$,
2. $(\sqrt{7} + j3)^5$,
3. $(\sqrt{11} - j7)^{-3}$,
4. $(\sqrt{6} + j9)^{\frac{1}{3}}$,
5. $\Im m\{je^{-j\frac{\pi}{7}} + e^{j\frac{2\pi}{5}}\}$.

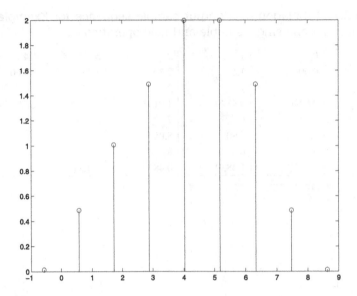

FIGURE 2.68

Stem plot of the output sample sequence for Example 2.41 after using the sample-and-hold operation

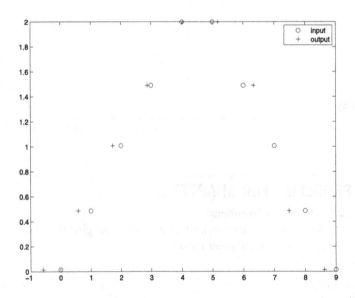

FIGURE 2.69

Stem plot of the input and output sample sequences for Example 2.41 on the same plot

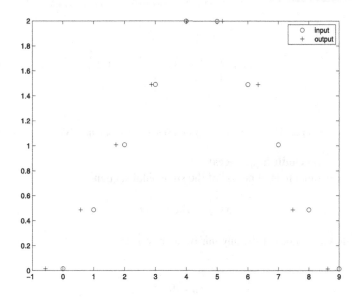

FIGURE 2.70

Stem plot of the input and output sample sequences for Example 2.41 after compensating for the delay in the sample-and-hold operator

Problem 2.2. (Representation of Sinusoidal Signals)
Define $x(t)$ as

$$x(t) = 7\cos\left(\omega_0 t + \frac{2\pi}{5}\right) + 6\cos\left(\omega_0 t + \frac{2\pi}{3}\right). \tag{2.494}$$

Express $x(t)$ in the form $x(t) = A\cos(\omega_0 t + \phi)$ by finding the numerical values of A and ϕ.

Problem 2.3. (Even and Odd Parts)
Determine the even and odd parts of the following sequences:

1.

$$x(n) = \left\{-3, -2, 0, \underset{\uparrow}{1}, 4, 5, 2\right\}, \tag{2.495}$$

2.

$$x(n) = 7\delta(n+2) + \delta(n+1) - 3\delta(n) + 4\delta(n-1) + 9\delta(n-2) - 2\delta(n-3), \tag{2.496}$$

3.

$$x(n) = \left\{ -5, 4, 3, \ \underset{\uparrow}{6}, \ -5, 0, 1 \right\},$$ (2.497)

4.

$$x(n) = \delta(n) - 2\delta(n-1) + 3\delta(n-2) - 4\delta(n-3).$$ (2.498)

Problem 2.4. (Periodic Sequences)
Determine the fundamental period of the sinusoidal sequences

$$x(n) = A\cos(\omega_0 n)$$ (2.499)

for the following values of the angular frequency ω_0:

1.

$$\omega_0 = 0.15\pi,$$ (2.500)

2.

$$\omega_0 = 0.26\pi,$$ (2.501)

3.

$$\omega_0 = 0.19\pi,$$ (2.502)

4.

$$\omega_0 = 0.85\pi.$$ (2.503)

Problem 2.5. (Periodic Sequences)
Determine if the following sequences are periodic and determine the fundamental period of the sequences if they are periodic:

1.

$$x(n) = 2.0\cos(0.17\pi n + 0.25),$$ (2.504)

2.

$$x(n) = \cos(0.28\pi n - 0.15),$$ (2.505)

3.

$$x(n) = \sin(0.62\pi n),$$ (2.506)

4.

$$x(n) = \cos(0.85\pi n - 0.33\pi).$$ (2.507)

Problem 2.6. (Phasor Representation)
Express the following discrete time signals in the form

$$x_k(n) = A_k \cos(\omega_k n + \phi_k), \quad k = 1, 2. \tag{2.508}$$

Specify the values of A_k, ω_k, and ϕ_k for each case. **(Hint)** You may need to use the trigonometric identity

$$\sin(\theta) = \cos\left(\theta - \frac{\pi}{2}\right). \tag{2.509}$$

1.

$$x_1(n) = 3\cos\left(0.57\pi n - \frac{3\pi}{7}\right) - 2\sin\left(0.57\pi n + \frac{2\pi}{3}\right); \tag{2.510}$$

2.

$$x_2(n) = 7\sin\left(0.37\pi n - \frac{2\pi}{3}\right) - 5\cos\left(0.37\pi n + \frac{3\pi}{5}\right). \tag{2.511}$$

Problem 2.7. (Poles and Zeros)
In this problem, the degrees of the numerator and denominator polynomials are different, so there should be zeros (or poles) at $z = 0$ or $z = \infty$. Determine the poles and zeros of the following filters:

1.

$$y(n) = 0.324y(n-1) + 0.125y(n-2) - 2.0x(n-1), \tag{2.512}$$

2.

$$y(n) = 0.3136y(n-2) + 2.0x(n-2), \tag{2.513}$$

3.

$$y(n) = -0.4y(n-1) + 0.16y(n-2) - 2.11x(n-2), \tag{2.514}$$

4.

$$y(n) = 0.4y(n-1) + 2.0x(n-1) + 2.0x(n-2). \tag{2.515}$$

Problem 2.8. (Impulse Response)
The transfer function of a discrete time system is given by

$$H(z) = \frac{0.3880z^3 + 1.1640z^2 + 1.1640z + 0.3880}{z^3 + 1.2210z^2 + 0.7344z + 0.1485}. \tag{2.516}$$

Find the impulse response $h(n)$ for this system for $0 \le n \le \infty$.

Problem 2.9. (Output of LTI System from System Function)
This system function for a LTI system is

$$H(z) = \frac{0.3591z^2 + 0.7182z + 0.3591}{z^2 + 0.162z - 0.2745}.$$
(2.517)

Find the output from this system for $0 \le n \le \infty$ when the input is

$$x(n) = 2.0\left[1.0 - (0.3)^n\right]u(n).$$
(2.518)

Problem 2.10. (Region of Convergence)
Determine the region of convergence for the following Z Transforms such that the unit circle ($|z| = 1$) is within the region of convergence:
1.

$$H(z) = \frac{z^2 + 3.5z + 10}{(z + 0.65)(z - 1.7)},$$
(2.519)

2.

$$H(z) = \frac{1.2188z^2 + 2.4376z + 1.2188}{(z + 1.112)(z - 2.35)},$$
(2.520)

3.

$$H(z) = \frac{0.4829z^3 + 1.4195z^2 + 1.4195z + 0.4829}{(z + 0.6948 - 0.5232j)(z + 0.6948 + 0.5232j)(z + 0.2094)},$$
(2.521)

4.

$$H(z) = \frac{1.27z^2 + 2.54z + 1.27}{(z + 1.569 - 0.429j)(z + 1.569 + 0.429j)}.$$
(2.522)

Problem 2.11. (Region of Convergence)
Determine the region of convergence for the following Z Transforms such that the unit circle ($|z| = 1$) is within the region of convergence:
1.

$$H(z) = \frac{z^2 - 0.5z - 14}{z^2 - 3.65z - 1.4},$$
(2.523)

2.

$$H(z) = \frac{0.6169z^2 + 1.2337z + 0.6169}{z^2 + 1.1454z + 0.4683},$$
(2.524)

3.

$$H(z) = \frac{10z^2 - 25z + 10}{z^2 + 2.65z + 1.3},$$
(2.525)

4.

$$H(z) = \frac{0.685z^2 + 1.37z + 0.685}{z^2 + 3.683z + 3.156}. \tag{2.526}$$

Problem 2.12. (Region of Convergence)
Determine the region of convergence for the following Z Transforms such that the unit circle ($|z| = 1$) is within the region of convergence:

1.

$$H(z) = \frac{0.6944 + 1.3888z^{-1} + 0.6944z^{-2}}{1.0 + 0.9182z^{-1} - 0.3327z^{-2}}, \tag{2.527}$$

2.

$$H(z) = \frac{1.1007z^2 - 2.2014z + 1.1007}{z^2 + 2.2530z + 1.1496}, \tag{2.528}$$

3.

$$H(z) = \frac{0.7097 + 1.4194z^{-1} + 0.7097z^{-2}}{1.0 - 2.1612z^{-1} + 4.0000z^{-2}}, \tag{2.529}$$

4.

$$H(z) = \frac{0.5997z^2 + 1.1993z + 0.5997}{z^2 - 1.9358z - 1.4629}. \tag{2.530}$$

Problem 2.13. (Region of Convergence)
Determine the region of convergence for the following Z Transforms such that the unit circle ($|z| = 1$) is within the region of convergence. **(Hint)** It may be useful to use the Matlab "roots" routine for this problem.

1.

$$
\begin{aligned}
H(z) &= \frac{B(z)}{A(z)}, \\
B(z) &= 0.0125 - 0.0872z^{-1} + 0.2615z^{-2} - 0.4359z^{-3} \\
&\quad + 0.4359z^{-4} - 0.2615z^{-5} + 0.0872z^{-6} - 0.0125z^{-7}, \\
A(z) &= 1.0 + 1.8211z^{-1} + 0.4776z^{-2} - 0.4250z^{-3} \\
&\quad - 0.3395z^{-4} - 0.6098z^{-5} - 0.0704z^{-6} \\
&\quad - 0.2599z^{-7};
\end{aligned} \tag{2.531}
$$

2.

$$
\begin{aligned}
H(z) &= \frac{B(z)}{A(z)}, \\
B(z) &= 0.0219 + 0.1534z^{-1} + 0.4601z^{-2} + 0.7668z^{-3} \\
&\quad + 0.7668z^{-4} + 0.4601z^{-5} + 0.1534z^{-6} + 0.0219z^{-7},
\end{aligned}
$$

$$A(z) = 1.0 + 1.9091z^{-1} + 0.6549z^{-2} - 0.5947z^{-3}$$
$$- 0.2251z^{-4} + 0.1059z^{-5} - 0.0160z^{-6}$$
$$- 0.0296z^{-7}; \tag{2.532}$$

3.

$$H(z) = \frac{B(z)}{A(z)},$$
$$B(z) = 0.0490 - 0.3429z^{-1} + 1.0288z^{-2} - 1.7147z^{-3}$$
$$+ 1.7147z^{-4} - 1.0288z^{-5} + 0.3429z^{-6} - 0.0490z^{-7},$$
$$A(z) = 1.0 + 2.4121z^{-1} + 4.7517z^{-2} + 4.2623z^{-3}$$
$$+ 1.1504z^{-4} - 2.3309z^{-5} - 3.4767z^{-6}$$
$$- 1.4977z^{-7}. \tag{2.533}$$

Problem 2.14. (Region of Convergence)
Determine the region of convergence for the following Z Transforms such that the unit circle ($|z| = 1$) is within the region of convergence:

1.

$$H(z) = \frac{0.6944 + 1.3888z^{-1} + 0.6944z^{-2}}{1.0 + 0.9182z^{-1} - 0.3327z^{-2}}, \tag{2.534}$$

2.

$$H(z) = \frac{1.1007z^2 - 2.2014z + 1.1007}{z^2 + 2.2530z + 1.1496}, \tag{2.535}$$

3.

$$H(z) = \frac{0.7097 + 1.4194z^{-1} + 0.7097z^{-2}}{1.0 - 2.1612z^{-1} + 4.0000z^{-2}}, \tag{2.536}$$

4.

$$H(z) = \frac{0.5997z^2 + 1.1993z + 0.5997}{z^2 - 1.9358z - 1.4629}. \tag{2.537}$$

Problem 2.15. (Inverse Z Transform)
The partial fraction expansion of the Z Transform for a discrete time sequence is given by

$$Y(z) = \frac{2.2594}{z + 0.4500} - \frac{0.099}{z - 0.5500}$$
$$+ \frac{1.0843}{z - 0.3210} - 2.9331. \tag{2.538}$$

Determine the discrete time function $y(n)$ $\forall\, 0 \le n \le \infty$ corresponding to this Z Transform.

Problem 2.16. (Inverse Z Transform)
Find a bounded sequence such that $|x(n)| < \infty$, $-\infty \le n \le \infty$ for the following
Z Transform:

$$X(z) = \frac{4.4000z^2 - 6.1100z}{(z - 2.15)(z - 0.81)}. \qquad (2.539)$$

Problem 2.17. (Impulse Response)
The system function for a discrete time system is given by

$$H(z) = \frac{0.0653z^2 + 0.1306z + 0.0653}{(z - 0.81)(z + 0.375)}. \qquad (2.540)$$

Determine the impulse response for the system $h(n)$, $0 \le n \le \infty$.

Problem 2.18. (System Response)
The difference equation representation for a discrete time system is given by

$$\begin{aligned} y(n) &= 0.1462x(n) - 0.2925x(n-1) + 0.1462x(n-2) \\ &\quad + 0.5y(n-1) - 0.0850y(n-2). \end{aligned} \qquad (2.541)$$

Find the output from this system for $0 \le n \le \infty$ when the input is

$$x(n) = 2.5(-0.47)^n u(n). \qquad (2.542)$$

Problem 2.19. (System Response)
The system response for a discrete time system is given by

$$H(z) = \frac{0.1593z^2 + 0.3187z + 0.1593}{(z + 0.472)(z - 0.567)}. \qquad (2.543)$$

Find the output for this system for $0 \le n \le \infty$ when the input is

$$x(n) = 2.5(0.76)^n u(n). \qquad (2.544)$$

Problem 2.20. (System Response)
1. The difference equation representation for a discrete time system is given by

$$\begin{aligned} y(n) &= 0.2943x(n) + 0.5754x(n-1) + 0.2943x(n-2) \\ &\quad + 0.0111y(n-1) - 0.3171y(n-2). \end{aligned} \qquad (2.545)$$

Find the output from this system for $0 \le n \le \infty$ when the input is

$$x(n) = 1.75(-0.87)^n u(n). \qquad (2.546)$$

2. The system transfer function for a discrete time system is given by

$$H(z) = \frac{0.2284z^2 + 0.4567z + 0.2284}{z^2 + 0.1690z - 0.2555}. \qquad (2.547)$$

Find the output from this system for $0 \le n \le \infty$ when the input is

$$x(n) = 2.35 \left[1 - (0.246)^n \right] u(n). \tag{2.548}$$

Problem 2.21. (System Response)
The system transfer function for a discrete time system is given by

$$H(z) = \frac{5.0840z^3 - 1.9255z^2 - 0.1281z + 0.1946}{(z + 0.7120)(z - 0.5930)(z - 0.4610)}. \tag{2.549}$$

Find the output for this system for $0 \le n \le \infty$ when the input is

$$x(n) = 2.25(-0.582)^n u(n). \tag{2.550}$$

Problem 2.22. (System Response)
The system transfer function for a discrete time system is given by

$$H(z) = \frac{5.0840z^3 - 1.9255z^2 - 0.1281z + 0.1946}{(z + 0.7120)(z - 0.5930)(z - 0.4610)}. \tag{2.551}$$

Find the output for this system for $0 \le n \le \infty$ when the input is

$$x(n) = 2.25(-0.582)^n u(n). \tag{2.552}$$

Problem 2.23. (System Response)
The system response for a discrete time system is given by

$$H(z) = \frac{0.1593z^2 + 0.3187z + 0.1593}{(z + 0.472)(z - 0.567)}. \tag{2.553}$$

Find the output for this system for $0 \le n \le \infty$ when the input is

$$x(n) = 2.5(0.76)^n u(n). \tag{2.554}$$

Problem 2.24. (Stability Analysis)
Determine if the following system transfer functions represent bounded-input, bounded-output (BIBO) stable, causal systems. Give a reason for each answer.

1.

$$H(z) = \frac{z^3 - 2.773z^2 - 2.3818z}{(z + 0.4605)\,(z - 0.7122)\,(z + 0.8146)}; \tag{2.555}$$

2.

$$H(z) = \frac{0.24\,(z - 0.321 - 1.423j)\,(z - 0.321 + 1.423j)}{(z + 0.8124 + 0.800j)\,(z + 0.8124 - 0.800j)}; \tag{2.556}$$

3.

$$H(z) = \frac{0.45\,(z - 0.550 - 0.836j)\,(z - 0.550 + 0.836j)}{(z + 0.6605 + 0.7883j)\,(z + 0.6605 - 0.7883j)}; \tag{2.557}$$

4.

$$H(z) = \frac{0.64\,(z+2.7)\,(z-2.0)}{(z-1.782)\,(z+0.6146)}.$$ (2.558)

Problem 2.25. (Stability Region)

Determine the stability region in terms of the coefficients of the system transfer function for the causal system

$$H(z) = \frac{b_0 + b_1 z^{-1} + b_2 z^{-2}}{1.0 + a_1 z^{-1} + a_2 z^{-2}}$$ (2.559)

by computing its poles and restricting them to be inside the unit circle.

Problem 2.26. (Sampling)

Given the discrete time sinusoid

$$x(n) = 2\sin(0.318\pi n + 0.6\pi)$$ (2.560)

and the sampling frequency $F_s = 3{,}000$ Hz,

1. Determine the corresponding continuous time sinusoid in the frequency range $0 < F_s < \frac{F_s}{2}$ that has the same sample sequence;
2. Determine two other continuous time sinusoids that have the same sample sequence.

Problem 2.27. (Sampling)

A continuous time signal $x(t)$ is given by

$$\begin{aligned}x(t) &= \cos(600\pi t + 0.6\pi) + 2.0\cos(850\pi t + 0.3\pi)\\ &\quad + 1.5\cos(1200\pi t - 0.4\pi) + \cos(2450\pi t + 0.2\pi)\\ &\quad + 0.75\cos(3725\pi t - 0.1\pi).\end{aligned}$$ (2.561)

The signal $x(t)$ is sampled at a 2.1 kHz rate, and the samples are passed through an ideal low pass filter with a cutoff frequency of 950 Hz to reconstruct the continuous time output signal $y(t)$. Determine the frequency components present in the reconstructed output signal.

Problem 2.28. (Sampling)

Consider the sampling of the band pass signal $x(t)$ whose spectrum is given in Fig. 2.71. Determine the minimum sampling frequency F_s to avoid aliasing. Assume that $x(t)$ is a real valued signal.

Problem 2.29. (Sampling)

A digital signal is given by

$$x(n) = 4\cos(0.419\pi n + 0.253\pi) + \cos(0.728\pi n - 0.319\pi).$$ (2.562)

Assume that this signal is played out through a D/A converter whose sampling rate is 4 kHz. Determine the resulting continuous time signal that will be heard if it is played through an audio system?

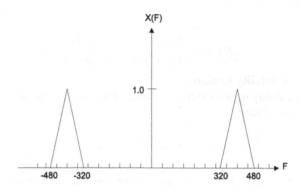

FIGURE 2.71

Plot of Spectrum for Problem 2.28

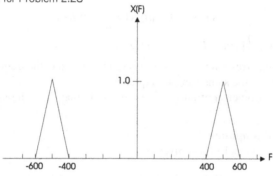

FIGURE 2.72

Plot of spectrum for Problem 2.30

Problem 2.30. (Sampling Band Pass Signals)
Consider the sampling of the band pass signal $x(t)$ whose spectrum is given in
Fig. 2.72. Determine the minimum sampling frequency F_s to avoid aliasing. Assume
that $x(t)$ is a real valued signal.

Problem 2.31. (Sampling)
Consider the continuous time function given by

$$x(t) = 3.0\cos(70\pi t + 0.6\pi) + 1.5\cos(82\pi t) + 2.5\cos(245\pi t). \qquad (2.563)$$

Assume that $x(t)$ is sampled using a sampling frequency of 200 samples per second.
Write an equation for the reconstructed continuous time signal if the sampled output,
$x(n)$, is passed through an ideal digital-to-analog (D/A) converter.

Problem 2.32. (Sampling)
A continuous time signal $x(t)$ is composed of a linear combination of sinusoidal
signals with frequencies at 0.3, 0.5, 1.2, 2.15, and 3.5 kHz. The signal $x(t)$ is sampled

at a 2.0 kHz rate, and the samples are passed through an ideal low pass filter with a cutoff frequency of 0.9 kHz to reconstruct the continuous time output signal $y(t)$. Determine the frequency components present in the reconstructed output signal.

Problem 2.33. (Sampling)

Given the discrete time sinusoid

$$x(n) = 2\cos(0.2\pi n + 0.1\pi) \tag{2.564}$$

and the sampling frequency $F_s = 2$ kHz,

1. Determine the corresponding continuous time sinusoid in the frequency range $0 < F_s < \frac{F_s}{2}$ that has the same sample sequence;
2. Determine two other continuous time sinusoids that have the same sample sequence.

Problem 2.34. (Sampling)

Given the discrete time sinusoid

$$x(n) = 2\sin(0.318\pi n + 0.6\pi) \tag{2.565}$$

and the sampling frequency $F_s = 3$ kHz,

1. Determine the corresponding continuous time sinusoid in the frequency range $0 < F < \frac{F_s}{2}$ that has the same sample sequence;
2. Determine two other continuous time sinusoids that have the same sample sequence.

Problem 2.35. (Difference Equations)

The zeros for a discrete time system are given by

$$
\begin{aligned}
q(1) &= 1.0000, \\
q(2) &= 0.6555 - 0.7552j, \\
q(3) &= 0.6555 + 0.7552j, \\
q(4) &= 0.3685 + 0.9296j, \\
q(5) &= 0.3685 - 0.9296j, \\
q(6) &= 0.4232 + 0.9060j, \\
q(7) &= 0.4232 - 0.9060j.
\end{aligned}
\tag{2.566}
$$

The poles are given by

$$
\begin{aligned}
p(1) &= 0.3316 + 0.9362j, \\
p(2) &= 0.3316 - 0.9362j, \\
p(3) &= 0.2722 + 0.9194j, \\
p(4) &= 0.2722 - 0.9194j,
\end{aligned}
$$

$$
\begin{aligned}
p(5) &= 0.0098 + 0.8014j, \\
p(6) &= 0.0098 - 0.8014j, \\
p(7) &= -0.4270.
\end{aligned}
\tag{2.567}
$$

The overall system gain is given by

$$
G = 0.1205.
\tag{2.568}
$$

Write one or more difference equations with only real coefficients that can be used to compute the output $y(n)$ for an input sequence $x(n)$. It is okay to use first order and/or second order sections to implement the filter?

Problem 2.36. (System Function)
A linear, time-invariant, discrete time system is described by the following difference equation:

$$
\begin{aligned}
y(n) &= 0.2533x(n) + 0.9487x(n-1) + 1.3930x(n-2) \\
&\quad + 0.9487x(n-3) + 0.2533x(n-4) - 1.3298y(n-1) \\
&\quad - 1.2940y(n-2) - 0.4373y(n-3) - 0.1994y(n-4). \tag{2.569}
\end{aligned}
$$

1. Determine the system function $H(z)$ for this system. Express $H(z)$ as a ratio of polynomials in z^{-1} (negative powers of z).
2. Obtain an expression for $H(\omega)$, the frequency response of the system. You should obtain this expression using $H(z)$ from above.

Problem 2.37. (Stability Analysis)
Determine if the following system transfer functions represent bounded-input, bounded-output (BIBO) stable, causal systems. Give a reason for each answer.

1.

$$
H(z) = \frac{z^3 - 2.773z^2 - 2.3818z}{(z+0.4605)\,(z-0.7122)\,(z+0.8146)};
\tag{2.570}
$$

2.

$$
H(z) = \frac{0.24\,(z-0.321-1.423j)\,(z-0.321+1.423j)}{(z+0.8124+0.800j)\,(z+0.8124-0.800j)};
\tag{2.571}
$$

3.

$$
H(z) = \frac{0.45\,(z-0.550-0.836j)\,(z-0.550+0.836j)}{(z+0.6605+0.7883j)\,(z+0.6605-0.7883j)};
\tag{2.572}
$$

4.

$$
H(z) = \frac{0.64\,(z+2.7)\,(z-2.0)}{(z-1.782)\,(z+0.6146)}.
\tag{2.573}
$$

Problem 2.38. (Difference Equations)

The transfer function for a discrete time system is given by

$$H(z) = \frac{0.4123(z-1)(z-0.9691-0.2467j)(z-0.9691+0.2467j)}{(z-0.1076)(z-0.6075-0.5899j)(z-0.6075+0.5899j)}. \quad (2.574)$$

Write one or more difference equations with only real coefficients that can be used to compute the output $y(n)$ for an input sequence $x(n)$.

2.31 MATLAB PROBLEMS FOR CHAPTER 2

Problem 2.39. (Sampling Sinusoidal Signals)

The following Matlab script generates a discrete time signal and makes a plot of it. Assume that the samples plotted are samples from a continuous time signal. Derive a formula for the continuous time signal corresponding to the samples plotted by this Matlab script.

```
dt = 1/200;
tt = -1: dt :1;
Fo = 3;
xx = 200*real( exp(j*2*pi*Fo*(tt - 0.35)));
%
plot(tt, xx, '+'), grid
title('SECTION of a SINUSOID'), xlabel('TIME (sec)')
```

Problem 2.40. (Complex Exponential Signals)

Define $s(t)$ as

$$s(t) = 10e^{j\frac{3\pi}{4}}e^{j23\pi t}. \quad (2.575)$$

1. Use Matlab to make a plot of $s_r(t) = \Re e\{s(t)\}$. Pick a range of values that will include exactly four periods of the signal.
2. Use Matlab to make a plot of $q(t) = \Im m\{\dot{s}(t)\}$, where the dot means differentiation with respect to time t. Again plot four cycles of the signal.

Problem 2.41. (Generating Sequences)

The Matlab functions "ones" and "zeros" can be used to generate a discrete time unit step function and an impulse. For example, 20 samples of a step function can be generated by the following Matlab code.

```
un = ones(1,20);
```

The indices for arrays and sequences for Matlab must be positive, finite integers. Therefore, it is a good idea to generate the corresponding time sequence to use with plots. In this case, the following Matlab script will generate a normalized time sequence with 20 samples.

```
tn = 0:19;
```

The "stem" function should be used to plot discrete time sequences instead of the "plot" function which is used for continuous time functions. You can plot the above step function using the Matlab script

```
stem(tn, un);
title(' Plot Title Here');
xlabel(' Label X axis Here');
ylabel(' Label Y axis Here');
```

Often, you want to generate a delayed step function so that you can observe the performance of a system just prior to the application of the unit step. The following Matlab script can be used to generate and plot 20 samples of a step function when it has been delayed by 10 samples:

```
un=[zeros(1,10), ones(1,20)];
tn = 0:29;
stem(tn, un);
```

A unit impulse delayed by 10 samples can be generated using the following Matlab code:

```
impn = [zeros(1,10), 1, zeros(1,19)];
tn = 0:29;
stem(tn, impn);
```

You can generate samples of a sinusoidal sequence using similar Matlab script. For example, the following Matlab script can be used to generate 20 samples of a sinusoidal sequence:

```
tn = 0:19;
xn = 5.0*sin(0.33*pi*tn + 0.75*pi);
stem(tn, xn);
```

Use the appropriate Matlab functions to generate the following sequences. Use the "stem" function to plot each of these. Turn in your Matlab code and the resulting plots.

1.

$$x(n) = \begin{cases} 0 & \text{for} \quad 0 \leq n \leq 9, \\ -1 & \text{for} \quad 10 \leq n \leq 19, \\ 0 & \text{for} \quad 20 \leq n \leq 29, \\ 1 & \text{for} \quad 30 \leq n \leq 39; \end{cases} \qquad (2.576)$$

2.

$$x(n) = \begin{cases} 0 & \text{for} \quad 0 \leq n \leq 19, \\ 1 & \text{for} \quad 20 \leq n \leq 39, \\ 0 & \text{for} \quad 40 \leq n \leq 49, \\ -1 & \text{for} \quad 50 \leq n \leq 69, \\ 0 & \text{for} \quad 70 \leq n \leq 79; \end{cases} \qquad (2.577)$$

3. 40 values of

$$x(n) = 5.0 * \sin(0.37\pi n - 0.47\pi). \tag{2.578}$$

Problem 2.42. (Generating Sequences)

1. Write a Matlab program to generate and display a random signal of length 75 whose elements are uniformly distributed in the interval $[-4, 4]$.
   (**Hint**) See the Matlab function "rand".
2. Write a Matlab program to generate and display a normal (Gaussian) random signal of length 75 whose elements are normally distributed with zero mean and a variance of 3.
   (**Hint**) See the Matlab function "randn". Also, the variance is modified by multiplying the output sequence by the square root of the desired variance.

Problem 2.43. (Inverse Z Transforms)

Use the Matlab program "residuez" to obtain the partial fraction expansion of the following Z Transforms and use the results to obtain the inverse Z Transform. Assume that the region of convergence includes the unit circle.

1.

$$X(z) = \frac{2z^2 - 4z + 2}{z^2 + 0.611z + 0.257}; \tag{2.579}$$

2.

$$X(z) = \frac{1.5z^2 - 1.5z}{z^2 + 1.5z - 0.675}. \tag{2.580}$$

Problem 2.44. (System Function)

A linear time-invariant filter is described by the difference equation

$$\begin{aligned} y(n) \;=\; & 0.1372x(n) - 0.2521x(n-1) + 0.1372x(n-2) \\ & - 0.8153y(n-1) - 0.4061y(n-2). \end{aligned} \tag{2.581}$$

1. Determine the system function $H(z)$ for this system. Express $H(z)$ as a ratio of polynomials in z^{-1} (negative powers of z) and also as a ratio of polynomials in positive powers of z.
2. Obtain an expression for $H(\omega)$, the frequency response of this system, from $H(z)$.
3. Use the Matlab function "freqz" to compute 100 values of $H(\omega)$ for $0 \le \omega \le \pi$. Use the Matlab functions "abs" and "angle" to compute the magnitude and phase for your values of $H(\omega)$. Turn in the magnitude and phase plots along with your Matlab code.

Problem 2.45. (Poles and Zeros)

Use the Matlab command *zplane* to plot the poles and zeros of the following transfer functions. You can use the form

```
zplane(b,a)
```

where the numerator coefficients are in the array b and the denominator coefficients are in the array a. You should also put titles and labels on your plots. Turn in your Matlab code and all plots.

1.

$$H(z) = \frac{B(z)}{A(z)} \tag{2.582}$$

where

$$
\begin{aligned}
B(z) &= 0.4985z^5 + 2.4240z^4 + 4.7817z^3 + 4.7817z^2 \\
&\quad + 2.4240z + 0.4985, \\
A(z) &= z^5 + 3.5579z^4 + 5.2635z^3 + 3.9391z^2 \\
&\quad + 1.4551z + 0.1929;
\end{aligned} \tag{2.583}
$$

2.

$$H(z) = \frac{0.1881z^4 - 0.3657z^2 + 0.1881}{z^4 + 0.5185z^2 + 0.3509}. \tag{2.584}$$

Problem 2.46. (Inverse Z Transform)
The Matlab function *residuez* can be used to obtain the partial fraction expansion of a Z Transform. This is useful in obtaining the inverse Z Transform. You can obtain information on how to use this function by using Matlab help.

```
help residuez
```

Use the Matlab *residuez* routine to obtain the partial fraction expansion of the following Z Transforms and then use the output to obtain the corresponding inverse Z transform:

1.

$$Y(z) = \frac{4.0000z^3 + 1.1700z^2 - 0.7722z - 0.1616}{z^3 + 0.3900z^2 - 0.3861z - 0.1616}, \tag{2.585}$$

2.

$$Y(z) = \frac{5.1720z^3 - 4.5899z^2 - 1.5222z}{z^3 - 0.0560z^2 - 0.2655z - 0.0465}. \tag{2.586}$$

Problem 2.47. (Poles and Zeros)
Use the Matlab command "zplane" to plot the poles and zeros of the following transfer functions. You can use the form

```
zplane(b,a)
```

where the numerator coefficients are in the array b and the denominator coefficients are in the array a. You should also put titles and labels on your plots. Turn in your Matlab code and all plots.

1.

$$H(z) = \frac{B(z)}{A(z)} \tag{2.587}$$

where

$$
\begin{aligned}
B(z) &= 0.1053z^5 + 0.4495z^4 + 0.8340z^3 + 0.8340z^2 \\
&\quad + 0.4495z + 0.1053, \\
A(z) &= z^5 - 0.5718z^4 - 1.2165z^3 + 0.0505z^2 \\
&\quad - 0.2684z + 0.2286. \tag{2.588}
\end{aligned}
$$

2.

$$H(z) = \frac{0.2929z^4 - 0.5858z^2 + 0.2929}{z^4 + 0.1716}. \tag{2.589}$$

Problem 2.48. (Region of Convergence)
Determine the region of convergence for the following Z Transforms such that the unit circle ($|z| = 1$) is within the region of convergence.
(Hint) It may be useful to use the Matlab "roots" routine for this problem.

1.

$$H(z) = \frac{2.4920z^4 - 8.6995z^3 + 11.1466z^2 - 6.1626z + 1.2236}{z^4 - 2.1576z^3 + 1.3933z^2 - 0.7313z + 0.0944}; \tag{2.590}$$

2.

$$H(z) = \frac{0.1101z^4 - 0.4402z^3 + 0.6604z^2 - 0.4402z + 0.1101}{z^4 - 1.5600z^3 + 0.8587z^2 + 0.5923z - 0.4508}; \tag{2.591}$$

3.

$$H(z) = \frac{0.3942z^4 + 1.4948z^3 + 2.2036z^2 + 1.4948z + 0.3942}{z^4 + 3.8329z^3 + 7.9217z^2 + 7.7857z + 3.5963}; \tag{2.592}$$

4.

$$H(z) = \frac{0.2679z^4 + 0.0029z^3 - 0.5266z^2 + 0.0029z + 0.2679}{z^4 + 0.7497z^3 + 0.3041z^2 + 0.2681z + 0.3175}. \tag{2.593}$$

Problem 2.49. (Inverse Z Transform)
The Matlab function *residuez* can be used to obtain the partial fraction expansion of a Z Transform. This is useful in obtaining the inverse Z Transform. You can obtain information on how to use this function by using the Matlab help utility as follows:

```
help residuez
```

Use the Matlab *residuez* function to obtain the partial fraction expansion of the following Z Transforms and then use the output from this function to obtain the corresponding inverse Z transform using the constraint that

$$|y(n)| < R < \infty \; \forall \; -\infty \leq n \leq \infty \tag{2.594}$$

where R is some large positive real number and $y(n)$ is the inverse Z Transform of $Y(z)$.

1.

$$Y(z) = \frac{0.3225 + 0.9675z^{-1} + 0.9675z^{-2} + 0.3225z^{-3}}{1.0 + 1.9209z^{-1} + 1.4806z^{-2} + 0.6592z^{-3} + 0.0995z^{-4}}; \tag{2.595}$$

2.

$$Y(z) = \frac{0.8107 - 3.2427z^{-1} + 4.8640z^{-2} - 3.2427z^{-3} + 0.8107z^{-4}}{1.0 - 3.1399z^{-1} + 4.1314z^{-2} - 2.8402z^{-3} + 1.0059z^{-4} - 0.1464z^{-5}} \tag{2.596}$$

Problem 2.50. (Region of Convergence)
Determine the region of convergence for the following Z Transforms such that the unit circle ($|z| = 1$) is within the region of convergence.
(Hint) It may be useful to use the Matlab "roots" routine for this problem.

1.

$$H(z) = \frac{B(z)}{A(z)},$$

$$B(z) = 0.0125 - 0.0872z^{-1} + 0.2615z^{-2} - 0.4359z^{-3}$$
$$+ 0.4359z^{-4} - 0.2615z^{-5} + 0.0872z^{-6} - 0.0125z^{-7},$$

$$A(z) = 1.0 + 1.8211z^{-1} + 0.4776z^{-2} - 0.4250z^{-3} - 0.3395z^{-4}$$
$$- 0.6098z^{-5} - 0.0704z^{-6} - 0.2599z^{-7}; \tag{2.597}$$

2.

$$H(z) = \frac{B(z)}{A(z)},$$

$$B(z) = 0.0219 + 0.1534z^{-1} + 0.4601z^{-2} + 0.7668z^{-3}$$
$$+ 0.7668z^{-4} + 0.4601z^{-5} + 0.1534z^{-6} + 0.0219z^{-7},$$

$$A(z) = 1.0 + 1.9091z^{-1} + 0.6549z^{-2} - 0.5947z^{-3} - 0.2251z^{-4}$$
$$+ 0.1059z^{-5} - 0.0160z^{-6} - 0.0296z^{-7}; \tag{2.598}$$

3.

$$H(z) = \frac{B(z)}{A(z)},$$

$$\begin{aligned}
B(z) &= 0.0490 - 0.3429z^{-1} + 1.0288z^{-2} - 1.7147z^{-3} \\
&\quad + 1.7147z^{-4} - 1.0288z^{-5} + 0.3429z^{-6} - 0.0490z^{-7}, \\
A(z) &= 1.0 + 2.4121z^{-1} + 4.7517z^{-2} + 4.2623z^{-3} + 1.1504z^{-4} \\
&\quad - 2.3309z^{-5} - 3.4767z^{-6} - 1.4977z^{-7}.
\end{aligned} \tag{2.599}$$

Problem 2.51. (Poles and Zeros)

Use the Matlab command "zplane" to plot the poles and zeros of the following transfer functions. You can use the form

```
zplane(b,a)
```

where the numerator coefficients are in the array b and the denominator coefficients are in the array a. You should also put titles and labels on your plots. Turn in your Matlab code and all plots.

1.

$$H(z) = \frac{0.1053z^5 + 0.4495z^4 + 0.8340z^3 + 0.8340z^2 + 0.4495z + 0.1053}{z^5 - 0.5718z^4 - 1.2165z^3 + 0.0505z^2 - 0.2684z + 0.2286}; \tag{2.600}$$

2.

$$H(z) = \frac{0.2929z^4 - 0.5858z^2 + 0.2929}{z^4 + 0.1716}. \tag{2.601}$$

Frequency Domain Analysis 3

3.1 FREQUENCY ANALYSIS OF DISCRETE TIME SIGNALS

The frequency analysis of discrete time signals is an important part of the analysis of linear, shift invariant, discrete time systems. Most of the discrete time signals of practical interest can be decomposed into a sum of sinusoidal signal components. A signal, represented with such a decomposition, is said to be represented in the frequency domain. A periodic signal can be decomposed into a sum of sinusoidal components, each with a frequency that is a multiple of the frequency of the fundamental sinusoidal signal, where the fundamental sinusoidal signal has a period that is the same as the period for the periodic signal. Each of the individual sinusoidal signals may have a unique magnitude and phase. This type of decomposition is called a Fourier series. The class of signals that are not periodic but have finite energy can be decomposed into a Fourier transform. These decompositions are important because a sinusoidal signal is an eigenfunction for a linear, shift invariant system such that the output for such a system with a sinusoidal input will be a sinusoidal output with the same frequency but possibly with different magnitude and phase.

3.2 FREQUENCY RESPONSE CHARACTERISTICS

This section discusses the frequency response of linear, shift invariant, discrete time systems. The frequency response for a linear, shift invariant, discrete time system, can be determined by applying a unit magnitude, zero phase, sinusoidal sequence with arbitrary frequency to the system as a chosen input. The frequency response, at that arbitrary frequency, is the ratio of the output to the input for the chosen input. This concept can be used to determine the frequency response of a linear, shift invariant, discrete time system.

The system transfer function for a causal, linear, shift invariant discrete time system has the following form:

$$H(z) = \frac{Y(z)}{X(z)} = \frac{\displaystyle\sum_{k=0}^{L} b(k)z^{-k}}{1.0 + \displaystyle\sum_{k=1}^{L} a(k)z^{-k}}. \tag{3.1}$$

Digital Signal Processing. DOI: 10.1016/B978-0-12-804547-3.00003-6

159

Eq. (3.1) represents an FIR discrete time system if all of the $a(k)$ are equal to zero. Otherwise, it represents an IIR discrete time system. The frequency response of the corresponding system can be obtained by applying a unit magnitude, sinusoidal signal with zero phase and arbitrary radial frequency, ω, as an input. The corresponding output is the frequency response for the system at the arbitrary frequency ω. The input signal, thus selected, can be represented as

$$x(n) = e^{j\omega n}. \tag{3.2}$$

Since the system is linear and shift invariant, the corresponding output will be a frequency dependent constant, multiplied by the input,

$$y(n) = H(\omega)e^{j\omega n}. \tag{3.3}$$

$H(\omega)$, which is frequency dependent, does not depend on the independent variable, n. It is also the frequency response for the discrete time system with system transfer function $H(z)$.

It is easy to show this by considering the general form for the difference equation for a causal, discrete time system

$$y(n) + \sum_{k=1}^{L} a(k)y(n-k) = \sum_{k=0}^{L} b(k)x(n-k). \tag{3.4}$$

If the output is assumed to be

$$y(n) = H(\omega)e^{j\omega n}, \tag{3.5}$$

then

$$y(n-k) = H(\omega)e^{j\omega(n-k)}. \tag{3.6}$$

Thus,

$$H(\omega)e^{j\omega n} + \sum_{k=1}^{L} a(k)H(\omega)e^{j\omega(n-k)} = \sum_{k=0}^{L} b(k)e^{j\omega(n-k)}, \tag{3.7}$$

$$H(\omega)e^{j\omega n}\left\{ 1.0 + \sum_{k=1}^{L} a(k)e^{-j\omega k} \right\} = e^{j\omega n}\left\{ \sum_{k=0}^{L} b(k)e^{-j\omega k} \right\}. \tag{3.8}$$

This equation can be solved for $H(\omega)$ to obtain

$$H(\omega) = \frac{\displaystyle\sum_{k=0}^{L} b(k)e^{-j\omega k}}{1.0 + \displaystyle\sum_{k=1}^{L} a(k)e^{-j\omega k}}. \tag{3.9}$$

This result is equivalent to evaluating the system transfer function on the unit circle where $z = e^{j\omega}$. Thus, the frequency response for the discrete time system with system

FIGURE 3.1

General discrete time system with input $e^{j\omega n}$

transfer function, $H(z)$, can be obtained by evaluating $H(z)$ on the unit circle where $z = e^{j\omega}$:

$$H(\omega) = H(z)|_{z=e^{j\omega}} = \frac{\displaystyle\sum_{k=0}^{L} b(k)e^{-j\omega k}}{1.0 + \displaystyle\sum_{k=1}^{L} a(k)e^{-j\omega k}}. \tag{3.10}$$

Fig. 3.1 shows a discrete time system with input $e^{j\omega}$. Observe that the output is $H(\omega)e^{j\omega}$. This output was verified in Eq. (3.10) as the output for a sinusoidal input with unit magnitude, zero phase, and arbitrary frequency to a causal, linear, shift invariant, discrete time system with all initial conditions equal to zero. This output can be obtained by multiplying the transfer function evaluated at the frequency of the sinusoidal input by the sinusoidal input. In this case, the input $e^{j\omega}$ is called an eigenfunction of the system because the output for this input is the input multiplied by a constant. Note that $H(\omega)$ is considered to be a constant since it is not a function of the independent variable n, even though it is a function of ω, which is a parameter in this case.

Example 3.1 illustrates the use of this concept to determine the frequency response for an FIR discrete time system.

EXAMPLE 3.1
(FIR Discrete Time System Frequency Response)

Problem:
Determine the frequency response for the FIR discrete time system with impulse response given by

$$h(n) = -0.0110\delta(n) + 0.1637\delta(n-1) + 0.6947\delta(n-2)$$
$$+ 0.1637\delta(n-3) - 0.0110\delta(n-4). \tag{3.11}$$

Solution:
The coefficients for an FIR discrete time system are equal to the corresponding values of the discrete time impulse response for the system. Thus, the correspond-

ing difference equation for the FIR discrete time system is

$$y(n) = -0.0110x(n) + 0.1637x(n-1) + 0.6947x(n-2)$$
$$+0.1637x(n-3) - 0.0110x(n-4). \tag{3.12}$$

The frequency response for the discrete time system can be determined by using a unit magnitude sinusoidal signal with zero phase and arbitrary frequency as the input to the system. The frequency response at the arbitrary frequency is the output divided by the corresponding input. Thus, if the input is chosen to be

$$x(n) = e^{j\omega n}. \tag{3.13}$$

Then

$$x(n-1) = e^{j\omega(n-1)} = e^{-j\omega}x(n),$$
$$x(n-2) = e^{j\omega(n-2)} = e^{-j2\omega}x(n),$$
$$x(n-3) = e^{j\omega(n-3)} = e^{-j3\omega}x(n),$$
$$x(n-4) = e^{j\omega(n-4)} = e^{-j4\omega}x(n). \tag{3.14}$$

The output, for the chosen input, is given by

$$y(n) = \left[-0.0110 + 0.1637e^{-j\omega} + 0.6947e^{-j2\omega}\right]e^{j\omega n}$$
$$\left[+0.1637e^{-j3\omega} - 0.0110e^{-j4\omega}\right]e^{j\omega n}. \tag{3.15}$$

The frequency response, $H(\omega)$, is the ratio of the output to the chosen input. Thus,

$$H(\omega) = \frac{y(n)}{x(n)} = \frac{y(n)}{e^{j\omega n}}, \tag{3.16}$$

$$H(\omega) = -0.0110 + 0.1637e^{-j\omega} + 0.6947e^{-j2\omega}$$
$$+ 0.1637e^{-j3\omega} - 0.0110e^{-j4\omega}. \tag{3.17}$$

$H(\omega)$ can also be written in the form

$$H(\omega) = e^{-2j\omega}\left\{0.6947 - 0.0110\left[e^{2j\omega} + e^{-2j\omega}\right]\right\}$$
$$+ e^{-2j\omega}\left\{0.1637\left[e^{j\omega} + e^{-j\omega}\right]\right\}, \tag{3.18}$$

$$H(\omega) = e^{-2j\omega}\{0.6947 + 0.3274\cos(\omega) - 0.0220\cos(2\omega)\}. \tag{3.19}$$

The magnitude and phase of $H(\omega)$, at some particular value of ω, can be obtained by substituting the value of ω into Eq. (3.19) for $H(\omega)$. Eq. (3.20) shows the

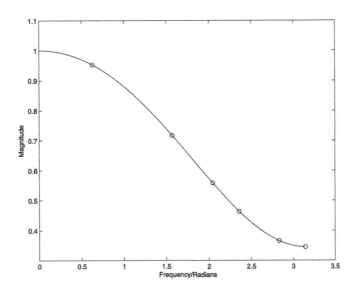

FIGURE 3.2

Magnitude spectrum for FIR filter

computation of the frequency response for ω equal to 0.2π, 0.5π, 0.65π, 0.75π, 0.9π, and π:

$$
\begin{aligned}
H(0.2\pi) &= e^{-0.4j\pi}\{0.6947 + 0.3274\cos(0.2\pi) - 0.0220\cos(0.4\pi)\} \\
&= 0.9528e^{-0.4j\pi}, \\
H(0.5\pi) &= e^{-j\pi}\{0.6947 + 0.3274\cos(0.5\pi) - 0.0220\cos(\pi)\} \\
&= 0.7194e^{-j\pi}, \\
H(0.65\pi) &= e^{-1.3j\pi}\{0.6947 + 0.3274\cos(0.65\pi) - 0.0220\cos(1.3\pi)\} \\
&= 0.5617e^{-1.3j\pi}, \\
H(0.75\pi) &= e^{-1.5j\pi}\{0.6947 + 0.3274\cos(0.75\pi) - 0.0220\cos(1.5\pi)\} \\
&= 0.4659e^{-1.5j\pi}, \\
H(0.9\pi) &= e^{-1.8j\pi}\{0.6947 + 0.3274\cos(0.9\pi) - 0.0220\cos(1.8\pi)\} \\
&= 0.3682e^{-1.8j\pi}, \\
H(\pi) &= e^{-2j\pi}\{0.6947 + 0.3274\cos(\pi) - 0.0220\cos(2\pi)\} \\
&= 0.3480. \tag{3.20}
\end{aligned}
$$

Fig. 3.2 shows the magnitude response of the system with the values computed in Eq. (3.20) marked with an o. Fig. 3.3 shows the corresponding phase response with the values computed in Eq. (3.20) marked with an o.

End of the Example

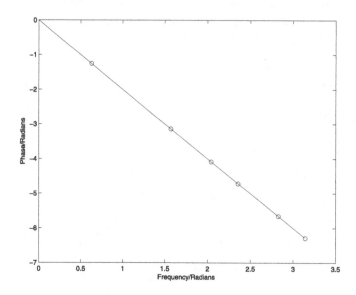

FIGURE 3.3

Phase spectrum for FIR filter

Example 3.2 illustrates the computation of the frequency response for an IIR discrete time system.

EXAMPLE 3.2
(IIR Discrete Time System Frequency Response)

Problem:
Determine the frequency response for the IIR discrete time system with the following system transfer function:

$$H(z) = \frac{0.6283z^2 - 1.2566z + 0.6283}{z^2 - 1.1804z + 0.4816}. \tag{3.21}$$

Solution:
The frequency response for the discrete time system with system transfer function given by $H(z)$ can be determined by evaluating $H(z)$ on the unit circle where $z = e^{j\omega}$ as follows:

$$H(e^{j\omega}) = H(z)|_{z=e^{j\omega}} = \frac{0.6283e^{2j\omega} - 1.2566e^{j\omega} + 0.6283}{e^{2j\omega} - 1.1804e^{j\omega} + 0.4816}. \tag{3.22}$$

Specific values of ω can be used in Eq. (3.22) to obtain the frequency response of the corresponding discrete time system at any given frequency. Eq. (3.23) shows the computation of the frequency response for ω equal to 0.2π, 0.5π, 0.65π, 0.75π, 0.9π, and π:

$$
\begin{aligned}
H(0.2\pi) &= \frac{0.6283(e^{0.4j\pi})^2 - 1.2566e^{0.4j\pi} + 0.6283}{(e^{0.4j\pi})^2 - 1.1804e^{0.4j\pi} + 0.4816} \\
&= -0.0470 + 0.7848j = 0.7862\angle 1.6306, \\
H(0.5\pi) &= \frac{0.6283(e^{j\pi})^2 - 1.2566e^{j\pi} + 0.6283}{(e^{j\pi})^2 - 1.1804e^{j\pi} + 0.4816} \\
&= -0.0470 + 0.7848j = 0.9747\angle 0.4138, \\
H(0.65\pi) &= \frac{0.6283(e^{1.3j\pi})^2 - 1.2566e^{1.3j\pi} + 0.6283}{(e^{1.3j\pi})^2 - 1.1804e^{1.3j\pi} + 0.4816} \\
&= 0.9283 + 0.2314j = 0.9567\angle 0.2443, \\
H(0.75\pi) &= \frac{0.6283(e^{1.5j\pi})^2 - 1.2566e^{1.5j\pi} + 0.6283}{(e^{1.5j\pi})^2 - 1.1804e^{1.5j\pi} + 0.4816} \\
&= 0.9374 + 0.1542j = 0.9500\angle 0.1631, \\
H(0.9\pi) &= \frac{0.6283(e^{1.8j\pi})^2 - 1.2566e^{1.8j\pi} + 0.6283}{(e^{1.8j\pi})^2 - 1.1804e^{1.8j\pi} + 0.4816} \\
&= 0.9432 + 0.0583j = 0.9450\angle 0.0618, \\
H(\pi) &= \frac{0.6283(e^{2.0j\pi})^2 - 1.2566e^{2.0j\pi} + 0.6283}{(e^{2.0j\pi})^2 - 1.1804e^{2.0j\pi} + 0.4816} \\
&= 0.9441 + 0.0000j = 0.9441\angle 0.0000. \quad (3.23)
\end{aligned}
$$

Fig. 3.4 shows the magnitude response of the system with the values computed in Eq. (3.23) marked with an o. Fig. 3.5 shows the corresponding phase response with the values computed in Eq. (3.23) marked with an o. Note that the magnitude of the frequency response is zero at $\omega = 0$. The plot shows that the phase is equal to zero at this value of ω as well.

End of the Example

3.3 FREQUENCY RESPONSE ESTIMATES USING POLE–ZERO PLOTS

Pole–zero plots can be used to help estimate the frequency response of discrete time systems. For example, it is usually desirable to have the magnitude of a low pass filter approach or equal to zero as the frequency approaches the Nyquist frequency ($\pm\pi$ for normalized frequencies). Therefore, we expect to find the zeros to be located at or near $z = -1$ for low pass filters. The location of the poles will then determine

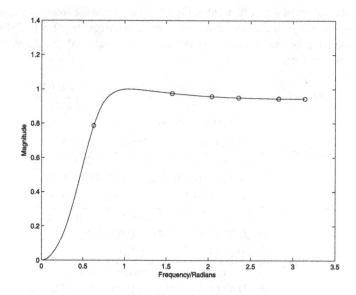

FIGURE 3.4

Magnitude plot of $H(\omega)$ for Example 3.2

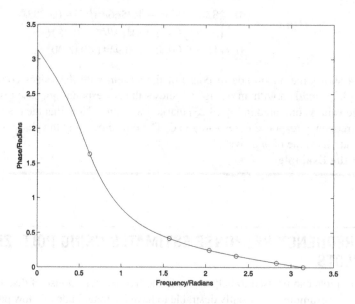

FIGURE 3.5

Phase plot of $H(\omega)$ for Example 3.2

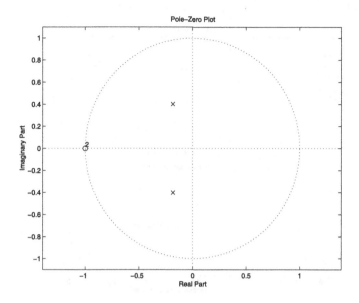

FIGURE 3.6

Pole–zero plot of $H(z)$ for low pass filter with cutoff frequency $\omega_c = 0.6\pi$

the value of the cutoff frequency which is typically defined as the point where the magnitude gain is equal to $\frac{\sqrt{2}}{2}$ or the power gain is equal to 0.5. For example, Fig. 3.6 gives the pole–zero plot for a second order low pass filter with cutoff frequency of $\omega_c = 0.6\pi$. The zeros are both located at $z = -1$ and the poles are located at

$$p(1) = -0.1848 + 0.4021j, \quad p(2) = p^*(1) = -0.1848 - 0.4021j. \quad (3.24)$$

Fig. 3.7 gives the magnitude and phase response plots for this filter. We also normally normalize the gain of the low pass filter to be equal to one at $\omega = 0$ or where $z = 1$. In this case, the second order low pass filter can be generally represented as

$$H(z) = \frac{A(z+1)^2}{z^2 - 2\Re e\{p(1)\}z + |p(1)|^2}. \quad (3.25)$$

It follows that

$$A = \frac{1 - 2\Re e\{p(1)\} + |p(1)|^2}{4}. \quad (3.26)$$

On the other hand, it is typical that the gain for a high pass filter is desired to be at or near zero at $\omega = 0$. Therefore, we expect the zeros of a high pass filter to be clustered around $z = 1$. The locations of the poles then determine the cutoff frequency. Fig. 3.8 gives the pole–zero plot for a second order high pass filter with a cutoff frequency of $\omega_c = 0.6\pi$. In this case, the poles are located in the same place as

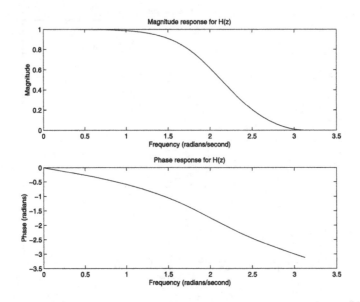

FIGURE 3.7

Magnitude and phase plots of $H(z)$ for low pass filter with cutoff frequency $\omega_c = 0.6\pi$

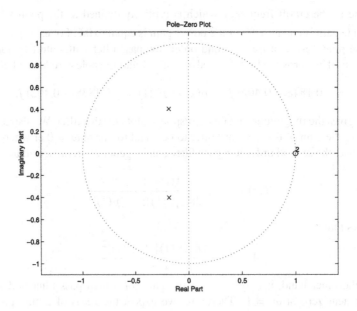

FIGURE 3.8

Pole–zero plot of $H(z)$ for high pass filter with cutoff frequency $\omega_c = 0.6\pi$

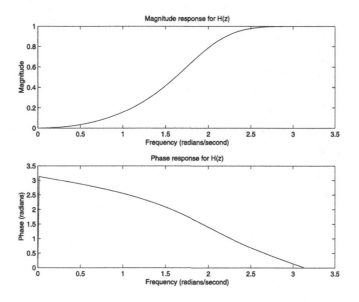

FIGURE 3.9

Magnitude and phase plots of $H(z)$ for high pass filter with cutoff frequency $\omega_c = 0.6\pi$

those for the low pass filter above with the same cutoff frequency. However, the zeros are located at $z = 1$ as desired. Fig. 3.9 gives the magnitude and phase response plots for this filter. We also normally normalize the gain of the high pass filter to be equal to one at $\omega = \pi$ or $z = -1$.

The gain for a band stop filter is desired to be at or near zero for the stop frequency which is normally given by $\omega_s = 0.5(\omega_1 + \omega_2)$ where ω_1 and ω_2 are the low and high cutoff frequencies, respectively. The locations of the poles then determine the cutoff frequencies. Fig. 3.10 gives a pole–zero plot for a fourth order, band stop, filter with cutoff frequencies of $\omega_1 = 0.6\pi$ and $\omega_2 = 0.75\pi$. There are two zeros located at

$$r(1) = r(3) = -0.5373 + 0.8434j \tag{3.27}$$

and two zeros located at

$$r(2) = r(4) = -0.5373 - 0.8434j. \tag{3.28}$$

The poles are located at

$$
\begin{aligned}
p(1) &= -0.5810 + 0.6372j, \\
p(2) &= -0.5810 - 0.6372j, \\
p(3) &= -0.3187 + 0.7678j, \\
p(4) &= -0.3187 - 0.7678j.
\end{aligned}
\tag{3.29}
$$

Fig. 3.11 gives the magnitude and phase response plots for this filter.

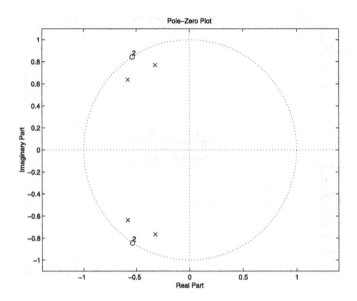

FIGURE 3.10

Pole–zero plot of $H(z)$ for band stop filter with cutoff frequencies at $\omega_1 = 0.6\pi$ and $\omega_2 = 0.75\pi$

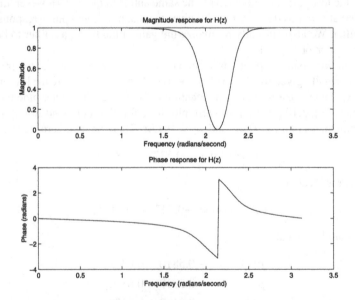

FIGURE 3.11

Magnitude and phase plots of $H(z)$ for band stop filter with cutoff frequency $\omega_1 = 0.6\pi$ and $\omega_2 = 0.75\pi$

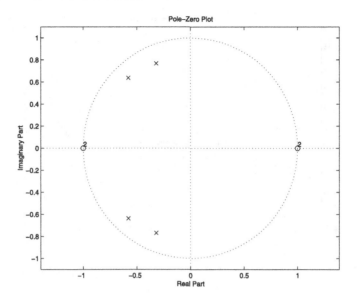

FIGURE 3.12

Pole–zero plot of $H(z)$ for band pass filter with cutoff frequencies at $\omega_1 = 0.6\pi$ and $\omega_2 = 0.75\pi$

The gain for a band pass filter is desired to be at or near zero for both $\omega = 0$ and $\omega = \pi$. Therefore, we expect to find zeros of the transfer function at both $z = 1$ and at $z = -1$. The locations of the poles then determine the cutoff frequencies. Fig. 3.12 gives a pole–zero plot for a fourth order band pass filter with cutoff frequencies of $\omega_1 = 0.6\pi$ and $\omega_2 = 0.75\pi$. There are two zeros at $z = 1$ and two zeros at $z = -1$. Since the critical frequencies for this example are the same as those for the band stop filter, the poles are located in the same place as there are for the band stop filter above. Fig. 3.13 gives the frequency magnitude and phase plots for this filter.

3.3.1 DIGITAL RESONATORS

A digital resonator is a special type of band pass filter with a pair of complex poles located near the unit circle. The name resonator refers to the fact that the filter has a rather large magnitude response in the vicinity of the pole location. A discrete time system can be designed to resonate with a resonant peak at ω_0, by selecting a pair of complex conjugate poles at

$$p_{1,2} = re^{\pm j\omega_0}. \tag{3.30}$$

The value of r should be selected to have a magnitude close to 1.0. Although there are several options for placing the zeros, one choice is to put one zero at $z = -1$ and one zero at $z = 1$ for each pair of complex poles. Another choice is to place both zeros at $z = 0$.

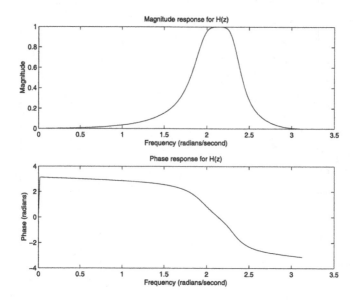

FIGURE 3.13

Magnitude and phase plots of $H(z)$ for band pass filter with cutoff frequency $\omega_1 = 0.6\pi$ and $\omega_2 = 0.75\pi$

EXAMPLE 3.3
(A Digital Resonator)

This example involves the design of a resonator with resonating frequency at $\omega = 0.7\pi$. We choose a radius of $r = 0.95$ for placement of the poles for the resonator. Fig. 3.14 gives a pole–zero plot of $H(z)$ for the resonator with the zeros located at $z = \pm 1$ and the poles at

$$p(1) = 0.95e^{j0.7\pi} = -0.5584 + 0.7686j, \qquad (3.31)$$

$$p(2) = p^*(1) = -0.5584 - 0.7686j. \qquad (3.32)$$

Fig. 3.15 gives the frequency response of the digital resonator with the zeros at $z = \pm 1$.

Fig. 3.16 gives a pole–zero plot of $H(z)$ for the resonator with the zeros located at $z = 0$ and the poles located in the same place at

$$p(1) = 0.95e^{j0.7\pi} = -0.5584 + 0.7686j, \qquad (3.33)$$

$$p(2) = p^*(1) = -0.5584 - 0.7686j. \qquad (3.34)$$

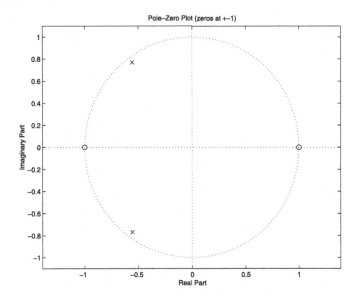

FIGURE 3.14

Pole–zero plot of $H(z)$ for digital resonator with zeros at $z = \pm 1$, $\omega = 0.7\pi$

Fig. 3.17 gives the frequency response of the digital resonator with the zeros at $z = 0$. Note that the peak in the magnitude response occurs for both cases at $\omega = 0.7\pi$. However, the peak is larger for the case with the zeros located at $z = \pm 1$.

End of the Example

3.4 FREQUENCY FILTERING

Many practical applications of digital signal processing involve the processing of a signal contaminated with noise. Often the noise, or a large portion of it, has a frequency distribution that is different from that of the desired signal. It may be practical to use a filter to suppress the noise while having a minimal impact on the magnitude and/or quality of the desired signal. For example, we know the normal range of frequencies for the human voice. We may be interested in designing a system that passes the frequencies associated with normal human speech and suppresses frequencies outside the range of human speech.

Fig. 3.18 gives a block diagram of system to process continuous time signals using a digital signal processing (DSP) system. The system includes an analog-to-digital converter (ADC), a digital-to-analog converter (DAC), a digital signal processor (DSP), and a clock. The ADC may include an analog, antialiasing filter and the DAC

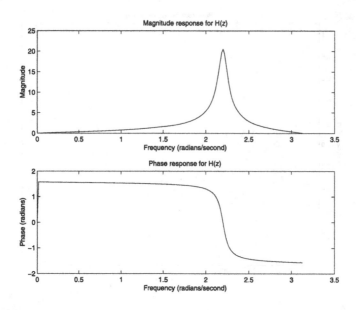

FIGURE 3.15

Frequency response of $H(z)$ for digital resonator with zeros at $z = \pm 1$, $\omega = 0.7\pi$

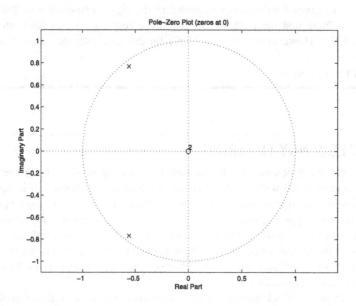

FIGURE 3.16

Pole–zero plot of $H(z)$ for digital resonator with zeros at $z = 0$, $\omega = 0.7\pi$

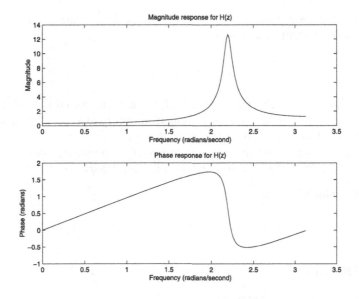

FIGURE 3.17

Frequency response of $H(z)$ for digital resonator with zeros at $z = 0$, $\omega = 0.7\pi$

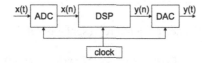

FIGURE 3.18

Block diagram of the overall system

may include low pass or bandpass filter to suppress noise with a frequency distribution outside of the desired frequency range for the signal of interest.

EXAMPLE 3.4
(Processing a Sample Sequence)

This example involves the analysis of a discrete time system used to filter an analog signal. The DSP has a clock rate of 250 MHz and the ADC and DCA run at 25 kHz. The DSP can perform 10000 operations per sample so it is able to implement simple digital filters. The desired -3 dB cutoff frequency for the low pass filter is 4 kHz.

We use normalized frequencies for digital filters with the sampling interval normalized to be equal to one unit ($T_s = 1.0/25000 = 4.0 \times 10^{-5}$ seconds) in this case. The corresponding Nyquist frequency of $\frac{\pi}{T_s}$ is normalized to π. Thus, the normalized cutoff frequency is

$$\omega_c = \frac{8000\pi}{25000} = 0.32\pi. \tag{3.35}$$

The functions in Matlab drop the factor of π. Thus, the digital filter should be designed in Matlab using a normalized cutoff frequency of $\omega_c = 0.32$. A second order filter will be used for this example ($order = 2$).

A second order digital filter with the desired cutoff frequency can be designed using the Matlab "butter" function and the following Matlab script.

Matlab Script 3.1.

```
wc = 0.32;
order = 2;
[b,a] = butter(order, wc);
```

End of the Script

The returned filter coefficients are

$$\begin{aligned} b &= \quad 0.1453 \quad 0.2906 \quad 0.1453 \\ a &= \quad 1.0 \quad -0.671 \quad 0.2523 \end{aligned} \tag{3.36}$$

The corresponding difference equation representation of the digital filter is

$$\begin{aligned} y(n) \quad - \quad & 0.671y(n-1) + 0.2523y(n-2) = 0.1453x(n) \\ & + 0.2906x(n-1) + 0.1453x(n-2). \end{aligned} \tag{3.37}$$

Assume that the analog input is given by

$$x(t) = [\cos(6000\pi t) + \cos(18000\pi t)]\, u(t). \tag{3.38}$$

Further, assume that the cosine signal at 6000π is the signal and the cosine signal at 18000π is undesirable noise. Matlab can be used to obtain samples of the input sequence using the following Matlab script:

Matlab Script 3.2.

```
w1 = 6000*pi;
w2 = 18000*pi;
fs = 25000;
dt = 1/fs;
nn = dt*(0:49);   % Generate 50 samples
x = cos(w1*nn) + cos(w2*nn);
```

End of the Script

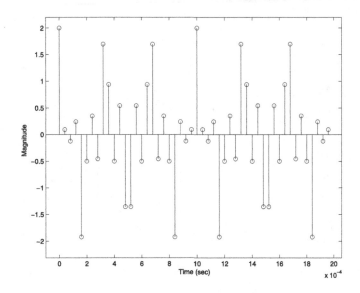

FIGURE 3.19

Plot of 50 samples of the input sequence for Example 3.4

Fig. 3.19 gives a plot of the input sequence $x(n)$.
   The Matlab *filter* function can be used to obtain the output.

```
y = filter(b, a, x);
```

Fig. 3.20 gives a plot of the output sequence after the suppression of the cosine sequence at 6000π by the filter, $y(n)$.
End of the Example

3.5 FREQUENCY DOMAIN SAMPLING

Previously, the discrete time Fourier Transform for a discrete time aperiodic signal $x(n)$ was determined to be given by

$$X(\omega) = X(z)|_{z=e^{j\omega}} \tag{3.39}$$

where

$$X(z) = \sum_{n=-\infty}^{\infty} x(n)z^{-n}. \tag{3.40}$$

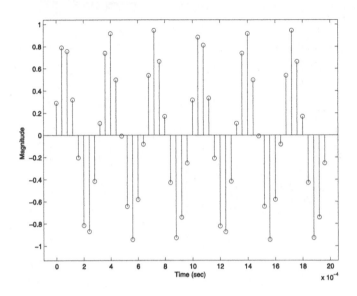

FIGURE 3.20

Plot of 50 samples of the output sequence for Example 3.4

If the region of convergence for $X(z)$ includes the unit circle, then $X(z)$ can be sampled on the unit circle at N equally spaced points separated in angle by

$$\omega_1 = \frac{2\pi}{N}. \tag{3.41}$$

The discrete Fourier series (DFS) representation of the sequence $x(n)$. For this case, it is given by

$$\tilde{X}(k) = X(z)|_{z=e^{j\frac{2\pi k}{N}}}, \quad k = 0, \pm 1, \pm 2, \ldots, \tag{3.42}$$

$$\tilde{X}(k) = \sum_{n=-\infty}^{\infty} x(n) e^{-j\frac{2\pi kn}{N}}. \tag{3.43}$$

It is easy to show that $\tilde{X}(k)$ is periodic since

$$e^{j\frac{2\pi(k+mN)}{N}} = e^{j\frac{2\pi k}{N}}, \quad m = 0, \pm 1, \pm 2, \ldots \tag{3.44}$$

Thus, $\tilde{X}(k)$ is periodic with period N. The inverse DFS (IDFS) is defined as

$$\tilde{x}(n) = \frac{1}{N} \sum_{k=0}^{N-1} \tilde{X}(k) e^{j\frac{2\pi kn}{N}}. \tag{3.45}$$

It follows directly that the IDFS is also periodic since the DFS is sampled and periodic.

3.6 THE DISCRETE FOURIER TRANSFORM

Consider the case where $x(n)$ is a finite sequence defined over the interval $0 \le n \le N - 1$. Then

$$X(z) = \sum_{n=0}^{N-1} x(n)z^{-n}. \tag{3.46}$$

If $X(z)|_{z=e^{j\omega}}$ is frequency sampled at $\omega = \frac{2\pi k}{N} \ \forall \ 0 \le k \le N - 1$, then we obtain the transform

$$\tilde{X}(k) = \sum_{n=0}^{N-1} x(n)e^{-j\frac{2\pi kn}{N}}. \tag{3.47}$$

The inverse transform is given by

$$\tilde{x}(n) = \frac{1}{N} \sum_{k=0}^{N-1} \tilde{X}(k)e^{j\frac{2\pi kn}{N}}. \tag{3.48}$$

We call this pair the Discrete Fourier Transform (DFT). It is important to emphasize here that $\tilde{x}(n)$ is periodic and is given by

$$\tilde{x}(n) = \sum_{m=-\infty}^{\infty} x(n - mN), \quad m = 0, \pm 1, \pm 2, \ldots \tag{3.49}$$

Thus, $\tilde{x}(n)$ is a linear combination of the original signal $x(n)$ and its infinite replicas, each shifted by multiples of $\pm N$. We can recover the original sequence by only using the values of $\tilde{x}(n)$ for $0 \le n \le N - 1$. This is equivalent to multiplying $\tilde{x}(n)$ by a rectangular window defined as

$$R_N(n) = \begin{cases} 1 & \text{for } 0 \le n \le N - 1, \\ 0 & \text{otherwise.} \end{cases} \tag{3.50}$$

3.7 DISCRETE FOURIER TRANSFORM EXAMPLES

This section provides examples of the computation of the DFT for some finite, discrete time signals.

EXAMPLE 3.5
(Fourier Transform of a Sample Sequence)

Problem:
Compute the Fourier transform of the following signal

$$x(n) = 0.0234\delta(n) + 0.0194\delta(n-1) + 0.9144\delta(n-2)$$
$$+ 0.0194\delta(n-3) + 0.0234\delta(n-4) \tag{3.51}$$

using

$$X(\omega) = \sum_{n=-\infty}^{\infty} x(n)e^{-j\omega n}.$$

Solution:
We can write

$$X(\omega) = 0.0234 + 0.0194e^{-j\omega} + 0.9144e^{-2j\omega} + 0.0194e^{-3j\omega} + 0.0234e^{-4j\omega}, \tag{3.52}$$

$$X(\omega) = e^{-2j\omega}\left[0.9144 + 0.0194\left(e^{j\omega} + e^{-j\omega}\right) + 0.0234\left(e^{2j\omega} + e^{-2j\omega}\right)\right], \tag{3.53}$$

$$X(\omega) = e^{-2j\omega}[0.9144 + 0.0388\cos(\omega) + 0.0468\cos(2\omega)]. \tag{3.54}$$

Fig. 3.21 gives the magnitude and phase plots for $X(\omega)$.
End of the Example

3.8 CONVOLUTION USING THE DFT

The periodicity of the reconstructed DFT has an impact when using the DFT for convolution. We illustrate this problem with an example.

EXAMPLE 3.6
(Convolution Using the DFT)

Consider the two discrete time signals given by

$$x_1(n) = \{ 4, \ 4, 4, 4, 4, 4, 4, 4\},$$
$$\uparrow$$
$$x_2(n) = \{ 0, \ 1, 2, 3, 4, 5, 6, 7\}. \tag{3.55}$$
$$\uparrow$$

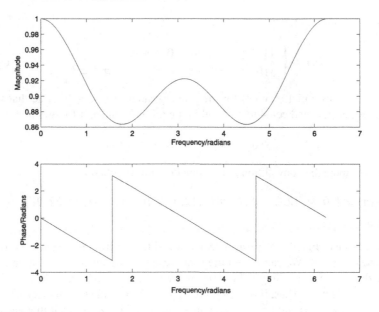

FIGURE 3.21

Magnitude and phase plots for Example 3.5

If we take the DFT of each, we have

$$X_1(k) = \{ \underset{\uparrow}{32}, \ 0, 0, 0, 0, 0, 0, 0 \},$$

$$X_2(k) = \{ \underset{\uparrow}{28.0}, \ -4.0 + 9.657j, \ -4.0 + 4.0j, \ -4.0 + 1.657j,$$

$$-4.0, \ -4.0 - 1.657j, \ -4.0 - 4.0j, \ -4.0 - 9.657j \}. \quad (3.56)$$

We form the element by element product of the two DFTs to obtain

$$Y_1(k) = X_1(k)X_2(k), \quad 0 \leq k \leq N - 1. \quad (3.57)$$

We then obtain what is called the circular convolution of the two sequences by taking the IDFT of $Y_1(k)$. Thus, we have

$$Y_1(k) = \{ \underset{\uparrow}{896}, \ 0, 0, 0, 0, 0, 0, 0 \}. \quad (3.58)$$

The corresponding IDFT is given by

$$y_1(n) = \begin{cases} 112 & \text{for} \quad 0 \leq n \leq 7, \\ y_1(n-8m) & \text{for} \quad m = 0, \pm 1, \pm 2, \pm 3, \ldots \end{cases} \tag{3.59}$$

Fig. 3.22 gives a plot of $y_1(n)$ for the primary range $0 \leq n \leq 7$. Note that since $y_1(n)$ is periodic and equal to 112 within the primary range, it follows that

$$y_1(n) = 112, \quad -\infty \leq n \leq \infty. \tag{3.60}$$

If we compute the convolution in the time domain, we obtain

$$y_2(n) = \{\ 0,\ 4, 12, 24, 40, 60, 84, 112, 112, 108, 100, 88, 72, 52, 28\}. \tag{3.61}$$
$$\uparrow$$

Fig. 3.23 gives the plot of $y_2(n)$. Clearly, it is different from the circular convolution result, $y_1(n)$. We can obtain the same result as the linear convolution in the time domain by expanding both $x_1(n)$ and $x_2(n)$ to accommodate the size of the linear convolution. Thus, if the length of $x_1(n)$ is L_1 and the length of $x_2(n)$ is L_2, then we should expand both of the two sequences with zeros such that each has length

$$L = L_1 + L_2 - 1. \tag{3.62}$$

Both L_1 and L_2 are equal to 8 for our example. Thus, we need to add 7 zeros to $x_1(n)$ and $x_2(n)$ to obtain

$$x_3(n) \quad = \quad \{\ 4,\ 4, 4, 4, 4, 4, 4, 4, 0, 0, 0, 0, 0, 0, 0\},$$
$$\uparrow$$

$$x_4(n) \quad = \quad \{\ 0,\ 1, 2, 3, 4, 5, 6, 7, 0, 0, 0, 0, 0, 0, 0\}. \tag{3.63}$$
$$\uparrow$$

The DFT of $x_3(n)$ is given in Table 3.1. The DFT of $x_4(n)$ is given in Table 3.2. We then form the element by element product of the two DFTs and take the IDFT to obtain

$$y_3(n) = \{\ 0,\ 4, 12, 24, 40, 60, 84, 112, 112, 108, 100, 88, 72, 52, 28\}. \tag{3.64}$$
$$\uparrow$$

which is the same result as we obtained by computing the convolution directly in the time domain. Fig. 3.24 gives a plot of the linear convolution of the two sequences using the DFT.

End of the Example

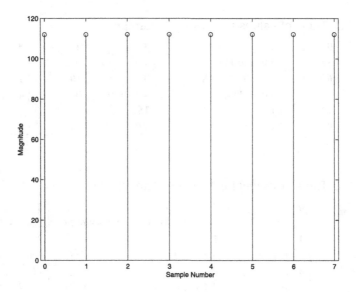

FIGURE 3.22

Circular convolution of $x_1(n)$ and $x_2(n)$

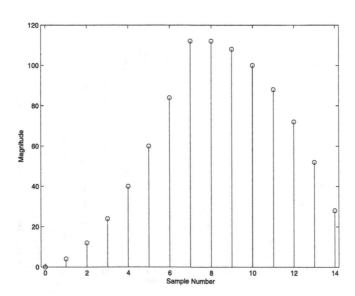

FIGURE 3.23

Linear convolution using convolution in the time domain

Table 3.1 Table of values for $X_3(k)$ for Example 3.6

k	0	1	2	3	4	5
$X_3(k)$	32	$4.0 - 20.11j$	0	$4.0 - 5.986j$	0	$4.0 - 2.673j$
k	6	7	8	9	10	11
$X_3(k)$	0	$4.0 - 0.7956j$	0	$4.0 + 0.7956j$	0	$4.0 + 2.673j$
k	12	13	14	15		
$X_3(k)$	0	$4.0 + 5.986j$	0	$4.0 + 20.11j$		

Table 3.2 Table of values for $X_4(k)$ for Example 3.6

k	0	1	2	3
$X_4(k)$	28	$-9.137 - 20.11j$	$-4.0 + 9.657j$	$2.380 - 5.986j$
k	4	5	6	7
$X_4(k)$	$-4.0 + 4.0j$	$3.277 - 2.673j$	$-4.0 + 1.657j$	$3.480 - 0.7956j$
k	8	9	10	11
$X_4(k)$	-4.0	$3.480 + 0.7956j$	$-4.0 - 1.657j$	$3.277 + 2.673j$
k	12	13	14	15
$X_4(k)$	$-4.0 - 4.0j$	$2.380 + 5.986j$	$-4.0 - 9.657j$	$-9.137 + 20.11j$

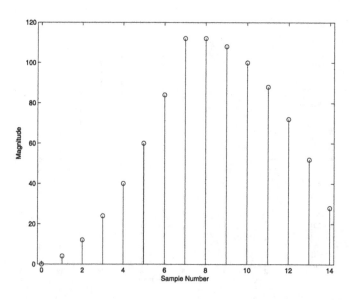

FIGURE 3.24

Linear convolution using the DFT

3.9 **FILTERING OF LONG DATA SEQUENCES**

In practical applications involving digital filtering, the input sequence $x(n)$ is often a very long sequence. On the other hand, the filter typically has a much smaller number of coefficients. For example, consider a sequence obtained by sampling a 4 second segment of speech at 8000 samples per second. Such a sequence would have 32,000 samples. A typical FIR filter may have 100 coefficients. A straightforward implementation would involve adding 99 zeros to the data sequence, adding 31,999 zeros to the filter, computing the DFT of both resulting sequences, forming the product of the DFTs, and finally computing the inverse DFT to obtain the output. This approach has at least two disadvantages:

1. The entire input sequence must be available in order to compute its DFT.
2. The DFTs for two sequences with 32,099 samples and one inverse DFT with 32,099 samples must be computed to obtain the output.

Actually, the signals would most likely be padded with zeros to a length of 32,768 (2^{15}) so that an FFT algorithm could be used to compute the DFTs and the inverse DFT.

Example 3.7 gives an example of the data access requirements for filtering a relatively long data sequence with a digital filter with a practical length of 100 filter coefficients.

EXAMPLE 3.7
(Filtering a Long Data Sequence)

This example considers the data access requirements for filtering a long data sequence. The input is assumed to have 32,000 samples and the filter length is assumed to be 100. Also, assume that the FFT block size will be 1024. Thus,

$$N = 1024,$$
$$M = 100,$$
$$L = N - M + 1 = 1024 - 100 + 1 = 925. \qquad (3.65)$$

The data access and computational requirements are as follows:

1. Add 924 zeros to the filter and compute the FFT of the result to obtain the 1024 point DFT of the filter.
2. Use 99 zeros and the first 925 samples of the input to form the first block of data.
3. Compute the FFT of the block of data to obtain the 1024 point DFT of the block of data.
4. Multiply the two DFTs together, point by point, to obtain the 1024 point DFT for a block of output.

5. Compute the inverse FFT of the DFT of the block of data and take the real part to obtain a 1024 sequence of output.
6. Discard the first 99 samples and save the last 925 samples of the output as 925 samples of the output sequence.
7. Form a new block of data using the last 99 samples of the previous block of data and 925 new samples of the input sequence if there are 925 or more samples of input data remaining.
   (a) If there are some samples left but not 925, then pad the block with zeros to make the final size equal to 1024 and go back to step 3.
   (b) If there are no samples left, then terminate the procedure.
8. Go back to step 3.

The number of blocks of data to be processed can be determined as

$$B = \left\lceil \frac{32000}{925} \right\rceil = \lceil 34.59 = 35. \tag{3.66}$$

End of the Example

There are two methods involving operations on blocks of data that can be used to overcome the disadvantages associated with using the DFT to filter long data sequences. Both of these approaches will require fewer complex multiplications and additions than the method described above. They are also more efficient than the direct computation using convolution or difference equations.

3.9.1 OVERLAP–SAVE METHOD

The overlap–save method involves the partitioning of the input into overlapping blocks. The choice of block size is somewhat arbitrary and should be chosen to minimize the total number of computations involved but it may also depend on available memory or allowable delay between the input and the corresponding output. Assume that the FIR filter has a length of M samples. Also assume that a block size of N is chosen for the DFTs and for the inverse DFTs. Normally, $N = 2^k$ for some integer k should be chosen to facilitate the use of an FFT algorithm. The blocks of input data would then be chosen as follows:

1. The first $M - 1$ samples of the first data block will be zeros. The next $L = N - M + 1$ samples will be the first L samples of the input data.
2. The first $M - 1$ samples of the subsequent data blocks will come from the last $M - 1$ data samples of the last data block. The next L samples will be the next L new samples of the data sequence.

In this way, each block will involve L new samples of the input sequence.

The output corresponding to each input block will be the last L samples of each output block. The first $M - 1$ samples are discarded.

EXAMPLE 3.8
(Overlap–Save)

Consider the use of the overlap–save procedure for filtering long sequences when the input sequence is a speech sample with a duration of 4 seconds and a sampling rate of 8000 samples per second. The FIR filter to be used has 100 filter coefficients. This example considers the total number of complex multiplication required to filter the sequence using the direct method and compares it with the number required for the overlap–save method. The number of complex multiplications required to compute an N point FFT is given by

$$K = N \log_2 N. \tag{3.67}$$

The total number of samples in the sequence is $N_1 = 4(8000) = 32000$. The minimum number of samples to use for the FFT in order to obtain a linear convolution result is

$$L = 32000 + 100 - 1 = 32099. \tag{3.68}$$

The next number that is a power of 2 is

$$N_2 = 2^{15} = 32768. \tag{3.69}$$

Thus, $32768 - 32000 = 768$ zeros need to be added to the end of the sample sequence and $32768 - 100 = 32668$ zeros need to be added to the end of the FIR filter. The FFT of both sequences must be computed and multiplied point by point to obtain the DFT of the output. Then, the inverse FFT (IFFT) must be computed to obtain the output. This means that the FFT must be computed twice and the IFFT must be computed once. The point by point multiplication of the two DFTs requires 32768 complex multiplications. The total number of complex multiplications is

$$K_1 = 3(32768)(15) + 32768 = 1507328. \tag{3.70}$$

If a block size of 256 (2^8) sample is used for the overlap–save method, then each time

$$L_2 = 256 - 100 + 1 = 157 \tag{3.71}$$

new samples will be used in each input block of data. It will require

$$B_1 = \left\lceil \frac{32000}{157} \right\rceil = \lceil 203.82 = 204 \tag{3.72}$$

blocks of data. The total number of complex multiplications required is

$$K_2 = 204\,[2(256)(8) + 256] + 256(8) = 889856. \tag{3.73}$$

It helps that the DFT of the filter only needs to be computed once. Thus, the computations for each block of data include the computation of the DFT for each block of data, the point by point multiplication of the two DFTs, and the computation of the inverse DFT of the product.

If a block size of 512 (2^9) is used for the overlap–save method, then each input block of data will use

$$L_3 = 512 - 100 + 1 = 413 \tag{3.74}$$

new samples. Thus,

$$B_3 = \left\lceil \frac{32000}{413} \right\rceil = \lceil 77.48 = 78 \tag{3.75}$$

blocks will be required. The total number of complex multiplications required is

$$K_2 = 78\,[2(512)(9) + 512] + 512(9) = 763392. \tag{3.76}$$

If a block size of 1024 (2^{10}) is used for the overlap–save method, then each block of data will use

$$L_3 = 1024 - 100 + 1 = 925 \tag{3.77}$$

new samples. Thus,

$$B_3 = \left\lceil \frac{32000}{925} \right\rceil = \lceil 34.59 = 35 \tag{3.78}$$

blocks will be required. The total number of complex multiplications required is

$$K_2 = 35\,[2(1024)(10) + 1024] + 1024(10) = 762880. \tag{3.79}$$

Similarly, if a block size of 2048 is used, then 17 blocks will be required and the number of complex multiplications required is 823296. Thus the smallest number of complex multiplications would be required when using a 1024 point FFT and 925 new samples per block.
End of the Example

3.9.2 OVERLAP–ADD METHOD

The overlap–add method also involves the partitioning of the input into overlapping blocks. The issues related to the choice of block size are the same as those for the choice of block size for the overlap–save method. Assume that the FIR filter has a length of M samples. Also assume that a block size of N samples is chosen for the DFTs and the inverse DFTs. Normally, the block size N should be chosen as $N = 2^k$ for some integer k to facilitate the use of an FFT algorithm. The blocks of input data are then chosen as follows:

1. The first $L = N - M + 1$ samples of the first data block will be the first L samples of the input sequence. The last $M - 1$ samples of the first data block will be zeros.
2. The first L samples of the subsequent data blocks will come from the next L data samples of the input sequence. The last $M - 1$ samples of the data block will be zeros.

In this way, each block will involve L new samples of the input sequence.

The output data sequence will be formed as follows:

1. The first N output samples will be put into the output sequence.
2. The first $M - 1$ samples of the new output block will be added to the last $M - 1$ samples of the output sequence which is the same as the last $M - 1$ samples of the last output block. The remaining L samples of the output block will form L new samples of the output sequence.

The savings for the overlap–add method are essentially the same as for the overlap–save method because each method uses $L = N - M + 1$ new samples for each block. The DFT sizes are the same and the number of complex multiplications required to multiply the two DFTs for each block are the same.

3.10 THE DISCRETE COSINE TRANSFORM

The Discrete Cosine Transform (DCT) is used extensively in some applications. It is the most widely used transform in a class of image coding systems known as transform coders. The DCT can be defined with either an even or an odd number of points. However, the even point DCT is more often used in practice so it is defined here.

Let $x(n)$ denote an N-point sequence that is zero outside of $0 \le n \le N - 1$. Create a new sequence, $y(n)$, which is related to $x(n)$ as follows:

$$y(n) = \begin{cases} x(n), & 0 \le n \le N - 1, \\ x(2N - 1 - n), & N \le n \le 2N - 1. \end{cases} \tag{3.80}$$

Then form a periodic sequence by repeating $y(n)$ every $2N$ points. The periodic sequence does not have any artificial discontinuities [7].

The $2N$-point DFT of $y(n)$ is given by

$$Y(k) = \sum_{n=0}^{2N-1} y(n) W_{2N}^{kn}, \quad 0 \le k \le 2N - 1 \tag{3.81}$$

where

$$W_{2N} = e^{-j(2\pi/2N)}. \tag{3.82}$$

The DFT of $y(n)$ can be defined in terms of $x(n)$ as follows:

$$Y(k) = \sum_{n=0}^{N-1} x(n) W_{2N}^{kn} + \sum_{n=N}^{2N-1} x(2N - 1 - n) W_{2N}^{kn}. \qquad (3.83)$$

This is equivalent to

$$Y(k) = \sum_{n=0}^{N-1} x(n) W_{2N}^{kn} + \sum_{n=0}^{N-1} x(n) W_{2N}^{k(2N-n)}. \qquad (3.84)$$

However, $W_{2N}^{2N} = 1$. Thus,

$$Y(k) = W_{2N}^{-k/2} \sum_{n=0}^{N-1} 2x(n) \cos(\pi k(2n + 1)/2N). \qquad (3.85)$$

The N-point DCT of $x(n)$ is defined as

$$C_x(k) = \begin{cases} W_{2N}^{k/2} Y(k), & 0 \leq k \leq N - 1, \\ 0, & \text{otherwise}. \end{cases} \qquad (3.86)$$

EXAMPLE 3.9
(DCT of a Sample Sequence)

Consider the 1-D sequence given by

$$x(n) = \begin{bmatrix} 0.0 & 0.5 & 1.0 & 1.5 & 2.0 & 2.5 & 3.0 & 3.5 \\ \uparrow \end{bmatrix}. \qquad (3.87)$$

The DCT for this sequence is given by

$$X(k) = [4.9497 \; -3.2212 \; 0.0 \; -0.3367 \; 0.0 \; -0.1005 \; 0.0 \; -0.0254].$$
$$(3.88)$$

Note that the DCT is real and has N points.
End of the Example

3.11 FAST FOURIER TRANSFORM ALGORITHMS

The basic idea behind the group of algorithms called the Fast Fourier Transform for computing the DFT is to successively decompose the DFT into smaller sized

DFTs to take advantage of the periodicity and symmetry of the complex exponentials.

3.11.1 DECIMATION IN TIME FFT ALGORITHM

Consider a sequence $x(n)$ with a length N that is a power of 2. A two-band polyphase decomposition of $x(n)$ can be used to express its Z Transform as

$$X(z) = X_0(z^2) + z^{-1}X_1(z^2) \tag{3.89}$$

where

$$X_0(z) \quad = \quad \sum_{n=0}^{\frac{N}{2}-1} x_0(n)z^{-n} = \sum_{n=0}^{\frac{N}{2}-1} x(2n)z^{-n},$$

$$X_1(z) \quad = \quad \sum_{n=0}^{\frac{N}{2}-1} x_1(n)z^{-n} = \sum_{n=0}^{\frac{N}{2}-1} x(2n+1)z^{-n}. \tag{3.90}$$

Observe that $X_0(z)$ is the Z Transform of the sequence $x(2n)$ (even samples) with length $\frac{N}{2}$, and $X_1(z)$ is the Z Transform of the sequence $x(2n+1)$ (odd samples) with length $\frac{N}{2}$.

Define

$$W_N = e^{-j\frac{2\pi}{N}}. \tag{3.91}$$

Then

$$W_N^k = e^{-j\frac{2\pi k}{N}} \tag{3.92}$$

and

$$W_{\frac{N}{2}} = e^{-j\frac{2\pi}{\frac{N}{2}}} = e^{-j\frac{4\pi}{N}} \tag{3.93}$$

which is periodic with period $\frac{N}{2}$ samples. If $X(z)$ is evaluated on the unit circle using N equally spaced samples, then the DFT of $x(n)$ will be obtained as

$$X(k) = X_{0<\frac{N}{2}>}(k) + W_N^k X_{1<\frac{N}{2}>}(k) \tag{3.94}$$

where $X_{0<\frac{N}{2}>}(k)$ and $X_{1<\frac{N}{2}>}(k)$ are $\frac{N}{2}$ point DFTs.

This process can be continued to further decompose each of the $\frac{N}{2}$ DFTs to two $\frac{N}{4}$ point DFTs until we finally have two point DFTs. The computational requirements for the DFT are traditionally stated in terms of the number of complex multiplications required. Using this decomposition approach the total number of complex multiplications and additions is

$$M = N \log_2 N. \tag{3.95}$$

In addition, the symmetry property for complex exponentials can be taken advantage of to obtain

$$W_N^{\frac{N}{2}+k} = -W_N^k \tag{3.96}$$

for a real sequence $x(n)$.

EXAMPLE 3.10

Consider the sequence given by

$$x = [1 \ \ 2 \ \ 2 \ \ 1 \ \ -1 \ \ -2 \ \ -2 \ \ -1]. \tag{3.97}$$

The 8-point, decimation in time, DFT is given by

$$W_8 = e^{-j\frac{2\pi}{8}} = e^{-j0.25\pi}. \tag{3.98}$$

It follows that

$$X(k) = 1 + 2W_8^k + 2W_8^{2k} + W_8^{3k} - W_8^{4k} - 2W_8^{5k} - 2W_8^{6k} - W_8^{7k}, \tag{3.99}$$

$$\begin{aligned} X(k) &= 1 + 2e^{-j0.25\pi k} + 2e^{-j0.25(2)\pi k} + e^{-j0.25(3)\pi k} - e^{-0.25(4)\pi k} \\ &\quad - 2e^{-j0.25(5)\pi k} - 2e^{-j0.25(6)\pi k} - e^{-j0.25(7)\pi k}, \end{aligned} \tag{3.100}$$

$$\begin{aligned} X(k) &= 1 + 2e^{-j0.25\pi k} + 2e^{-j0.5\pi k} + e^{-j0.75\pi k} - e^{-j\pi k} \\ &\quad - 2e^{-j1.25\pi k} - 2e^{-j1.5\pi k} - e^{-j1.75\pi k}. \end{aligned} \tag{3.101}$$

Now consider a time frequency decomposition, which is equivalent to a polyphase decomposition:

$$X(z) = X_0(z) + z^{-1}X_1(z). \tag{3.102}$$

The term $X_0(z)$ uses the even numbered samples, and the term $X_1(z)$ uses the odd numbered samples:

$$\begin{aligned} X_0(z) &= 1 + 2z^{-2} - z^{-4} - 2z^{-6}, \\ X_1(z) &= 2 + z^{-2} - 2z^{-4} - z^{-6}. \end{aligned} \tag{3.103}$$

It follows that

$$\begin{aligned} X_0(k) &= 1 + 2W_4^k - W_4^{2k} - 2W_4^{3k}, \\ X_1(k) &= 2 + W_4^k - 2W_4^{2k} - W_4^{3k} \end{aligned} \tag{3.104}$$

where

$$W_4 = e^{-j\frac{2\pi}{4}} = e^{-0.5\pi} = W_8^2. \tag{3.105}$$

In other words, $X_0(k)$ and $X_1(k)$ are both 4-point DFTs.

$$X_0(k) = 1 + 2e^{-j0.5\pi k} - e^{-j0.5(2)\pi k} - 2e^{-j0.5(3)\pi k},$$
$$X_1(k) = 2 + e^{-j0.5\pi k} - 2e^{-j0.5(2)\pi k} - e^{-j0.5(3)\pi k}, \qquad (3.106)$$

$$X_0(k) = 1 + 2e^{-j0.5\pi k} - e^{-j\pi k} - 2e^{-j1.5\pi k},$$
$$X_1(k) = 2 + e^{-j0.5\pi k} - 2e^{-j\pi k} - e^{-j1.5\pi k}, \qquad (3.107)$$

$$X_0(k) = \begin{cases} 0 \\ 2.0000 - 4.0000j \\ 0 \\ 2.0000 + 4.0000j \end{cases}, \qquad (3.108)$$

$$X_1(k) = \begin{cases} 0 \\ 4.0000 - 2.0000j \\ 0 \\ 4.0000 + 2.0000j \end{cases}. \qquad (3.109)$$

It is easy to verify that

$$X(k) = X_0(k) + W_8 X_1(k) = X_0(k) + e^{-j0.25\pi k} X_1(k). \qquad (3.110)$$

The process can be continued to obtain 4 two-point DFTs to implement $X(k)$.
End of the Example

3.11.2 DECIMATION IN FREQUENCY FFT ALGORITHM

The same idea that was used for the decimation in time FFT algorithm can be used to obtain a decimation in frequency algorithm by selecting the first $\frac{N}{2}$ points for $x_0(n)$ and the last $\frac{N}{2}$ points for $x_1(n)$. Then, the Z Transform of $x(n)$ can be written as

$$X(z) = X_0(z) + Z^{-\frac{N}{2}} X_1(z) \qquad (3.111)$$

where

$$X_0(z) = \sum_{n=0}^{\frac{N}{2}-1} x(n)z^{-n},$$
$$X_1(z) = \sum_{n=0}^{\frac{N}{2}-1} x\left(\frac{N}{2}+n\right)z^{-n}. \qquad (3.112)$$

The corresponding DFT can be obtained by evaluating $X(z)$ on the unit circle at N evenly spaced samples:

$$X(k) = \sum_{n=0}^{\frac{N}{2}-1} x(n)W_N^{nk} + W_N^{\frac{Nk}{2}} \sum_{n=0}^{\frac{N}{2}-1} x\left(\frac{N}{2}+n\right)W_N^{nk}. \qquad (3.113)$$

Note that

$$W_N^{\frac{Nk}{2}} = (-1)^k. \qquad (3.114)$$

Thus,

$$X(k) = \sum_{n=0}^{\frac{N}{2}-1} \left[x(n) + (-1)^k x\left(\frac{N}{2}+n\right)\right] W_N^{nk}. \qquad (3.115)$$

If k is even ($k = 2m$), then

$$X(2m) = \sum_{n=0}^{\frac{N}{2}-1} \left[x(n) + x\left(\frac{N}{2}+n\right)\right] W_N^{2mn}. \qquad (3.116)$$

If k is odd ($k = 2m + 1$), then

$$X(k) = \sum_{n=0}^{\frac{N}{2}-1} \left[x(n) - x\left(\frac{N}{2}+n\right)\right] W_N^{n(2m+1)}. \qquad (3.117)$$

The above expressions represent $\frac{N}{2}$-point DFTs of the following $\frac{N}{2}$ point sequences:

$$\begin{aligned} x_0(n) &= \left[x(n) + x\left(\frac{N}{2}+n\right)\right], \\ x_1(n) &= \left[x(n) - x\left(\frac{N}{2}+n\right)\right] W_N^n. \end{aligned} \qquad (3.118)$$

The decomposition using this approach can be continued until there are two point DFTs. The computational complexity is equivalent to that for the decimation in time FFT.

EXAMPLE 3.11
(Comparison of the DCT and the DFT)

This example compares the characteristics of the DCT and the DFT for a sample sequence. The following Matlab script can be used to generate 32 samples of a test sequence.

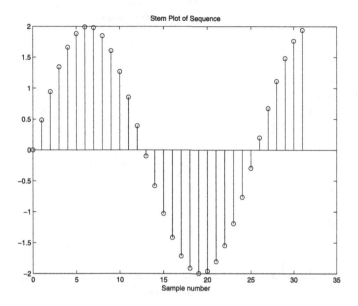

FIGURE 3.25

Stem plot of a test sequence for Example 3.11

Matlab Script 3.3.

```
n = 0:31;
dw = (pi + 3*pi/2)/32;
w = n*dw;
x = 2.0*sin(w);
print -dps examp3_10a.ps
```

End of the Script

Fig. 3.25 gives a stem plot of the resulting sequence. The Matlab FFT function can be used to compute the DFT and the Matlab DCT function can be used to compute the DCT as shown in the following Matlab script.

Matlab Script 3.4.

```
y1 = fft(x);
mag1 = abs(y1);
y2 = dct(x);
mag2 = abs(y2);
```

End of the Script

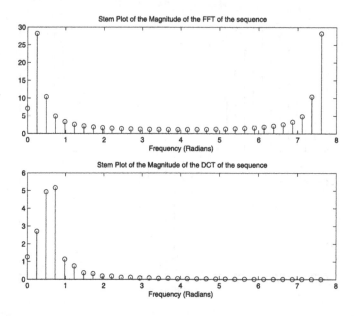

FIGURE 3.26

Stem plots of the magnitudes of the DFT and the DCT for the test sequence for Example 3.11

Fig. 3.26 gives stem plots for the magnitude of the DFT and the magnitude of the DCT for this sequence. Note the frequencies near the Nyquist frequency (normalized frequency of π radians) have a higher magnitude for the DFT but they are essentially zero for the DCT. The minimum magnitude occurs at the Nyquist frequency for both transforms. Note that the DFT has 32 complex values while the DCT has 32 real values. However, only need 32 complex values are needed to determine the DFT since the input sequence was real. The magnitude of the DFT has a maximum of 28 and a minimum of 1.1233 while the DCT has a maximum of 5.1561 and a minimum of 0.0017. The magnitude of the values of the DCT decrease to zero much faster than they do for the DFT. This is a demonstration of the compactness of the DCT relative to the DFT. This property makes the DCT useful for speech and image coding.

End of the Example

3.12 PROBLEMS FOR CHAPTER 3

Problem 3.1. (Frequency Representation)

The frequency response, $H(\omega)$, for a symmetric fifth order transfer function for an FIR filter with real coefficients has the specific values given in Table 3.3.

Determine $H(z)$.

Table 3.3 Frequency response values for Problem 3.1

Normalized Frequency	$H(\omega)$
0	1.0
0.25π	$-0.3104 + 0.7495j$
0.5π	$-0.2967 - 0.2967j$
0.75π	$0.1142 - 0.0473j$
π	0.0

Problem 3.2. (Filtering Long Sequences)
The length of a sampled speech signal is 31912 samples and the sampling rate is 8000 samples per second. The goal is to use the overlap–save filtering procedure to filter the sequence with an FIR digital filter that has 130 coefficients.

1. Determine the total number of complex multiplications required to filter the input sequence using an FFT algorithm directly without using the overlap–save procedure. The FFT size should be large enough to accommodate linear convolution of the two sequences.
2. Determine the total number of complex multiplications required to filter the sequence using the overlap–save procedure and a block size of 512 samples.
3. Determine the total number of complex multiplications required to filter the sequence using the overlap–save procedure and a block size of 1024 samples.
4. Determine the total number of complex multiplications required to filter the sequence using the overlap–save procedure and a block size of 2048 samples.

Problem 3.3. (The Discrete Fourier Transform)
The first seven values of the DFT of a 12 point sequence is given in Table 3.4. Determine the remaining 5 values of the DFT for this sequence.

Problem 3.4. (Overlap–Save Procedure)
The length of a sampled speech signal is 16360 samples and the sampling rate is 8000 samples per second. It is desired to use the overlap–save filtering procedure with an FIR digital filter with 43 coefficients. The block size should be 512 samples.

1. Determine the number of blocks of data that will be used in implementing the filter. Round any fraction up to the next whole number (assume that the last block will be padded with zeros to make it of the appropriate length).
2. Determine the number of zeros that must be added to the filter before taking the DFT. Give the appropriate Matlab statement to add the zeros and take the required FFT of the filter.
3. Determine the number of zeros that must be added as initial conditions at the beginning of the first data block. Determine the number of data values that need to be used in the first data block. Give the appropriate Matlab code to add the required zeros and take the FFT of the first data block.

Table 3.4 Values of the DFT for Problem 3.3

n	$\Re e\{X(\omega)\}$	$\Im m\{X(\omega)\}$
0	13.9575	0.0
1	−5.9474	−0.5465
2	−0.4746	−2.1078
3	−1.4974	1.1407
4	0.7233	1.5387
5	3.3181	−1.3810
6	−4.7579	0
7		
8		
9		
10		
11		

4. Determine which data values will be used in the second block of data. Give the appropriate Matlab code to put the data in an array for the second block of data and take the FFT of the second data block.
5. Determine the size of the first output block of data and write Matlab code to obtain the first output data block. Assume that you have the FFT of the filter as "ft1" and the FFT of the first block of data as "ft2".

Problem 3.5. (Discrete Time Fourier Transform)
Compute the Fourier transform of the following signals using

$$X(\omega) = \sum_{n=-\infty}^{\infty} x(n)e^{-j\omega n}. \tag{3.119}$$

1.

$$
\begin{aligned}
x(n) \;=\;&\; 0.0234\delta(n) + 0.0194\delta(n-1) + 0.9144\delta(n-2) \\
&+ 0.0194\delta(n-3) + 0.0234\delta(n-4); \tag{3.120}
\end{aligned}
$$

2.

$$x(n) = -0.0546\delta(n) + 0.6525\delta(n-2) + 0.6525\delta(n-4) - 0.0546\delta(n-6). \tag{3.121}$$

Problem 3.6. (Discrete Fourier Transform)
Two sequences are given below:

$$x_1(n) = \left\{ \begin{array}{c} 1, \, 2, 3, 1 \\ \uparrow \end{array} \right\}, \tag{3.122}$$

$$x_2(n) = \left\{\begin{array}{c} 4, \ 3, 2, 2 \\ \uparrow \end{array}\right\}. \qquad (3.123)$$

1. Compute the 4-point DFTs for the two sequences.
2. Compute the circular convolution of the two sequences by:
   (a) Multiplying the two 4-point DFTs point by point.
   (b) Computing the inverse DFT of the product.
3. Compute the 7-point DFTs of the two sequences (add three zeros at the end of each sequence).
4. Compute the linear convolution of the two sequences by:
   (a) Multiplying the two 7-point DFTs point by point.
   (b) Computing the inverse DFT of the product.

Problem 3.7. (Discrete Fourier Transform)
This problem involves the computation and comparison of the Fourier Transform of two sequences.

1. Determine the Fourier Transform of $X_1(\omega)$ for the signal given by

$$x_1(n) = \left\{\begin{array}{c} 1, 2, \ 3, \ 2, 1, 0 \\ \uparrow \end{array}\right\}. \qquad (3.124)$$

2. Determine the 6-point DFT, $X_2(k)$, of the sequence given by

$$\begin{aligned} x_2(n) \ &= \ 3\delta(n) + 2\delta(n-1) + \delta(n-2) \\ &+ \delta(n-4) + 2\delta(n-5). \end{aligned} \qquad (3.125)$$

3. Determine the relationship, if any, between $X_1(\omega)$ and $X_2(k)$.

Problem 3.8. (Filtering Long Sequences)
The length of a sampled speech signal is 47192 samples and the sampling rate is 8000 samples per second. We wish to use the overlap–add filtering procedure with an FIR digital filter with 135 coefficients.

1. Determine the total number of complex multiplications required to filter the input sequence using an FFT algorithm directly without using the overlap–add procedure. The DFT size should be large enough to accommodate linear convolution of the two sequences but it does not have to be a power of 2. Assume that you will not use an FFT algorithm for this part so the number of complex multiplications to compute a DFT of size N will be N^2.
2. Repeat part 1 of this problem with the DFT size extended so that it is a power of 2 and an FFT algorithm is used to compute the DFT.
3. Determine the total number of complex multiplications required to filter the sequence using the overlap–add procedure and a block size of 512 samples.

Table 3.5 Tabulation of the sequence for Problem 3.9

n	0	1	2	3	4	5
s(n)	0.0	1.0	2.0	3.0	2.0	1.0
n	6	7	8	9	10	11
s(n)	0.0	−1.0	−2.0	−3.0	−2.0	−1.0

4. Determine the total number of complex multiplications required to filter the sequence using the overlap–add procedure and a block size of 1024 samples.
5. Determine the total number of complex multiplications required to filter the sequence using the overlap–add procedure and a block size of 2048 samples.

3.13 MATLAB PROBLEMS FOR CHAPTER 3

Problem 3.9. (Frequency Domain Interpolation)
The values of a sample sequence are given in Table 3.5. The corresponding DFT (frequency representation) for this sequence can be computed using the following Matlab script

```
s = [0.0 1.0 2.0 3.0 2.0 1.0 0.0 -1.0 -2.0 -3.0 -2.0 -1.0];
sfft = fft(s);
display(sfft)
```

1. Use the *Frequency Domain Interpolation* procedure to interpolate this sequence to a length of 24 samples with the sampling interval 0.5 times the original sampling interval (one interpolated sample between each original sample).
2. Plot the original sample sequence.
3. Plot the interpolated sample sequence.

Problem 3.10. (Sampling)
Consider the continuous time function given by

$$x(t) = 2.0\cos(60\pi t + 0.6\pi) + 3.0\cos(95\pi t) + 1.5\cos(230\pi t). \qquad (3.126)$$

1. Determine the equation for the corresponding discrete time sequence $x(n)$ using a sampling frequency of 200 samples per second.
2. Write a Matlab script to obtain 128 samples of $x(n)$ starting at $n = 0$. Turn in the listing for your Matlab script and the stem plot of the sampled sequence.
3. Use the Matlab script *fft* to compute the FFT of the sequence $x(n)$. Turn in your Matlab script to compute the FFT.
4. Compute an estimate of the frequency content of $x(n)$ as the magnitude of the FFT for $x(n)$. Turn in your Matlab script and a plot of the magnitude of the FFT for $x(n)$.
5. Repeat parts (a) through (d) for a sampling frequency of 300 samples per second.

Table 3.6 Tabulation of the sequence for Problem 3.12

n	0	1	2	3	4	5
s(n)	0.0	1.0	2.0	3.0	2.0	1.0
n	6	7	8	9	10	11
s(n)	0.0	−1.0	−2.0	−3.0	−2.0	−1.0

Problem 3.11. (Representation of Discrete Time Sequences)
Consider the continuous time function given by

$$x(t) = \cos(20\pi t) + \cos(45\pi t). \tag{3.127}$$

1. Determine the equation for the corresponding discrete time sequence $x(n)$ using a sampling frequency of 100 samples per second.
2. Write a Matlab script to obtain 128 samples of $x(n)$ starting at $n = 0$. Turn in the listing for your Matlab script and the stem plot of the sampled sequence.
3. Use the Matlab script *fft* to compute the FFT of the sequence $x(n)$. Turn in your Matlab script to compute the FFT.
4. Compute an estimate of the frequency content of $x(n)$ as the magnitude of the FFT of $x(n)$. Turn in your Matlab script and a stem plot of the magnitude of the FFT.
5. Use the Matlab script *spectrum* with appropriate input parameters to obtain an improved estimate of the spectral content of $x(n)$. Turn in a list of the parameters you used for the *spectrum* script whether you used the default values or specified your own values. State why you used the default values if you used them. Turn in your Matlab script to use the Matlab *spectrum* script and a stem plot of the resulting spectrum.
6. The spectral estimate from the Matlab *spectrum* script should be better than the one obtained as the magnitude of the FFT. Write a summary of your results and explain why the Matlab *spectrum* script gave a better estimate of the spectrum.

Problem 3.12. (Frequency Domain Interpolation)
The values of a sample sequence are given in Table 3.6. The corresponding DFT (frequency representation) for this sequence can be computed using the following Matlab script:

```
s = [0.0 1.0 2.0 3.0 2.0 1.0 0.0 -1.0 -2.0 -3.0 -2.0 -1.0];
sfft = fft(s);
display(sfft)
```

1. Use the *Frequency Domain Interpolation* procedure to interpolate this sequence to a length of 24 samples with the sampling interval 0.5 times the original sampling interval (one interpolated sample between each original sample).
2. Plot the original sample sequence.
3. Plot the interpolated sample sequence.

Table 3.7 FFT for sequence for Problem 3.13

k	X(k)
0	0
1	$0 - 6.8284j$
2	0
3	$0 + 1.1716j$
4	0
5	$0 - 1.1716j$
6	0
7	$0 + 6.8284j$

Problem 3.13. (Frequency Domain Interpolation)
The FFT for a sample sequence, $x(n)$, is given in Table 3.7. You are required to interpolate the sequence by a factor of 2 by adding zeros as appropriate in the FFT for $x(n)$ as given in Table 3.7 and then taking its inverse to obtain the interpolated output $y(n)$.

1. Give a table similar to Table 3.7 for the FFT of the interpolated output $y(n)$.
2. Use the Matlab function *ifft* as appropriate to obtain the original sequence $x(n)$. Turn in your Matlab script and a stem plot of the sequence $x(n)$.
3. Use the Matlab function *ifft* as appropriate to obtain the interpolated sequence $y(n)$. Turn in your Matlab script and a stem plot of the output sequence $y(n)$.

Problem 3.14. (Filtering Long Sequences)
The length of a sampled speech signal is 31912 samples and the sampling rate is 8000 samples per second. The goal is to use the overlap–save filtering procedure with an FIR digital filter with 130 coefficients and a block size to be 1024 samples.

1. Determine the number of blocks of data that will be used in implementing the filter. Round any fraction up to the next whole number (assume that the last block will be padded with zeros to make it of the appropriate length).
2. Determine the number of zeros that must be added to the filter before taking the DFT. Give the appropriate Matlab statement to add the zeros and take the required FFT of the filter.
3. Determine the number of zeros that must be added as initial conditions at the beginning of the first data block. Determine the number of data values that need to be used in the first data block. Give the appropriate Matlab script to add the required zeros and take the FFT of the first data block.
4. Determine which data values will be used in the second block of data. Give the appropriate Matlab script to put the data in an array for the second block of data and take the FFT of the second data block.
5. Determine the size of the first output block of data and write a Matlab script to obtain the first output data block. Assume that you have the FFT of the filter as "ft1" and the FFT of the first block of data as "ft2".

Problem 3.15. (Frequency Response)
Use the Matlab command "freqz" and other Matlab script as necessary to compute
and plot 201 values of the frequency response and the phase response of the following
transfer functions for the normalized frequency range $-\pi \leq \omega \leq \pi$. You should put
titles and labels on your plots. Turn in your Matlab script and all plots.

1.

$$H(z) = \frac{0.1053z^5 + 0.4495z^4 + 0.8340z^3 + 0.8340z^2 + 0.4495z + 0.1053}{z^5 - 0.5718z^4 - 1.2165z^3 + 0.0505z^2 - 0.2684z + 0.2286};$$

(3.128)

2.

$$H(z) = \frac{0.2929z^4 - 0.5858z^2 + 0.2929}{z^4 + 0.1716}.$$

(3.129)

Problem 3.16. (Stabilization of a System Transfer Function)
The system transfer function for an unstable discrete time system is given by

$$H(z) = \frac{B(z)}{A(z)}$$

(3.130)

where

$$B(z) = 0.3588 + 1.7938z^{-1} + 3.5876z^{-2} + 3.5876z^{-3} + 1.7938z^{-4} + 0.3588z^{-5}$$

(3.131)

and

$$A(z) = 1.0000 + 2.7549z^{-1} + 3.9560z^{-2} + 2.7070z^{-3} + 0.9336z^{-4} + 0.1286z^{-5}.$$

(3.132)

Required:
1. Determine the system transfer function for a stable discrete time system that has
   the same magnitude response as the one given above in Eq. (3.130). Give a detailed
   description of what you did to obtain the stable system transfer function.
2. Use the Matlab *freqz* function to compute 200 values of the magnitude response of
   the original system transfer function as given in Eq. (3.130). Turn in your Matlab
   script and the plots that you generated.
3. Use the Matlab *freqz* function to compute 200 values of the magnitude response
   of your modified, stable system transfer function. Turn in your Matlab script and
   the plots that you generated.

Hints:
1. The Matlab *roots* function can be used to determine the roots of a polynomial.
2. The Matlab *abs* function can be used to determine the absolute value of a constant
   or a matrix as appropriate.

3. The Matlab *help* function can be used to display the comments at the beginning of a Matlab "m" file. This is usually information on how to use a function or a script file. For example, the Matlab statement

```
help roots
```

will result in a display of information on how to use the Matlab *roots* function.

Problem 3.17. (Filtering Long Sequences)
The length of a sampled speech signal is 47192 samples and the sampling rate is 8000 samples per second. We wish to use the overlap–add filtering procedure with an FIR digital filter with 123 coefficients. We want the block size to be 1024 samples. Assume that the input has been placed in an array x and the filter coefficients have been placed in an array b for all parts of this problem. Also, assume that you will store the output in an array y.

1. Determine the number of blocks of data that will be used in implementing the filter. Round any fraction up to the next whole number (assume that the last block will be padded with zeros to make it of the appropriate length).
2. Determine the number of zeros that must be added to the filter before computing the DFT. Give the appropriate Matlab statement to add the required data and zeros to the first data block and then compute the required DFT of the filter.
3. Determine the number of zeros that must be added to the first block of data. Determine the number of data values that need to be used in the block of data. Give the appropriate Matlab script to add the required zeros and compute the DFT of the first data block.
4. Determine the number of output values that will be generated for the first block of data and write a Matlab script to obtain the first output data block and store the appropriate data values from this output data block into the proper location in the output array y. Assume that you have the DFT of the filter as "ft1" and the DFT of the first block of data as "ft2".
5. Determine which data values and how many zeros will be used in the second block of data. Give the appropriate Matlab script to put the data and the zeros in an array for the second block of data and compute the DFT of the second data block.
6. Determine number of output values from the second output data block should be inserted and/or added to the output array y. Write a Matlab script compute the output for the second data block insert and/or add the appropriate values in the output array y.

Design of Digital Filters

4

There are numerous applications that require the selective removal of signals with frequencies in one or more ranges while selectively permitting signals with frequencies in one or more different ranges. This process is generally called filtering. Filters can be implemented by using continuous time systems or by using discrete time system. This text is particularly oriented to discrete time systems so the discussions are confined to filters to be implemented using discrete time systems, which are typically called digital filters.

This chapter presents several methods to design digital filters to be implemented using linear, shift invariant, discrete time systems. The methods used to design digital filters can also be used to design discrete time systems for other applications such as digital control, tracking, etc. The methods use information about the desired system such as the desired impulse response or the desired frequency response to design the desired filter. There are numerous computer software programs to design digital filters. Matlab, which is used for many of the examples presented in this text, is one such program. The goal of the discussions on digital filter design, as presented in this text, is to provide the reader with knowledge to facilitate the use of the available software packages, such as Matlab, to meet design goals and/or specifications for the practical design and implementation of digital filters.

4.1 FILTER SPECIFICATIONS

The design procedures for FIR and IIR digital filters involve the determination of digital filter coefficients to meet the desired filter characteristics or specifications. The desired filter characteristics are typically given in the frequency domain in terms of the desired magnitude response and/or the desired phase response. The filter design problem involves the determination of a realizable system transfer function, $H(z)$, for a filter with frequency specifications approximately equal to the desired frequency characteristics.

The design of FIR filters is based upon the approximation of the specified magnitude response, which can be done in either the frequency domain or in the time domain using the desired impulse response. It is practical to design FIR filters with linear phase so this desirable property is typically added to the design specifica-

Digital Signal Processing. DOI: 10.1016/B978-0-12-804547-3.00004-8

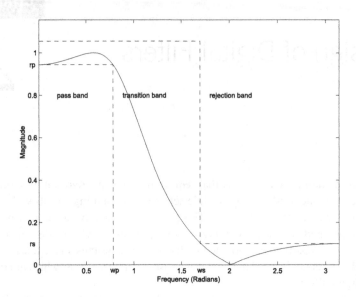

FIGURE 4.1

Specifications for a low pass filter

tions [8]. The design of IIR digital filters, based upon prototypes for continuous time filters, is a very popular method.

The typical approach to providing the specifications for a digital filter involves defining critical frequencies for the band edges, the permitted pass band attenuation and the desired stop band suppression for the desired filter. An ideal low pass filter passes all frequencies below a designated cutoff frequency (pass band) without distortion or attenuation. It blocks the transmission of all frequencies above the designated cutoff frequency (stop or rejection band). Practical low pass filters permit attenuation or distortion of frequencies, within specified limits, in the pass band and attenuates or suppresses frequencies, within specified limits, in the stop band. It is also practical to designate a transition band between the end (or beginning) of the pass band and the beginning (or end) of the stop band.

Fig. 4.1 gives an example of the specifications for a low pass filter in terms of:

1. The cutoff frequency wp,
2. The rejection frequency wr,
3. The pass band ripple rp, and
4. The stop band ripple rs.

The filter in Fig. 4.1 was designed using the following Matlab parameters (The normalized frequency and the actual values for the other parameters are provided within the parentheses.):

1. $wp = 0.25$ $(wp = 0.25\pi)$,
2. $order = 2$,

3. $rp = 0.5$ ($rp = \pm 0.5$ dB),

4. $rs = 20$ ($rs = -20$ dB).

Note that the Matlab filter design functions use scaled specifications for the critical frequencies. For example, the Nyquist frequency, π, is scaled to 1.0 for use in Matlab. Thus, the parameter $wp = 0.25$ represents a normalized cutoff frequency of $wp = 0.25\pi$.

4.2 CAUSALITY CONSTRAINTS

An ideal filter is considered to have a specified, nonzero magnitude for one or more bands of frequencies and is considered to have zero magnitude for one or more bands of frequencies. On the other hand, practical implementation constraints require that a filter be causal. Inputs can be delayed for the implementation of a discrete time system that only uses input samples (FIR filter, for example). The delay itself leads to a phase change which is inconsistent with the frequency response specification for an ideal filter. On the other hand, a discrete time system that uses previous values of the output for inputs cannot use output values that have not been computed (IIR filter, for example). This restriction also leads to a different type of phase change and it is also inconsistent with the frequency response specification for an ideal filter.

The causality constraint can be stated in the form of the requirement that the impulse response

$$h(n) = 0 \quad \forall\, n < 0 \tag{4.1}$$

where

$$H(z) = \sum_{n=-\infty}^{\infty} h(n)z^{-n}. \tag{4.2}$$

The discussions for this text will only deal with causal implementations. This leads to the designation

$$H(z) = \sum_{n=0}^{\infty} h(n)z^{-n}. \tag{4.3}$$

The Paley–Wiener Theorem [4] states if $h(n) = 0 \;\; \forall\, n < 0$ and $h(n)$ has finite energy, then

$$\int_{\omega=-\pi}^{\pi} |\ln |H(\omega)|| < \infty. \tag{4.4}$$

If $H(\omega)$ is zero over any finite band of frequencies, then this integral is not finite. Thus, the Paley–Wiener Theorem leads to the conclusion that the magnitude function for a given causal filter $|H(\omega)|$ can be zero at some frequencies but it cannot be zero over any finite band of frequencies. It follows that practical filters, which are required to be causal, cannot have ideal frequency specifications.

4.3 SYMMETRY AND ANTISYMMETRY FOR FIR FILTERS

An FIR filter of length M with input $x(n)$ and output $y(n)$ can be described by the difference equation [4]

$$y(n) = \sum_{k=0}^{M-1} b(k)x(n-k) \qquad (4.5)$$

where the $b(k)$ are filter coefficients. Note that the difference equation for the FIR filter uses the current input $x(n)$ and delayed input samples $x(n-k)$ but does not use any delayed samples of the output $y(n-k)$. Alternatively, the filter can be expressed in terms of the convolution summation as

$$y(n) = \sum_{k=0}^{M-1} h(k)x(n-k) \qquad (4.6)$$

where the $h(k)$ are the samples of the filter's impulse response. Eqs. (4.5) and (4.6) can be shown to be the same by equating $h(k)$ and $b(k)$. Thus, the filter coefficients for an FIR filter are equivalent to the samples of the corresponding filter's impulse response.

An FIR filter can also be represented in terms of its system transfer function

$$H(z) = \sum_{k=0}^{M-1} h(k)z^{-k} = \sum_{k=0}^{M-1} b(k)z^{-k}. \qquad (4.7)$$

Its corresponding frequency response can be obtained by evaluating $H(z)$ on the unit circle where $z = e^{j\omega}$. Thus,

$$H(\omega) = \sum_{k=0}^{M-1} b(k)e^{-j\omega k}. \qquad (4.8)$$

$H(\omega)$ can be written in terms of its magnitude and phase components where

$$H(\omega) = A(\omega)\Phi(\omega). \qquad (4.9)$$

The term $A(\omega)$ is the magnitude (or amplitude) response of $H(\omega)$ and the term $\Phi(\omega)$ is the corresponding phase response.

An FIR filter can be designed to have linear phase. Linear phase is characterized by the phase of the frequency response for the filter being proportional to the frequency such that [9]

$$\Phi(\omega) = -M\omega \pm \text{ possible jumps of } \pi. \qquad (4.10)$$

The phase will have a jump of $\pm\pi$ whenever the frequency response has a zero crossing (positive to negative or negative to positive). A zero for the frequency response occurs when its real and imaginary parts are both equal to zero.

Table 4.1 Filter coefficients for a high pass FIR filter

n	0	1	2	3	4
$b(n)$	−0.0034	−0.0100	0.0921	−0.2408	0.3209
n	5	6	7	8	
$b(n)$	−0.2408	0.0921	−0.0100	−0.0034	

An FIR filter has linear phase if its unit sample response satisfies the condition [4]

$$h(n) = \pm h(M - 1 - n), \quad \forall \; 0 \leq n \leq M - 1. \tag{4.11}$$

If the linear phase condition is imposed on the system transfer function for the FIR filter as given in Eq. (4.7), then if M is odd [4]

$$H(z) \;=\; z^{-(M-1)/2} H_1(z), \quad \forall \; M \text{ odd, where}$$

$$H_1(z) \;=\; \left\{ b\left(\frac{M-1}{2}\right) + \sum_{k=0}^{(M-3)/2} b(k) \left[z^{(M-1-2k)/2} \pm z^{-(M-1-2k)/2} \right] \right\}. \tag{4.12}$$

If M is even, then

$$H(z) \;=\; z^{-(M-1)/2} H_2(z), \quad \forall \; M \text{ even, where}$$

$$H_2(z) \;=\; \sum_{k=0}^{(M/2)-1} b(k) \left[z^{(M-1-2k)/2} \pm z^{-(M-1-2k)/2} \right]. \tag{4.13}$$

This concept is demonstrated by Example 4.1.

EXAMPLE 4.1
(Linear Phase FIR Filter (Odd))

Consider the FIR filter with system transfer function given by

$$H(z) \;=\; -0.0034 - 0.0100z^{-1} + 0.0921z^{-2} - 0.2408z^{-3} + 0.3209z^{-4}$$
$$- 0.2408z^{-5} + 0.0921z^{-6} - 0.0100z^{-7} - 0.0034z^{-8}. \tag{4.14}$$

This system transfer function has 9 coefficients ($M = 9$, which is odd). The corresponding filter coefficients are also given in Table 4.1.

Fig. 4.2 shows the stem plot of the coefficients in $H(z)$.

Note that the filter coefficients, as shown in Fig. 4.2 and in Table 4.1, are symmetric such that

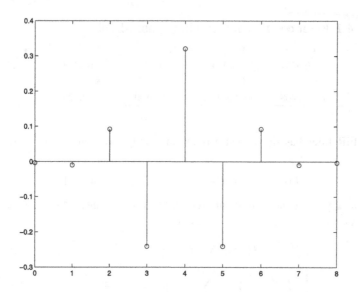

FIGURE 4.2

Stem plot for the coefficients in $H(z)$ for Example 4.1

$$
\begin{aligned}
b(0) &= b(8), \\
b(1) &= b(7), \\
b(2) &= b(6), \\
b(3) &= b(5).
\end{aligned}
\tag{4.15}
$$

Thus, the system transfer function is consistent with Eq. (4.12) and can be written as

$$
\begin{aligned}
H(z) &= z^{-4}\left[b(4) + b(0)\left(z^4 + z^{-4}\right) + b(1)\left(z^3 + z^{-3}\right)\right] \\
&\quad + z^{-4}\left[b(2)\left(z^2 + z^{-2}\right) + b(3)\left(z^1 + z^{-1}\right)\right], \\
H(z) &= z^{-4}\left[0.3209 - 0.0034\left(z^4 + z^{-4}\right) - 0.0100\left(z^3 + z^{-3}\right)\right] \\
&\quad + z^{-4}\left[0.0921\left(z^2 + z^{-2}\right) - 0.2408\left(z^1 + z^{-1}\right)\right].
\end{aligned}
\tag{4.16}
$$

The frequency response for $H(z)$ can be determined by evaluating $H(z)$ on the unit circle in the complex plane where $z = e^{j\omega}$. The trigonometric identities

$$
\begin{aligned}
e^{j\omega} &= \cos(\omega) + j\sin(\omega), \\
e^{-j\omega} &= \cos(\omega) - j\sin(\omega)
\end{aligned}
\tag{4.17}
$$

can be used to determine that

$$\left[z^4 + z^{-4}\right]\Big|_{z=e^{j\omega}} = e^{4j\omega} + e^{-4j\omega}$$
$$= \cos(4\omega) + \sin(4\omega) + \cos(4\omega) - \sin(4\omega)$$
$$= 2\cos(4\omega). \tag{4.18}$$

Similarly, the trigonometric identities in Eq. (4.17) can be used to show that

$$\left[z^3 + z^{-3}\right]\Big|_{z=e^{j\omega}} = 2\cos(3\omega),$$
$$\left[z^2 + z^{-2}\right]\Big|_{z=e^{j\omega}} = 2\cos(2\omega),$$
$$\left[z + z^{-1}\right]\Big|_{z=e^{j\omega}} = 2\cos(\omega). \tag{4.19}$$

It follows that

$$H(\omega) = e^{-4j\omega} [0.3209 - 2(0.0034)\cos(4\omega) - 2(0.0100)\cos(3\omega)]$$
$$+ e^{-4j\omega} [2(0.0921)\cos(2\omega) - 2(0.2408)\cos(\omega)], \tag{4.20}$$

$$H(\omega) = e^{-4j\omega} [0.3209 - 0.0068\cos(4\omega) - 0.0200\cos(3\omega)]$$
$$+ e^{-4j\omega} [0.1842\cos(2\omega) - 0.4816\cos(\omega)]. \tag{4.21}$$

Fig. 4.3 gives the magnitude and phase plots for the normalized frequency response for $H(z)$. Note the jump in phase by π radians just after the normalized frequency 0.6566 ($\Omega = 0.209\pi$) radians/sample. This is due to a zero crossing of the frequency response at the corresponding normalized frequency.
End of the Example

Example 4.2 gives an example of a linear phase FIR filter with an even number of filter coefficients.

EXAMPLE 4.2
(Linear Phase FIR Filter (Even))

The system transfer function for an FIR filter is given by

$$H(z) = -0.0004 - 0.0168z^{-1} - 0.0180z^{-2} + 0.1425z^{-3} + 0.3927z^{-4}$$
$$+ 0.3927z^{-5} + 0.1425z^{-6} - 0.0180z^{-7} - 0.0168z^{-8} - 0.0004z^{-9}. \tag{4.22}$$

This system transfer function has 10 coefficients ($M = 10$, which is even). The corresponding filter coefficients are also given in Table 4.2. Fig. 4.4 shows the stem plot of the coefficients in $H(z)$. Fig. 4.4 and Table 4.2 show that the filter

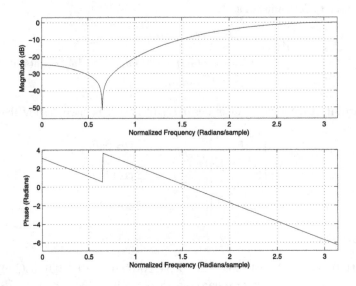

FIGURE 4.3

Magnitude and phase plots for $H(z)$ for Example 4.1

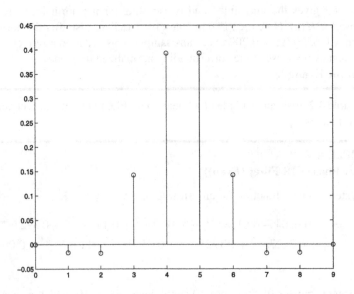

FIGURE 4.4

Stem plot for the coefficients in $H(z)$ for Example 4.2

coefficients are symmetric such that

$$
\begin{aligned}
b(0) &= b(9),\\
b(1) &= b(8),\\
b(2) &= b(7),\\
b(3) &= b(6),\\
b(4) &= b(5).
\end{aligned}
\tag{4.23}
$$

It follows that the system transfer function is consistent with Eq. (4.13) and can be written as

$$
H(z) = z^{-4.5} H_2(z)
\tag{4.24}
$$

where

$$
H_2(z) = \sum_{k=0}^{4} b(k)\left[z^{(9-2k)/2} + z^{-(9-2k)/2} \right].
\tag{4.25}
$$

Thus,

$$
\begin{aligned}
H(z) =\ & z^{-4.5}\left\{ b(0)\left[z^{4.5}+z^{-4.5}\right] + b(1)\left[z^{3.5}+z^{-3.5}\right]\right.\\
& \left.+ b(2)\left[z^{2.5}+z^{-2.5}\right]\right\}\\
& + z^{-4.5}\left\{ b(3)\left[z^{1.5}+z^{-1.5}\right] + b(4)\left[+z^{0.5}+z^{-0.5}\right]\right\},
\end{aligned}
\tag{4.26}
$$

$$
\begin{aligned}
H(z) =\ & z^{-4.5}\left\{ -0.0004\left[z^{4.5}+z^{-4.5}\right] - 0.0168\left[z^{3.5}+z^{-3.5}\right]\right\}\\
& + z^{-4.5}\left\{ -0.0180\left[z^{2.5}+z^{-2.5}\right] + 0.1425\left[z^{1.5}+z^{-1.5}\right]\right\}\\
& + z^{-4.5}\left\{ 0.3927\left[+z^{0.5}+z^{-0.5}\right]\right\}.
\end{aligned}
\tag{4.27}
$$

The trigonometric identities in Eq. (4.17) can be used to determine

$$
\begin{aligned}
\left[z^{4.5}+z^{-4.5}\right]\Big|_{z=e^{j\omega}} &= 2\cos(4.5\omega),\\
\left[z^{3.5}+z^{-3.5}\right]\Big|_{z=e^{j\omega}} &= 2\cos(3.5\omega),\\
\left[z^{2.5}+z^{-2.5}\right]\Big|_{z=e^{j\omega}} &= 2\cos(2.5\omega),\\
\left[z^{1.5}+z^{-1.5}\right]\Big|_{z=e^{j\omega}} &= 2\cos(1.5\omega),\\
\left[z^{0.5}+z^{-0.5}\right]\Big|_{z=e^{j\omega}} &= 2\cos(0.5\omega).
\end{aligned}
\tag{4.28}
$$

It follows that

$$
\begin{aligned}
H(\omega) =\ & e^{-j4.5\omega}\left[-0.0008\cos(4.5\omega) - 0.0337\cos(3.5\omega) - 0.0360\cos(2.5\omega)\right]\\
& + e^{-j4.5\omega}\left[0.2850\cos(1.5\omega) + 0.7854\cos(0.5\omega)\right].
\end{aligned}
\tag{4.29}
$$

Table 4.2 Filter coefficients for a low pass FIR filter

n	0	1	2	3	4
$b(n)$	−0.0004	−0.0168	−0.0180	0.1425	0.3927
n	5	6	7	8	9
$b(n)$	0.3927	0.1425	−0.0180	−0.0168	−0.0004

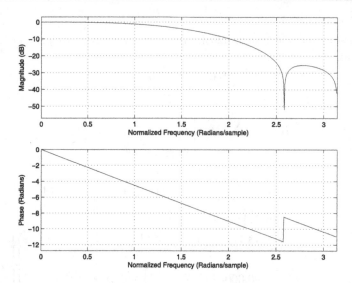

FIGURE 4.5

Magnitude and phase plots for $H(z)$ for Example 4.2

Fig. 4.5 gives the magnitude and phase plots for the normalized frequency response for $H(z)$. Note the jump in phase by π radians just after the normalized frequency 2.5833 ($\Omega = 0.8223\pi$). This is due to a zero crossing of the frequency response at the corresponding normalized frequency.
End of the Example

4.4 WINDOW BASED FIR FILTER DESIGN

It is common to design linear phase, frequency selective filters such as low pass, high pass, band pass, and band stop FIR digital filters using samples of the impulse response of the corresponding ideal continuous time filter. The impulse response of the corresponding ideal continuous time filter is typically infinite in duration due to sharp transitions between the pass and stop bands. Thus, the ideal impulse response, $h_d(n)$, generally has a range of $-\infty \le n \le \infty$. Filters with an impulse response having a range of $-\infty \le n \le \infty$ are not physically realizable. The impulse response for the desired ideal filter must be truncated to estimate the finite duration discrete time impulse response for a realizable FIR filter. This can be accomplished by multiplying the ideal impulse response by a finite duration window. The filter can then be approximated by using the product of the chosen window and impulse response of the corresponding ideal, continuous time filter.

The system transfer function for a FIR digital filter is given by

$$H(z) = \sum_{k=-L_1}^{L_2} b(k)z^{-k}. \tag{4.30}$$

The filter coefficients for a FIR filter, $b(k)$, are also the samples of the discrete time impulse response with

$$b(k) = h(k) \quad \forall -L_1 \le k \le L_2. \tag{4.31}$$

The desired frequency response, $H_d(\omega)$, can be defined over the interval $-\pi \le \omega \le \pi$, which is the normalized, primary frequency interval for a digital filter. Then, the impulse response of the desired ideal filter can be obtained as the inverse Fourier Transform of the desired continuous time filter as specified in the frequency domain:

$$h_d(n) = \frac{1}{2\pi} \int_{-\pi}^{\pi} H_d(\omega)e^{j\omega n} d\omega. \tag{4.32}$$

The following subsections will illustrate this concept by designing examples of low pass and high pass digital filters.

4.4.1 LOW PASS FILTER DESIGN

An ideal low pass filter can be defined such that

$$\begin{aligned} H_d(\omega) &= 1 \quad \text{for } 0 \le |\omega| \le \omega_c, \\ H_d(\omega) &= 0 \quad \text{for } \omega_c < |\omega| \le \pi. \end{aligned} \tag{4.33}$$

Fig. 4.6 gives a normalized frequency plot of an ideal low pass filter with $\omega_c = 0.67\pi$.

The inverse Fourier Transform of $H_d(\omega)$ can be determined using the following equation:

$$h_d(n) = \frac{1}{2\pi} \int_{-\omega_c}^{\omega_c} e^{j\omega n} d\omega, \tag{4.34}$$

$$h_d(n) = \frac{\sin(\omega_c n)}{\pi n} \quad \forall n \ne 0,$$

$$h_d(0) = \frac{\omega_c}{\pi}. \tag{4.35}$$

A causal approximation to the desired filter can be obtained by truncating $h_d(n)$ to use only samples for $-M \le n \le M$. The truncated filter coefficients can then be delayed by M samples to obtain a causal sequence.

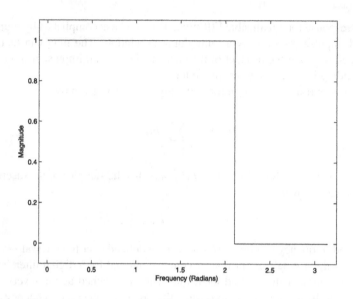

FIGURE 4.6

Example of an ideal low pass filter ($\omega_c = 0.67\pi$)

The truncated impulse response, $h_d(n)$, can also be delayed by M samples by multiplying $H_d(\omega)$ by $e^{-j\omega M}$ prior to computing its inverse Fourier Transform according to the time shift property of the Fourier Transform:

$$H_{d2}(\omega) = e^{-j\omega M} H_d(\omega), \tag{4.36}$$

$$
\begin{aligned}
H_{d2}(\omega) &= e^{-j\omega M} \quad \text{for } 0 \le |\omega| \le \omega_c, \\
H_{d2}(\omega) &= 0 \quad \text{for } \omega_c < |\omega| \le \pi.
\end{aligned}
\tag{4.37}
$$

The inverse Fourier transform of $H_{d2}(\omega)$ will be the delayed sequence, $h_{d2}(n)$. Thus,

$$h_{d2}(n) = \frac{1}{2\pi} \int_{-\omega_c}^{\omega_c} e^{-j\omega M} e^{j\omega n} d\omega, \tag{4.38}$$

$$h_{d2}(n) = \frac{1}{2\pi} \int_{-\omega_c}^{\omega_c} e^{j\omega(n-M)} d\omega, \tag{4.39}$$

$$
\begin{aligned}
h_{d2}(n) &= \frac{\sin\left[\omega_c\,(n-M)\right]}{\pi\,(n-M)} \quad \text{for } n \ne M, \\
h_{d2}(M) &= \frac{\omega_c}{\pi} \quad \text{for } n = M.
\end{aligned}
\tag{4.40}
$$

Example 4.3 further illustrates this design approach through the design of a low pass filter as an example.

EXAMPLE 4.3
(Low Pass Filter Design)

This example involves the design a low pass filter with a normalized cutoff frequency of $\omega_c = 0.67\pi$ with 41 coefficients. The parameter M for this example can be determined as

$$M = \frac{41 - 1}{2} = 20. \tag{4.41}$$

The filter coefficients for $0 \le n \le 19$ can be computed with the use of Eq. (4.40) and are given by

$$b(n) = \frac{\sin\left[0.67\pi\,(n - 20)\right]}{\pi\,(n - 20)}. \tag{4.42}$$

The filter coefficient for $M = 20$ is given

$$b(20) = \frac{0.67\pi}{\pi} = 0.67. \tag{4.43}$$

The filter coefficients, $b(n)$, are symmetric. Therefore,

$$b(n) = b(40 - n) \quad \text{for } 21 \le n \le 40. \tag{4.44}$$

The magnitude response should have a value of 1.0 at $\omega = 0.0$ for a low pass filter. Thus,

$$|H(0)| = 1.0. \tag{4.45}$$

However, the impulse response has been truncated. Therefore, the magnitude of $H(0)$, for the truncated sequence, will normally not be equal to 1.0 as a result. The filter coefficients can be normalized to obtain the coefficients for a new filter $b_1(n)$ so that $H_1(0) = 1.0$, to improve the performance of the filter. Thus, the modified filter coefficients for the filter with $H_1(0) = 1$ can be determined by

$$b_1(n) = \frac{b(n)}{\displaystyle\sum_{k=0}^{40} b(k)}. \tag{4.46}$$

Fig. 4.7 gives the a stem plot of the impulse response for the filter with coefficients $b_1(n)$. Fig. 4.8 gives the magnitude and phase response plots for the filter with coefficients $b_1(n)$. Note that the phase is linear for the pass band and linear for the stop band except for jumps of π where the frequency response has a zero crossing.
End of the Example

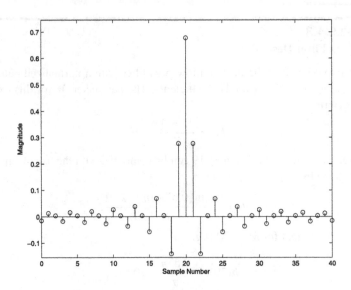

FIGURE 4.7

Impulse response for the low pass filter in Example 4.3

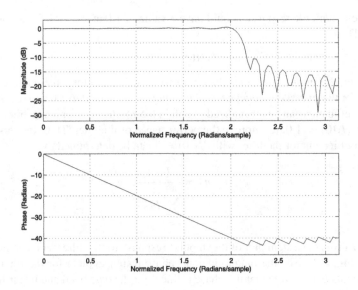

FIGURE 4.8

Magnitude and phase responses for the low pass filter in Example 4.3

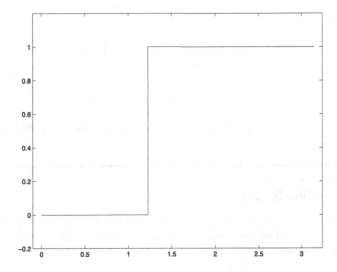

FIGURE 4.9

Example of an ideal high pass filter ($\omega_c = 0.391\pi$)

4.4.2 HIGH PASS FILTER DESIGN

The magnitude response for an ideal high pass filter can be defined such that

$$\begin{aligned} H_d(\omega) &= 0 \quad \text{for } 0 \le |\omega| \le \omega_c, \\ H_d(\omega) &= 1 \quad \text{for } \omega_c < |\omega| \le \pi. \end{aligned} \tag{4.47}$$

Fig. 4.9 gives a normalized frequency plot of an ideal high pass filter with a normalized cutoff frequency $\omega_c = 0.391\pi = 1.2284$.

The same procedure that we used for the design of the low pass filter, in Example 4.3, can be used to design a practical high pass filter. The frequency specification for a high pass filter shifted in time by M samples can be given as

$$\begin{aligned} H_{d2}(\omega) &= 0 \quad \text{for } 0 \le |\omega| \le \omega_c, \\ H_{d2}(\omega) &= e^{-j\omega M} \quad \text{for } \omega_c < |\omega| \le \pi. \end{aligned} \tag{4.48}$$

Thus, if $n \ne M$,

$$h_{d2}(n) = \frac{1}{2\pi} \int_{-\pi}^{-\omega_c} e^{j\omega(n-M)} d\omega + \frac{1}{2\pi} \int_{\omega_c}^{\pi} e^{j\omega(n-M)} d\omega, \tag{4.49}$$

$$h_{d2}(n) = \frac{1}{2\pi} \left[\frac{e^{j\omega(n-M)}}{j(n-M)} \right]\Big|_{-\pi}^{-\omega_c} + \frac{1}{2\pi} \left[\frac{e^{j\omega(n-M)}}{j(n-M)} \right]\Big|_{\omega_c}^{-\pi}, \tag{4.50}$$

$$h_{d2}(n) = -\frac{\sin\left[\omega_c(n-M)\right]}{\pi(n-M)} \quad \forall\, n \ne M. \tag{4.51}$$

If $n = M$, then

$$h_{d2}(n) = \frac{1}{2\pi} \int_{-\pi}^{-\omega_c} d\omega + \frac{1}{2\pi} \int_{\omega_c}^{\pi} d\omega, \qquad (4.52)$$

$$h_{d2}(M) = \frac{\pi - \omega_c}{\pi} \quad \text{for} \ n = M. \qquad (4.53)$$

Example 4.4 presents an example of the design of a high pass filter using this approach.

EXAMPLE 4.4
(High Pass Filter Design)

Problem:
Design a high pass FIR filter with a normalized cutoff frequency of $\omega_c = 0.391\pi$ with 41 coefficients.

Solution:
The value of M, for this case, can be determined as

$$M = \frac{41 - 1}{2} = 20. \qquad (4.54)$$

The filter coefficients for $0 \le n \le 19$ are given by

$$b(n) = -\frac{\sin\left[0.391\pi \ (n - 20)\right]}{\pi \ (n - 20)}, \qquad (4.55)$$

$$b(20) = \frac{\pi - 0.391\pi}{\pi} = 1.000 - 0.391 = 0.609. \qquad (4.56)$$

The impulse response is symmetric. Therefore,

$$b(n) = b(40 - n) \quad \text{for} \ 21 \le n \le 40. \qquad (4.57)$$

The magnitude of the frequency response for the filter, at $\omega = \pi$, is expected to be 1.0 because the design is for a high pass filter. However, the impulse response for the ideal, high pass filter, has been truncated so the magnitude of $H(\pi)$ will normally not be equal to 1.0 for the frequency response for the truncated impulse response. The truncated impulse response can be normalized to obtain the coefficients for a new filter $b_2(n)$ so that $H_2(\pi) = 1.0$ to improve the performance of the filter. It follows that

$$b_2(n) = \frac{b(n)}{\displaystyle\sum_{k=0}^{40} (-1)^k b(k)}. \qquad (4.58)$$

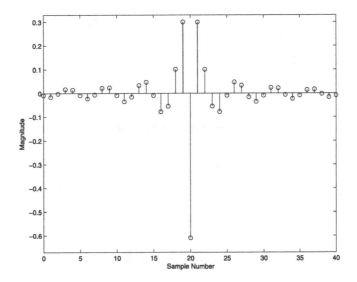

FIGURE 4.10

Impulse response for the high pass filter in Example 4.4

Fig. 4.10 gives the a stem plot of the impulse response for the filter with coefficients $b_2(n)$.

Fig. 4.11 gives the magnitude and phase plots for the filter with coefficients $b_2(n)$.

End of the Example

4.4.3 OTHER FREQUENCY SELECTIVE FILTERS

The procedure to use the normalized, truncated impulse response for the corresponding ideal frequency selective filter, to design low pass and high pass filters, was illustrated in Sections 4.4.1 and 4.4.2. This procedure can also be used to design band pass filters, band stop filters and other frequency selective filters. Table 4.3 gives the equations for the truncated impulse responses for ideal low pass and high pass filters with $2M + 1$ coefficients shifted by M samples. These equations can be used for the design of causal, low pass and high pass FIR filters [9].

Table 4.4 gives the impulse responses for ideal filters for $2M + 1$ coefficients shifted by M samples, which can be used for the design of causal, band pass and band stop FIR filters [9].

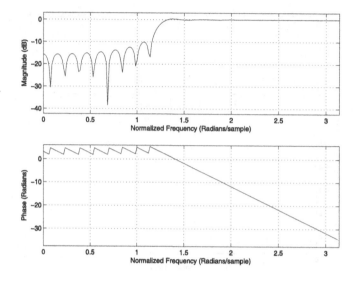

FIGURE 4.11

Magnitude and phase responses for the high pass filter in Example 4.4

Table 4.3 The impulse responses for low pass and high pass ideal filters for $2M + 1$ coefficients shifted by M samples

Filter Type	Shifted Causal Impulse Responses
Lowpass	$\begin{cases} \frac{\sin[\omega_c(n-M)]}{\pi(n-M)} & \forall\, n \neq M, \\ \frac{\omega_c}{\pi} & \text{for } n = M. \end{cases}$
Highpass	$\begin{cases} \frac{-\sin[\omega_c(n-M)]}{\pi(n-M)} & \forall\, n \neq M, \\ \frac{\pi-\omega_c}{\pi} & \text{for } n = M. \end{cases}$

Table 4.4 The impulse responses for band pass and band stop ideal filters for $2M + 1$ coefficients shifted by M samples

Filter Type	Shifted Causal Impulse Responses
Band Stop	$\begin{cases} \frac{-\sin[\omega_2(n-M)]}{\pi(n-M)} + \frac{\sin[\omega_1(n-M)]}{\pi(n-M)} & \forall\, n \neq M, \\ \frac{\pi-\omega_2+\omega_1}{\pi} & \text{for } n = M. \end{cases}$
Band Pass	$\begin{cases} \frac{\sin[\omega_2(n-M)]}{\pi(n-M)} - \frac{\sin[\omega_1(n-M)]}{\pi(n-M)} & \forall\, n \neq M, \\ \frac{\omega_2-\omega_1}{\pi} & \text{for } n = M. \end{cases}$

Table 4.5 Frequency domain characteristics for window functions [8]

Type of Window	Width of Main Lobe	Peak Sidelobe Level (dB)
Rectangular	$\frac{4\pi}{N}$	-13
Bartlett	$\frac{8\pi}{N}$	-25
Hanning	$\frac{8\pi}{N}$	-31
Hamming	$\frac{8\pi}{N}$	-41
Blackman	$\frac{12\pi}{N}$	-57

4.5 FIXED WINDOW FUNCTIONS

Truncation of the impulse response is the same as using a rectangular window such that

$$w(n) = 1 \quad \forall - M \leq n \leq M. \tag{4.59}$$

The corresponding number of filter coefficients will be

$$N = 2M + 1. \tag{4.60}$$

Many tapered windows have been proposed by various authors to improve the performance of window based FIR filter designs. The four most commonly used tapered windows of length $N = 2M + 1$ are [8,9]:

$$(\text{Bartlett}) \quad w(n) = 1 - \frac{|n|}{M+1} \quad \forall - M \leq n \leq M, \tag{4.61}$$

$$(\text{Hanning}) \quad w(n) = 0.5 \left[1.0 + \cos\left(\frac{2\pi n}{2M+1}\right) \right] \quad \forall - M \leq n \leq M, \tag{4.62}$$

$$(\text{Hamming}) \quad w(n) = 0.54 + 0.46 \cos\left(\frac{2\pi n}{2M+1}\right) \quad \forall - M \leq n \leq M, \tag{4.63}$$

$$(\text{Blackman}) \quad w(n) = 0.42 + 0.5 \cos\left(\frac{2\pi n}{2M+1}\right)$$
$$+ \quad 0.08 \cos\left(\frac{4\pi n}{2M+1}\right) \quad \forall - M \leq n \leq M. \tag{4.64}$$

These windows contribute different properties to the corresponding FIR filter in terms of the width of the main lobe, the relative magnitude of the side lobes, the minimum stop band attenuation and the transition width. Table 4.5 gives frequency domain characteristics the rectangular window function and for the window functions described in Eqs. (4.61) through (4.64).

The width of the main lobe represents the width of the transition region between the pass band and the stop band. The peak sidelobe level represents the attenuation in the stop band. Thus, Table 4.5 illustrates the tradeoff between the width of the transition region and the stop band attenuation for the given window functions.

4.5.1 WINDOW BASED FIR FILTER DESIGN EXAMPLES

The design of FIR digital filters using the product of the impulse response for an ideal filter and a finite duration window was discussed in Sect. 4.4. The Matlab function *fir1* uses the window based design approach to design FIR filters. The type of window to be used can be specified as an input parameter for the function. The details on the use of the function can be obtained by using the Matlab Help Utility. Example 4.5 illustrates the use of the window based design procedure to design a band stop digital filter.

EXAMPLE 4.5
(FIR Band Stop Filter Design)

This example involves the design of a band stop digital filter with 31 coefficients using the window method. Both rectangular and Hamming windows are used and the results are compared. The goal is to design a FIR, linear phase, digital filter approximating the ideal frequency response given by

$$H_d(\omega) = \begin{cases} 1 & \text{for} \quad |\omega| \leq \frac{\pi}{7}, \\ 0 & \text{for} \quad \frac{\pi}{7} < |\omega| < \frac{\pi}{2}, \\ 1 & \text{for} \quad \frac{\pi}{2} \leq |\omega| \leq \pi. \end{cases} \tag{4.65}$$

The filter coefficients for the rectangular window design can be determined by computing the inverse Fourier transform of $H_d(\omega)$ and then using a 31 length rectangular window to truncate the impulse response:

$$\begin{aligned} h_d(n) &= DFT^{-1}\{H_d(\omega)\} \\ &= \frac{1}{2\pi} \int_{-\pi}^{\pi} H_d(\omega)e^{j\omega n} d\omega \\ &= \frac{1}{2\pi} \left(\int_{-\pi}^{-\frac{\pi}{2}} e^{j\omega n} d\omega + \int_{-\frac{\pi}{7}}^{\frac{\pi}{7}} e^{j\omega n} d\omega + \int_{\frac{\pi}{2}}^{\pi} e^{j\omega n} d\omega \right). \end{aligned} \tag{4.66}$$

Evaluating the above integrals and using Euler's identities, this reduces to

$$h_d(n) = \begin{cases} \frac{1}{\pi n} \left(\sin \frac{\pi n}{7} - \sin \frac{\pi n}{2} \right) & n \neq 0, \\ 1 - \frac{5\pi}{14} & n = 0. \end{cases} \tag{4.67}$$

Matlab Script 4.1 can be used to determine the 31 coefficients from the impulse response derived above.

Matlab Script 4.1.

```
% Impulse Response Filter Design
w1 = pi/7;
w2 = pi/2;
```

```
n1 = -15:-1;
b2neg = (1/pi)*(sin(w1*(n1))./(n1) - sin(w2*(n1))./(n1));
n2 = 1:15;
b2pos = (1/pi)*(sin(w1*(n2))./(n2) - sin(w2*(n2))./(n2));
b216 = 1-5/14;
b2 = [b2neg b216 b2pos];
b2 = b2/sum(b2);
```

End of the Script

The Matlab function *fir1* can be used to design an FIR filter using the window method. Matlab Script 4.2 can be used to obtain the same filter, as the one designed in Matlab Script 4.1 design, using the Matlab *fir1* function with a rectangular window.

Matlab Script 4.2.

```
% FIR Filter Design Using the fir1 Matlab Function
%    with a rectangular window
order = 30;
wn = [1/7 1/2];
w = window(@rectwin, order+1);
b1 = fir1(order,wn,'stop',w);
```

End of the Script

Fig. 4.12 shows the impulse responses of the band stop filter using the rectangular window using Matlab Script 4.1 (using the +) and with the use of the Matlab *fir1* function as shown in Matlab Script 4.2 (using the *o* on the same plot for comparison. The normalized peak error was of the order of 10^{-16} and the average mean square error is of the order of 10^{-32}. Thus the filter coefficients are essentially the same.

Fig. 4.13 shows the magnitude and phase response of the band stop filter using the rectangular window. The magnitude response for the desired filter, with the stop band attenuation set at -41 dB, is also shown using a dashed line on the magnitude response plot for comparison.

The Hamming window is the default window for the Matlab *fir1* function. Therefore, the corresponding filter can be designed using the Matlab *fir1* function as shown in Matlab Script 4.3.

Matlab Script 4.3.

```
% FIR Filter Design Using a Hamming Window
N = 30;
wc = [1/7 1/2];
b4 = fir1(N,wc,'stop');
```

End of the Script

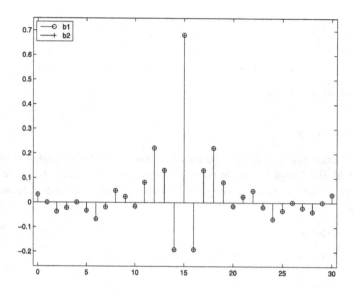

FIGURE 4.12

Impulse responses for the two FIR filter designs using the rectangular window for comparison

Fig. 4.14 shows the stem plot of the impulse response of the band stop filter using the Hamming window with the Matlab *fir1* function. Fig. 4.15 shows the magnitude and phase responses for the corresponding FIR filter designed using the Matlab fir1 function with the Hamming window. The magnitude response for the desired filter, with the stop band attenuation set at −41 dB, is also shown using a dashed line on the magnitude response plot for comparison with the magnitude and phase responses for the design using the rectangular window. The design using the Hamming window more closely meets the desired specifications regarding the stop band attenuation than does the design using the rectangular window. However, the transition band is wider with the Hamming window than it is with the rectangular window. This is consistent with the parameters listed in Table 4.5.
End of the Example

4.6 DESIGN OF FIR FILTERS USING MATLAB

The Signal Processing Toolbox for Matlab includes several Matlab functions for the design of FIR digital filters. This includes files to determine the filter order from given specifications and to determine the frequency response of a linear, discrete time system (filter).

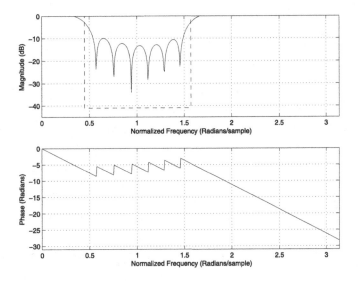

FIGURE 4.13

Magnitude and phase response of band stop filter using a rectangular window

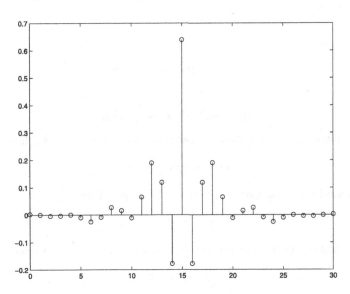

FIGURE 4.14

Impulse responses for the FIR filter design using the Hamming window

The parameters for the design of the filter are specified in decibels for the use of Matlab. For example, if the magnitude response for a low pass filter is desired to fall

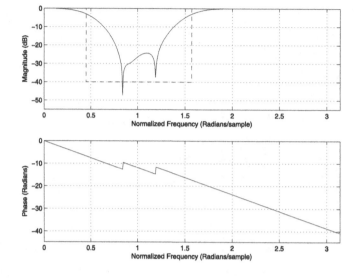

FIGURE 4.15

Magnitude and phase response of band stop filter using Matlab's *fir1* function and a Hamming window

within the range

$$1 - \delta_p \leq |H(\omega)| \leq 1 + \delta_p \qquad (4.68)$$

then

$$rp = -20 \log_{10}(1 - \delta_p) \quad \text{dB} \qquad (4.69)$$

If the magnitude response in the stop band is desired to not exceed δ_s, then

$$rs = -20 \log_{10}(\delta_s). \qquad (4.70)$$

The pass band covers the range of frequencies (low pass filter)

$$0 \leq \omega \leq wp. \qquad (4.71)$$

The stop band or rejection band covers the range of frequencies (low pass filter)

$$ws \leq \omega \leq \pi. \qquad (4.72)$$

The frequency band

$$wp < \omega < ws \qquad (4.73)$$

is called the transition band.

The following Matlab functions have been provided in the Matlab Signal Processing Toolbox for the design of FIR digital filters:

1. *fir1* – Design of FIR filters using the window method.
2. *fir2* – Design of FIR filters using the frequency sampling method.
3. *firls* – Design of linear phase FIR filters using the least squares criteria.
4. *fircls* – Design of linear phase FIR filters using a constrained, least squares criteria.
5. *firpm* – Design of FIR filters using the Parks–McClellan, optimal, equiripple, FIR filter design method [10].

The following Matlab functions have been provided in the Matlab Signal Processing Toolbox to help define the parameters needed to design an FIR filter to meet given specifications.

1. *firpmord* – Finds the approximate order N, normalized frequency band edges Fo, frequency band magnitudes Ao and weights W to be used by the *firpm* function.
2. *kaiserord* – Finds the order N, the normalized frequency band edges Wn, the Kaiser window beta parameter BTA and filter type FILTYPE to be used by the *fir1* function [11].

4.6.1 FILTER ORDER ESTIMATION

The *firpmord* Matlab function in the Signal Processing Toolbox can be used to determine the filter order for use with the corresponding *firpm* function to design an FIR filter using the Parks–McClellan optimal equiripple FIR filter design procedure [10]. Information on its use can be obtained using the Matlab *Help Utility*. The required inputs include:

1. *fedge* – a vector of the normalized frequency band edges,
2. *mval* – a vector of the corresponding magnitude values at the band edges,
3. *dev* – a vector specifying the maximum allowable deviation between the magnitude of the designed filter and the desired magnitude,
4. *Fs* – the sampling frequency.

If Fs is not specified, then it defaults to $Fs = 2$. Then, the frequency values range from 0 to 1 (normalized frequencies). The length of *fedge* is twice the length of *mval*, minus 2 (it must therefore be even). The first frequency band is assumed to start at zero, and the last one always ends at $Fs/2$.

The returned values are:

1. N – The order of the filter to be designed,
2. Fo – Normalized frequency band edges,
3. Ao – A real vector the same size as Fo which specifies the desired amplitude of the frequency response of the resultant filter,
4. *wts* – a vector specifying the weights to be used in each frequency band for the optimization of the filter coefficients.

The Matlab function *kaiserord* can be used to determine the order for an FIR filter designed using a Kaiser window [11]. The design parameters are:

1. *fedge* – a vector of the normalized frequency band edges,
2. *mval* – a vector of the corresponding magnitude values at the band edges,

3. *dev* – a vector specifying the maximum allowable deviation between the magnitude of the designed filter and the desired magnitude,

4. *Fs* – The sampling interval for the design.

The parameters for *kaiserord* have the same form as the parameters for *firpmord*. The returned output values are:

1. *N* – The order of the desired filter,

2. *Wn* – The normalized band edges for the filter design,

3. *beta* – The Kaiser window design parameter,

4. *type* – The filter type.

4.7 COMPARISON OF FIR FILTER DESIGNS

This section compares the results for the design of an FIR digital filter using the window based method with a Hamming window, the window based method using a Kaiser window, the use of the Matlab function *firpm* to design the filter using the Parks–McClellan method, the use of the *firls* function to design the filter using the least squares method, and the use of the *fircls* function to design the filter using the constrained, least squares method. Example 4.6 presents the comparison of the results for the designs using the indicated methods.

EXAMPLE 4.6
(Comparison of FIR Filter Designs)

Problem:
A FIR digital filter is desired to meet the following specifications:

- Sampling frequency – 8192 Hz
- Pass band – 0 to 3000 Hz
- Stop band – 3500 to 4096 Hz
- Pass band ripple – $\delta_p = 0.1087 = \pm 1$ dB
- Stop band ripple – $\delta_s = 0.0178 = -35$ dB

Design an FIR filter to meet the desired specifications using

1. The window method with a Hamming window,

2. The window method using a Kaiser window,

3. The Parks–McClellan method using the Matlab *firpm* function,

4. The least squares method using the Matlab *firls* function,

5. The constrained, least squares method using the Matlab *fircls* function, and

6. The frequency sampling method using the Matlab *fir2* function.

Solution using Hamming Window and *fir1*:
The requirement is to design an FIR filter using a Hamming window to meet the specifications. The type of window determines the stop band ripple (δ_s) and the

number of coefficients determines the width of the transition region. The peak side lobe attenuation for the Hamming window is -43 dB and the approximate transition width of the main lobe is $\frac{8\pi}{N}$ where N is the length of the filter (also the window length).

The required length of the filter can be computed by solving for N:

$$d\omega = \omega_s - \omega_c,$$

$$N = \left\lceil \frac{8\pi}{d\omega} \right\rceil \tag{4.74}$$

where

$$\omega_c = \frac{(2\pi)(3000)}{8192},$$

$$\omega_s = \frac{(2\pi)(3500)}{8192}. \tag{4.75}$$

$d\omega$ is the transition width of the main lobe and $\lceil \cdot \rceil$ is the *ceil* function (round up). The order of the filter is therefore

$$\omega_s = 0.8545\pi,$$

$$\omega_c = 0.7324\pi,$$

$$d\omega = 0.1221\pi,$$

$$N = \left\lceil \frac{8\pi}{0.1221\pi} \right\rceil$$

$$= 66. \tag{4.76}$$

The filter designed with the order equal to 66 should have a stop band attenuation of 43 dB. The order can be reduced to meet the less strict specifications for the stop ban attenuation to be 35 dB. It was determined by experimentation that a filter with order equal to 40 could meet the specifications. The following Matlab script can be used to design the require filter. Remember that all frequencies are normalized by π within Matlab. Thus, a cutoff frequency in Matlab given as $\omega_c = 0.5$ actually implies that the cutoff frequency ω_c has been set to 0.5π within Matlab.

Matlab Script 4.4.

```
% Design using fir1 with a Hamming window
wcd = 2*3000/8192;
wc = 2*3200/8192;
wr = 2*3500/8192;
delp = 0.1087;
dels = 0.0178;
dw = (wr - wcd);
% n1 = ceil(8/dw);
```

```
n1 = 40;
win = window(@hamming, n1+1);
b1 = fir1(n1, wc, win);
```

End of the Script

Fig. 4.16 shows the magnitude response of the FIR filter using the Hamming window with the order set to 40. The specifications for the desired filter are also shown on the plot using dashed lines.

Solution using Kaiser Window and *fir1*:

Several parameters need to be determined for the design using a Kaiser window and the *fir1* function. The Matlab function *kaiserord* can be used to compute the parameters.

Matlab Script 4.5.

```
[N,Wn,beta,type] = kaiserord(fedge, mval, dev);
```

End of the Script

The parameter, *fedge*, is the vector of pass band and stop band frequencies, the parameter, *mval*, is the magnitude of the ideal filter, and the parameter, *dev*, is a vector of maximum deviations allowable. The function computes the order of the filter, the new cutoff frequency, the *beta* parameter used for the Kaiser window and the type of the filter.

The following Matlab script was used to design the FIR filter using a Kaiser window.

Matlab Script 4.6.

```
% Design using fir1 with a Kaiser window
wc = 2*3000/8192;
wr = 2*3500/8192;
fedge = [wc wr];
mval = [1 0];
dev = [delp dels];
[n2,Wn,beta,type] = kaiserord(fedge, mval, dev);
b2 = fir1(n2, Wn, type, kaiser(n2+1,beta));
```

End of the Script

Fig. 4.17 shows the magnitude and phase responses for the FIR filter using the Kaiser window. The specifications for the desired filter are also shown on the plot using dashed lines.

Solution using *firpm*:

The Matlab function *firpmord* can be used to determine the normalized frequency band edges *Fo*, the frequency band magnitudes *Ao* and the weights *wt3* to be used with the *firpm* Matlab function. Once the parameters have been determined, the Matlab function *firpm* can be used to design the filter. The filter designed with the parameters obtained using *firpmord* did not meet the specifications with the initial values used for *dev* and *dels* as specified by the problem:

$$dev = [delp \quad dels]. \tag{4.77}$$

The filter did meet specifications when the parameters *dev* and *dels* were multiplied by 0.75. The following Matlab script was used to design the FIR filter using the Matlab *firpm* function.

Matlab Script 4.7.

```
% Design using firpm
wc = 2*3000/8192;
wr = 2*3500/8192;
fedge = [wc wr];
mval = [1.0 0.0];
dev = [0.75*delp 0.75*dels];
% dev = [delp dels];
[n3, Fo, Ao, wt3] = firpmord(fedge, mval, dev);
b3 = firpm(n3, Fo, Ao, wt3);
```

End of the Script

Fig. 4.18 shows the magnitude response of the FIR filter using the Matlab *firpm* function.

Solution using *firls*:

The order, frequency band edges, desired amplitudes and weights need to be specified in order to use the Matlab function *fircls* to design a FIR filter using the least squares design method. The order of the filter was chosen by trial and error to be 36 to meet the specifications. The following Matlab script was used to design the filter using the *firls* Matlab function.

Matlab Script 4.8.

```
% Design using firls
wc = 2*3000/8192;
wr = 2*3500/8192;
n4 = 36;
fpts = [0.0 wc wr 1.0];
mag = [1.0 1.0 0.75*dels 0.75*dels];
wt4 = [1 delp/dels];
wt4 = [1 1];
b4 = firls(n4, fpts, mag, wt4);
```

End of the Script

Fig. 4.19 shows the magnitude response of the FIR filter using the least squares method.

Solution using *fircls*:

The order, frequency band edges, desired amplitudes and weights need to be specified in order to use the Matlab function *fircls* to design a FIR filter using the constrained least squares design method. The order of the filter was chosen by trial and error to be 30 to meet the specifications. The following Matlab script was used to design the filter using the Matlab *fircls* function.

Matlab Script 4.9.

```
% Design using fircls
wc = 2*3100/8192;
wcd = 2*3000/8192;
wr = 2*3350/8192;
wrd = 2*3500/8192;
n5 = 30;
fedge = [0.0 wc wr 1.0];
A = [1.0 0.5 0.0];
up = [1.0+delp 0.5+delp dels];
lo = [1.0-delp 0.5-delp -dels];
b5 = fircls(n5, fedge, A, up, lo);
```

End of the Script

Fig. 4.20 shows the magnitude response of the FIR filter using the constrained least squares method.

Solution using *fir2*:

The Matlab *fir2* function can be used to design an arbitrary shaped FIR filter using the frequency sampling method. Equal length vectors, *F* and *A*, are used to specify the frequencies and magnitudes for the filer. The first value in *F* must be 0 and the last value in *F* must be 1, which corresponds to half the sample rate. The numbers in *F* must be given in increasing order. The following Matlab script was used to design the filter using the Matlab *fir2* function.

Matlab Script 4.10.

```
% Design using fir2
wc = 2*3100/8192;
wcd = 2*3000/8192;
wr = 2*3300/8192;
wrd = 2*3500/8192;
n6 = 46;
f6 = [0 wcd 0.5*(wc+wr) wr 1.0];
A = [1 1 0.5 0.5*dels 0.5*dels];
b6 = fir2(n6, f6, A);
```

End of the Script

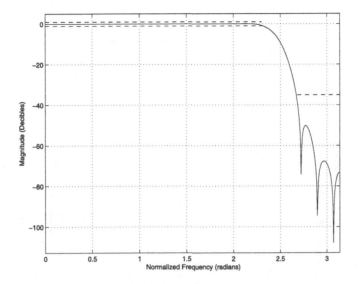

FIGURE 4.16

Magnitude response for the FIR filter using the Hamming window

Fig. 4.21 shows the magnitude response of the FIR filter designed using the Matlab *fir2* function.

Table 4.6 gives a comparison of the order of the for each of the filters. The design parameters for all of the designs were adjusted so that the filters designed would meet the specifications. The FIR filter designed using the *Hamming* window has the smallest transition width. However, it has the second highest order. The frequency sampling design using the Matlab *fir2* function required the highest order to meet the specifications. The design using the Matlab *firpm* function required the smallest order to meet the specifications. Generally, the choice of the best FIR filter design approach for a particular application involves design decisions regarding the filter order, the transition width, the pass band ripple, and the stop band ripple required to meet specifications for the filter.

End of the Example

4.8 DESIGN BY SIMULATING THE DIFFERENTIATOR

Digital filters can be designed by modeling the differentiator to transform analog filter prototypes into digital filters. The analog filter prototypes can be represented in the Laplace Transform domain using the Laplace variable s. The differentiator may be

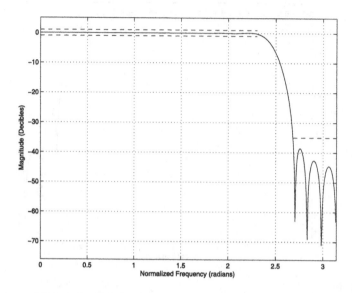

FIGURE 4.17

Magnitude response for the FIR filter using the Kaiser window

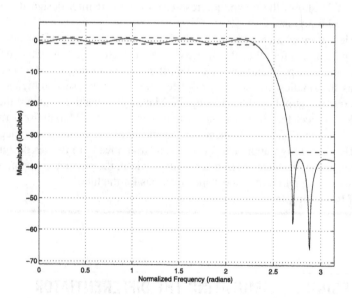

FIGURE 4.18

Magnitude response for the design using *firpm*

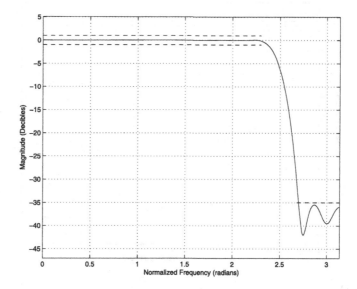

FIGURE 4.19

Design using *firls* for Example 4.6

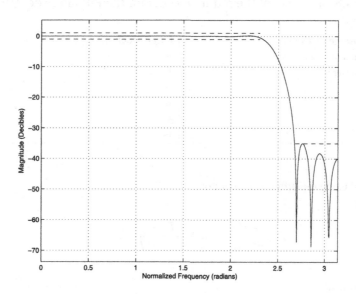

FIGURE 4.20

Design using *fircls* for Example 4.6

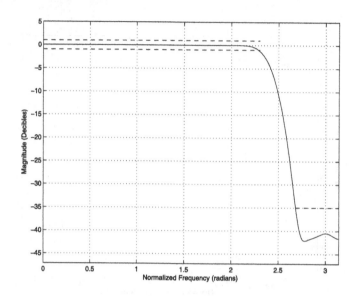

FIGURE 4.21

Design using *fir2* for Example 4.6

Table 4.6 Summary of the required order for the FIR filters designed for Example 4.6

Filter design method	Filter order
fir1 (Hamming window)	40
fir1 (Kaiser window)	31
firpm	20
firls	36
fircls	30
fir2	46

modeled using the difference operator

$$\frac{dy(t)}{dt}\bigg|_{t=nT} = \frac{y(nT) - y(nT - T)}{T}, \tag{4.78}$$

or equivalently,

$$s \Rightarrow \frac{1 - z^{-1}}{T}. \tag{4.79}$$

This approach can be continued in this manner to determine the second derivative as

$$\left.\frac{d^2 y(t)}{dt}\right|_{t=nT} = \frac{d}{dt}\left[\left.\frac{dy(t)}{dt}\right]\right|_{t=nT} = \frac{y^2(nT) - 2y(nT - T) + y(nT - 2T)}{T},$$

(4.80)

or equivalently,

$$s^2 \Rightarrow \frac{\left(1 - z^{-1}\right)^2}{T}.$$

(4.81)

This approach works reasonably well when the sampling interval is relatively small. However, other more specialized techniques with improved performance are available for digital filter design. These more specialized techniques take advantage of the special properties of filters to provide transfer function approximations that improve performance over the differentiator model discussed above. Some of these techniques will be discussed later in this chapter.

The consequences of mapping from the S plane to the Z plane using [4]

$$s \Rightarrow \frac{1 - z^{-1}}{T},$$

(4.82)

or equivalently,

$$z \Rightarrow \frac{1}{1 - sT}$$

(4.83)

can be considered by evaluating the expressions for the differentiator in the complex frequency plane where

$$s = j\Omega$$

(4.84)

and

$$z = e^{j\omega}.$$

(4.85)

Thus,

$$z \Rightarrow \frac{1}{1 - j\omega T} = \frac{1}{1 + \omega^2 T^2} + j\frac{\omega T}{1 + \omega^2 T^2}.$$

(4.86)

As ω goes from $-\infty$ to ∞, the corresponding locus of points in the Z plane is a circle of radius $\frac{1}{2}$ with center at $z = \frac{1}{2}$. Points in the left half of the s-plane are mapped to points inside this circle and points in the right half of the s-plane are mapped to the outside of this circle. This means that this mapping is not suitable for mapping analog filters into digital filters because the stable poles in the S plane are mapped into a circle or radius $\frac{1}{2}$ with center at $\frac{1}{2}$. However, this mapping can yield practical results if the mapping is a confined to the design of low pass filters and band pass filters with relatively low critical frequencies.

4.9 IMPULSE INVARIANT DESIGN

The goal of the Impulse Invariant design approach is to obtain a digital filter with the same response to a digital impulse as the corresponding continuous time filter has to an analog impulse. The procedure is to form the system transfer function for the digital system from the Z Transform of samples of the impulse response for the continuous time system. This approach will be illustrated by Example 4.7.

EXAMPLE 4.7

(An Impulse Invariant FIR System Design)

The goal for Example 4.7 is to design an impulse invariant digital system corresponding to the continuous time system with system transfer function given by

$$H(s) = \frac{s + 0.1}{(s + 0.1)^2 + 5}. \tag{4.87}$$

The continuous time system transfer function, $H(s)$, can be expanded into partial fractions to obtain

$$H(s) = \frac{0.5}{s + 0.1 - \sqrt{5}j} + \frac{0.5}{s + 0.1 + \sqrt{5}j}. \tag{4.88}$$

The corresponding inverse Laplace Transform gives the impulse response as

$$h(t) = 0.5e^{-(0.1 + 2.2361j)t}u(t) + 0.5e^{-(0.1 - 2.2361j)t}u(t). \tag{4.89}$$

If $h(t)$ is sampled at $t = nT$, the sampled impulse response is given by

$$h(n) = h(t)|_{t=nT} = 0.5e^{-(0.1 + 2.2361j)nT}u(n) + 0.5e^{-(0.1 - 2.2361j)nT}u(n). \tag{4.90}$$

The relationship between the digital system frequency response and the analog system impulse response is given by [4]

$$X(f) = Fs \sum_{k=-\infty}^{\infty} X_a\left[(f - k)F_s\right] \tag{4.91}$$

where

$$Fs = \frac{1}{T}. \tag{4.92}$$

Thus, the Z Transform of the sampled sequence should be multiplied by the sampling interval T to obtain the Z Transform for the impulse invariant impulse response. Thus,

$$H(z) = T \left[\frac{0.5}{1 - e^{-(0.1-2.2361 j)T} z^{-1}} + \frac{0.5}{1 - e^{-(0.1+2.2361 j)T} z^{-1}} \right]. \quad (4.93)$$

The two terms can be combined to obtain

$$H(z) = T \left[\frac{1 - e^{-0.1T} \cos(2.2361T) z^{-1}}{1 - 2e^{-0.1T} \cos(2.2361T) z^{-1} + e^{-0.2T} z^{-2}} \right]. \quad (4.94)$$

End of the Example

Example 4.8 provides another example of the use of the impulse invariant design approach to design a digital filter.

EXAMPLE 4.8
(Impulse Invariant Design Example Two)

Problem:
The system transfer function for a continuous time system is given by

$$H(s) = \frac{5}{s+5}. \quad (4.95)$$

Use the impulse invariant approach with a sampling interval of $T = 0.06$ to determine the transfer function for an impulse invariant discrete time system $H_{IP}(z)$.

Solution:
The discrete time system $H_{IP}(z)$ can be determined from the impulse response of the original filter:

$$\begin{aligned} h(t) &= L^{-1}\{H(s)\}, \\ h(t) &= 5e^{-5t} u(t) \end{aligned} \quad (4.96)$$

where L^{-1} is the inverse Laplace transform. The discrete time impulse response is derived from the sampled $h(t)$

$$\begin{aligned} h(n) &= h(t)|_{t=nT}, \\ h(n) &= 5Te^{-5nT} u(n), \\ h(n) &= 5Te^{-(5T)n} u(n). \end{aligned} \quad (4.97)$$

Taking the Z transform of $h(n)$

$$\begin{aligned} H_{IP}(z) &= \frac{5T}{1 - e^{-5T} z^{-1}}, \\ H_{IP}(z) &= \frac{0.3000}{1 - 0.7408 z^{-1}}. \end{aligned} \quad (4.98)$$

Matlab can be used to verify the results:

```
Ts = 0.06;
Fs = 1/Ts;
bs = [5];
as = [1 5];
[BZ,AZ] = impinvar(bs,as,Fs);
```

The results are

$$
\begin{aligned}
Bz &= 0.3000, \\
Az &= 1.0000 \quad -0.7408 ,
\end{aligned}
\tag{4.99}
$$

which is the same result.
End of the Example

4.10 DESIGN USING THE BILINEAR TRANSFORMATION

The bilinear transformation is very popular for designing digital filters from analog
filter prototypes. It was inspired by the trapezoidal rule for simulating integration:

$$
y(n) = \frac{(x(n) + x(n-1))\,T}{2} + y(n-1).
\tag{4.100}
$$

The Z Transform of both sides can be computed to obtain

$$
Y(z) = \frac{\left(X(z) + z^{-1}X(z)\right)T}{2} + z^{-1}Y(z).
\tag{4.101}
$$

The corresponding system transfer function is given by

$$
G(z) = \frac{Y(z)}{X(z)} = \frac{\left(1 + z^{-1}\right)T}{2\left(1 - z^{-1}\right)}.
\tag{4.102}
$$

Integration for a continuous time system can be represented by the Laplace Transform

$$
G(s) = \frac{1}{s}.
\tag{4.103}
$$

Thus, the expression for the transformation is given by

$$
\frac{1}{s} \rightarrow \frac{\left(1 + z^{-1}\right)T}{2\left(1 - z^{-1}\right)},
\tag{4.104}
$$

or

$$
s \Rightarrow \frac{2\left(1 - z^{-1}\right)}{T\left(1 + z^{-1}\right)} = \frac{2(z-1)}{T(z+1)}.
\tag{4.105}
$$

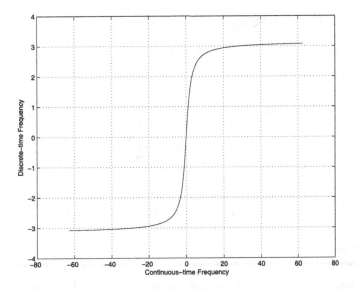

FIGURE 4.22

A plot of the mapping of continuous time frequencies to discrete time frequencies for the bilinear transformation

The bilinear transformation maps the continuous time frequencies for the range $-\infty \leq \Omega \leq \infty$ into the range of discrete time frequencies from $-\pi \leq \omega \leq \pi$. Fig. 4.22 shows this mapping for the range of continuous time frequencies $-20\pi \leq \Omega \leq 20\pi$.

Example 4.9 illustrates the design of an IIR filter with the use of the bilinear transformation.

EXAMPLE 4.9
(Bilinear Transformation IIR Filter Design)

Use the bilinear transformation to obtain a digital system to simulate

$$H(s) = \frac{2.0s}{s + 10.0}. \tag{4.106}$$

The bilinear transformation can be used to substitute for s in Eq. (4.106) to obtain

$$H(z) = \frac{2\left(\frac{2(z-1)}{T(z+1)}\right)}{\left(\frac{2(z-1)}{T(z+1)}\right) + 10}. \tag{4.107}$$

Simplifying $H(z)$, we obtain

$$H(z) = \frac{\left(\frac{4}{2+10T}\right)(z-1)}{z + \left(\frac{10T-2}{10T+2}\right)}. \tag{4.108}$$

If T is set to 1, the result is

$$H(z) = \frac{0.3333(z-1)}{z + 0.6667}. \tag{4.109}$$

If T is set to 0.25, the result is

$$H(z) = \frac{0.8889(z-1)}{z + 0.1111}. \tag{4.110}$$

End of the Example

The performance of the bilinear transformation can be evaluated by evaluating it on the unit circle in the Z plane where $z = e^{j\omega}$:

$$F(z) = \left.\frac{2(z-1)}{T(z+1)}\right|_{z=e^{j\omega}} = \frac{2\left(e^{j\omega}-1\right)}{T\left(e^{j\omega}+1\right)} = \frac{2e^{j0.5\omega}\left(e^{j0.5\omega}-e^{-j0.5\omega}\right)}{T\left(e^{j0.5\omega}+e^{-j0.5\omega}\right)}, \tag{4.111}$$

$$F(\omega) = j\Omega = j\frac{2}{T}\tan\left(\frac{\omega}{2}\right). \tag{4.112}$$

Thus, we see that the analog frequency space is distorted by the following relationship when using the bilinear transformation:

$$\Omega \Rightarrow \frac{2}{T}\tan\left(\frac{\omega}{2}\right). \tag{4.113}$$

4.11 DESIGN USING ANALOG FILTER PROTOTYPES

The bilinear transformation is often used to obtain digital filters from analog filter prototypes. The normal procedure is to first convert the critical frequencies in the continuous time domain into the corresponding frequencies in the digital domain after using the bilinear transformation. For example, analog low pass filters with a cutoff frequency of Ω_c and a magnitude of

$$|H(\Omega_c)| = \frac{1}{1+\epsilon^2} \tag{4.114}$$

can be designed using

$$H(s)H(-s) = \frac{1}{1 + \epsilon^2 \left(\frac{s}{\Omega_c}\right)^{2n}} \qquad (4.115)$$

where $H(s)$ is composed of the poles in the left half of the s plane. For example, with $n = 2$ and $\epsilon = 1$

$$H(s) = \frac{1}{\left(\frac{s}{\Omega_c}\right)^2 + \sqrt{2}\left(\frac{s}{\Omega_c}\right) + 1}. \qquad (4.116)$$

The first step in designing a digital filter using an analog prototype is to prewarp the cutoff frequency so that the digital filter and the analog filter will have the same cutoff frequency. Thus,

$$\Omega_c \rightarrow \frac{2}{T} \tan\left(\frac{\omega_c}{2}\right) \qquad (4.117)$$

where Ω_c is the cutoff frequency for the analog filter and ω_c is the cutoff frequency for the digital filter. Then, the bilinear transformation is used such that

$$s \rightarrow \frac{2}{T}\left(\frac{z-1}{z+1}\right). \qquad (4.118)$$

Note that the substitutions for Ω_c and s for the ratio $\frac{s}{\Omega_c}$ results in the factor $\frac{2}{T}$ being divided out everywhere. Therefore, the procedure can be simplified by replacing Ω_c using

$$\Omega_c \rightarrow \tan\left(\frac{\omega_c}{2}\right) \qquad (4.119)$$

and then replacing s using

$$s \rightarrow \left(\frac{z-1}{z+1}\right). \qquad (4.120)$$

Mitra provides a procedure for designing a digital filter using an analog prototype [8].

4.12 FREQUENCY TRANSFORMATIONS

A common practice for the design of IIR digital filters is to convert an analog filter to a digital filter using a transformation. The specifications for the digital filter are converted to specifications to the selected analog filter prototype and then the transformation is used to obtain the digital filter [8]. This approach has been widely used for several reasons:

1. Advanced techniques for the design of analog filters are readily available.
2. Analog filter design techniques normally involve closed form solutions.

Table 4.7 Frequency transformations for analog low pass and high pass filters [4]

Type of Transformation	Transformation	Band edge frequencies of new filter
Low pass	$s \rightarrow \dfrac{\Omega_p s}{\Omega_p'}$	Ω_p'
High pass	$s \rightarrow \dfrac{\Omega_p \Omega_p'}{s}$	Ω_p'

3. Extensive tables are available for analog filter design.
4. Many real world applications require digital simulation of analog filters.

The design procedure may involve using the transformation on the analog filter prototype and then using the bilinear transformation on the transformed analog filter. Alternately, the procedure may involve using the bilinear transformation to design a normalized low pass filter ($\Omega_c = 1$) and then using a spectral transformation in the Z plane to design the desired filter.

4.13 FREQUENCY TRANSFORMATION PROCEDURES

There are two procedures that can be used to transform low pass filters to low pass, high pass, band pass, and band stop digital filters [8].

1. One procedure starts with a low pass analog filter prototype, uses continuous time to continuous time frequency transformation and then uses the bilinear transformation to map the continuous time filter to a corresponding discrete time filter.
2. The other procedure starts with a low pass digital filter and uses discrete time to discrete time frequency transformations to map the digital low pass filter to the desired filter.

4.13.1 CONTINUOUS TIME TRANSFORMATION PROCEDURE

The design approach for the use of continuous time to continuous time frequency transformations follows [8]:

1. Use the appropriate frequency transformation to obtain the corresponding analog filter prototype for the desired kind of filter as given in Tables 4.7 and 4.8.
2. Prewarp the specified discrete time frequencies for the desired digital filter to the corresponding continuous time frequencies using the equation

$$\Omega_k' = \tan\left(\frac{\omega_k}{2}\right). \tag{4.121}$$

3. Use the transformation

$$s \rightarrow \frac{z-1}{z+1} \tag{4.122}$$

to obtain the corresponding discrete time filter.

Table 4.8 Frequency transformations for analog band pass and band stop filters [4]

Type of Transformation	Transformation	Band edge frequencies of new filter
Band pass	$s \rightarrow \Omega_p \left[\frac{s^2 + \Omega_l \Omega_u}{s(\Omega_u - \Omega_l)} \right]$	Ω_l, Ω_u
Band stop	$s \rightarrow \Omega_p \left[\frac{s(\Omega_u - \Omega_l)}{s^2 + \Omega_l \Omega_u} \right]$	Ω_l, Ω_u

Table 4.7 gives frequency transformations for continuous time low pass and high pass filters. The prototype low pass filter has a band edge frequency of Ω_p for each transformation.

Table 4.8 gives frequency transformations for continuous time low pass and high pass filters. The prototype low pass filter has a band edge frequency of Ω_p for each transformation. Example 4.10 gives and example using this procedure.

EXAMPLE 4.10

The system transfer function for the Butterworth prototype low pass continuous time filter with band edge frequency Ω_c and band pass ripple equal to -3 dB ($\epsilon = 1.0$) is given by

$$H(s) = \frac{1.0}{\left(\frac{s}{\Omega}\right)^2 + \sqrt{2}\left(\frac{s}{\Omega}\right) + 1}. \qquad (4.123)$$

The goal is to design a discrete time filter with a band pass edge $\omega_c = 0.678\pi$ and band pass ripple equal to -3 dB. The first step is to prewarp the desired discrete time frequency to obtain

$$\Omega'_c = \tan\left(\frac{0.678\pi}{2}\right) = 1.8055. \qquad (4.124)$$

Then use the transformation

$$s \rightarrow \frac{z-1}{z+1} \qquad (4.125)$$

to obtain

$$H(z) = \frac{1.0}{\left(\frac{\left[\frac{z-1}{z+1}\right]}{1.8055}\right)^2 + \sqrt{2}\left(\frac{\left[\frac{z-1}{z+1}\right]}{1.8055}\right) + 1}. \qquad (4.126)$$

$H(z)$ can be simplified to obtain

$$H(z) = \frac{2.5534(z+1)^2}{6.8134z^2 + 4.5199z + 1.7065}. \qquad (4.127)$$

$H(z)$ can be simplified to standard form by dividing each of the filter coefficients by 6.8134 to obtain

$$H(z) = \frac{0.4785z^2 + 0.9569z + 0.4785}{z^2 + 0.6634z + 0.2505}, \quad (4.128)$$

or

$$H(z) = \frac{0.4785 + 0.9569z^{-1} + 0.4785z^{-2}}{1.0 + 0.6634z^{-1} + 0.2505z^{-2}}. \quad (4.129)$$

The Matlab *butter* function uses this approach to design digital filters. The filter designed in this example can be verified through the use of the Matlab *butter* function to design the same filter. The following Matlab script can be used for the design.

Matlab Script 4.11.

```
order = 2;
wc = 0.678;
[b2, a2] = butter(order, wc);
```

End of the Script

The resulting filter coefficients are give by

```
b2 =
    0.4785    0.9569    0.4785

a2 =
    1.0000    0.6634    0.2505
```

which is the same result obtained using the design procedure for this example.
End of the Example

4.13.2 DISCRETE TIME TRANSFORMATION PROCEDURE

A frequency transformation can be used to obtain a high pass filter, a band stop filter or a band pass filter from a low pass filter. For example, the following frequency transformation can be used to transform a low pass filter to a high pass filter [8]:

$$z^{-1} \Rightarrow -\frac{z^{-1} + \alpha}{1 + \alpha z^{-1}},$$

$$\alpha = -\frac{\cos\left(0.5\left(\omega_c + \hat{\omega}_c\right)\right)}{\cos\left(0.5\left(\omega_c - \hat{\omega}_c\right)\right)} \quad (4.130)$$

where Z is the Z Transform variable in the low pass filter, ω_c is the normalized cutoff frequency for the low pass filter, and $\hat{\omega}_c$ is the cutoff frequency for the high pass filter.

Table 4.9 Digital frequency transformations for low pass and high pass digital filters [4]

Type of Transformation	Transformation	Parameters
Low pass	$Z^{-1} \rightarrow \dfrac{z^{-1}-\alpha}{1-\alpha z^{-1}}$	$\hat{\omega}_p =$ edge frequency of new filter $\alpha = \dfrac{\sin[(\omega_p - \hat{\omega}_p)/2]}{\sin[(\omega_p + \hat{\omega}_p)/2]}$
High pass	$Z^{-1} \rightarrow -\dfrac{z^{-1}+\alpha}{1+\alpha z^{-1}}$	$\hat{\omega}_p =$ edge frequency of new filter $\alpha = -\dfrac{\cos[(\omega_p + \hat{\omega}_p)/2]}{\cos[(\omega_p - \hat{\omega}_p)/2]}$ $\omega_l =$ lower edge frequency $\omega_u =$ upper edge frequency $a_1 = 2\alpha K/(K+1)$

Table 4.10 Digital frequency transformations for band pass and band stop digital filters [4]

Type of Transformation	Transformation	Parameters
Band pass	$Z^{-1} \rightarrow -\dfrac{z^{-2}-a_1 z^{-1}+a_2}{a_2 z^{-2}-a_1 z^{-1}+1}$	$a_2 = (K-1)/(K+1)$ $\alpha = \dfrac{\cos[(\omega_u + \omega_l)/2]}{\cos[(\omega_u - \omega_l)/2]}$ $K = \cot\left(\dfrac{\omega_u - \omega_l}{2}\right)\tan\left(\dfrac{\omega_p}{2}\right)$ $\omega_l =$ lower edge frequency $\omega_u =$ upper edge frequency
Band stop	$Z^{-1} \rightarrow \dfrac{z^{-2}-a_1 z^{-1}+a_2}{a_2 z^{-2}-a_1 z^{-1}+1}$	$a_1 = 2\alpha/(K+1)$ $a_2 = (1-K)/(1+K)$ $\alpha = \dfrac{\cos[(\omega_u + \omega_l)/2]}{\cos[(\omega_u - \omega_l)/2]}$ $K = \tan\left(\dfrac{\omega_u - \omega_l}{2}\right)\tan(\dfrac{\omega_p}{2})$

Table 4.9 gives frequency transformations for low pass to low pass and low pass to high pass high pass digital filters. The prototype low pass filter has a band edge frequency of ω_p for each transformation.

Table 4.10 gives frequency transformations for low pass to band pass and low pass to band stop digital filters. The prototype low pass filter has a band edge frequency of ω_p for each transformation.

The design approach for the use of discrete time to discrete time frequency transformations follows [8]:

1. Begin with the design of a low pass discrete time filter with band edge ω_p.
2. Transform any specifications in terms of continuous time frequencies to the corresponding discrete time frequencies using the equation

$$\omega_k = \frac{\Omega_k}{F_s} = T\Omega_k \qquad (4.131)$$

where F_s is the sampling frequency and T is the sampling interval.

3. Use the appropriate discrete time to discrete time frequency transformation to obtain the corresponding discrete time filter for the desired kind of filter as given in Table 4.9 or Table 4.10 as appropriate.

EXAMPLE 4.11

A second order low pass filter with a 3 dB attenuation at the normalized cutoff frequency $\omega_c = 0.35\pi$ has a system transfer function given by

$$H(z) = \frac{0.1675 + 0.3350z^{-1} + 0.1675z^{-2}}{1.0000 - 0.5570z^{-1} + 0.2270z^{-2}}. \tag{4.132}$$

The goal is to design a second order high pass filter $H_{HP}(z)$ with a 3 dB attenuation cutoff frequency of $\widehat{\omega}_c = 0.57\pi$ using a low pass to high pass spectral transformation

$$z^{-1} = -\frac{z^{-1} + \alpha}{1 + \alpha z^{-1}},$$

$$\alpha = -\frac{\cos\left[0.5(\omega_c + \widehat{\omega}_c)\right]}{\cos\left[0.5(\omega_c - \widehat{\omega}_c)\right]}. \tag{4.133}$$

Solution: We first determine the value for α. Let

$$d1 = \frac{\omega_c + \widehat{\omega}_c}{2} = 0.5(0.35\pi + 0.57\pi) = 0.46\pi = 1.4451,$$

$$d2 = \frac{\omega_c - \widehat{\omega}_c}{2} = 0.5(0.35\pi - 0.57\pi) = -0.11\pi = -0.1332, \tag{4.134}$$

$$\alpha = -\frac{\cos(1.4451)}{\cos(-0.1332)} = -0.1332. \tag{4.135}$$

We can now write the system transfer function $H_{HP}(z)$ for the resulting high pass filter using the transformation with the value of α inserted. The transformation with the numeric value for α inserted is

$$z^{-1} = -\frac{z^{-1} - 0.1332}{1 - 0.1332z^{-1}}. \tag{4.136}$$

Using the transformation on $H(z)$, we obtain

$$H_{HP}(z) = \frac{B(z)}{A(z)} \tag{4.137}$$

where

$$B(z) = 0.1675 - 0.3350 \left[-\frac{z^{-1} - 0.1332}{1 - 0.1332z^{-1}} \right]$$

$$+ 0.1675 \left[-\frac{z^{-1} - 0.1332}{1 - 0.1332z^{-1}} \right]^2, \tag{4.138}$$

$$A(z) = 1.0000 + 0.5570 \left[-\frac{z^{-1} - 0.1332}{1 - 0.1332z^{-1}} \right]$$

$$+ 0.2270 \left[-\frac{z^{-1} - 0.1332}{1 - 0.1332z^{-1}} \right]^2. \tag{4.139}$$

We can write

$$H(z) = \frac{N(z)}{D(z)} \tag{4.140}$$

where

$$N(z) = 0.1675 \left[1 - 0.1332z^{-1} \right]^2$$

$$- 0.3350 \left[z^{-1} - 0.1332 \right] \left[1 - 0.1332z^{-1} \right]$$

$$+ 0.1675 \left[z^{-1} - 0.1332 \right]^2,$$

$$D(z) = \left[1 - 0.1332z^{-1} \right]^2 + 0.5570 \left[z^{-1} - 0.1332 \right] \left[1 - 0.1332z^{-1} \right]$$

$$+ 0.2270 \left[z^{-1} - 0.1332 \right]^2. \tag{4.141}$$

Simplifying, we obtain

$$H(z) = \frac{0.2151 - 0.4302z^{-1} + 0.2151z^{-2}}{0.9298 + 0.2400z^{-1} + 0.1705z^{-2}}, \tag{4.142}$$

$$H(z) = \frac{0.2313 - 0.4626z^{-1} + 0.2313z^{-2}}{1.0000 + 0.2581z^{-1} + 0.1834z^{-2}}. \tag{4.143}$$

End of the Example

The Matlab *butter* function and the following Matlab script to design the same filter.

Matlab Script 4.12.

```
order = 2;
wc = 0.57;
[b, a] = butter(order, wc, 'high');
```

End of the Script

The design using the Matlab *butter* function provided the following filter coefficients.

$$b = [0.2313 \quad -0.4626 \quad 0.2313]$$
$$a = [1.0000 \quad 0.2581 \quad 0.1834] \tag{4.144}$$

Thus, the results are the same.

4.14 DESIGN OF IIR FILTERS USING MATLAB

The Signal Processing Toolbox for Matlab includes several m-files for the design of IIR digital filters. This includes files to determine the filter order from given specifications. The Matlab Signal Processing Toolbox provides the following m-files to design digital filters based upon the use of the bilinear transformation and analog filter prototypes.

Matlab Script 4.13.

```
[b, a] = butter(n2. wn1);  % Butterworth filter prototype
[b, a] = cheby1(n2, rp, wn2);  % Chebyshev type 1 prototype
[b, a] = cheby2(n3, rs, wn3);  % Chebyshev type 2 prototype
[b, a] = ellip(n4, rp, rs, wn4) % Elliptic filter prototype
```

End of the Script

In each case, the parameter *b* is the vector of returned numerator coefficients and the parameter *a* is the vector of returned denominator coefficients.

The parameters for specification of the filter are as follows:

1. wp – The normalized pass band edge frequency
2. ws – The normalized stop band edge frequency
3. rp – The pass band ripple in dB
4. rs – The stop band ripple in dB

The returned parameters are:

1. N – The order of the filter to meet the specifications
2. wn – The normalized frequency to use in the design for the pass band edge frequency

The filter order for IIR filters can be determined using the Matlab m-files corresponding to the Matlab design m-files.

Matlab Script 4.14.

```
[N, wn] = buttord(wp, ws, rp, rs);
[N, wn] = cheb1ord(wp, ws, rp, rs);
[N, wn] = cheb2ord(wp, ws, rp, rs);
[N, wn] = ellipord(wp, ws, rp, rs);
```

End of the Script

EXAMPLE 4.12

Consider the design example where the specifications for a desired IIR digital filter are given by

Matlab Script 4.15.

```
wp = 0.651;
ws = 0.713;
rp = 1;
rs = 40;
```

End of the Script

   The goal is to determine the order required for each of the filter types supported by Matlab. The following Matlab script can be used to determine the order for each of the filter types.

Matlab Script 4.16.

```
wp = 0.651;
ws = 0.713;
rp = 1;
rs = 40;
[n1, wn1] = buttord(wp, ws, rp, rs);
[n2, wn2] = cheb1ord(wp, ws, rp, rs);
[n3, wn3] = cheb2ord(wp, ws, rp, rs);
[n4, wn4] = ellipord(wp, ws, rp, rs);
```

End of the Script

   The results are

$$
\begin{aligned}
n1 &= 23, & wn1 &= 0.66, \\
n2 &= 9, & wn2 &= 0.651, \\
n3 &= 9, & wn3 &= 0.713, \\
n4 &= 5, & wn4 &= 0.651.
\end{aligned}
\tag{4.145}
$$

The order required for the Butterworth filter is much larger than the order required for the other filters. The elliptic filter requires the lowest order.
End of the Example

   Example 4.13 provides an example using the Matlab functions *buttord* and *butter* to design an IIR filter to meet given specifications.

EXAMPLE 4.13

(IIR Filter Design using the Matlab *butter* Function)

Problem:

An IIR digital filter is desired to meet the following specifications:

- Sampling frequency – 8192 Hz
- Pass band – 0 to 3000 Hz
- Stop band – 3500 to 4096 Hz
- Pass band ripple – $\delta_p = 0.1087$
- Stop band ripple – $\delta_s = 0.0178$

Use the Matlab function *buttord* to determine the order of the desired filter and design an IIR filter using the Matlab function *butter* to design a filter to meet the specifications.

Solution:

The pass band and stop band ripples need to be computed in dB in order to design this filter. Therefore,

$$
\begin{aligned}
r_p &= -20\log(1.0 - \delta_p), \\
r_s &= -20\log(\delta_s). \tag{4.146}
\end{aligned}
$$

The Matlab function *buttord* can be used to determine the order of the IIR filter and to compute the new 3 dB frequency. The parameters for the function are the normalized pass band and stop band edges and the pass band and stop band ripples in dB. The following Matlab script can be used to determine the parameters and to design the IIR filter using the Matlab *butter* function.

Matlab Script 4.17.

```
fs = 8192;
delt = 2/fs;
wp = delt*2800;
ws = delt*3100;
delp = 0.1087;
dels = 0.0178;
rp = -20*log10(1.0 - delp);
rs = -20*log10(dels);
[order1, wc1] = buttord(wp, ws, rp, rs);
[b1, a1] = butter(order1, wc1);
```

End of the Script

Fig. 4.23 shows the magnitude response of the IIR filter designed using the Butterworth filter prototype. The specifications are shown on the filter using the hashed line. A 16th order filter was required to meet the specifications, and the 3 dB normalized cutoff frequency was $wc1 = 0.6964\pi$.

End of the Example

Example 4.14 illustrates the design of a filter to meet the specifications given in Example 4.13 using the Matlab functions *cheb1ord* and *cheby1*.

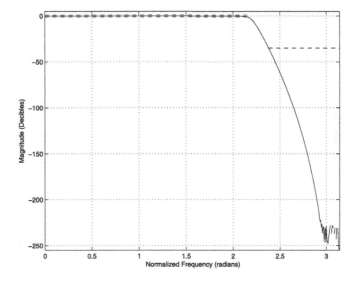

FIGURE 4.23

Magnitude plot for Example 4.13

EXAMPLE 4.14
(Design Example using Matlab Functions *cheb1ord* and *cheby1*)

Problem:
An IIR digital filter is desired to meet the following specifications:

- Sampling frequency – 8192 Hz
- Pass band – 0 to 3000 Hz
- Stop band – 3500 to 4096 Hz
- Pass band ripple – $\delta_p = 0.1087$
- Stop band ripple – $\delta_s = 0.0178$

Use the Matlab function *cheb1ord* to determine the order of the desired filter and design an IIR filter using the Matlab function *cheby1* to design a filter to meet the specifications.

Solution:
The Matlab *cheb1ord* function can be used to determine the order of the desired filter and the filter to meet the desired specifications can be designed using the Matlab *cheby1* function. The Matlab function *cheb1ord* can be used in a similar manner to the way that the Matlab function *buttord* was used in Example 4.13 to determine the IIR filter parameters. However, the Chebyshev type 1 filter characteristic is determined by the pass band ripple and not by the stop band ripple. The

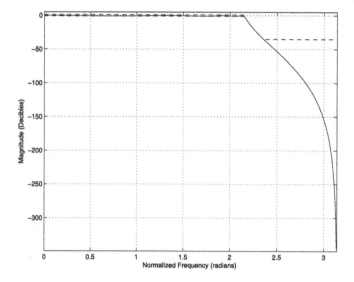

FIGURE 4.24

Magnitude plot for Example 4.14

following Matlab script can be used to determine the parameters and to design the required filter.

Matlab Script 4.18.

```
fs = 8192;
delt = 2/fs;
wp = delt*2800;
ws = delt*3100;
delp = 0.1087;
dels = 0.0178;
rp = -20*log10(1.0 - delp);
rs = -20*log10(dels);
[order2, wc2] = cheb1ord(wp, ws, rp, rs);
[b2, a2] = cheby1(order2, rp, wc2);
```

End of the Script

Fig. 4.24 shows the magnitude response of the IIR filter designed using the Chebyshev Type 1 filter prototype. The specifications are shown on the filter using the hashed line. A seventh order filter was required to meet the specifications and the 3 dB normalized cutoff frequency was $wc1 = 0.6836\pi$.

End of the Example

4.15 **PROBLEMS FOR CHAPTER 4**

Problem 4.1. (Design using Windows)
Use the windows design method to design a linear phase FIR digital filter with frequency response approximately equal to the ideal frequency response given by

$$H_d(\omega) = \begin{cases} 0 & \text{for} \quad |\omega| \le \frac{\pi}{3}, \\ 1 & \text{for} \quad \frac{\pi}{3} < |\omega| < \frac{2\pi}{3}, \\ 0 & \text{for} \quad \frac{2\pi}{3} \le |\omega| \le \pi. \end{cases} \qquad (4.147)$$

1. Determine the coefficients for linear phase filter with 39 filter coefficients based on the use of the window method with a Hamming window. You are required to
   (a) Develop an equation for the samples of the impulse response for $H_d(\omega)$.
   (b) Develop an equation for the appropriate Hamming window to be used to compute the 39 filter coefficients.
   (c) Multiply the two equations and make any adjustment in the size of the coefficients to obtain the final filter coefficients.
2. Determine and plot the magnitude response and the phase response for the filter you designed.
3. Write a Matlab script using the Matlab *fir1* function to design a digital filter to meet the specifications given above.
4. Determine and plot the magnitude response and the phase response for the filter you designed using the Matlab *fir1* function.
5. Compare your coefficients to the coefficients obtained with the Matlab *fir1* function.

Problem 4.2. (FIR Filter Design)
Design an FIR, linear phase, digital filter approximating the ideal frequency response given by

$$H_d(\omega) = \begin{cases} 0 & \text{for} \quad |\omega| \le \frac{\pi}{7}, \\ 1 & \text{for} \quad \frac{\pi}{7} < |\omega| < \frac{\pi}{3}, \\ 0 & \text{for} \quad \frac{\pi}{3} \le |\omega| \le \pi. \end{cases} \qquad (4.148)$$

Determine the coefficients for a 5 coefficient filter based on the use of the window method with the use of a Hanning window. **(Hint)** The equation for the Hanning window shifted by M samples in time is given by

$$W(n) = 0.5 + 0.5 \cos\left[2\pi(n - M)/N\right] \qquad (4.149)$$

where N is the number of samples in the window and

$$M = \frac{N - 1}{2}. \qquad (4.150)$$

Problem 4.3. (FIR Filter Design)
An FIR digital filter is desired to meet the following specifications:

- Sampling frequency – 8192 Hz
- Pass band – 2500 to 4096 Hz
- Stop band – 0 to 2000 Hz
- Pass band ripple = ±1 dB
- Stop band ripple = −30 dB

Select a Kaiser window for the design of the required filter. Determine appropriate parameters to use the *kaiserord* Matlab function to determine the order of the filter.

```
[N,Wn,beta,type] = kaiserord(F,A,DEV)
```

Determine F, A, and DEV. F is a vector of band edge frequencies in Hz, in ascending order between 0 and half the sampling frequency Fs. A is a vector of 0s and 1s specifying the desired function's amplitude on the bands defined by F. DEV is a vector of maximum deviations or ripples allowable for each band.

Problem 4.4. (Frequency Sampling Method)
Design a linear phase FIR filter of length $N = 37$ that has a symmetric unit sample response and a frequency response that satisfies the condition

$$|H(\omega)|_{\omega=\frac{2\pi k}{37}} = \begin{cases} 0 & \text{for } 0 \le k \le 11, \\ 0.71 & \text{for } k = 12, \\ 1 & \text{for } 13 \le k \le 23, \\ 0.71 & \text{for } k = 24, \\ 0 & \text{for } 25 \le k \le 36. \end{cases} \qquad (4.151)$$

1. Determine the 37 filter coefficients.
2. Plot the magnitude response and the phase response for your filter.
3. Determine if the magnitude response of your filter matches the desired magnitude response at the specified frequencies.

Problem 4.5. (Parks–McClellan Optimal Equiripple Design)
An FIR digital filter is desired to meet the following specifications:

- Sampling frequency – 4096 Hz
- Pass band – 0 to 925 Hz
- Stop band – 1150 to 2048 Hz
- Pass band ripple – $\delta_p = 0.1088$
- Stop band ripple – $\delta_s = 0.0175$
- The number of coefficients should not exceed 31.

1. Determine appropriate parameters for the filter using the *firpmord* Matlab routine and design the desired filter using the Matlab *firpm* FIR filter design routine. Use appropriate weights to ensure that the filter meets the desired specifications.
2. Determine and plot the magnitude response and the phase response for the filter you designed.

3. Make a table containing the following entries for the filter:
   (a) Minimum number of filter coefficients to meet the filter specifications
   (b) Maximum ripple in the stop band
   (c) Actual pass band edge frequency (where the gain last reaches $1 - \delta_p$)
   (d) Actual stop band edge frequency (where the gain first reaches δ_s).

Problem 4.6. (Least Squares Optimal Design)
An FIR digital filter is desired to meet the following specifications:

- Sampling frequency – 4096 Hz
- Pass band – 0 to 925 Hz
- Stop band – 1150 to 2048 Hz
- Pass band ripple – $\delta_p = 0.1088$
- Stop band ripple – $\delta_s = 0.0175$
- The number of coefficients should not exceed 31.

1. Determine appropriate parameters and design the desired filter using the Matlab *firls* FIR filter design function. Use appropriate weights to ensure that the filter meets the desired specifications.
2. Determine and plot the magnitude and phase responses of the filter.
3. Make a table containing the following entries for the filter:
   (a) Minimum number of filter coefficients to meet the filter specifications
   (b) Maximum ripple in the stop band
   (c) Actual pass band edge frequency (where the gain last reaches $1 - \delta_p$)
   (d) Actual stop band edge frequency (where the gain first reaches δ_s).

Problem 4.7. (Bilinear Transformation)
The continuous time transfer function for the third order Butterworth low pass filter with cutoff frequency, $\Omega_c = 1.0$ and with attenuation at the cutoff frequency equal to 3 dB is given by

$$H(s) = \frac{1}{s^3 + 2s^2 + 2s + 1}. \tag{4.152}$$

1. Use the bilinear transformation

$$s \Rightarrow \frac{z-1}{z+1} \tag{4.153}$$

   along with appropriate frequency transformations and prewarping to design a high pass digital filter with normalized cutoff frequency $\omega_c = 0.39\pi$ with attenuation at the cutoff frequency equal to 3 dB.
2. Use the Matlab function *freqz* to verify that your digital filter has the correct magnitude response. Determine the actual attenuation at $\omega_c = 0.39\pi$.

260 CHAPTER 4 Design of Digital Filters

Problem 4.8. (IIR Filter Design)

The continuous time transfer function for the Butterworth low pass filter with cutoff frequency, $\Omega_c = 1.0$ and with attenuation at the cutoff frequency equal to 3 dB is given by

$$H(s) = \frac{1}{s^3 + 2s^2 + 2s + 1}. \qquad (4.154)$$

1. Use the bilinear transformation

$$s \Rightarrow \frac{2(z-1)}{T(z+1)} \qquad (4.155)$$

along with prewarping

$$\Omega_c \Rightarrow \frac{2}{T}\tan\left(\frac{\omega}{2}\right) \qquad (4.156)$$

and the appropriate frequency transformations to design a high pass digital filter with normalized cutoff frequency $\omega_c = 0.362\pi$ with attenuation at the cutoff frequency equal to 3 dB.
2. Use the Matlab function *freqz* to verify that your digital filter has the correct magnitude response.

Problem 4.9. (IIR Filter Design)

A second order type 1 Chebyshev low pass IIR digital filter with a 1.0 dB attenuation cutoff frequency at $\omega_c = 0.281\pi$ has a system transfer function given by

$$H(z) = \frac{0.1243\left(1 + z^{-1}\right)^2}{1.0000 - 0.8544z^{-1} + 0.4122z^{-2}}. \qquad (4.157)$$

Design a back stop filter $H_{BS}(z)$ with a center frequency of $\omega_s = 0.65\pi$ and a bandwidth of 0.20π by using a low pass to band stop spectral transformation. Use Matlab function *cheby1* with appropriate inputs to determine if the filter you designed is essentially equal to the corresponding filter designed using *cheby1*.

Problem 4.10. (IIR Filter Design)

A third order low pass filter with a 1 dB attenuation at the normalized cutoff frequency $\omega_1 = 0.319\pi$ has a system transfer function given by

$$H(z) = \frac{0.0405 + 0.1215z^{-1} + 0.1215z^{-2} + 0.0405z^{-3}}{1.0000 - 1.4686z^{-1} + 1.1684z^{-2} - 0.3759z^{-1}}. \qquad (4.158)$$

Design a third order low pass filter $H_{LP}(z)$ with a 1 dB attenuation cutoff frequency of $\omega_2 = 0.615\pi$ using a low pass to low pass spectral transformation. Use Matlab to plot the magnitude responses of the both low pass filters on the same figure using different line types.

Problem 4.11. (IIR Filter Design)
The system transfer function for a continuous time system is given by

$$H(s) = \frac{6(s-1)}{s(s-2)}. \tag{4.159}$$

1. Use the impulse invariant approach with a sampling interval of $T = 0.05$ to determine the transfer function for an impulse invariant discrete time system $H_{IP}(z)$.
2. Use the bilinear transformation

$$s \Rightarrow \frac{2(z-1)}{T(z+1)} \tag{4.160}$$

to determine the transfer function for a discrete time system $H_{BT}(z)$.
3. Use Matlab to plot 200 samples of the frequency response of the original continuous system $H(\omega)$, the frequency response of the impulse invariant system $H_{IP}(z)$ and the frequency response of the system $H_{BT}(z)$ on the same figure using different line types.
4. Summarize your thoughts on which design procedure gave the best results and give reasons for your thoughts.

Problem 4.12. (IIR Filter Design)
Design an IIR digital filter with frequency response specifications given by

- Pass band cutoff frequency = 6.25 kHz
- Stop band frequency = 8.5 kHz
- Pass band ripple = 1.0 dB
- Stop band attenuation = 45 dB
- Sampling frequency = 32 kHz

Determine the order of the required digital filter and the critical normalized digital frequencies required to design

1. A Butterworth type digital filter using the Matlab *butter* function
2. A Chebyshev type 1 digital filter using the Matlab *cheby1* function
3. A Chebyshev type 2 digital filter using the Matlab *cheby2* function
4. An Elliptical type digital filter using the Matlab *ellip* function

Show all calculation or provide the Matlab code used to obtain your answers.

Problem 4.13. (IIR Filter Design)
Design an IIR digital filter with frequency response specifications given by

- Pass band cutoff frequency = 8 kHz
- Stop band frequency = 9 kHz
- Pass band ripple = 0.5 dB
- Stop band attenuation = 40 dB
- Sampling frequency = 44 kHz

Determine the order of the required digital filter and the critical normalized digital frequencies required to design

1. A Butterworth type digital filter using the Matlab *butter* function
2. A Chebyshev type 1 digital filter using the Matlab *cheby1* function
3. A Chebyshev type 2 digital filter using the Matlab *cheby2* function
4. An Elliptical type digital filter using the Matlab *ellip* function

Show all calculation or provide the Matlab code used to obtain your answers.

Problem 4.14. (IIR Filter Design)
A fourth order Butterworth filter has a cutoff frequency of $\Omega_c = 200\pi$ rad/s.

1. Determine the zeros and poles of the analog transfer function for the filter.
2. Determine the pass band and stop band frequencies if we want 1 dB ripple in the pass band and 40 dB attenuation in the stop band.
3. Assume that we want to design a digital filter using the analog Butterworth filter transfer function and the bilinear transformation with a sampling frequency $F_s = 1$ kHz. Determine the zeros and poles in the Z plane for the corresponding digital filter.

Problem 4.15. (IIR Filter Design)
A second order low pass type 1 Chebyshev IIR digital filter with a 1.0 dB attenuation cutoff frequency at $\omega_c = 0.37\pi$ has a system transfer function given by

$$H(z) = \frac{0.2794\left(1 + z^{-1}\right)^2}{1.0000 - 0.0599z^{-1} + 0.3139z^{-2}}. \qquad (4.161)$$

Design a fourth order band pass filter $H_{BP}(z)$ with a center frequency of $\hat{\omega}_p = 0.67\pi$ and a bandwidth of 0.20π by using a low pass to band pass spectral transformation. Use Matlab to plot the magnitude responses of the low pass filter and the band pass filter on the same figure using different line types. You can use the Matlab *hold* function to put more than one plot on the same figure.

Problem 4.16. (IIR Filter Design)
A second order low pass filter with a 3 dB attenuation at the normalized cutoff frequency $\omega_c = 0.5\pi$ has a system transfer function given by

$$H(z) = \frac{0.2929 + 0.5858z^{-1} + 0.2929z^{-2}}{1.0000 + 0.1716z^{-2}}. \qquad (4.162)$$

Design a second order high pass filter $H_{HP}(z)$ with a 3 dB attenuation cutoff frequency of $\hat{\omega}_c = 0.385\pi$ using a low pass to high pass spectral transformation. Use Matlab to plot the magnitude responses of the both filters on the same figure using different line types.

Problem 4.17. (IIR Filter Design)
The system transfer function for a continuous time system is given by

$$H(s) = \frac{4s + 2}{s^2 + 6s + 8}. \tag{4.163}$$

1. Use the impulse invariant approach with a sampling interval of $T = 0.05$ to determine the transfer function for an impulse invariant discrete time system $H_{IP}(z)$.
2. Use the bilinear transformation

$$s \Rightarrow \frac{2(z - 1)}{T(z + 1)} \tag{4.164}$$

to determine the transfer function for a discrete time system $H_{BT}(z)$.
3. Use Matlab to plot 200 samples of the frequency response of the original continuous system $H(\omega)$, the frequency response of the impulse invariant system $H_{IP}(z)$ and the frequency response of the system $H_{BT}(z)$ on the same figure using different line types.
4. Summarize your thoughts on which design procedure gave the best results and give reasons for the results.

Problem 4.18. (IIR Filter Design)
The continuous time transfer function for the Butterworth low pass filter with cutoff frequency, $\Omega_c = 1.0$ and with attenuation at the cutoff frequency equal to 3 dB is given by

$$H(s) = \frac{1}{s^3 + 2s^2 + 2s + 1}. \tag{4.165}$$

1. Use the bilinear transformation

$$s \Rightarrow \frac{z - 1}{z + 1} \tag{4.166}$$

along with appropriate frequency transformations and prewarping to design a low pass digital filter with normalized cutoff frequency $\omega_c = 0.45\pi$ with attenuation at the cutoff frequency equal to 3 dB.
2. Use the Matlab function *freqz* to verify that your digital filter has the correct magnitude response.

Problem 4.19. (IIR Filter Design)
A second order low pass type 1 Chebyshev IIR digital filter with a 1.0 dB attenuation cutoff frequency at $\omega_c = 0.37\pi$ has a system transfer function given by

$$H(z) = \frac{0.1930 \left(1 + z^{-1}\right)^2}{1.0000 - 0.4773z^{-1} + 0.3435z^{-2}}. \tag{4.167}$$

Design a fourth order bandpass filter $H_{BP}(z)$ with a center frequency of $\hat{\omega}_p = 0.55\pi$ and a bandwidth of 0.20π by using a low pass to band pass spectral transformation. Use Matlab to plot the magnitude responses of the low pass filter and the band pass filter on the same figure using different line types. You can use the Matlab *hold* function to put more than one plot on the same figure.

Problem 4.20. (IIR Filter Design)
A second order low pass filter with a 3 dB attenuation at the normalized cutoff frequency $\omega_c = 0.61\pi$ has a system transfer function given by

$$H(z) = \frac{0.4020 + 0.8039z^{-1} + 0.4020z^{-2}}{1.0000 + 0.4068z^{-1} + 0.2010z^{-2}}. \tag{4.168}$$

Design a second order low pass filter $H_{LP}(z)$ with a 3 dB attenuation cutoff frequency of $\hat{\omega}_c = 0.41\pi$ using a low pass to low pass spectral transformation. Use Matlab to plot the magnitude responses of the both low pass filters on the same figure using different line types.

Problem 4.21. (IIR Filter Design)
The system transfer function for a continuous time system is given by

$$H(s) = \frac{5}{s+5}. \tag{4.169}$$

1. Use the impulse invariant approach with a sampling interval of $T = 0.06$ to determine the transfer function for an impulse invariant discrete time system $H_{IP}(z)$.
2. Use the bilinear transformation

$$s \Rightarrow \frac{2(z-1)}{T(z+1)} \tag{4.170}$$

to determine the transfer function for a discrete time system $H_{BT}(z)$.
3. Use Matlab to plot 200 samples of the frequency response of the original continuous system $H(\omega)$, the frequency response of the impulse invariant system $H_{IP}(z)$ and the frequency response of the system $H_{BT}(z)$ on the same figure using different line types.
4. Summarize your thoughts on which design procedure gave the best results and give reasons for the results.

Problem 4.22. (IIR Filter Design)
A second order low pass filter with a 3 dB attenuation at the normalized cutoff frequency $\omega_c = 0.35\pi$ has a system transfer function given by

$$H(z) = \frac{0.1675 + 0.3350z^{-1} + 0.1675z^{-2}}{1.0000 - 0.5570z^{-1} + 0.2270z^{-2}}. \tag{4.171}$$

We want to design a second order high pass filter $H_{HP}(z)$ with a 3 dB attenuation cutoff frequency of $\hat{\omega}_c = 0.57\pi$ using a low pass to high pass spectral transformation

$$z^{-1} = \frac{z^{-1} + \alpha}{1 + \alpha z^{-1}},$$

$$\alpha = -\frac{\cos(\omega_c + \widehat{\omega}_c)}{\cos(\omega_c - \widehat{\omega}_c)}. \tag{4.172}$$

1. Determine the value of α to use in the transformation.
2. Write the system transfer function $H_{HP}(z)$ for the resulting high pass filter using the transformation with the value of α inserted. You do not need to simplify it to determine the numeric value of the filter coefficients.

Problem 4.23. (Digital Filter Design)
The continuous time transfer function for the second order Butterworth low pass filter with cutoff frequency $\Omega_c = 1.0$ and with attenuation at the cutoff frequency equal to 3 dB is given by

$$H(s) = \frac{1}{s^2 + \sqrt{2}s + 1}. \tag{4.173}$$

We want to design a second order low pass filter with a 3 dB cutoff frequency of 1260 Hz for a system with a sampling frequency of 4000 Hz. Use the bilinear transformation

$$s \Rightarrow \frac{z-1}{z+1} \tag{4.174}$$

along with appropriate frequency transformations and prewarping to design the required filter.

Problem 4.24. (IIR Filter Design)
The system transfer function for a continuous time system is given by

$$H(s) = \frac{1.161(s + 1.95)}{(s + 1.71)(s + 0.982)}. \tag{4.175}$$

Assume that the sampling interval $T = 0.085$ for this problem. Use the impulse invariant approach to determine the transfer function for an impulse invariant discrete time system $H_{IP}(z)$ based upon $H(s)$.

Problem 4.25. (IIR Filter Design using Transformations)
A second order low pass filter with a normalized cutoff frequency at $\omega_c = 0.319\pi$ has a system transfer function given by

$$H(z) = \frac{0.1517 + 0.2323z^{-1} + 0.1517z^{-2}}{1.0 - 0.8063z^{-1} + 0.4430z^{-2}}. \tag{4.176}$$

We want to design a second order high pass filter $H_{HP}(z)$ with a cutoff frequency of $\widehat{\omega}_c = 0.681\pi$ using a low pass to high pass spectral transformation

$$z^{-1} = -\frac{z^{-1} + \alpha}{1 + \alpha z^{-1}},$$

$$\alpha = -\frac{\cos\left(\frac{\omega_c+\widehat{\omega_c}}{2}\right)}{\cos\left(\frac{\omega_c-\widehat{\omega_c}}{2}\right)}. \tag{4.177}$$

1. Determine the value of α to use in the transformation.
2. Determine the system transfer function for the high pass filter ($H_{HP}(z)$) with the filter coefficients represented as equations involving α as a parameter with $a(0) = 1$. Determine the filter coefficients $b(0), b(1), b(2), a(1),$ and $a(2)$ in terms of α. You do not need to simplify the transfer function beyond that point.

Problem 4.26. (IIR Filter Design)
The system transfer function for a continuous time system is given by

$$H(s) = \frac{s^2 + 5.1s + 2}{(s+2.5)(s+1.6)}. \tag{4.178}$$

Assume that the sampling interval $T = 0.25$ for this problem. Use the impulse invariant approach to determine the transfer function for an impulse invariant discrete time system $H_{IP}(z)$ based upon $H(s)$.

Problem 4.27. (IIR Filter Design)
A second order low pass filter with a normalized pass band edge frequency at $\omega_c = 0.25\pi$ has a system transfer function given by

$$H(z) = \frac{0.0996z^2 + 0.1131z + 0.0996}{z^2 - 1.1634z + 0.5566}. \tag{4.179}$$

We want to design a second order low pass filter $H_{LP}(z)$ with a normalized pass band edge frequency at $\widehat{\omega_c} = 0.38\pi$ using a low pass to low pass spectral transformation

$$z^{-1} \Rightarrow \frac{z^{-1}-\alpha}{1-\alpha z^{-1}},$$

$$\alpha = \frac{\sin[(\omega_p - \widehat{\omega}_p)/2]}{\sin[(\omega_p + \widehat{\omega}_p)/2]} \tag{4.180}$$

where $\widehat{\omega}_p$ is the desired pass band edge frequency for the new filter and ω_p is the pass band edge for the original low pass filter.

1. Determine the value of α to use in the transformation.
2. Write the system transfer function $H_{LP}(z)$ for the resulting low pass filter using the transformation with the value of α inserted. You do not need to simplify it to determine the numeric value of the filter coefficients.

Problem 4.28. (IIR Filter Design)
The continuous time transfer function for the Butterworth low pass filter with cutoff frequency, $\Omega_c = 1.0$ and with attenuation at the cutoff frequency equal to 3 dB is

given by

$$H(s) = \frac{1}{s^3 + 2s^2 + 2s + 1}. \tag{4.181}$$

1. Use the bilinear transformation

$$s \Rightarrow \frac{2(z-1)}{T(z+1)} \tag{4.182}$$

along with prewarping

$$\Omega_c \Rightarrow \frac{2}{T} \tan\left(\frac{\omega}{2}\right) \tag{4.183}$$

and the appropriate frequency transformations to design a low pass digital filter with normalized cutoff frequency $\omega_c = 0.712\pi$ with attenuation at the cutoff frequency equal to 3 dB.

2. Use the Matlab function *freqz* to verify that your digital filter has the correct magnitude response.

Problem 4.29. (Spectral Transformation Design)
A second order low pass IIR digital filter with a pass band ripple of 1.0 dB, a stop band ripple of 45 dB, and a cutoff frequency at $\omega_c = 0.510\pi$ has a system transfer function given by

$$H(z) = \frac{0.3191 + 0.6313z^{-1} + 0.3191z^{-2}}{1.0 + 0.1073z^{-1} + 0.3170z^{-2}}. \tag{4.184}$$

Design a bandpass filter $H_{BP}(z)$ with a center frequency of $\omega_s = 0.42\pi$ and a bandwidth of 0.25π by using a low pass to band pass spectral transformation. Use Matlab function *ellip* with appropriate inputs to determine if the filter you designed is essentially equal to the corresponding filter designed using *ellip*.

Problem 4.30. (Spectral Transformation Design)
A third order low pass filter with a 1 dB attenuation at the normalized cutoff frequency $\omega_1 = 0.247\pi$ has a system transfer function given by

$$H(z) = \frac{0.0204 + 0.0612z^{-1} + 0.0612z^{-2} + 0.0204z^{-3}}{1.0000 - 1.8831z^{-1} + 1.5142z^{-2} - 0.4679z^{-3}}. \tag{4.185}$$

Design a third order high pass filter $H_{HP}(z)$ with a 1 dB attenuation cutoff frequency of $\omega_2 = 0.575\pi$ using a low pass to high pass spectral transformation. Compare your filter coefficients with the corresponding filter designed using the Matlab *cheby1* function.

Problem 4.31. (Impulse Invariant Design)
The system transfer function for a continuous time system is given by

$$H(s) = \frac{2s^2 + 11s + 21.5}{s^3 + 6.5s^2 + 12.5s + 7}. \tag{4.186}$$

1. Use the impulse invariant approach with a sampling interval of $T = 0.25$ to determine the transfer function for an impulse invariant discrete time system $H_{IP}(z)$ to simulate $H(s)$.
2. Plot and compare the impulse responses of the continuous time system $H(s)$ and the impulse invariant system $H_{IP}(z)$.

Problem 4.32. (Bilinear Transformation Design)
The system transfer function for a continuous time system is given by

$$H(s) = \frac{2s^2 + 11s + 21.5}{s^3 + 6.5s^2 + 12.5s + 7}. \tag{4.187}$$

1. Use the bilinear transformation

$$s \Rightarrow \frac{2(z-1)}{T(z+1)} \tag{4.188}$$

with a sampling interval of $T = 0.25$ to determine the transfer function for a discrete time system $H_{BT}(z)$ to simulate $H(s)$.
2. Plot and compare the impulse responses of the continuous time system $H(s)$ and the bilinear transformation system $H_{BT}(z)$.

Problem 4.33. (IIR Filter Design)
This problem involves the design of an IIR digital filter with frequency response specifications given by

- Pass band cutoff frequency = 4.97 kHz
- Stop band frequency = 5.86 kHz
- Pass band ripple = 1.0 dB
- Stop band attenuation = 45 dB
- Sampling frequency = 16 kHz

Determine the order of the required digital filter and the critical normalized digital frequencies required to design

1. A Butterworth type digital filter using the Matlab *butter* function
2. A Chebyshev type 1 digital filter using the Matlab *cheby1* function
3. A Chebyshev type 2 digital filter using the Matlab *cheby2* function
4. An Elliptical type digital filter using the Matlab *ellip* function

Show all calculation or provide the Matlab script used to obtain your answers.

Problem 4.34. (IIR Filter Design)
The continuous time transfer function for the Butterworth low pass filter with cutoff frequency $\Omega_c = 1.0$ and with attenuation at the cutoff frequency equal to 3 dB is given by

$$H(s) = \frac{1}{s^3 + 2s^2 + 2s + 1}. \tag{4.189}$$

1. Use the bilinear transformation with prewarping and the appropriate frequency transformations to design a high pass digital filter with normalized cutoff frequency $\omega_c = 0.419\pi$ with attenuation at the cutoff frequency equal to 3 dB.
2. Use the Matlab function *butter* with appropriate inputs to determine if your filter coefficients computed using this function are the same as the coefficients you computed.
3. Use the Matlab function *freqz* to verify that your digital filter has the correct magnitude response.

Problem 4.35. (IIR Filter Design)
A second order type 2 Chebyshev low pass IIR digital filter with a back stop frequency edge of $\omega_s = 0.401\pi$ and a back stop ripple of -20 dB has a system transfer function given by

$$H(z) = \frac{0.1463 + 0.1710z^{-1} + 0.1463z^{-2}}{1.0000 - 0.8489z^{-1} + 0.3124z^{-2}}. \tag{4.190}$$

1. Design a back stop filter $H_{BS}(z)$ with a center frequency of $\omega_s = 0.659\pi$ and a bandwidth of 0.25π by using a low pass to band stop spectral transformation.
2. Use the Matlab function *cheby2* with appropriate inputs to determine if the filter you designed is essentially equal to the corresponding filter designed using *cheby2*.
3. Use the Matlab function *freqz* to verify that your digital filter has the correct magnitude response.

Problem 4.36. (IIR Filter Design)
A third order low pass filter with a 1 dB attenuation at the normalized cutoff frequency $\omega_1 = 0.617\pi$ and an attenuation of -30 dB in the stop band, has a system transfer function given by

$$H(z) = \frac{0.2898 + 0.7419z^{-1} + 0.7419z^{-2} + 0.2898z^{-3}}{1.0 + 0.5141z^{-1} + 0.6186z^{-2} - 0.0694z^{-3}}. \tag{4.191}$$

1. Design a third order low pass filter $H_{LP}(z)$ with a 1 dB attenuation cutoff frequency of $\omega_2 = 0.384\pi$ using a low pass to low pass spectral transformation.
2. Use the Matlab function *ellip* with appropriate inputs to determine if the filter you designed is essentially equal to the corresponding filter designed using *ellip*.
3. Use the Matlab function *freqz* to verify that your digital filter has the correct magnitude response.

Problem 4.37. (IIR Filter Design)
The system transfer function for a continuous time system is given by

$$H(s) = \frac{20(s - 1)}{(s + 4)(s + 3)}. \tag{4.192}$$

1. Use the impulse invariant approach with a sampling interval of $T = 0.05$ to determine the transfer function for an impulse invariant discrete time system $H_{IP}(z)$ using $H(s)$.
2. Use the bilinear transformation

$$s \Rightarrow \frac{2(z-1)}{T(z+1)} \qquad (4.193)$$

with $T = 0.05$ to determine the system transfer function for a discrete time system $H_{BT}(z)$ using $H(s)$.
3. Use Matlab to plot 100 samples of the frequency response for normalized radial frequencies in the interval

$$0.0 \leq \omega \leq 0.5\pi \qquad (4.194)$$

(a) For the original continuous system $H(\omega)$,
(b) For the impulse invariant system $H_{IP}(z)$, and
(c) For the bilinear transformation system $H_{BT}(z)$
on the same figure using different line types. Note that this is required only for given frequency range rather than for the entire frequency response.
4. Summarize your thoughts on which design procedure gave the best results and give reasons for your thoughts.

Problem 4.38. (IIR Filter Design)
Design an IIR digital filter with frequency response specifications given by

- Stop band edge frequency = 5.25 kHz
- Pass band edge frequency = 7.35 kHz
- Pass band ripple = ±1.5 dB
- Stop band attenuation = 45 dB
- Sampling frequency = 32 kHz

Determine the order of the required digital filter and the critical normalized digital frequencies required to design

1. A Butterworth type digital filter using the Matlab *butter* function
2. A Chebyshev type 1 digital filter using the Matlab *cheby1* function
3. A Chebyshev type 2 digital filter using the Matlab *cheby2* function
4. An Elliptical type digital filter using the Matlab *ellip* function
5. Write a Matlab script to design filters using each of the four Matlab functions using appropriate input parameters including the order you determined from Problems 5a through 5d
6. Use the Matlab function *freqz* to verify that your designs meet the desired specifications

Show all calculation or provide the Matlab code used to obtain your answers.

Problem 4.39. (IIR Filter Design)

The continuous time transfer function for the Butterworth low pass filter with cutoff frequency $\Omega_c = 1.0$ and with attenuation at the cutoff frequency equal to 3 dB is given by

$$H(s) = \frac{1}{s^3 + 2s^2 + 2s + 1}. \tag{4.195}$$

Use the bilinear transformation

$$s \Rightarrow \frac{2(z-1)}{T(z+1)} \tag{4.196}$$

along with prewarping

$$\Omega_c \Rightarrow \frac{2}{T} \tan\left(\frac{\omega}{2}\right) \tag{4.197}$$

and the appropriate frequency transformations to design a low pass digital filter with normalized cutoff frequency $\omega_c = 0.712\pi$ with attenuation at the cutoff frequency equal to 3 dB.

4.16 MATLAB PROBLEMS FOR CHAPTER 4

Problem 4.40. (Design using Windows)
Use the windows design method to design a linear phase FIR digital filter with frequency response approximately equal to the ideal frequency response given by

$$H_d(\omega) = \begin{cases} 0 & for \quad |\omega| \le \frac{\pi}{3} \\ 1 & for \quad \frac{\pi}{3} < |\omega| < \frac{2\pi}{3} \\ 0 & for \quad \frac{2\pi}{3} \le |\omega| \le \pi \end{cases} \tag{4.198}$$

1. Determine the coefficients for linear phase filter with 39 filter coefficients based on the use of the window method with a Hamming window. You are required to
   (a) Develop an equation for the samples of the impulse response for $H_d(\omega)$,
   (b) Develop an equation for the appropriate Hamming window to be used to compute the 39 filter coefficients.
   (c) Multiply the two equations and make any adjustment in the size of the coefficients to obtain the final filter coefficients.
2. Determine and plot the magnitude response and the phase response for the filter you designed.
3. Write a Matlab script using the Matlab *fir1* function to design a digital filter to meet the specifications given above.
4. Determine and plot the magnitude response and the phase response for the filter you designed using the Matlab *fir1* function.
5. Compare your coefficients to the coefficients obtained with the Matlab *fir1* function.

Problem 4.41. (Frequency Sampling Method)
Design a linear phase FIR filter of length $N = 37$ that has a symmetric unit sample response and a frequency response that satisfies the condition

$$|H(\omega)|_{\omega = \frac{2\pi k}{37}} = \begin{cases} 0 & for \quad 0 \le k \le 11 \\ 0.71 & for \quad k = 12 \\ 1 & for \quad 13 \le k \le 23 \\ 0.71 & for \quad k = 24 \\ 0 & for \quad 25 \le k \le 36 \end{cases} \quad (4.199)$$

1. Determine the 37 filter coefficients.
2. Plot the magnitude response and the phase response for your filter.
3. Determine if the magnitude response of your filter matches the desired magnitude response at the specified frequencies.

Problem 4.42. (Parks–McClellan Optimal Equiripple Design)
A FIR digital filter is desired to meet the following specifications:

- Sampling frequency – 4096 Hz
- Pass band – 0 to 925 Hz
- Stop band – 1150 to 2048 Hz
- Pass band ripple – $\delta_p = 0.1088$
- Stop band ripple – $\delta_s = 0.0175$
- The number of coefficients should not exceed 31.

1. Determine appropriate parameters for the filter using the *firpmord* Matlab routine and design the desired filter using the Matlab *firpm* FIR filter design routine. Use appropriate weights to ensure that the filter meets the desired specifications.
2. Determine and plot the magnitude response and the phase response for the filter you designed.
3. Make a table containing the following entries for the filter:
   (a) Minimum number of filter coefficients to meet the filter specifications
   (b) Maximum ripple in the stop band
   (c) Actual pass band edge frequency (where the gain last reaches $1 - \delta_p$)
   (d) Actual stop band edge frequency (where the gain first reaches δ_s).

Problem 4.43. (Least Squares Optimal Design)
A FIR digital filter is desired to meet the following specifications:

- Sampling frequency – 4096 Hz
- Pass band – 0 to 925 Hz
- Stop band – 1150 to 2048 Hz
- Pass band ripple – $\delta_p = 0.1088$
- Stop band ripple – $\delta_s = 0.0175$
- The number of coefficients should not exceed 31.

1. Determine appropriate parameters and design the desired filter using the Matlab *firls* FIR filter design routine. Use appropriate weights to ensure that the filter meets the desired specifications.
2. Determine and plot the magnitude and phase responses of the filter.
3. Make a table containing the following entries for the filter:
   (a) Minimum number of filter coefficients to meet the filter specifications
   (b) Maximum ripple in the stop band
   (c) Actual pass band edge frequency (where the gain last reaches $1 - \delta_p$)
   (d) Actual stop band edge frequency (where the gain first reaches δ_s).

Problem 4.44. (Least Squares Optimal Design)
A FIR digital filter is desired to meet the following specifications:

- Order = 32;
- Sampling frequency = 8192 Hz
- Pass band 1 = 0 to 935 Hz
- Stop band = 1350 to 2215 Hz
- Pass band 2 = 2630 to 4096 Hz
- Pass band ripple = ±2 dB
- Stop band ripple = −35 dB

Determine appropriate parameters to design the desired filter using the Matlab *firls* FIR filter design function. Use appropriate weights to ensure that the filter meets the desired specifications.

Problem 4.45. This problem involves the design of an IIR digital filter with frequency response specifications given by

- Pass band cutoff frequency = 4.97 kHz.
- Stop band frequency = 5.86 kHz.
- Pass band ripple = 1.0 dB.
- Stop band attenuation = 45 dB.
- Sampling frequency = 16 kHz.

Determine the order of the required digital filter and the critical normalized digital frequencies required to design

1. A Butterworth type digital filter using the Matlab *butter* function.
2. A Chebyshev type 1 digital filter using the Matlab *cheby1* function.
3. A Chebyshev type 2 digital filter using the Matlab *cheby2* function.
4. An Elliptical type digital filter using the Matlab *ellip* function.

Show all calculation or provide the Matlab code used to obtain your answers.

Problem 4.46. (Design using Windows)

Design a linear phase FIR digital filter with frequency response approximately equal to the ideal frequency response given by

$$H_d(\omega) = \begin{cases} 1 & \text{for} \quad |\omega| \le \frac{\pi}{4} \\ 0 & \text{for} \quad \frac{\pi}{4} < |\omega| < \frac{3\pi}{4} \\ 1 & \text{for} \quad \frac{3\pi}{4} \le |\omega| \le \pi \end{cases} \qquad (4.200)$$

1. Determine the coefficients for linear phase filter with 35 filter coefficients based on the use of the window method with a Hanning window.
2. Determine and plot the magnitude response and the phase response for the filter you designed.
3. Repeat parts (a) and (b) using a Blackman window.
4. Compare the filters designed to the ideal filter and discuss the relative merits of each in terms of the width of the transition region and the suppression in the stop bands.

Problem 4.47. (Frequency Sampling Method)
Design a linear phase FIR filter of length $N = 35$ that has a symmetric unit sample response and a frequency response that satisfies the condition

$$|H(\omega)|_{\omega = \frac{2\pi k}{25}} = \begin{cases} 1 & \text{for} \quad 0 \le k \le 4 \\ 0.7 & \text{for} \quad k = 5 \\ 0 & \text{for} \quad 6 \le k \le 11 \\ 0.7 & \text{for} \quad k = 12 \\ 1 & \text{for} \quad 13 \le k \le 17 \end{cases} \qquad (4.201)$$

1. Determine the 35 filter coefficients.
2. Plot the magnitude response and the phase response for your filter.
3. Determine if the magnitude response of your filter matches the desired magnitude response at the specified frequencies.

Problem 4.48. (Parks–McClellan Optimal Equiripple Design)
A FIR digital filter is desired to meet the following specifications:

- Sampling frequency – 8192 Hz
- Stop band 1 – 0 to 100 Hz
- Pass band – 400 to 2600 Hz
- Stop band 2 – 3000 to 4096 Hz
- Pass band ripple – $\delta_p = 0.1088$
- Stop band ripple – $\delta_s = 0.0175$
- The number of coefficients should not exceed 35.

1. Determine appropriate parameters for the filter using the *firpmord* Matlab routine and design the desired filter using the Matlab *firpm* FIR filter design routine. Use appropriate weights to ensure that the filter meets the desired specifications.

2. Determine and plot the magnitude response and the phase response for the filter you designed.
3. Make a table containing the following entries for the filter:
   (a) Minimum number of filter coefficients to meet the filter specifications
   (b) Maximum ripple in the stop band
   (c) Actual pass band edge frequency (where the gain last reaches $1 - \delta_p$)
   (d) Actual stop band edge frequency (where the gain first reaches δ_s).

Problem 4.49. (Least Squares Optimal Design)
Use the filter design parameters given for Problem 3 for Problem 4.

1. Determine appropriate parameters and design the desired filter using the Matlab *firls* FIR filter design routine. Use appropriate weights to ensure that the filter meets the desired specifications.
2. Determine and plot the magnitude and phase responses of the filter.
3. Make a table containing the following entries for the filter:
   (a) Minimum number of filter coefficients to meet the filter specifications
   (b) Maximum ripple in the stop band
   (c) Actual pass band edge frequency (where the gain last reaches $1 - \delta_p$)
   (d) Actual stop band edge frequency (where the gain first reaches δ_s).

Problem 4.50. A FIR digital filter is desired to meet the following specifications:

- Sampling frequency – 8192 Hz
- Pass band – 0 to 2800 Hz
- stop band – 3300 to 4096 Hz
- Pass band ripple – $\delta_p = 0.1088$
- Stop band ripple – $\delta_s = 0.0175$

Determine which of the following techniques can best meet the requirement using the least number of coefficients.

1. Window based design method,
2. Frequency sampling design method,
3. Parks–McClellan optimal equiripple design,
4. Least Squares Optimal Design

Determine the design method to use, the minimum number of coefficients to meet the required specifications and the actual filter coefficients using that method. Finally, verify that your design meets the requirements by evaluation of the magnitude of the frequency response at the critical frequencies for your filter design.

Implementation of Discrete Time Systems

This chapter focuses on the realization of discrete time systems which can be done either in software, in hardware, or partially in software and partially in software. The simplest structures for the realization of discrete time systems are called direct form realizations. However, these simple realizations often do not provide good performance because of limitations due to the use of quantization of the system parameters and errors due to the use of finite precision arithmetic for the computations. There are other structures that offer advantages regarding quantization effects, software or hardware efficiency, or computational efficiency. The cascade, parallel and lattice structures are of particular importance. Therefore, it is of interest to consider the alternative forms for the realization of digital filters and the impact of using finite precision arithmetic for the implementation. We consider several different realizations for discrete time systems in this chapter. We cover the effects of the use of finite precision arithmetic in Chap. 6.

5.1 STRUCTURES FOR THE REALIZATION OF FIR SYSTEMS

The convolution of two discrete time sequences can be represented as an FIR digital filter with one of the sequences serving as the input and the other one serving as the filter coefficients,

$$y(n) = \sum_{k=0}^{L} h(k)x(n-k). \tag{5.1}$$

This is equivalent to a difference equation involving one sequence $h(k)$ which can be considered to be the sequence of FIR filter coefficients and a second sequence $x(n)$ which can be considered to be the input sequence. Thus, we have

$$y(n) = \sum_{k=0}^{L} b(k)x(n-k) \tag{5.2}$$

Digital Signal Processing. DOI: 10.1016/B978-0-12-804547-3.00005-X

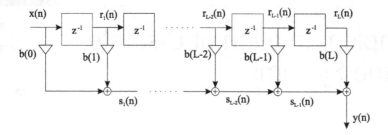

FIGURE 5.1

Block diagram for the direct realization of an FIR discrete time system

where $b(k) = h(k)$. The equivalent system transfer function has only a numerator polynomial,

$$H(z) = \sum_{k=0}^{L} b(k)z^{-k}. \tag{5.3}$$

Furthermore, the unit sample impulse response for the system is formed from the shifted coefficients of the difference equation:

$$h(n) = \sum_{k=0}^{L} b(k)\delta(n-k) = \begin{cases} b(n) & 0 \le n \le L, \\ 0 & \text{otherwise.} \end{cases} \tag{5.4}$$

5.1.1 DIRECT FORM STRUCTURES

Consider the system transfer function for the FIR discrete time system as given by

$$H(z) = \sum_{k=0}^{L} b(k)z^{-k}. \tag{5.5}$$

This FIR discrete time system can be represented by the block diagram shown in Fig. 5.1. The corresponding equations to implement the system are given by

$$
\begin{aligned}
s_k(n) &= b(k)r_k(n) + s_{k-1}(n) \ \forall \ 0 \le k \le L, \\
r_k(n) &= r_{k-1}(n-1) \ \forall \ 0 \le k \le L, \\
s_{-1}(n) &= 0.0, \\
r_0(n) &= x(n), \\
y(n) &= s_L(n).
\end{aligned} \tag{5.6}
$$

This iterative set of equations is a convenient form to use for the software implementation of an FIR filter because the equations can be implemented in a *for* loop.

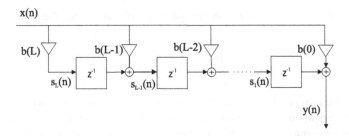

FIGURE 5.2

Block diagram for the direct implementation of an FIR digital filter

An alternate, transpose direct form for realizing the FIR filter, can be obtained by writing $H(z)$ in the form

$$H(z) = \left[\cdots \left[b(L)X(z)z^{-1} + b(L-1)X(z) \right] z^{-1} + \cdots + b(0)X(z) \right]. \quad (5.7)$$

For example, this form for a third order filter can be written as

$$H(z) = [[[[b(3)z^{-1} + b(2)]z^{-1}] + b(1)]z^{-1} + b(0)]. \quad (5.8)$$

Fig. 5.2 gives the corresponding block diagram representation for a 1-D FIR system with order equal to L. A set of iterative equations can be developed to implement the structure in Fig. 5.2 as follows:

$$
\begin{aligned}
s_L(n) &= b(L)x(n), \\
s_k(n) &= b(k)x(n) + s_{k+1}(n-1) \ \forall \ (L-1) \geq k \geq 1, \\
y(n) &= b(0)x(n) + s_1(n-1).
\end{aligned}
\quad (5.9)
$$

This alternative iterative set of equations is also a convenient form to use for the software implementation of an FIR filter because the equations can be implemented in a *for* loop.

5.1.2 CASCADE FORM STRUCTURES

It is simple to factor the system function into second order FIR system transfer functions so that

$$H(z) = \prod_{m=1}^{M} H_m(z) \quad (5.10)$$

where

$$H_m(z) = b_{m0} + b_{m1}z^{-1} + b_{m2}z^{-2} \ \forall 1 \leq m \leq M. \quad (5.11)$$

The original filter coefficient b_0 can be equally distributed among the M filter sections such that

$$b_0 = \prod_{m=1}^{M} b_{m0} = b_{10}b_{20} \cdots b_{M0} \qquad (5.12)$$

or it may be assigned to a single filter section. Considerations regarding the scaling of the coefficients may determine the distribution of the original b_0. For example, it may be implemented as a scaling adjustment after all of the other computations have been completed. The values of all of the b_{m0} coefficients would be equal to 1 in this special case.

The system can be implemented using the following set of difference equations:

$$
\begin{aligned}
g_0(n) &= x(n), \\
g_m(n) &= b_{m0}g_{m-1}(n) + b_{m1}g_{m-1}(n-1) \\
&\quad + b_{m2}g_{m-1}(n-2) \ \forall \ 1 \le m \le M, \\
y(n) &= g_M(n).
\end{aligned} \qquad (5.13)
$$

The cascade structure also provides a convenient form for the implementation of the FIR filter using software because the above equations can conveniently be implemented using a *for* loop. Implementing the filter as a cascade of second order or possibly fourth order filters can provide a reduction of the roundoff errors due to the quantization of the filter coefficients. This will be discussed in more detail later in Chap. 6.

5.1.3 LINEAR PHASE FIR FILTER

Many of the FIR filters of interest are designed with linear phase. The filter coefficients are symmetric for a linear phase FIR filter such that

$$
\begin{aligned}
H(z) &= \sum_{k=0}^{L} b(k)z^{-k}, \\
b(L-k) &= b(k) \ \forall \ 0 \le k \le \left\lfloor \frac{L}{2} \right\rfloor.
\end{aligned} \qquad (5.14)
$$

Consider the linear phase FIR filter with system transfer function given by

$$H(z) = \sum_{n=0}^{L} b(n)z^{-n} \qquad (5.15)$$

where $b(k) = b(L-k)$. If L is even, we can write

$$H(z) = \sum_{n=0}^{M-1} b(n)\left[z^{-n} + z^{-(L-n)}\right] + b(M)z^{-M} \qquad (5.16)$$

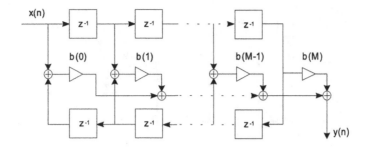

FIGURE 5.3

Block diagram for a linear phase FIR digital filter

where $M = L/2$. If L is odd, we can write

$$H(z) = \sum_{n=0}^{M-1} b(n) \left[z^{-n} + z^{-(L-n)} \right]. \qquad (5.17)$$

Fig. 5.3 gives a block diagram of the linear phase FIR filter when L is even where we have taken advantage of the symmetry in $H(z)$ to reduce the number of multiplications as indicated in Eq. (5.16). The corresponding difference equation is

$$
\begin{aligned}
y(n) \;=\; & b(0)\,[x(n) + x(n - L)] \\
& + b(1)\,[x(n - 1) + x(n - L + 1)] \\
& + \cdots + b(M - 1)\,[x(n - M + 1) + x(n - M - 1)] \\
& + b(M)x(M).
\end{aligned}
\qquad (5.18)
$$

EXAMPLE 5.1
(FIR Filter Design Example One)

The following Matlab script can be used to design an eight order, FIR, high pass filter with relative cutoff frequency equal to 0.34 using the *fir1* Matlab function.

Matlab Script 5.1.

```
order = 8;
wc = 0.34;
type = 'high'
b = fir1(order, wc, type);
```

End of the Script

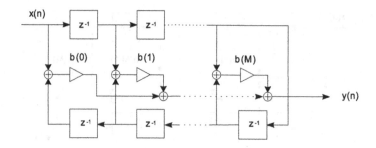

FIGURE 5.4

Block diagram for a linear phase FIR digital filter

The resulting coefficients are

```
b =
  Columns 1 through 6
    0.0057     0.0014    -0.0721    -0.2399     0.6559    -0.2399

  Columns 7 through 9
   -0.0721     0.0014     0.0057
```

The corresponding difference equation is

$$
\begin{aligned}
y(n) \;=\;\; & 0.0057x(n) + 0.0014x(n-1) - 0.0721x(n-2) \\
& - 0.2399x(n-3) + 0.6559x(n-4) - 0.2399x(n-5) \\
& - 0.0721x(n-6) + 0.0014x(n-7) \\
& + 0.0057x(n-8).
\end{aligned}
\tag{5.19}
$$

This can be rewritten as

$$
\begin{aligned}
y(n) \;=\;\; & 0.0057\,[x(n) + x(n-8)] + 0.0014\,[x(n-1) + x(n-7)] \\
& - 0.0721\,[x(n-2) + x(n-6)] - 0.2399\,[x(n-3) + x(n-5)] \\
& + 0.6559x(n-4).
\end{aligned}
\tag{5.20}
$$

Note that the original difference equation required 9 multiplications and 8 additions or subtractions. The modified difference equation required 5 multiplications and 8 additions or subtractions. Thus, 4 multiplications were spared with the modified computational structure.

End of the Example

The case when L is odd will now be considered. Fig. 5.4 gives a block diagram of the linear phase FIR filter when L is odd where we have taken advantage of the symmetry in $H(z)$ to reduce the number of multiplications as indicated in Eq. (5.17). The corresponding difference equation is

$$
\begin{aligned}
y(n) \;=\; & b(0)\,[x(n)+x(n-L)] \\
& + b(1)\,[x(n-1)+x(n-L+1)] \\
& + \cdots + b(M)\,[x(n-M)+x(n-M-1)].
\end{aligned}
\qquad (5.21)
$$

EXAMPLE 5.2
(FIR Filter Design Example Two)

The following Matlab script can be used to design a seventh order low pass filter with relative cutoff frequency equal to 0.673 using the *fir1* Matlab function.

Matlab Script 5.2.

```
order = 7;
wc = 0.673;
b = fir1(order, wc);
```

End of the Script

The resulting coefficients are

```
b =

  Columns 1 through 6
    0.0065    -0.0268    -0.0040    0.5244    0.5244    -0.0040

  Columns 7 through 8
    -0.0268    0.0065
```

The corresponding difference equation is

$$
\begin{aligned}
y(n) \;=\; & 0.0065x(n) - 0.0268x(n-1) - 0.0040x(n-2) + 0.5244x(n-3) \\
& + 0.5244x(n-4) - 0.0040x(n-5) - 0.0268x(n-6) \\
& + 0.0065x(n-7).
\end{aligned}
\qquad (5.22)
$$

This can be rewritten as

$$
\begin{aligned}
y(n) \;=\; & 0.0065[x(n)+x(n-7)] - 0.0268[x(n-1)+x(n-6)] \\
& - 0.0040[x(n-2)+x(n-5)] \\
& + 0.5244[x(n-3)+x(n-4)].
\end{aligned}
\qquad (5.23)
$$

Note that the original difference equation required 8 multiplications and 7 additions or subtractions. The modified difference equation required 4 multiplications and 7 additions or subtractions. Thus, 4 multiplications were spared with the modified computational structure.
End of the Example

5.1.4 POLYPHASE FIR FILTER REALIZATION

The polyphase realization is a parallel decomposition of a FIR digital filter based on the decomposition of the filter in multiple powers of z. We will discuss the polyphase FIR realization in this section. The polyphase FIR filter realization can provide advantages in computational efficiency when used for decimation or interpolation of discrete time signals. We will discuss these advantages later in Chap. 7.

Consider the general system transfer function for an FIR digital filter as given by

$$H(z) = \sum_{k=0}^{L} b(k) z^{-k}. \tag{5.24}$$

We can partition $H(z)$ such that

$$
\begin{aligned}
H(z) &= \sum_{m=1}^{M} z^{m-1} \sum_{k=0}^{K} b(n) z^{n}, \\
n &= Mk + (m-1), \\
K &= \left\lceil \frac{L}{M} \right\rceil
\end{aligned}
\tag{5.25}
$$

where some of the $b(n)$ will be equal to zero if $L \neq KM$.

This can be illustrated by an example using an eight order FIR digital filter (9 coefficients) [8] with $M = 2$.

EXAMPLE 5.3
(Polyphase FIR Polyphase Filter Example One)

The order of the filter, with $L = 8$, is 9 for this example. Therefore,

$$H(z) = \sum_{k=0}^{8} b(k) z^{-k}. \tag{5.26}$$

It follows that

$$K = \left\lceil \frac{8}{2} \right\rceil = 4. \tag{5.27}$$

Thus, $H(z)$ can be written in the form

$$
\begin{aligned}
H(z) &= \sum_{m=1}^{2} z^{m-1} \sum_{k=0}^{4} b(n) z^{n}, \\
n &= 2k + (m-1).
\end{aligned}
\tag{5.28}
$$

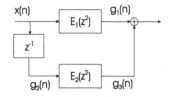

FIGURE 5.5

Polyphase FIR implementation using 2 filters

It follows that $H(z)$ can be written as

$$
\begin{aligned}
H(z) &= \left[b(0) + b(2)z^{-2} + b(4)z^{-4} + b(6)z^{-6} + b(8)z^{-8} \right] \\
&\quad + z^{-1} \left[b(1) + b(3)z^{-2} + b(5)z^{-4} + b(7)z^{-6} \right].
\end{aligned} \tag{5.29}
$$

The coefficient $b(9) = 0$ for this case. Two digital filters can be designed using the coefficients of $H(z)$ as follows:

$$
\begin{aligned}
E_0(z) &= b(0) + b(2)z^{-1} + b(4)z^{-2} + b(6)z^{-3} + b(8)z^{-4}, \\
E_1(z) &= b(1) + b(3)z^{-1} + b(5)z^{-2} + b(7)z^{-3}.
\end{aligned} \tag{5.30}
$$

The transfer functions for these two filters can be used to write $H(z)$ in the form

$$
H(z) = E_0(z^2) + z^{-1}E_1(z^2). \tag{5.31}
$$

Fig. 5.5 gives a system block diagram of the resulting implementation using these two filters. The corresponding difference equations are

$$
\begin{aligned}
g_1(n) &= b(0)x(n) + b(2)x(n-2) + b(4)x(n-4) \\
&\quad + b(6)x(n-6) + b(8)x(n-8), \\
g_2(n) &= x(n-1), \\
g_3(n) &= b(1)g_2(n) + b(3)g_2(n-2) + b(5)g_2(n-4) + b(7)g_2(n-6), \\
y(n) &= g_1(n) + g_3(n).
\end{aligned} \tag{5.32}
$$

End of the Example

The transfer function, $H(z)$, can be decomposed in a similar way for the case with $M = 3$.

EXAMPLE 5.4

(Polyphase FIR Polyphase Filter Example Two)

Example 5.4 considers the decomposition of a ninth order FIR filter into 3 polyphase FIR filters:

$$K = \left\lceil \frac{8}{3} \right\rceil = 3, \tag{5.33}$$

$$H(z) = \sum_{m=1}^{3} z^{m-1} \sum_{k=0}^{3} b(n)z^n,$$

$$n = 3k + (m-1). \tag{5.34}$$

Thus, $H(z)$ can be written in the form

$$\begin{aligned} H(z) = & \; b(0) + b(3)z^{-3} + b(6)z^{-6} \\ & + z^{-1}\left[b(1) + b(4)z^{-3} + b(7)z^{-6}\right] \\ & + z^{-2}\left[b(2) + b(5)z^{-3} + b(8)z^{-6}\right]. \end{aligned} \tag{5.35}$$

Again, filters $E_0(z)$, $E_1(z)$, and $E_2(z)$ can be defined such that

$$H(z) = E_0(z^3) + z^{-1}E_1(z^3) + z^{-2}E_2(z^3). \tag{5.36}$$

Fig. 5.6 gives a system block diagram of the resulting implementation using three filters. The corresponding difference equations are:

$$\begin{aligned} g_1(n) &= x(n-1), \\ g_2(n) &= g_1(n-1), \\ s_1(n) &= b(0)x(n) + b(3)x(n-3) + b(6)x(n-6), \\ s_2(n) &= b(1)g_1(n) + b(4)g_1(n-3) + b(7)g_1(n-6), \\ s_3(n) &= b(2)g_2(n) + b(5)g_2(n-3) + b(8)g_2(n-6), \\ y(n) &= s_1(n) + s_2(n) + s_3(n). \end{aligned} \tag{5.37}$$

End of the Example

5.2 FIR LATTICE IMPLEMENTATION

Lattice filters are used in many applications including digital speech processing and in the implementation of adaptive filters [4]. We will show how to implement FIR filters using a lattice structure in this section. Consider the general FIR filter with system transfer function given by

$$H(z) = \sum_{k=0}^{L} b(k)z^{-k}. \tag{5.38}$$

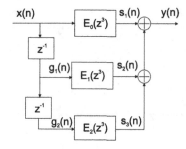

FIGURE 5.6

Polyphase FIR implementation using 3 filters

We can also write this as

$$H(z) = b(0)H_m(z) \qquad (5.39)$$

where

$$H_L(z) = 1.0 + \sum_{k=1}^{L} \beta(k)z^{-k} \qquad (5.40)$$

and

$$\beta(k) = \frac{b(k)}{b(0)}. \qquad (5.41)$$

The filter $H_L(z)$ can be considered to be related to prediction where the estimate of $\widehat{x(n)}$ is given by

$$\widehat{x(n)} = -\sum_{k=1}^{L} \beta(k)x(n-k). \qquad (5.42)$$

Then $y(n)$ can be considered to be the error obtained from subtracting the estimate of $x(n)$ from its current value:

$$y(n) = x(n) - \widehat{x(n)} = x(n) + \sum_{k=1}^{L} \beta(k)x(n-k). \qquad (5.43)$$

The block diagram for the FIR lattice section is given in Fig. 5.7. We can obtain a single-section FIR lattice filter by letting $g_0(n) = f_0(n)$ or by connecting both inputs to $x(n)$ as shown in Fig. 5.8. The corresponding difference equations for the single-section FIR lattice filter are then given by

$$\begin{aligned} f_1(n) &= y(n) = f_0(n) + K_1 g_0(n-1), \\ g_1(n) &= K_1 f_0(n) + g_0(n-1). \end{aligned} \qquad (5.44)$$

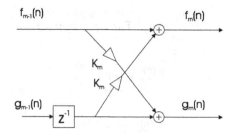

FIGURE 5.7

FIR lattice filter section

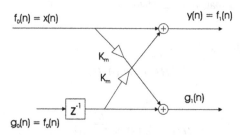

FIGURE 5.8

Block diagram for single section FIR lattice filter

Substituting $x(n)$ for $f_0(n)$ and $g_0(n)$, we obtain

$$
\begin{aligned}
f_1(n) &= y(n) = x(n) + K_1 x(n-1), \\
g_1(n) &= K_1 x(n) + x(n-1).
\end{aligned}
\tag{5.45}
$$

We see that the output $f_1(n)$ is the output for an FIR filter with system transfer function

$$
F_1(z) = \left[1.0 + K_1 z^{-1}\right] X(z).
\tag{5.46}
$$

We can equate the system in Eq. (5.46) to the one in Eq. (5.40) if we set $K_1 = b(1)$. Similarly, the output $g_1(n)$ is the output for an FIR filter with system transfer function

$$
G_1(z) = K_1 + z^{-1}.
\tag{5.47}
$$

We can obtain a second order lattice FIR filter by using a second section with the outputs from the first section used as the inputs to the second section as shown in Fig. 5.9. We can write difference equations for this second order section as follows.

$$
\begin{aligned}
f_2(n) &= y(n) = f_1(n) + K_2 g_1(n-1), \\
g_2(n) &= K_2 f_1(n) + g_1(n-1),
\end{aligned}
$$

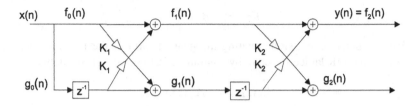

FIGURE 5.9

Second order FIR lattice filter

$$
\begin{aligned}
f_1(n) &= f_0(n) + K_1 g_0(n-1), \\
g_1(n) &= K_1 f_0(n) + g_0(n-1).
\end{aligned}
\tag{5.48}
$$

However, we know that the inputs are connected so that $g_0(n) = f_0(n) = x(n)$. If we substitute for $f_1(n)$ and $g_1(n)$ in the equation for $y(n)$ ($y(n) = f_2(n)$), we obtain

$$
y(n) = x(n) + K_1 x(n-1) + K_2 [K_1 x(n-1) + x(n-2)],
$$
$$
y(n) = x(n) + K_1 (1 + K_2) x(n-1) + K_2 x(n-2).
\tag{5.49}
$$

In a similar way, we can substitute for $f_1(n)$ and $g_1(n)$ in the equation for $g_2(n)$ to obtain

$$
g_2(n) = K_2 x(n) + K_1 (1 + K_2) x(n-1) + K_2 x(n-2).
\tag{5.50}
$$

If we take the Z Transform of both sides of the equation for $y(n)$, we obtain

$$
Y(z) = \left[1.0 + K_1 (1 + K_2) z^{-1} + K_2 z^{-2} \right] X(z),
\tag{5.51}
$$

or

$$
H_2(z) = 1.0 + K_1 (1 + K_2) z^{-1} + z^{-2}.
\tag{5.52}
$$

We can equate Eq. (5.52) with the general FIR system transfer function in Eq. (5.40) for $m = 2$ to solve for K_1 and K_2. We have

$$
H_2(z) = 1.0 + \beta(1) z^{-1} + \beta(2) z^{-2}.
\tag{5.53}
$$

Thus,

$$
\begin{aligned}
\beta(2) &= K_2, \\
\beta(1) &= K_1 (1 + K_2), \\
\beta(0) &= 1.0.
\end{aligned}
\tag{5.54}
$$

It follows that

$$
K_1 = \frac{\beta(1)}{1 + \beta(2)},
$$

$$K_2 = \beta(2). \tag{5.55}$$

We can derive expressions relating the general form of the FIR filter with the corresponding FIR lattice structure by continuing in this way [4]. We define

$$\begin{aligned} F_m(z) &= A_m(z)X(z), \\ G_m(z) &= B_m(z)X(z). \end{aligned} \tag{5.56}$$

Note that $A_m(z)$ and $B_m(z)$ have the same coefficients but in reverse order. Therefore, we have the relationship

$$B_m(z) = z^{-m}A_m(z^{-1}). \tag{5.57}$$

The input $x(n)$ is connected to both inputs for the first FIR lattice section so that

$$f_0(n) = g_0(n) = x(n). \tag{5.58}$$

It follows directly that

$$\begin{aligned} F_0(z) &= G_0(z), \\ A_0(z) &= B_0(z) = 1. \end{aligned} \tag{5.59}$$

We can use the computational structure for the general FIR lattice section m as given in Fig. 5.7 to obtain equations for $F_m(z)$ and $G_m(z)$ as follows:

$$\begin{aligned} F_m(z) &= F_{m-1}(z) + z^{-1}K_m G_{m-1}(z), \\ G_m(z) &= K_m F_{m-1}(z) + z^{-1}G_{m-1}(z). \end{aligned} \tag{5.60}$$

We can then use Eq. (5.56) to obtain

$$\begin{aligned} A_m(z) &= A_{m-1}(z) + z^{-1}K_m B_{m-1}(z), \\ B_m(z) &= K_m A_{m-1}(z) + z^{-1}B_{m-1}(z). \end{aligned} \tag{5.61}$$

Thus, a lattice section can be described by the following matrix equation:

$$\begin{bmatrix} A_m(z) \\ B_m(z) \end{bmatrix} = \begin{bmatrix} 1 & z^{-1}K_m \\ K_m & z^{-1} \end{bmatrix} \begin{bmatrix} A_{m-1}(z) \\ B_{m-1}(z) \end{bmatrix}. \tag{5.62}$$

It is simple to determine $A_L(z)$ from $F_L(z)$ for an Lth order FIR filter since

$$F_L(z) = H_L(z)X(z) = A_L(z)X(z). \tag{5.63}$$

It follows that K_L is the coefficient for z^{-L} in $H_L(z)$. We can then determine $B_L(z)$ from $A_L(z)$ as

$$B_L(z) = z^{-L}A_L(z^{-1}). \tag{5.64}$$

We can then find $A_{L-1}(z)$ using Eq. (5.61),

$$A_{L-1}(n) = \frac{A_L(z) - K_L B_L(z)}{1 - K_L^2}. \tag{5.65}$$

The lattice coefficient K_{L-1} is the coefficient of $z^{-(L-1)}$ in $A_{L-1}(z)$. We can continue in this way until we determine all of the lattice coefficients.

EXAMPLE 5.5
(FIR Lattice Filter Example One)

The system function for an FIR digital filter is given by

$$H(z) = 0.1671 + 0.1622z^{-1} - 0.2127z^{-2} - 0.0887z^{-3}. \tag{5.66}$$

We wish to implement $H(z)$ using a FIR lattice structure. We must first normalize $H(z)$ by dividing it by $b(0)$ to obtain $A_3(z)$:

$$A_3(z) = \frac{0.1671 + 0.1622z^{-1} - 0.2127z^{-2} - 0.0887z^{-3}}{0.1671}, \tag{5.67}$$

$$A_3(z) = 1.0 + 0.9707z^{-1} - 1.2729z^{-2} - 0.5308z^{-3}. \tag{5.68}$$

We then have K_3 as the coefficient for z^{-3},

$$K_3 = -0.5308. \tag{5.69}$$

We can obtain $B_3(z)$ as

$$B_3(z) = z^{-3}A_3(z^{-1}), \tag{5.70}$$

$$A_3(z^{-1}) = 1.0 + 0.9707z - 1.2729z^2 - 0.5308z^3, \tag{5.71}$$

$$B_3(z) = -0.5308 - 1.2729z^{-1} + 0.9707z^{-2} + z^{-3}. \tag{5.72}$$

Then, we have

$$A_2(z) = \frac{A_3(z) - K_3 B_3(z)}{1 - K_3^2}, \tag{5.73}$$

$$A_2(z) = 1.0 + 0.4107z^{-1} - 1.0549z^{-2}. \tag{5.74}$$

It follows from the use of Eq. (5.55) that

$$K_2 = -1.0549,$$
$$K_1 = \frac{0.4107}{1 - 1.0549} = -7.4862. \tag{5.75}$$

Thus,

$$\begin{aligned}
K_1 &= -7.4862, \\
K_2 &= -1.0549, \\
K_3 &= -0.5308.
\end{aligned} \tag{5.76}$$

The corresponding difference equations are given by:

$$\begin{aligned}
f_0(n) &= x(n), \\
g_0(n) &= f_0(n), \\
f_1(n) &= f_0(n) - 7.4862 g_0(n-1), \\
g_1(n) &= -7.4862 f_0(n) + g_0(n-1), \\
f_2(n) &= f_1(n) - 1.0549 g_1(n-1), \\
g_2(n) &= -1.0549 f_1(n) + g_1(n-1), \\
f_3(n) &= f_2(n) - 0.5308 g_2(n-1), \\
g_3(n) &= -0.5308 f_2(n) + g_2(n-1), \\
y(n) &= 0.1671 f_3(n).
\end{aligned} \tag{5.77}$$

End of the Example

Note that the denominator Eq. (5.65) has the term

$$1 - K_m^2. \tag{5.78}$$

We see that we have a problem and the algorithm fails if $K_m = 1$. This occurs for linear phase FIR filters with symmetric coefficients. One approach to the solution of this problem for linear phase filters is to partition $H_m(z)$ into two system transfer functions and implement them in cascade. One of the system transfer functions can include the roots outside and/or on the unit circle. The other one can include roots inside and/or on the unit circle. It may be necessary to form a third system transfer function including the roots on the unit circle if there are two or more zeros on the unit circle.

EXAMPLE 5.6
(FIR Lattice Filter Example Two)

The system transfer function for an FIR digital filter is given by

$$\begin{aligned}
H(z) = &-0.0078 + 0.0645z^{-1} + 0.4433z^{-2} + 0.4433z^{-3} \\
&+ 0.0645z^{-4} - 0.0078z^{-5}.
\end{aligned} \tag{5.79}$$

We wish to determine the K_m parameters for a lattice implementation of this FIR filter. We first normalize $H(z)$ by dividing it by $b(0)$ to obtain

$$H_5(z) = \frac{H(z)}{-0.0078},$$

$$H_5(z) = 1.0 - 8.2886z^{-1} - 57.0092z^{-2} - 57.0092z^{-3}$$
$$- 8.2886z^{-4} + z^{-5}. \tag{5.80}$$

It follows that $K_5 = 1.0$. The algorithm fails because $1 - K_5^2 = 0.0$. This FIR filter can be split into a second order filter and a third order filter. The roots of $H_5(z)$ are given by

$$q(1) = 13.0108,$$
$$q(2) = -3.5145,$$
$$q(3) = -1.0000,$$
$$q(4) = -0.2845,$$
$$q(5) = 0.0769. \tag{5.81}$$

We can form $S_1(z)$ using $q(1)$ through $q(2)$ and form $S_2(z)$ using $q(3)$ and $q(5)$. Using this approach, we have

$$H(z) = (-0.0078)S_1(z)S_2(z),$$
$$S_1(z) = 1.0000 - 9.4963z^{-1} - 45.7267z^{-2},$$
$$S_2(z) = 1.0 + 1.2077z^{-1} + 0.1858z^{-2} - 0.0219z^{-3}. \tag{5.82}$$

The parameter K_2 for $S_1(z)$ can be obtained as the coefficient for z^{-2} in $S_1(z)$. The other parameter for the second order filter, $S_1(z)$, can be obtained using the procedure used in Example 5.5. Thus, the K parameters for $S_1(z)$ are

$$K_{21} = 0.2123,$$
$$K_{21} = -45.7267. \tag{5.83}$$

The parameter K_3 for S_2 can be obtained as the coefficient for z^{-3} in $S_2(z)$. Then we can determine K_2 and K_1 using the procedure used in Example 5.5. Thus, the K parameters for $S_2(z)$ are

$$K_{12} = 1.0,$$
$$K_{22} = 0.2123,$$
$$K_{32} = -0.0219. \tag{5.84}$$

The corresponding difference equations are given by:

$$
\begin{aligned}
f_{01}(n) &= x(n), \\
g_{01}(n) &= f_{01}(n), \\
f_{11}(n) &= f_{01}(n) + 0.2123 g_{01}(n-1), \\
g_{11}(n) &= 0.2123 f_{01}(n) + g_{01}(n-1), \\
f_{21}(n) &= f_{11}(n) - 45.7267 g_{11}(n-1), \\
g_{21}(n) &= -45.7267 f_{11}(n) + g_{11}(n-1), \\
f_{02}(n) &= f_{21}(n), \\
g_{02}(n) &= f_{02}(n), \\
f_{12}(n) &= f_{02}(n) + g_{02}(n), \\
g_{12}(n) &= f_{02}(n) + g_{02}(n), \\
f_{22}(n) &= f_{12}(n) + 0.2123 g_{12}(n-1), \\
g_{22}(n) &= 0.2123 f_{12}(n) + g_{12}(n-1), \\
f_{32}(n) &= f_{22}(n) - 0.0219 g_{22}(n-1), \\
g_{32}(n) &= -0.0219 f_{22}(n) + g_{22}(n-1), \\
y(n) &= -0.0078 f_{32}(n). \quad\quad (5.85)
\end{aligned}
$$

Where $x(n)$ is the input and $y(n)$ is the output. Note that $g_{21}(n)$ and $g_{32}(n)$ are not used in subsequent equations for this structure and could have been left out of the set of difference equations.
End of the Example

5.3 CASCADE IMPLEMENTATION OF FIR FILTERS

A given system transfer function for a discrete time system

$$
H(z) = \frac{B(z)}{A(z)} = \frac{\displaystyle\sum_{k=0}^{L} b(k)}{1.0 + \displaystyle\sum_{k=1}^{L} a(k)} \quad\quad (5.86)
$$

can be represented as a product of first order or higher order system transfer functions such that

$$
H(z) = \prod_{m=1}^{M} H_m(z). \quad\quad (5.87)
$$

The $a(k) = 0.0$ for an FIR filter. The system transfer function coefficients for a linear, shift invariant system are real constants. We can factor $H(z)$ such that

$$H(z) = B(z) = b(0) \prod_{k01}^{L} (z - q_k).$$

(5.88)

The q_k are real or they occur in complex conjugate pairs. We can form each $H_m(z)$ using one or more real q_k or complex conjugate pairs of q_k so that the coefficients of all of the $H_m(z)$ are also real. A standard practice for partitioning $H(z)$ into second order sections is to:

1. Obtain the roots of the numerator polynomial $B(z)$ for the FIR digital filter such that

$$H(z) = B(z) = b(0) \prod_{m=1}^{M} (z - q_m).$$

(5.89)

2. Order the q_m in increasing (or decreasing) magnitude.
3. Combine the complex conjugate pairs in increasing (or decreasing) magnitude to form second order system transfer functions

$$H_k(z) = (z - q_m)(z - q_m^*) = z^2 - 2\Re(q_m)z + \left|q_m^2\right|.$$

(5.90)

4. Pair real q_m in increasing (or decreasing) magnitude pairs so that the q_m closest in magnitude are paired to form second order system transfer functions

$$H_k(z) = (z - q_m)(z - q_{m+1}) = z^2 - (q_m + q_{m+1})z + q_m q_{m+1}.$$

(5.91)

5. If the number of real roots is odd, then form a first order section using the q_m that is left after the above second order sections have been formed.

Pairing real roots closest in magnitude has the benefit of reducing rounding error when the coefficients are quantized.

The coefficient $b(0)$ can be used as a multiplying constant for $H_1(z)$ (or another $H_m(z)$) or it can be distributed among the second order sections. For example, we can determine

$$A = |b(0)|^{\frac{1}{M}}.$$

(5.92)

Then, we form

$$\widehat{H_m(z)} = AH_m(z).$$

(5.93)

It follows that

$$H(z) = \prod_{m=1}^{M} \widehat{H_m(z)} = A^M \prod_{m=1}^{M} H_m(z)$$

(5.94)

if $b(0)$ is positive and

$$H(z) = - \prod_{m=1}^{M} \widehat{H_m(z)} = A^M \prod_{m=1}^{M} H_m(z) \qquad (5.95)$$

if $b(0)$ is negative.

We can illustrate this concept using an example.

EXAMPLE 5.7
(Cascade FIR Filter)

Consider the FIR filter with system transfer function

$$
\begin{aligned}
H(z) \ = \ & -0.0049 + 0.0077z^{-1} + 0.0069z^{-2} - 0.1057z^{-3} \\
& + 0.5960z^{-4} + 0.5960z^{-5} - 0.1057z^{-6} + 0.0069z^{-7} \\
& + 0.0077z^{-8} - 0.0049z^{-9}.
\end{aligned}
\qquad (5.96)
$$

We can obtain the q_m which are the roots of $H(z)$ as

$$
\begin{aligned}
q_1 &= 3.5923, \\
q_2 &= 1.0395 + 3.2775j, \\
q_3 &= 1.0395 - 3.2775j, \\
q_4 &= -3.2602, \\
q_5 &= -1.0000, \\
q_6 &= -0.3067, \\
q_7 &= 0.0879 + 0.2772j, \\
q_8 &= 0.0879 - 0.2772j, \\
q_9 &= 0.2784.
\end{aligned}
\qquad (5.97)
$$

Note that the q_m have been arranged in decreasing magnitude order. Using the procedure described above, we can form

$$
\begin{aligned}
H_1(z) &= (z - q_2)(z - q_3) = z^2 - 2.0790z + 11.8225, \\
H_2(z) &= (z - q_7)(z - q_8) = z^2 - 0.1759z + 0.0846, \\
H_3(z) &= (z - q_1)(z - q_4) = z^2 - 0.3321z - 11.7114, \\
H_4(z) &= (z - q_6)(z - q_9) = z^2 + 0.0284z - 0.0854, \\
H_5(z) &= (z - q_5) = z + 1.0000,
\end{aligned}
\qquad (5.98)
$$

$$A = (0.0049)^{\frac{1}{5}} = 0.3454. \qquad (5.99)$$

The corresponding difference equations with the use of A are given by

$$
\begin{aligned}
s_1(n) &= -0.3454x(n) + 0.7182x(n-1) - 4.0838x(n-2), \\
s_2(n) &= 0.3454s_1(n) - 0.0607s_1(n-1) + 0.0292s_1(n-2), \\
s_3(n) &= 0.3454s_2(n) - 0.1147s_2(n-1) - 4.0454x_2(n-2), \\
s_4(n) &= 0.3454s_3(n) + 0.0098s_3(n-1) - 0.0295s_3(n-2), \\
y(n) &= s_5(n) = 0.3454s_4(n) + 0.3454s_4(n-1).
\end{aligned}
\tag{5.100}
$$

We multiplied $H_1(z)$ by -1 because $b(0)$ was negative.
End of the Example

5.4 STRUCTURES FOR THE REALIZATION OF IIR SYSTEMS

There are several different realizations for IIR digital filters. The system transfer function for IIR digital filters has a denominator polynomial. This denominator polynomial results in feedback of previously computed values of the output for the computation of a given output sample. This feedback provides advantages over FIR digital filters in terms of the required filter order to meet filter specifications for frequency selective filters. However, FIR filters typically have better phase characteristics for particular applications. The use of feedback can also lead to an unstable filter and poor performance due to quantization of the filter coefficients and the use of finite precision arithmetic for the computations. The choice of the most appropriate realization for a particular application may vary depending on the application requirements. Therefore, it is appropriate to study the various realizations and their properties.

IIR digital filters are typically implemented as second order sections because second order sections provide better performance in terms of roundoff errors when using finite precision arithmetic for the required computations. Therefore, we primarily discuss the realizations in terms of second order sections. However, the concepts presented are general and can be easily extended to higher order sections.

The system function for the IIR digital filter can be written as

$$
H(z) = \frac{\displaystyle\sum_{k=0}^{L} b(k)z^{-k}}{1.0 + \displaystyle\sum_{k=1}^{L} a(k)z^{-k}}
\tag{5.101}
$$

where $a(k)$ and $b(k)$ are the filter coefficients. We assume that the input is $x(n)$ and the output is $y(n)$. Therefore, we can write

$$H(z) = \frac{\displaystyle\sum_{n=0}^{\infty} y(n)z^{-n}}{\displaystyle\sum_{n=0}^{\infty} x(n)z^{-n}} = \frac{Y(z)}{X(z)}. \tag{5.102}$$

It follows that

$$Y(z) = H(z)X(z). \tag{5.103}$$

5.4.1 DIRECT FORM STRUCTURE

We can write the system function in the form

$$H(z) = H_1(z)H_2(z) \tag{5.104}$$

where

$$H_1(z) = \frac{G(z)}{X(z)} = \frac{1}{1.0 + \displaystyle\sum_{k=1}^{L} a(k)z^{-k}} \tag{5.105}$$

and

$$H_2(z) = \frac{Y(z)}{G(z)} = \sum_{k=0}^{L} b(k)z^{-k}. \tag{5.106}$$

The difference equation associated with $H_1(z)$ is

$$g(n) = x(n) - \sum_{k=1}^{L} a(k)g(n-k). \tag{5.107}$$

The difference equation associated with $H_2(z)$ is

$$y(n) = \sum_{k=0}^{L} b(k)g(n-k). \tag{5.108}$$

Fig. 5.10 shows a block diagram of a typical second order section of an IIR filter using a direct form realization.

The difference equations for this second order section are

$$
\begin{aligned}
g(n) &= x(n) - a(1)g(n-1) - a(2)g(n-2), \\
y(n) &= b(0)g(n) + b(1)g(n-1) + b(2)g(n-2).
\end{aligned} \tag{5.109}
$$

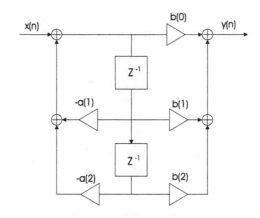

FIGURE 5.10

Direct form realization of a second order IIR digital filter

5.4.2 TRANSPOSE DIRECT FORM STRUCTURE

We can use the result

$$Y(z) = H(z)X(z) \qquad (5.110)$$

to obtain an expression of the form

$$Y(z) = b(0)X(z) + \sum_{k=1}^{L} [b(k)X(z) - a(k)Y(z)] z^{-k}. \qquad (5.111)$$

The corresponding difference equation is given by

$$Y(z) = b(0)x(n) + \sum_{k=1}^{L} \left[b(k)x(n-k) - a(k)y(n-k) \right]. \qquad (5.112)$$

Fig. 5.11 gives the block diagram for a second order IIR filter using this approach. The corresponding difference equation for this second order filter is

$$
\begin{aligned}
y(n) \;=\; & b(0)x(n) + b(1)x(n-1) - a(1)y(n-1) \\
& + b(2)x(n-2) - a(2)y(n-2). \qquad (5.113)
\end{aligned}
$$

Fig. 5.12 gives the block diagram for a fourth order IIR filter using this approach. The corresponding difference equation for this fourth order filter is

$$
\begin{aligned}
y(n) \;=\; & b(0)x(n) + b(1)x(n-1) - a(1)y(n-2) \\
& + b(2)x(n-2) - a(2)y(n-2) + b(3)x(n-3) - a(3)y(n-3) \\
& + b(4)x(n-4) - a(4)y(n-4). \qquad (5.114)
\end{aligned}
$$

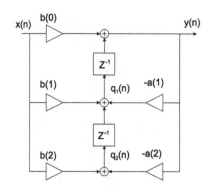

FIGURE 5.11

Transpose direct form realization of a second order IIR digital filter

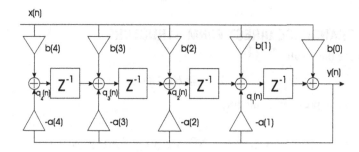

FIGURE 5.12

Block diagram for a fourth order IIR system

5.4.3 CASCADE FORM IIR REALIZATION

The use of second order sections for the implementation of digital filters is very popular because it reduces the roundoff error when fixed point arithmetic is used. Roundoff error also affects the frequency response so that a high order filter implemented using second order sections (SOS) typically gives better performance than the implementation of the same filter using a single difference equation. Second order sections are used because the poles and zeros typically occur in complex conjugate pairs. Real poles and zeros can also be paired to obtain the SOS.

The procedure for partitioning $H(z)$ into second order sections for the IIR filter is similar to the procedure for FIR filters. However, both zeros and poles are involved. The procedure below has been modified from the one for FIR digital filters to make it appropriate for IIR digital filters:

1. Obtain the roots of the numerator polynomial $B(z)$ and the denominator polynomial $A(z)$ for the IIR digital filter such that

$$H(z) = \frac{B(z)}{A(z)} = \frac{b(0) \prod\limits_{m=1}^{M} (z - q_m)}{\prod\limits_{m=1}^{M} (z - p_m)}. \tag{5.115}$$

2. Order the q_m and the p_m in increasing (or decreasing) magnitude.
3. Combine the complex conjugate pairs in increasing (or decreasing) magnitude to form second order system transfer functions. The magnitude of the corresponding poles or zeros have second priority to combining complex conjugates. A pair of complex conjugate poles may be combined with a pair of real zeros or vice versa to obtain a second order section. If both poles and zeros are complex conjugates, then the second order section can be represented as

$$H_k(z) = \frac{(z - q_m)(z - q_m^*)}{(z - p_m)(z - p_m^*)} = \frac{z^2 - 2\Re(q_m)z + |q_m^2|}{z^2 - 2\Re(p_m)z + |p_m^2|}. \tag{5.116}$$

If a complex pair of poles is combined with real pair of zeros, then the second order section can be represented as

$$H_k(z) = \frac{(z - q_m)(z - q_{m+1})}{(z - p_m)(z - p_m^*)} = \frac{z^2 - (q_m + q_{m+1})z + q_m q_{m+1}}{z^2 - 2\Re(p_m)z + |p_m^2|}. \tag{5.117}$$

We can represent the case for a real pair of poles and a complex pair of zeros in a similar way.

4. Pair real q_m and real q_m in increasing (or decreasing) magnitude pairs so that the q_m and p_m closest in magnitude are paired to form second order system transfer functions

$$H_k(z) = \frac{(z - q_m)(z - q_{m+1})}{(z - p_m)(z - p_{m+1})} = \frac{z^2 - (q_m + q_{m+1})z + q_m q_{m+1}}{z^2 - (p_m + p_{m+1})z + p_m p_{m+1}}. \tag{5.118}$$

5. If the number of real poles is odd, then form a first order section using the p_m and q_m that is left after the above second order sections have been formed where it should be noted that some or all of the q_m may be zeros.

Pairing real poles and zeros closest in magnitude has the benefit of reducing rounding error when the coefficients are quantized.

The coefficient $b(0)$ can be used as a multiplying constant for $H_1(z)$ (or another $H_m(z)$) or it can be distributed among the second order sections. For example, we can determine

$$A = |b(0)|^{\frac{1}{M}}. \tag{5.119}$$

Then, we form

$$\widehat{H_m(z)} = A \prod_{m=1}^{M} H_m(z). \qquad (5.120)$$

It follows that

$$H(z) = \prod_{m=1}^{M} \widehat{H_m(z)} = A^M \prod_{m=1}^{M} H_m(z) \qquad (5.121)$$

if $b(0)$ is positive and

$$H(z) = -\prod_{m=1}^{M} \widehat{H_m(z)} = A^M \prod_{m=1}^{M} H_m(z) \qquad (5.122)$$

if $b(0)$ is negative.

EXAMPLE 5.8
(Second Order Section)

Consider the system transfer function

$$H(z) = \frac{0.3038 - 0.0034z^{-1} - 0.5989z^{-2} - 0.0034z^{-3} + 0.3038z^{-4}}{1.0 - 0.9061z^{-1} + 0.1651z^{-2} - 0.3150z^{-3} + 0.3903z^{-4}}. \quad (5.123)$$

The zeros of $H(z)$ are located at

$$
\begin{aligned}
q(1, 1) &= -0.9936 + 0.1131j, \\
q(2, 1) &= -0.9936 - 0.1131j, \\
q(3, 1) &= 0.9992 + 0.0394j, \\
q(4, 1) &= 0.9992 - 0.0394j. \qquad (5.124)
\end{aligned}
$$

The poles are located at

$$
\begin{aligned}
p(1, 1) &= -0.3458 + 0.6338j, \\
p(2, 1) &= -0.3458 - 0.6338j, \\
p(3, 1) &= 0.7988 + 0.3326j, \\
p(4, 1) &= 0.7988 - 0.3326j, \qquad (5.125)
\end{aligned}
$$

$$A = (0.3038)^{\frac{1}{2}} = 0.5512. \qquad (5.126)$$

Fig. 5.13 gives the Z plane plot for $H(z)$.

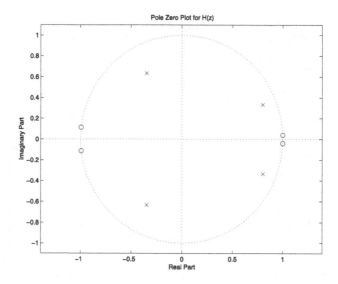

FIGURE 5.13

Z plane plot for $H(z)$ for Example 5.8

Two second order sections can be formed by pairing the complex conjugate pairs of poles and zeros such that

$$\widehat{H_1(z)} = \frac{0.5512(z + 0.9936 - 0.1131\,j)(z + 0.9936 + 0.1131\,j)}{(z + 0.3458 - 0.6338\,j)(z + 0.3458 + 0.6338\,j)}, \qquad (5.127)$$

$$\widehat{H_2(z)} = \frac{0.5512(z + 0.9992 - 0.0394\,j)(z + 0.9992 - 0.0394\,j)}{(z - 0.7988 - 0.3326\,j)(z - 0.7988 + 0.3326\,j)}. \qquad (5.128)$$

The system transfer functions for the two SOS are

$$H_1(z) = \frac{0.5512 + 1.0954z^{-1} + 0.5512z^{-2}}{1.0 - 1.5977z^{-1} + 0.7488z^{-2}},$$

$$H_2(z) = \frac{0.5512 - 1.1016z^{-1} + 0.5512z^{-2}}{1.0 + 0.6915z^{-1} + 0.5213z^{-2}}. \qquad (5.129)$$

Figs. 5.14 and 5.15 give the Z plane plots for the two SOS. The corresponding difference equations are given by

$$\begin{aligned}
g(n) &= 0.5512x(n) + 1.0954x(n-1) + 0.5512x(n-2) \\
&\quad + 1.5977g(n-1) - 0.7488g(n-2), \\
y(n) &= 0.5512g(n) - 1.1016g(n-1) + 0.5512g(n-2) \\
&\quad - 0.6915y(n-1) - 0.5213y(n-2).
\end{aligned} \qquad (5.130)$$

End of the Example

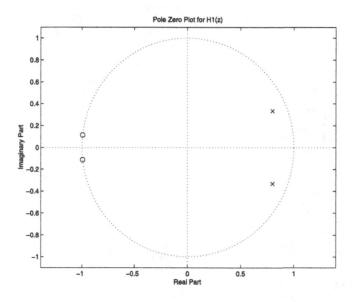

FIGURE 5.14

Z plane plot for $H_1(z)$

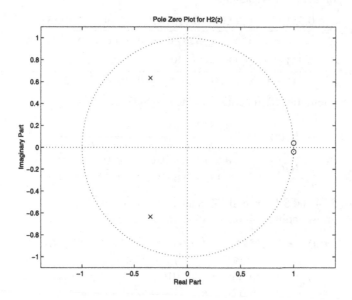

FIGURE 5.15

Z plane plot $H_2(z)$

5.4.4 PARALLEL IIR FORM STRUCTURE

The parallel implementation of IIR digital filters using parallel second order sections (SOS) will be discussed in this section. The IIR digital filter can be implemented in parallel using SOS form by

1. Partitioning the system transfer function into partial fractions,
2. Combining partial fractions with complex conjugate coefficients to form second order system transfer functions, and
3. Combining partial fractions with real coefficients to form second order sections.

We can partition $\frac{H(z)}{z}$ into partial fractions to obtain

$$\frac{H(z)}{z} = \frac{C_0}{z} + \sum_{k=1}^{L} \frac{r_k}{z - p_k}. \tag{5.131}$$

It follows that

$$H(z) = C_0 + \sum_{k=1}^{L} \frac{r_k z}{z - p_k}. \tag{5.132}$$

Two partial fractions with complex conjugate coefficients can be combined to form a single, second order section. Assume that $p_{k+1} = p_k^*$ and $r_{k+1} = r_k^*$. A second order system transfer function can be formed as

$$H_m(z) = \frac{r_k z}{z - p_k} + \frac{r_k^* z}{z - p_k^*}, \tag{5.133}$$

$$H_m(z) = \frac{(r_k z)(z - p_k^*)}{(z - p_k)(z - p_k^*)}$$

$$+ \frac{(r_k^* z)(z - p_k)}{(z - p_k)(z - p_k^*)}, \tag{5.134}$$

$$H_m(z) = \frac{(r_k + r_k^*)z^2 - (p_k^* r_k + p_k r_k^*)z}{z^2 - (p_k + p_k^*)z + p_k p_k^*}, \tag{5.135}$$

$$H_m(z) = \frac{b_m(0)z^2 + b_m(1)z}{z^2 + a_m(1)z + a_m(2)} \tag{5.136}$$

where

$$\begin{aligned}
b_m(0) &= r_k + r_k^* = 2\Re(r_k), \\
b_m(1) &= -(p_k^* r_k + p_k r_k^*) = -2\Re(p_k r_k^*), \\
a_m(1) &= -(p_k + p_k^*) = -2\Re(p_k), \\
a_m(2) &= p_k p_k^* = |p_k|^2.
\end{aligned} \tag{5.137}$$

It also follows that

$$H_m(z) = \frac{b_m(0) + b_m(1)z^{-1}}{1.0 + a_m(1)z^{-1} + a_m(2)z^{-2}}. \tag{5.138}$$

Example 5.9 gives an example of this procedure.

EXAMPLE 5.9
(Parallel Second Order Section)

The system transfer functions for a digital filter is given by

$$H(z) = \frac{B(z)}{A(z)},$$

$$B(z) = 0.0684 + 0.0773z^{-1} + 0.1312z^{-2}$$
$$+ 0.0773z^{-3} + 0.0684z^{-4},$$

$$A(z) = 1.0000 - 1.6903z^{-1} + 2.0315z^{-2}$$
$$- 1.2334z^{-3} + 0.4242z^{-4}. \tag{5.139}$$

We want to implement $H(z)$ using parallel second order sections in the form

$$H(z) = C_0 + H_1(z) + H_2(z) \tag{5.140}$$

where $H_1(z)$ and $H_2(z)$ are SOS and C_0 is a real constant.
We begin by partitioning $H(z)$ into partial fractions.

$$H(z) = 0.1613 + \frac{r_1}{z - p_1} + \frac{r_2}{z - p_2}$$
$$+ \frac{r_3}{z - p_3} + \frac{r_4}{z - p_4} \tag{5.141}$$

where

$$r_1 = -0.0228 + 0.0620j,$$
$$r_2 = -0.0228 - 0.0620j,$$
$$r_3 = -0.0236 - 0.3776j,$$
$$r_4 = -0.0236 + 0.3776j,$$
$$p_1 = 0.3177 + 0.8816j,$$
$$p_2 = 0.3177 - 0.8816j,$$
$$p_3 = 0.5275 + 0.4525j,$$
$$p_4 = 0.5275 - 0.4525j. \tag{5.142}$$

We observe that $r_2 = r_1^*$ and $p_2 = p_1^*$. Therefore, we can form an SOS using the partial fractions involving $r_1, r_2, p_1,$ and p_2. It follows that

$$H_1(z) = \frac{b_1(0) + b_1(1)z^{-1}}{1.0 + a_1(1)z^{-1} + a_1(2)z^{-2}} \tag{5.143}$$

where

$$
\begin{aligned}
b_1(0) &= 2\Re(r_1) = -0.0456, \\
b_1(1) &= -2\Re(p_1 r_1^*) = -0.0949, \\
a_1(1) &= -2\Re(p_1) = -0.6354, \\
a_1(2) &= |p_1|^2 = 0.8782.
\end{aligned} \tag{5.144}
$$

Thus,

$$H_1(z) = \frac{-0.0456 - 0.0949z^{-1}}{1.0 - 0.6354z^{-1} + 0.8782z^{-2}}. \tag{5.145}$$

We also observe that $r_4 = r_3^*$ and $p_4 = p_3^*$. Therefore, we can form an SOS using the partial fractions involving $r_3, r_4, p_3,$ and p_4. It follows that

$$H_2(z) = \frac{b_2(0) + b_2(1)z^{-1}}{1.0 + a_2(1)z^{-1} + a_2(2)z^{-2}} \tag{5.146}$$

where

$$
\begin{aligned}
b_2(0) &= 2\Re(r_3) = -0.0473, \\
b_2(1) &= -2\Re(p_3 r_3^*) = 0.3667, \\
a_2(1) &= -2\Re(p_3) = -1.0550, \\
a_2(2) &= |p_3|^2 = 0.4830.
\end{aligned} \tag{5.147}
$$

Thus,

$$H_2(z) = \frac{-0.0473 + 0.3667z^{-1}}{1.0 - 1.0550z^{-1} + 0.4830z^{-2}}. \tag{5.148}$$

We also observe that

$$C_0 = 0.1613. \tag{5.149}$$

It follows that

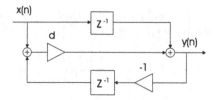

FIGURE 5.16

A computational structure for a first order all pass IIR digital filter

$$H(z) = 0.1613 + \frac{-0.0456 - 0.0949z^{-1}}{1.0 - 0.6354z^{-1} + 0.8782z^{-2}}$$
$$+ \frac{-0.0473 + 0.3667z^{-1}}{1.0 - 1.0550z^{-1} + 0.4830z^{-2}}. \tag{5.150}$$

End of the Example

5.5 ALL PASS DIGITAL FILTERS

The all pass digital filter has a system transfer function of the form

$$H(z) = \frac{z^{-m} D_m(z^{-1})}{D_m(z)} = \frac{B(z)}{A(z)}. \tag{5.151}$$

Stability requirements dictate that the roots of $D_m(z)$ all have magnitudes less than 1. We can develop a computational structure that can take advantage of the redundancy between the numerator polynomial $B(z)$ and the denominator polynomial $A(z)$.

We begin our discussion with the first order, all pass filter which can be represented as

$$H(z) = \frac{d + z^{-1}}{1.0 + dz^{-1}}. \tag{5.152}$$

Fig. 5.16 gives a computational structure that can implement $H(z)$ using one multiplier instead of two. The corresponding difference equation is given by

$$y(n) = d\left[x(n) - y(n-1)\right] + x(n-1). \tag{5.153}$$

Note that the negation of $y(n - 1)$ can be implemented in an adder instead of by multiplying $y(n)$ by -1. Thus, only one multiplication is needed to implement the first order all pass filter.

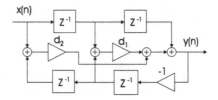

FIGURE 5.17

A computational structure for a second order all pass IIR digital filter

The second order all pass IIR digital filter can be represented as

$$H(z) = \frac{d_2 + d_1 z^{-1} + z^{-2}}{1.0 + d_1 z^{-1} + d_2 z^{-2}}. \tag{5.154}$$

Fig. 5.17 gives a computational structure that can implement $H(z)$ using two multipliers instead of four. The corresponding difference equation is given by

$$y(n) = d_2 \left[x(n) - y(n-2) \right] + d_1 \left[x(n-1) - y(n-1) \right] + x(n-2). \tag{5.155}$$

This concept can be extended to any arbitrary order all pass digital filter to reduce the number of required multiplications.

5.6 ALL POLE IIR LATTICE STRUCTURE

The FIR Lattice structure implements the all zero FIR digital filter. The difference equation for the FIR filter with $b(0)$ normalized to 1 is given by

$$y(n) = x(n) + \sum_{k=1}^{m} b(k)x(n-k). \tag{5.156}$$

If we replace $b(k)$ with $\alpha(k)$, we have

$$y(n) = x(n) + \sum_{k=1}^{m} \alpha(k)x(n-k). \tag{5.157}$$

If we exchange inputs and outputs in the above equation, we obtain

$$x(n) = y(n) + \sum_{k=1}^{m} \alpha(k)y(n-k). \tag{5.158}$$

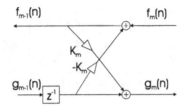

FIGURE 5.18

Basic cell for the IIR all pole lattice filter

Solving for $y(n)$, we obtain

$$y(n) = x(n) - \sum_{k=1}^{m} \alpha(k)y(n-k),$$ (5.159)

which is the equation for an all pole IIR filter with system transfer function

$$H_m(z) = \frac{1}{1.0 + \sum_{k=1}^{m} \alpha(k)z^{-k}}.$$ (5.160)

This suggest that we can modify the FIR lattice structure to obtain an all pole IIR structure by making $f_{m-1}(n)$ the output and $f_m(n)$ the input.

Recall that the equations for cell m for the FIR lattice filter are

$$\begin{aligned} f_m(n) &= f_{m-1}(n) + K_m g_{m-1}(n-1), \\ g_m(n) &= K_m f_{m-1}(n) + g_{m-1}(n-1). \end{aligned}$$ (5.161)

If we solve for $f_{m-1}(n)$ as the output and $f_m(n)$ as the input, we obtain

$$\begin{aligned} f_{m-1}(n) &= f_m(n) - K_m g_{m-1}(n), \\ g_m(n) &= K_m f_{m-1}(n) + g_{m-1}(n) \end{aligned}$$ (5.162)

where the equation for $g_m(n)$ has not been changed. We use Eq. (5.162) to determine the computational structure for a basic section of an all pole IIR digital filter. Thus the basic cell for the all pole IIR filter is given in Fig. 5.18. We obtain the IIR filter by using the output $f_0(n)$ as a feedback input to $g_0(n)$. Thus, we have

$$g_0(n) = f_0(n).$$ (5.163)

We can obtain the lattice structure for a second order all pole IIR filter by connecting the inputs and outputs for two sections as shown in Fig. 5.19. This concept can be extended to obtain the lattice structure for any arbitrary order all pole IIR digital filter.

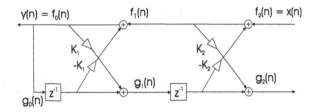

FIGURE 5.19

Lattice structure for a second order all pole IIR filter

5.7 IIR LATTICE STRUCTURE

We can use the IIR all pole lattice structure to obtain the IIR lattice structure for the general IIR digital filter. The general IIR filter $H(z)$ can be represented as

$$H(z) = H_1(z)H_2(z) \tag{5.164}$$

where

$$H_1(z) = \frac{G_0(z)}{X(z)} = \frac{1.0}{1.0 + \displaystyle\sum_{k=1}^{m} a(k)z^{-k}} \tag{5.165}$$

and

$$H_2(z) = \frac{Y(z)}{G_0(z)} = \sum_{k=0}^{m} b(k)z^{-k}. \tag{5.166}$$

Thus, we can use the output from the all pole filter with appropriate coefficients to obtain

$$y(n) = \sum_{k=1}^{m} \beta(k)g_0(n - k). \tag{5.167}$$

We can then solve for the values of $\beta(k)$ from the original values of $b(k)$.

Consider the second order IIR filter given by

$$H(z) = \frac{\displaystyle\sum_{k=0}^{2} b(k)z^{-k}}{1.0 + \displaystyle\sum_{k=1}^{2} a(k)z^{-k}}. \tag{5.168}$$

Fig. 5.20 gives the block diagram for a second order IIR lattice filter structure.

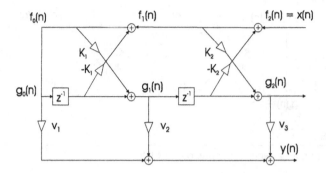

FIGURE 5.20

Lattice structure for a second order IIR filter

The lattice parameters for implementing $H(z)$ using this structure will now be obtained. The difference equations corresponding to the computational structure in Fig. 5.20 are given by

$$
\begin{aligned}
f_2(n) &= x(n), \\
g_2(n) &= K_2 f_1(n) + g_1(n-1), \\
f_1(n) &= f_2(n) - K_2 g_1(n-1), \\
g_1(n) &= K_1 f_0(n) + g_0(n-1), \\
f_0(n) &= f_1(n) - K_1 g_0(n-1), \\
g_0(n) &= f_0(n), \\
y(n) &= V_1 g_0(n) + V_2 g_1(n) + V_3 g_2(n). \quad (5.169)
\end{aligned}
$$

We can solve for $f_0(n)$ in terms of $x(n)$, $g_0(n-1)$, and $g_0(n-2)$ to obtain the equation for the output in terms of the input $x(n)$ where $g_0(n) = f_0(n)$. We have

$$
f_0(n) = f_1(n) - K_1 g_0(n-1). \quad (5.170)
$$

Substituting for $f_1(n)$, we have

$$
f_0(n) = g_0(n) = f_2(n) - K_2 g_1(n-1) - K_1 g_0(n-1). \quad (5.171)
$$

Since $x(n) = f_2(n)$, we can substitute for $g_1(n-1)$ to obtain

$$
g_0(n) = x(n) - K_1 \left[K_1 x(n-1) + g_0(n-2) \right] - K_1 g_0(n-1), \quad (5.172)
$$

$$
g_0(n) = x(n) - [K_1 + K_1 K_2] g_0(n-1) - K_2 g_0(n-2). \quad (5.173)
$$

The system equation for the all pole second order system is given by

$$
H_1(z) = \frac{G_0(z)}{X(z)} = \frac{1.0}{1.0 + a(1)z^{-1} + a(2)z^{-2}}. \quad (5.174)
$$

The corresponding difference equation for the output is

$$y(n) = x(n) - a(1)y(n-1) - a(2)y(n-2).$$ (5.175)

We can compare this with Eq. (5.173) for $g_0(n)$. The equations match is we set

$$
\begin{aligned}
y(n) &= g_0(n), \\
a(2) &= K_2, \\
a(1) &= K_1 + K_1 K_2.
\end{aligned}
$$ (5.176)

It follows that

$$
\begin{aligned}
K_2 &= a(2), \\
K_1 &= \frac{a(1)}{1.0 + a(2)}.
\end{aligned}
$$ (5.177)

Note that the above equations are very similar to the equations for the second order FIR lattice filter except that $b(1)$ is replaced by $a(1)$ and $b(2)$ is replace by $a(2)$. This suggest that the algorithm for obtaining the K lattice coefficients for the FIR filter can also be used to obtain the K lattice coefficients for the IIR lattice filter.

We now consider the derivation of the V parameters for the structure. We consider the transfer function relationships for $G_1(z)$ and $G_2(z)$. We see from Fig. 5.20 that

$$G_1(z) = \left(K_1 + z^{-1}\right) G_0(z),$$ (5.178)

$$G_2(z) = K_2 F_1(z) + z^{-1} G_1(z) = K_2 F_1(z) + z^{-1} \left(K_1 + z^{-1}\right) G_0(z).$$ (5.179)

We also can write

$$F_0(z) = G_0(z) = F_1(z) - K_1 z^{-1} G_0(z).$$ (5.180)

We can solve for $F_1(z)$ to obtain

$$F_1(z) = \left[1.0 + K_1 z^{-1}\right] G_0(z).$$ (5.181)

Thus,

$$G_2(z) = K_2 \left[1.0 + K_1 z^{-1}\right] G_0(z) + z^{-1} \left[K_1 + z^{-1}\right] G_0(z),$$ (5.182)

$$G_2(z) = \left[K_2 + (K_1 K_2 + K_1) z^{-1} + z^{-2}\right] G_0(z).$$ (5.183)

The output $y(n)$ is formed as the linear sum

$$y(n) = V_1 g_0(n) + V_2 g_1(n) + V_3 g_2(n).$$ (5.184)

Thus

$$Y(z) = V_1 G_0(z) + V_2 G_1(z) + V_3 G_2(z), \tag{5.185}$$

$$
\begin{aligned}
Y(z) &= V_1 G_0(z) + \left[V_2 \left(K_1 + z^{-1} \right) + V_3 K_2 \right] G_0(z) \\
&\quad \left[V_3 \left(K_1 + K_1 K_2 \right) G_0(z) z^{-1} + V_3 z^{-2} \right] G_0(z), \tag{5.186}
\end{aligned}
$$

$$
\begin{aligned}
Y(z) &= [(V_1 + K_1 V_2 + K_2 V_3)] G_0(z) \\
&\quad [(V_2 + K_1 V_3 + K_1 K_2 V_3)] G_0(z) z^{-1} \\
&\quad + V_3 G_0(z) z^{-2}. \tag{5.187}
\end{aligned}
$$

We can obtain an alternative expression for $Y(z)$ from the direct implementation of the second order IIR filter. Thus, we have

$$Y(z) = b(0) G_0(z) + b(1) z^{-1} G_0(z) + b(2) z^{-2} G_0(z). \tag{5.188}$$

Comparing the two equations, we obtain

$$
\begin{aligned}
b(2) &= V_3, \\
b(1) &= V_2 + K_1 V_3 + K_1 K_2 V_3, \\
b(0) &= V_1 + K_1 V_2 + K_2 V_3. \tag{5.189}
\end{aligned}
$$

We can solve for the V_k parameters to obtain

$$
\begin{aligned}
V_3 &= b(2), \\
V_2 &= b(1) - (K_1 K_2 + K_1) V_3, \\
V_1 &= b(0) - K_1 V_2 - K_2 V_3. \tag{5.190}
\end{aligned}
$$

We arranged the above equations with V_3 first because V_3 can be used to solve for V_2, and V_2 and V_3 can be used to solve for V_1.

We can implement an arbitrary order IIR filter by using second order sections. We can then implement each second order section using a second order lattice structure. The Matlab function *tf2latc* can be used to obtain the lattice parameters from the coefficients for the system transfer function for an FIR or an IIR digital filter.

EXAMPLE 5.10
(IIR Lattice)

The system transfer function for an IIR digital filter is given by

$$H(z) = \frac{0.4949 + 0.9842 z^{-1} + 0.4949 z^{-2}}{1.0 + 0.8918 z^{-1} + 0.4543 z^{-2}}. \tag{5.191}$$

We wish to determine the IIR lattice parameters for this filter (K and V parameters).

The K parameters can be determined as follows:

$$K_2 = a(2) = 0.4543,$$

$$K_1 = \frac{a(1)}{1.0 + a(2)} = 0.6132. \qquad (5.192)$$

The V parameters can be determined as follows:

$$V_3 = b(2) = 0.4949,$$

$$V_2 = b(1) - (K_1 K_2 + K_1) V_3 = 0.5428,$$

$$V_1 = b(0) - K_1 V_2 - K_2 V_3 = -0.0628. \qquad (5.193)$$

The filter $H(z)$ is a second order section. The corresponding difference equations are given by

$$
\begin{aligned}
f_2(n) &= x(n), \\
g_2(n) &= K_2 f_1(n) + g_1(n-1), \\
f_1(n) &= f_2(n) - K_2 g_1(n-1), \\
g_1(n) &= K_1 f_0(n) + g_0(n-1), \\
f_0(n) &= f_1(n) - K_1 g_0(n-1), \\
g_0(n) &= f_0(n), \\
y(n) &= V_1 g_0(n) + V_2 g_1(n) + V_3 g_2(n). \qquad (5.194)
\end{aligned}
$$

Using the Lattice parameters for $H(z)$, we have

$$
\begin{aligned}
f_2(n) &= x(n), \\
g_2(n) &= 0.4543 f_1(n) + g_1(n-1), \\
f_1(n) &= f_2(n) - 0.4543 g_1(n-1), \\
g_1(n) &= 0.6132 f_0(n) + g_0(n-1), \\
f_0(n) &= f_1(n) - 0.6132 g_0(n-1), \\
g_0(n) &= f_0(n), \\
y(n) &= -0.0628 g_0(n) + 0.5428 g_1(n) + 0.4949 g_2(n). \qquad (5.195)
\end{aligned}
$$

End of the Example

5.8 IMPLEMENTING IIR FILTERS USING MATLAB
5.8.1 GENERAL FILTER FUNCTION USING MATLAB

IIR digital filters can be implemented through the use of the general difference equation given by

$$y(n) = \sum_{k=0}^{L} b(k)x(n-k) - \sum_{k=1}^{L} a(k)y(n-k). \tag{5.196}$$

We can use Matlab to implement this difference equation for an arbitrary order IIR filter using the Matlab function *myfilter:scr* in Appendix C.2. This function shows the use of the dot-product Matlab function *dot* to compute the multiplication of appropriate filter coefficients by the current and delayed inputs and the delayed outputs in the difference equation representation. The dot-product is more efficient in performing these computations than it would be to use a *for* loop to perform the computations. The *myfilter* function is functionally equivalent to the Matlab *filter* function.

EXAMPLE 5.11
(Dot-Product Filter Implementation)

We can demonstrate the use of the *myfilter* function to filter a sample sequence. We will compare the results with the results from the Matlab *filter* function. The following Matlab script can be used to design and implement a filter and perform the required comparison.

Matlab Script 5.3.

```
order = 6;
wc = [0.1 0.65];
rp = 1.0;
rs = 40;
[b, a] = ellip(order, rp, rs, wc);
x = sampdata;
y1 = myfilter(b, a, x);
y2 = filter(b, a, x);
```

End of the Script

 Fig. 5.21 gives stem plots of the output from the Matlab *filter* function and the *myfilter* function for comparison. The outputs are essentially the same.
End of the Example

5.8.2 IMPLEMENTATION USING SECOND ORDER SECTIONS

The use of second order sections to implement IIR filters provides advantages such as lower roundoff errors, parallel implementation in hardware, etc. Example 5.12

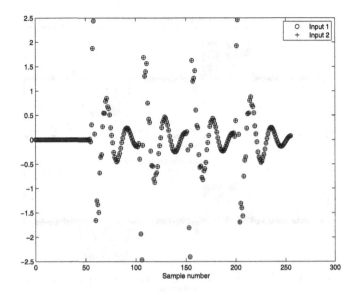

FIGURE 5.21

Stem plots for the outputs using the Matlab *filter* function and the stem plot of the corresponding output using the *myfilter* function for Example 5.11

provides an example of the use of second order sections to implement an IIR digital filter.

EXAMPLE 5.12
(IIR Second Order Section)

Matlab script *SOS:mat* in Appendix C.1 illustrates the decomposition of a fourth order IIR filter into second order sections. Fig. 5.22 gives comparison plots for the implementation using second order sections the implementation using the Matlab *filter* function.
End of Example

5.8.3 THE LINEAR PHASE IIR FILTER

Recall that one of the primary advantages of FIR filters is that FIR filters can have a linear phase which means that they can be implemented with no phase distortion since the linear phase only results in delay. However, IIR filters can also be implemented using a linear phase for applications does not require real-time. If the data is stored in memory or on a disk, the linear phase can be obtained by using the procedure depicted in Fig. 5.23. The block *Rev* in Fig. 5.23 means that the data is reversed in time. The delay z^{-N} is necessary because the sequence must be delayed before

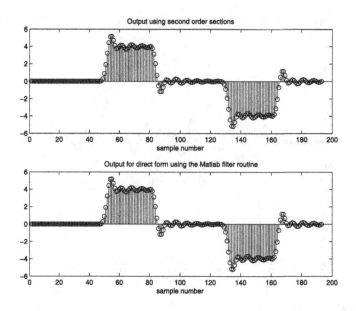

FIGURE 5.22

Comparison plots for IIR filter using second order sections and using the direct form

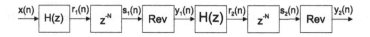

FIGURE 5.23

Procedure for linear phase using IIR filters

it can be reversed. Thus, the procedure involves filtering the data using the filter $H(z)$, delaying the output by N samples and reversing the delayed sequence in time, filtering the delayed and reversed output sequence $y_1(n)$ using the filter $H(z)$, and finally delaying and reversing the output sequence $r_2(n)$ to obtain the linear phase output $y_2(n)$.

This procedure uses the reverse and delay properties of Z Transforms for linear shift invariant system. If a sequence $x(n)$ has a Z Transform $X(z)$ such that

$$x(n) \Leftrightarrow X(z). \tag{5.197}$$

Then

$$x(n - N) \Leftrightarrow z^{-N}X(z) \tag{5.198}$$

and

$$x(-n) \Leftrightarrow X(z^{-1}). \tag{5.199}$$

It follows that

$$
\begin{aligned}
R_1(z) &= H(z)X(z), \\
S_1(z) &= z^N R_1(z) = z^{-N} H(z)X(z), \\
Y_1(z) &= S_1(-z) = z^{-N} H(z^{-1})X(z^{-1}), \\
R_2(z) &= H(z)Y_1(z) = z^{-N} H(z)H(z^{-1})X(z^{-1}), \\
S_2(z) &= z^{-N} R_2(z) = z^{-2N} H(z)H(z^{-1})X(z^{-1}), \\
Y_2(z) &= S_2(z^{-1}) = z^{-2N} H(z^{-1})H(z)X(z).
\end{aligned}
\tag{5.200}
$$

Note that the frequency representation of $y_2(n)$ is given by

$$
Y_2(\omega) = Y_2(z)|_{z=e^{j\omega}} = e^{-2j\omega N} H(-\omega)H(\omega)X(\omega).
\tag{5.201}
$$

We assume that the system is linear and shift invariant. It follows that

$$
H(-\omega) = H^*(\omega)
\tag{5.202}
$$

and

$$
Y_2(\omega) = e^{-2j\omega N} |H(\omega)|^2 X(\omega).
\tag{5.203}
$$

The term

$$
e^{-2j\omega N}
\tag{5.204}
$$

is a linear phase term corresponding to a delay of $2N$ sample intervals. Thus, the magnitude response of the resulting linear phase IIR filter is the square of the magnitude of the original filter.

EXAMPLE 5.13
(Linear Phase IIR Filter)

This example involves filtering a sample input sequence $x(n)$ with a fourth order IIR filter designed using the Matlab ellip filter design function. Matlab script C.3 in Appendix C.3 shows the implementation of an IIR filter using linear phase. Fig. 5.24 gives a stem plot of the linear phase output for this example.
End of the Example

The Matlab function *filtfilt* implements this procedure to obtain a linear phase IIR digital filter.

5.8.4 IMPLEMENTATION OF THE IIR LATTICE FILTER STRUCTURE

The IIR Lattice structure was presented in Sect. 5.7. We will discuss the use of Matlab to implement this structure in this section. We can use the Matlab function *tf2latc* to

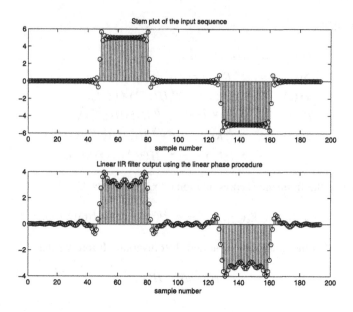

FIGURE 5.24

Stem plots for the IIR linear phase filter example

obtain the IIR lattice parameters from for an IIR filter. This is demonstrated in the following Matlab script.

Matlab Script 5.4.

```
% IIR lattice structure example (iirlatcex1)
x = sampdata;
wp = 0.3;
order = 4;
rp = 2.0;
rs = 45;
[b, a] = ellip(order, rp, rs, wp);
[sos, g] = tf2sos(b, a);
[n1, n2] = size(sos);
for k=1:n1
    bsos(1, :) = sos(k, 1:3);
    asos(1, :) = sos(k, 4:6);
    [kpram, vpram]  = tf2latc(bsos, asos);
    kp(:, k) = kpram;
    vp(:, k) = vpram;
end

% filter data
xin = x;
for k=1:n1
    kploc = kp(:, k);
```

```
      vploc = vp(:, k);
      y = latcfilt(kploc, vploc, xin);
      xin = y;
end
```

End of the Script

The Matlab *filter* function can now be used to compute the output to verify that the output is correct.

Matlab Script 5.5.

```
y2 = filter(b, a, x);
subplot(2,1,1)
stem(y1)
title('Output using second order lattice sections');
xlabel('sample number');
subplot(2,1,2)
stem(y2)
title('Output using the Matlab filter function');
xlabel('sample number');
print -dps iirlatc.ps
```

End of the Script

Fig. 5.25 gives a stem plot of the outputs for the IIR Lattice filter structure example.

5.8.5 IIR LATTICE REALIZATIONS USING SOS

EXAMPLE 5.14
(Cascade Second Order IIR Filter)

The system transfer function for an IIR digital filter is given by

$$H(z) = \frac{B(z)}{A(z)} \qquad (5.205)$$

where

$$
\begin{aligned}
B(z) &= 0.0503 - 0.1510z^{-2} + 0.1510z^{-4} - 0.0503z^{-6}, \\
A(z) &= 1.0 - 0.4974z^{-1} + 1.2361z^{-2} - 0.4458z^{-3} \\
&\quad + 0.7306z^{-4} - 0.1313z^{-5} + 0.1357z^{-6}. \qquad (5.206)
\end{aligned}
$$

We can use the Matlab function *tf2sos* to obtain the coefficients for the cascade implementation of $H(z)$ using second order lattice filter sections. We can determine the lattice parameters for each of the second order sections separately. The equations for obtaining the lattice parameters for an all pole IIR filter are the same equations for finding the lattice parameters for an FIR filter. The K and V param-

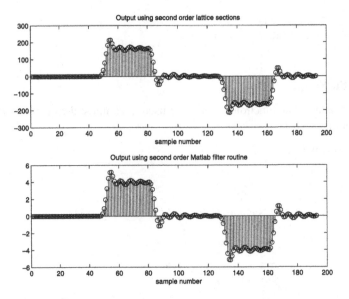

FIGURE 5.25

Stem plots of outputs for IIR lattice filter example

eters can be computed for each of the second order sections using the following equations for a second order filter:

$$
\begin{aligned}
K_1 &= \frac{a(1)}{1 + a(2)}, \\
K_2 &= a(2), \\
V_3 &= b(2), \\
V_2 &= b(1) - (K_1 K_2 + K_1)b(2), \\
V_1 &= b(0) - V_2 K_1 - V_3 K_2.
\end{aligned}
\tag{5.207}
$$

The following Matlab script can be used to obtain the lattice parameters for the implementation of $H(z)$ using three second order IIR lattice sections.

Matlab Script 5.6.

```
b = [0.0503 0 -0.1510 0 0.1510 0 -0.0503];
a = [1.0 -0.4974 1.2361 -0.4458 0.7306 -0.1313 0.1357];
[sos, gain] = tf2sos(b, a);
[K1, V1] = tf2latc(sos(1, 1:3), sos(1, 4:6));
[K2, V2] = tf2latc(sos(2, 1:3), sos(2, 4:6));
[K3, V3] = tf2latc(sos(3, 1:3), sos(3, 4:6));
```

End of the Script

The results are:

```
sos =
    1.0000    0.0000   -1.0000    1.0000   -0.1584    0.3214
    1.0000    2.0000    1.0000    1.0000    0.5142    0.6354
    1.0000   -2.0000    1.0000    1.0000   -0.8532    0.6644

gain =
    0.0503

K(1:2,1) =
   -0.1199
    0.3215

V(1:3,1) =
    1.3163
   -0.1656
   -1.0456

K(3:4,1) =
    0.3145
    0.6353

V(4:6,1) =
   -0.0851
    1.4751
    0.9780

K(5:6,1) =
   -0.5127
    0.6643

V(7:9,1) =
   -0.2359
   -1.1435
    0.9780
```

Fig. 5.26 gives the block diagram of the system showing the three second order sections.

The difference equations corresponding to the block diagram in Fig. 5.26 follow:

$$
\begin{aligned}
f_3(n) &= x(n), \\
g_3(n) &= K_2 f_2(n) + g_2(n-1), \\
f_2(n) &= f_3(n) - K_2 g_2(n-1), \\
g_2(n) &= K_1 f_1(n) + g_1(n-1), \\
f_1(n) &= f_2(n) - K_1 g_1(n-1), \\
g_1(n) &= f_1(n), \\
s_1(n) &= V_1 g_1(n) + V_2 g_2(n) + V_3 g_3(n), \quad (5.208)
\end{aligned}
$$

$$
\begin{aligned}
f_6(n) &= s_1(n), \\
g_6(n) &= K_4 f_5(n) + g_5(n-1), \\
f_5(n) &= f_6(n) - K_4 g_5(n-1), \\
g_5(n) &= K_3 f_4(n) + g_4(n-1), \\
f_4(n) &= f_5(n) - K_3 g_4(n-1), \\
g_4(n) &= f_4(n), \\
s_2(n) &= V_4 g_4(n) + V_5 g_5(n) + V_6 g_6(n), \qquad (5.209)
\end{aligned}
$$

$$
\begin{aligned}
f_3(n) &= x(n), \\
g_3(n) &= K_2 f_2(n) + g_2(n-1), \\
f_2(n) &= f_3(n) - K_2 g_2(n-1), \\
g_2(n) &= K_1 f_1(n) + g_1(n-1), \\
f_1(n) &= f_2(n) - K_1 g_1(n-1), \\
g_1(n) &= f_1(n), \\
s_3(n) &= V_7 g_7(n) + V_8 g_8(n) + V_9 g_9(n), \\
y(n) &= 0.0503 s_3(n). \qquad (5.210)
\end{aligned}
$$

End of the Example

5.9 STATE SPACE REALIZATION

The state space representation of a discrete time system involves an internal description of the system in the form of states as well as the description of the input/output relationship [2,12,13]. The mathematical equations describing the system, its input, and its output are divided into two parts:

1. A set of equations relating the next state variables to the current state variables and the current input signal
2. A second set of equations relating the current state variables and the current input to the output signal

The state variables provide information about the internal signals in the system. Therefore, the state space description can be considered to provide a more detailed description of the system dynamics than does the input–output relationship provided by the system transfer function or by the system difference equation.

The state space representation was developed for the analysis and design of both continuous time and discrete time control systems. However, it can also be used to analyze and design discrete time systems. We will limit our discussion to the state space representation of single input, single output discrete time systems in this text. We show that the state space representation can be used to analyze, design, and to develop computational structures for discrete time systems.

The assignment of state variables for the state space representation is somewhat arbitrary. Therefore, the state space representation for a given system is usually not

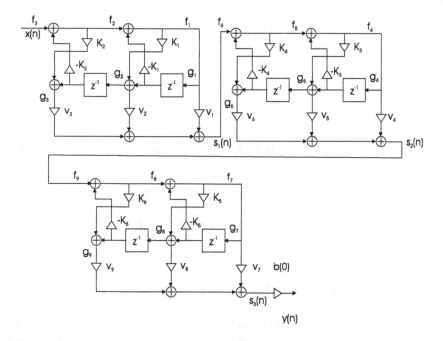

FIGURE 5.26

Block diagram of the system for Example 5.14

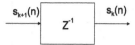

FIGURE 5.27

Standard assignment of state variables for a discrete time system

unique. The standard approach for discrete time systems is to assign the current state variables $s_k(n)$ to the outputs of the delays as shown in Fig. 5.27. The state equations are then written for the inputs to the delays which are the next states variables $s_k(n + 1)$. The state variables are used to form a column vector. The column vector for a state space representation with L states is given by

$$S(n) = \begin{bmatrix} s_1(n) \\ s_2(n) \\ \vdots \\ s_L(n) \end{bmatrix}.$$ (5.211)

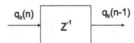

FIGURE 5.28

Modified assignment of state variables for a discrete time system

The state equations, in matrix form, for a single input sequence $x(n)$ and a single output sequence $y(n)$ can be represented in the form

$$
\begin{aligned}
S(n+1) &= AS(n) + Bx(n), \\
y(n) &= CS(n) + Dx(n).
\end{aligned}
\tag{5.212}
$$

If the system has L states, then A is an $L \times L$ matrix, B is an $L \times 1$ column vector, C is a $1 \times L$ row vector, and D is a 1×1 matrix (a scalar).

The next state variables in $S(n+1)$ are the values to be saved in the delays (memories) for a software or hardware implementation. However, it is more common to write difference equations to compute the output(s) for the current value of the index, n, rather than for the next value of the index, $n+1$. The next state equations can be modified to compute the states for the index n by making the substitution

$$
Q(n) =
\begin{bmatrix}
q_1(n) \\
q_2(n) \\
\vdots \\
q_L(n)
\end{bmatrix}
= S(n+1) =
\begin{bmatrix}
s_1(n+1) \\
s_2(n+1) \\
\vdots \\
s_L(n+1)
\end{bmatrix}.
\tag{5.213}
$$

In this way, the outputs for the difference equations still represent signals that need to be saved for later computations, and therefore should be stored in memory. This approach provides advantages for both software and hardware implementation of the discrete time system without changing the analytical basis for the state space representation. The modified assignment of state variables is shown in Fig. 5.28.

This has the effect of identifying the state variables as the inputs to the delays rather than the outputs from the delays as would be a convenient way to identify state variables for Eq. (5.212). The advantage of this substitution is that the difference equations directly compute the inputs to the delays which is more consistent for the actual implementation. The state space representation for the single input, single output discrete time system, with this change, is given by

$$
\begin{aligned}
Q(n) &= AQ(n-1) + Bx(n), \\
y(n) &= CQ(n-1) + Dx(n)
\end{aligned}
\tag{5.214}
$$

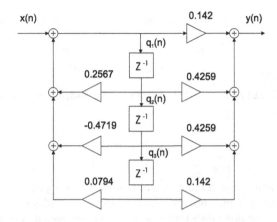

FIGURE 5.29

Direct form realization for Example 5.15

where $x(n)$ is the current input, $y(n)$ is the current output, A, B, C, and D are matrices of appropriate dimensions, and $Q(n)$ is a state vector given by

$$Q(n) = \begin{bmatrix} q_1(n) \\ q_2(n) \\ q_3(n) \\ \vdots \\ q_L(n) \end{bmatrix}. \tag{5.215}$$

Note that the A, B, C, and D for Eq. (5.214) are the same as the corresponding matrices for Eq. (5.212). The matrix A is called the state transition matrix. We will use Eq. (5.214) for the development of the state space representations in this text.

We can illustrate the concept of the state space representation using an example.

EXAMPLE 5.15
(Direct Form Representation)

This example involves the state space representation of a recursive discrete time system with the difference equation representation given by

$$\begin{aligned} y(n) &= 0.142x(n) + 0.4259x(n-1) + 0.4259x(n-2) \\ &+ 0.142x(n-3) + 0.2567y(n-1) \\ &- 0.4719y(n-2) + 0.0794y(n-3). \end{aligned} \tag{5.216}$$

The state space representation of this discrete time system should be in the form

$$\begin{aligned} Q(n) &= AQ(n-1)+Bx(n), \\ y(n) &= CQ(n-1)+Dx(n) \end{aligned} \tag{5.217}$$

where $A, B, C,$ and D are matrices of appropriate dimensions, and $Q(n)$ is a state vector given by

$$Q(n) = \begin{bmatrix} q_1(n) \\ q_2(n) \\ q_3(n) \end{bmatrix}. \tag{5.218}$$

Solution: We first obtain a block diagram or signal flow graph representation of the discrete time system. Then we can assign state variables to the inputs of the delays. This can be done directly from the difference equation or we can obtain the transfer function for the system as shown in Eq. (5.219):

$$H(z) = \frac{0.142 + 0.4259z^{-1} + 0.4259z^{-2} + 0.142z^{-3}}{1.0 - 0.2567z^{-1} + 0.4719z^{-2} - 0.794z^{-3}}. \tag{5.219}$$

Fig. 5.29 gives the block diagram for the direct form realization for $H(z)$. The corresponding state equations, with the state variables assigned to the inputs to the delays, are given by

$$\begin{aligned} q_1(n) &= 0.2567q_1(n-1) - 0.4719q_2(n-1) + 0.0794q_3(n-1) + x(n), \\ q_2(n) &= q_1(n-1), \\ q_3(n) &= q_2(n-1). \end{aligned} \tag{5.220}$$

The output equation is given by

$$y(n) = 0.142q_1(n). \tag{5.221}$$

The output equation is given in terms of the current value of the state variable $q_1(n)$, which is not in the desired form. We can obtain the correct form by substituting for $q_1(n)$ in the output equation. Thus, we obtain

$$\begin{aligned} y(n) &= 0.142\big[0.2567q_1(n-1) - 0.4719q_2(n-1) \\ &\quad + 0.0794q_3(n-1) + x(n)\big], \\ y(n) &= 0.4624q_1(n-1) + 0.3589q_2(n-1) + 0.1307q_3(n-1) \\ &\quad + 0.142x(n). \end{aligned} \tag{5.222}$$

The corresponding matrices for the state space representation are given by

$$A = \begin{bmatrix} 0.2567 & -0.4719 & 0.0794 \\ 1.0000 & 0000.0 & 0000.0 \\ 0.0000 & 1.0000 & 0000.0 \end{bmatrix},$$

$$B = \begin{bmatrix} 1.0000 \\ 0.0000 \\ 0.0000 \end{bmatrix},$$

$$C = [0.4624 \quad 0.3589 \quad 0.1533],$$

$$D = [0.142]. \tag{5.223}$$

We can use the Matlab function *tf2ss* to obtain the state space representation for the direct form as given in this example. The following Matlab script can be used to obtain this state space representation.

Matlab Script 5.7.

```
b = [0.142 0.4259 0.4259 0.142];
a = [1.0 -0.2567 0.4719 -0.0794];
[A B C D] = tf2ss(b,a);
```

End of the Script

The results for this discrete time system are given by

```
A =
    0.2567    -0.4719    0.0794
    1.0000          0         0
         0     1.0000         0

B =
    1
    0
    0

C =
    0.4624    0.3589    0.1533

D =
    0.1420
```

Note that this is consistent with the results obtained from the direct form block diagram as given in Eq. (5.223).
End of the Example

EXAMPLE 5.16
(Transpose Direct Form Representation)

We now consider the development of the state space realization of the recursive discrete time system used in Eq. (5.219) using the transpose direct form. We will develop the state space representation for the transpose direct form using the equa-

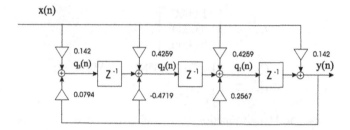

FIGURE 5.30

Block diagram representation of system for Example 5.16

tions

$$Q(n) = A_2Q(n-1) + B_2x(n),$$
$$y(n) = C_2Q(n-1) + D_2x(n). \tag{5.224}$$

Fig. 5.30 gives a block diagram for the transpose direct form representation for the discrete time system used in discrete time system. We can assign state variables to the inputs to the delays and write equations for the state equations and the output as follows:

$$y(n) = 0.142x(n) + q_1(n-1),$$
$$q_1(n) = 0.4259x(n) + q_2(n-1) + 0.2567y(n),$$
$$q_2(n) = 0.4259x(n) + q_3(n-1) - 0.4719y(n),$$
$$q_3(n) = 0.142x(n) + 0.0794y(n). \tag{5.225}$$

The state equations are given in terms of the feedback of the output $y(n)$. We can obtain the correct form by substituting for $y(n)$ in state equations for $q_1(n)$, $q_2(n)$, and $q_3(n)$. Thus, we obtain

$$q_1(n) = [0.4259 + 0.2567(0.142)]x(n) + q_2(n-1)$$
$$+ 0.2567q_1(n-1),$$
$$q_2(n) = [0.4259 - 0.4719(0.142)]x(n) + q_3(n-1)$$
$$- 0.4719q_1(n-1),$$
$$q_3(n) = [0.142 + 0.0794(0.142)]x(n) + 0.0794q_1(n-1). \tag{5.226}$$

It follows that

$$A_2 = \begin{bmatrix} 0.2567 & 1 & 0 \\ -0.4719 & 0 & 1 \\ 0.0794 & 0 & 0 \end{bmatrix},$$

$$B_2 = \begin{bmatrix} 0.4624 \\ 0.3589 \\ 0.1533 \end{bmatrix},$$
$$C_2 = [1 \quad 0 \quad 0],$$
$$D_2 = [0.142]. \tag{5.227}$$

The matrices for the state space representation for the transpose direct form are transposes of the corresponding matrices for the direct form. Thus, we can use the state space representation for the transpose direct form by using the Matlab function *tf2ss* followed by transposing the matrices. This can be accomplished by using the following Matlab script.

Matlab Script 5.8.

```
b = [0.142 0.4259 0.4259 0.142];
a = [1.0 -0.2567 0.4719 -0.0794];
[A B C D] = tf2ss(b,a);
A2 = transpose(A);
B2 = transpose(C);
C2 = transpose(B);
D2 = D;
```

End of the Script

The results are:

```
A2 =
    0.2567    1.0000         0
   -0.4719         0    1.0000
    0.0794         0         0

B2 =
    0.4624
    0.3589
    0.1533

C2 =
    1    0    0

D2 =
    0.1420
```

These results are consistent with the representation in Eq. (5.227).

The procedures presented in Examples 5.15 and 5.16 are general and can be used to obtain the state space representation corresponding to the direct form or the transpose direct form for any single input, single output, discrete time system.
End of the Example

5.9.1 DIAGONAL FORM FOR THE STATE TRANSITION MATRIX

The update of the states for the single input, single output, state space representation involves the multiplication of a matrix A and a column vector $Q(n)$ and the multiplication of a column vector by a scalar $x(n)$. Generally, the matrix A may be fully populated so that $(L+1) \times L$ multiplications and $L \times L$ additions are needed. However, a similarity transformation may be used to diagonalize A if it does not have any repeated eigenvalues [13]. If we diagonalize the state transition matrix A, then the update of the states will require $2 \times L$ multiplications and L additions. This can result in substantial computational savings for high order systems.

We will discuss a procedure to accomplish this in this section and provide examples for single input, single output, discrete time systems. We begin with the state space representation of a single input, single output system as given by

$$
\begin{aligned}
Q(n) &= AQ(n-1) + Bx(n), \\
y(n) &= CQ(n-1) + Dx(n).
\end{aligned} \tag{5.228}
$$

We can perform a similarity transformation such that

$$
Q(n) = PR(n) \tag{5.229}
$$

where P is a square matrix with an inverse. The corresponding state space representation using this similarity transformation then becomes

$$
\begin{aligned}
PR(n) &= APR(n-1) + Bx(n), \\
y(n) &= CPR(n-1) + Dx(n).
\end{aligned} \tag{5.230}
$$

We consider the special case where the matrix P is composed of the eigenvectors of the matrix A. If the eigenvalues of A are distinct, then P has an inverse P^{-1}. We can multiply the equation for the update of the state variables by P^{-1} to obtain

$$
\begin{aligned}
P^{-1}PR(n) &= P^{-1}APR(n-1) + P^{-1}Bx(n), \\
y(n) &= CPR(n-1) + Dx(n).
\end{aligned} \tag{5.231}
$$

It follows that

$$
\begin{aligned}
R(n) &= P^{-1}APR(n-1) + P^{-1}Bx(n), \\
y(n) &= CPR(n-1) + Dx(n).
\end{aligned} \tag{5.232}
$$

We define the matrix product

$$
J = P^{-1}AP. \tag{5.233}
$$

The matrix J is a diagonal matrix with the eigenvalues for the matrix A on its diagonal for the special case where the eigenvalues of A are distinct and the matrix P is composed of the eigenvectors of A. It follows for this case that

$$
\begin{aligned}
R(n) &= JR(n-1) + P^{-1}Bx(n), \\
y(n) &= CPR(n-1) + Dx(n).
\end{aligned}
\tag{5.234}
$$

5.9.2 COMPLEX EIGENVALUES

The coefficients for the system transfer function and also for the finite difference equations are real, in general. Thus, the eigenvalues of A are real or they occur in complex conjugate pairs. We refer to the element in row k_1 and column k_2 as $J(k_1, k_2)$. Suppose the eigenvalue at $J(k_1, k_1)$ is complex and eigenvalue $J(k_1 + 1, k_1 + 1)$ is its complex conjugate. The use of complex numbers during computations typically increases the computational complexity compared to the use of only real numbers for the computations. The multiplication of two complex numbers requires 4 multiplications and 2 additions.

We can use another similarity transformation to transform the equations for the update of the state variables so that only real numbers are needed. We consider a case for a second order system with complex poles to illustrate this procedure. We begin with the state space representation given in Eq. (5.234). If a second order system has two complex eigenvalues, then we can define the diagonal matrix J as

$$
J = \begin{bmatrix} \alpha + j\beta & 0.0 \\ 0.0 & \alpha - j\beta \end{bmatrix}
\tag{5.235}
$$

where $\alpha + j\beta$ and $\alpha - j\beta$ are the complex conjugate eigenvalues of A. Ogata [13] defines the matrix T as

$$
T = \begin{bmatrix} 0.5 & -0.5j \\ 0.5 & 0.5j \end{bmatrix}.
\tag{5.236}
$$

We can use the similarity transformation

$$
R = TV
\tag{5.237}
$$

to transform the matrices in the state equations so that only real numbers are needed. The inverse of T is given by

$$
T^{-1} = \begin{bmatrix} 1.0 & 1.0 \\ j & -j \end{bmatrix}.
\tag{5.238}
$$

The state equations using this similarity transformation are given by

$$
\begin{aligned}
TV(n) &= JTV(n-1) + P^{-1}Bx(n), \\
y(n) &= CPTV(n-1) + Dx(n).
\end{aligned}
\tag{5.239}
$$

We can multiply the state update equation by T^{-1} to obtain

$$
T^{-1}TV(n) = T^{-1}JTV(n-1) + T^{-1}P^{-1}Bx(n),
$$

$$y(n) = CPTV(n-1) + Dx(n). \qquad (5.240)$$

We can then write the state equations for the system in the following form using these results:

$$V(n) = T^{-1}JTV(n-1) + T^{-1}P^{-1}Bx(n),$$
$$y(n) = CPTV(n-1) + Dx(n). \qquad (5.241)$$

We define

$$U = T^{-1}JT. \qquad (5.242)$$

Thus,

$$U = \begin{bmatrix} 1.0 & 1.0 \\ j & -j \end{bmatrix} \begin{bmatrix} \alpha + j\beta & 0.0 \\ 0.0 & \alpha - j\beta \end{bmatrix} \begin{bmatrix} 0.5 & -0.5j \\ 0.5 & 0.5j \end{bmatrix}, \qquad (5.243)$$

$$U = \begin{bmatrix} (\alpha + j\beta) & (\alpha - j\beta) \\ j(\alpha + j\beta) & -j(\alpha - j\beta) \end{bmatrix} \begin{bmatrix} 0.5 & -0.5j \\ 0.5 & 0.5j \end{bmatrix},$$

$$U = \begin{bmatrix} \alpha & \beta \\ -\beta & \alpha \end{bmatrix}. \qquad (5.244)$$

Generally, the matrix J will be larger that a 2×2 matrix. We can modify the T matrix to accommodate higher order systems with larger state transition matrices. Generally, the eigenvalue at position $J(k_1, k_1)$ is real or complex. If it is real, then the corresponding element in the matrix T should be set to 1.0. If it is complex, then the eigenvalue at $J(k_1 + 1, k_1 + 1)$ is its complex conjugate. We can form the square matrix T such that

1. If the eigenvalue at $J(k, k)$ is real, then the diagonal element $T(k, k) = 1.0$.
2. If the eigenvalue at $J(k, k)$ is complex and eigenvalue $J(k+1, k+1)$ is its complex conjugate, then
   (a) $T(k, k) = 0.5$,
   (b) $T(k, k+1) = -j0.5$,
   (c) $T(k+1, k) = 0.5$,
   (d) $T(k+1, k+1) = j0.5$.
3. The elements of T are zeros otherwise.

Then, with

$$U = T^{-1}JT \qquad (5.245)$$

the corresponding state space equations are given by

$$V(n) = UV(n-1) + T^{-1}P^{-1}Bx(n),$$
$$y(n) = CPTV(n-1) + Dx(n). \qquad (5.246)$$

Let

$$A_3 = U, \tag{5.247}$$

$$B_3 = T^{-1}P^{-1}B, \tag{5.248}$$

$$C_3 = CPT, \tag{5.249}$$

and

$$D_3 = D. \tag{5.250}$$

We then have

$$\begin{aligned} V(n) &= A_3 V(n-1) + B_3 x(n), \\ y(n) &= C_3 T(n-1) + D_3 x(n). \end{aligned} \tag{5.251}$$

Note that A_3, B_3, C_3, and D_3 have real coefficients.

We now consider an example involving a system with complex eigenvalues.

EXAMPLE 5.17

(State Space Representation using Matlab)

We can use the following Matlab script to design a low pass digital filter.

Matlab Script 5.9.

```
order = 3;
wc = 0.45;
rp = 1.0;
rs = 40;
[b, a] = ellip(order, rp, rs, wc);
```

End of the Script

The corresponding filter coefficient vectors are given by

$$\begin{aligned} b &= [0.1261 \ 0.3012 \ 0.3012 \ 0.1261], \\ a &= [1.0000 \ -0.6333 \ 0.7300 \ -0.2421]. \end{aligned} \tag{5.252}$$

We can use the following Matlab script to compute the state space matrices and the corresponding P and J matrices.

Matlab Script 5.10.

```
[A B C D] = tf2ss(b, a);
[P J] = eig(A);
```

End of the Script

The results are given by

$$A = \begin{bmatrix} 0.6333 & -0.7300 & 0.2421 \\ 1.0000 & 0.0000 & 0.0000 \\ 0 & 1.0000 & 0.0000 \end{bmatrix}, \qquad (5.253)$$

$$B = \begin{bmatrix} 1.0000 \\ 0.0000 \\ 0.0000 \end{bmatrix}, \qquad (5.254)$$

$$C = [0.3811 \quad 0.2092 \quad 0.1566], \qquad (5.255)$$

$$D = [0.1261]. \qquad (5.256)$$

Matrix P, which is composed of the eigenvectors for A, is given by

$$P = \begin{bmatrix} (-0.4222 + 0.1385j) & (-0.4222 - 0.1385j) & 0.1350 \\ (0.0881 + 0.5510j) & (0.0881 - 0.5510j) & 0.3535 \\ 0.7008 & 0.7008 & 0.9256 \end{bmatrix}, \qquad (5.257)$$

$$J = \begin{bmatrix} (0.1257 + 0.7863j) & 0.0000 & 0.0000 \\ 0.0000 & (0.1257 - 0.7863j) & 0.0000 \\ 0.0000 & 0.0000 & 0.3819 \end{bmatrix}. \qquad (5.258)$$

We observe that there are two complex eigenvalues. Thus, matrix T is given by

$$T = \begin{bmatrix} 0.5000 & (0.0000 - 0.5000j) & 0.0000 \\ 0.5000 & (0.0000 + 0.5000j) & 0.0000 \\ 0.0000 & 0.0000 & 1.0000 \end{bmatrix}. \qquad (5.259)$$

It follows that

$$
\begin{aligned}
A_3 &= T^{-1}JT = T^{-1}P^{-1}APT \\
&= \begin{bmatrix} 0.1257 & 0.7863 & 0.0000 \\ -0.7863 & 0.1257 & 0.0000 \\ 0.0000 & 0.0000 & 0.3819 \end{bmatrix}, \qquad (5.260)
\end{aligned}
$$

$$B_3 = T^{-1}P^{-1}B = \begin{bmatrix} -2.0865 \\ -0.6798 \\ 1.5797 \end{bmatrix}, \qquad (5.261)$$

$$C_3 = CPT = [-0.0327 \quad 0.1681 \quad 0.2704], \qquad (5.262)$$

$$D_3 = D = [0.1261]. \qquad (5.263)$$

Note that all of the elements of these matrices are real.

Fig. 5.31 gives a stem plot of an input sequence. We use the Matlab function *filter* to filter the input sequence.

Matlab Script 5.11.

```
y1 = filter(b, a, x);
```

End of the Script

The output is $y1(n)$ and the input is $x(n)$. Fig. 5.32 gives a stem plot of the output sequence, $y1(n)$. The following Matlab script can be used to filter the input sequence $x(n)$ using the modified state space representation.

Matlab Script 5.12.

```
Vm1 = zeros(3,1);
m = length(x);
for k=1:m
    f = x(k);
    V = A3*Vm1 + B3*f;
    y2(k) = C3*Vm1 + D*f;
    Vm1 = V;
end
```

End of the Script

Fig. 5.33 gives a comparison stem plot of the two output sequences, $y1(n)$ and $y2(n)$. The normalized peak error using $y1(n)$ as the reference was 6.526300×10^{-16}. The normalized average mean square error using $y1(n)$ as the reference was 2.889576×10^{-32}.

End of the Example

5.10 PROBLEMS FOR CHAPTER 5

Problem 5.1. (IIR System Implementation)

Consider a pole–zero system with system function

$$H(z) = \frac{0.2929 \left(1 - 0.5e^{j\pi/4}z^{-1}\right)\left(1 - 0.5e^{-j\pi/4}z^{-1}\right)}{\left(1 - 0.8e^{j\pi/3}z^{-1}\right)\left(1 - 0.8e^{-j\pi/3}z^{-1}\right)}. \qquad (5.264)$$

1. Draw a block diagram for the regular direct form realization of the system using only real coefficients.
2. Write one or more difference equations to implement the system based upon your block diagram from part 1.
3. Draw a block diagram for the transpose direct form realization of the system using only real coefficients.

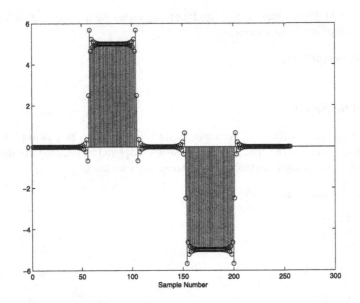

FIGURE 5.31

Stem plot of an input sequence

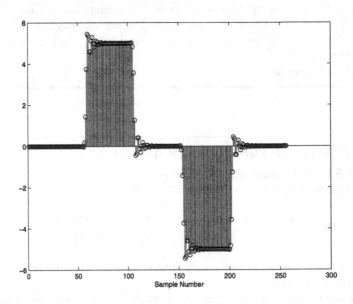

FIGURE 5.32

Stem plot of the output sequence $y1(n)$

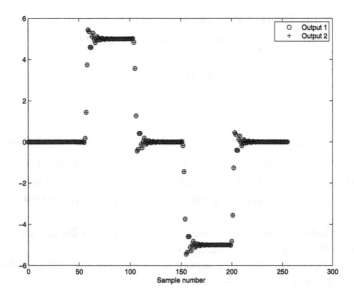

FIGURE 5.33

Stem plot of the output sequence $y1(n)$

4. Write one or more difference equations to implement the system based upon your block diagram from part 3.

Problem 5.2. (FIR Lattice System Implementation)

Consider the FIR lattice filter with coefficients

$$
\begin{aligned}
K_1 &= 0.654, \\
K_2 &= -0.549, \\
K_3 &= 0.483.
\end{aligned}
$$

(5.265)

1. Determine the impulse response of this system by tracing a unit impulse

$$ x(n) = \delta(n) $$

(5.266)

through the lattice structure.

2. Determine the system transfer function

$$ H(z) = b(0) + b(1)z^{-1} + b(2)z^{-2} + b(3)z^{-3}. $$

(5.267)

In other words, determine $b(0)$, $b(1)$, $b(2)$, and $b(3)$.

3. Draw the equivalent direct form implementation for this filter.

Problem 5.3. (Second Order Sections)
The system transfer function for an IIR digital filter is given by

$$H(z) = \frac{B(z)}{A(z)} \tag{5.268}$$

where

$$
\begin{aligned}
B(z) &= 0.0993 + 0.2919z^{-1} + 0.6264z^{-2} + 0.8893z^{-3} + 1.0398z^{-4} \\
&\quad + 0.8893z^{-5} + 0.6264z^{-6} + 0.2919z^{-7} + 0.0993z^{-8}, \\
A(z) &= 1.0 + 1.4837z^{-1} + 0.4873z^{-2} + 0.5524z^{-3} + 1.3204z^{-4} \\
&\quad + 0.4069z^{-5} - 0.2463z^{-6} + 0.2077z^{-7} + 0.2337z^{-8}. \tag{5.269}
\end{aligned}
$$

1. Use the Matlab routine *tf2sos* to obtain the coefficients for the cascade implementation of $H(z)$ using second order sections.
2. Develop appropriate difference equations for the implementation of the overall filter using the second order IIR sections.

Problem 5.4. (State Space Representation)
The system transfer function for a discrete time digital system is given by

$$H(z) = \frac{B(z)}{A(z)} \tag{5.270}$$

where

$$
\begin{aligned}
B(z) &= 0.1429 - 0.7147z^{-1} + 1.4295z^{-2} - 1.4295z^{-3} \\
&\quad + 0.7147z^{-4} - 0.1429z^{-5}, \\
A(z) &= 1.0 - 1.4104z^{-1} + 1.3328z^{-2} - 0.6331z^{-3} \\
&\quad + 0.1777z^{-4} - 0.0202z^{-5}. \tag{5.271}
\end{aligned}
$$

Develop a state space representation for the system using the form

$$Q(n) = \mathbf{A}Q(n-1) + \mathbf{B}x(n), \tag{5.272}$$

$$y(n) = \mathbf{C}Q(n-1) + \mathbf{D}x(n). \tag{5.273}$$

Clearly identify the $\mathbf{A}$, $\mathbf{B}$, $\mathbf{C}$, and $\mathbf{D}$ matrices for your implementation.

Problem 5.5. (Second Order IIR Lattice Sections)
The system transfer function for an IIR digital filter is given by

$$H(z) = \frac{B(z)}{A(z)} \tag{5.274}$$

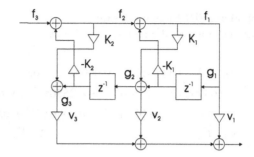

FIGURE 5.34

Block diagram for a second order IIR lattice section

where

$$
\begin{aligned}
B(z) &= 0.1203 + 0.5399z^{-1} + 1.1534z^{-2} + 1.4611z^{-3} \\
&\quad + 1.1534z^{-4} + 0.5399z^{-5} + 0.1203z^{-6}, \\
A(z) &= 1.0 + 1.0466z^{-1} + 1.8933z^{-2} + 0.7336z^{-3} \\
&\quad + 0.8238z^{-4} + 0.0399z^{-5} + 0.1721z^{-6}.
\end{aligned}
\tag{5.275}
$$

The results from using the Matlab *tf2sos* function to obtain the coefficients for a second order section implementation of $H(z)$ are given by

```
sos =
    1.0000    1.8854    1.0000    1.0000   -0.3513    0.2489
    1.0000    1.4166    1.0000    1.0000    0.5288    0.7336
    1.0000    1.1867    1.0000    1.0000    0.8691    0.9423

gain =
    0.1203
```

1. Fig. 5.34 gives a block diagram for a second order IIR lattice section. Determine the lattice parameters for each of the second order sections using the output from the Matlab *tf2sos* function as given above.
2. Develop a block diagram for the IIR lattice implementation of each of the second order sections.
3. Develop appropriate difference equations for the implementation of the overall filter using the second order IIR lattice sections.

Problem 5.6. (All Pole IIR Lattice Section)
The system function for an all pole IIR digital filter is given by

$$
H(z) = \frac{0.8795}{1.0 - 0.4672z^{-1} + 0.3912z^{-2} - 0.0445z^{-3}}.
\tag{5.276}
$$

Determine the lattice parameters for an all pole lattice implementation of this filter.

Problem 5.7. (Polyphase FIR Implementation)
The system transfer function for a digital filter is given by

$$
\begin{aligned}
H(z) \;=\; & -0.0023 - 0.0062z^{-1} + 0.0005z^{-2} + 0.0333z^{-3} \\
& + 0.0390z^{-4} - 0.0786z^{-5} - 0.2880z^{-6} + 0.6014z^{-7} \\
& - 0.2880z^{-8} - 0.0786z^{-9} + 0.0390z^{-10} + 0.0333z^{-11} \\
& + 0.0005z^{-12} - 0.0062z^{-13} - 0.0023z^{-14}. \tag{5.277}
\end{aligned}
$$

1. Develop a polyphase representation of $H(z)$ in the form

$$
H(z) = E_1(z^3) + z^{-1}E_2(z^3) + z^{-2}E_3(z^3). \tag{5.278}
$$

2. Write appropriate difference equations to implement the polyphase version of the system.

Problem 5.8. (Linear Phase FIR Filter)
The system transfer function for a digital filter is given by

$$
\begin{aligned}
H(z) \;=\; & -0.0019 + 0.0165z^{-1} - 0.0443z^{-2} - 0.0187z^{-3} \\
& + 0.5484z^{-4} + 0.5484z^{-5} - 0.0187z^{-6} - 0.0443z^{-7} \\
& + 0.0165z^{-8} - 0.0019z^{-9}. \tag{5.279}
\end{aligned}
$$

1. Draw a block diagram for the implementation of $H(z)$ using only 5 multipliers.
2. Write one or more difference equations to implement $H(z)$ using only 5 multiplications.

Problem 5.9. (FIR Lattice Structure)
The system transfer function for a FIR digital filter is given by

$$
\begin{aligned}
H(z) \;=\; & 0.7652 + 0.6177z^{-1} - 1.0587z^{-2} + 1.2362z^{-3} \\
& - 0.8202z^{-4} + 0.2598z^{-5}. \tag{5.280}
\end{aligned}
$$

1. Determine the K_m parameters for a lattice implementation of this FIR filter. Show the details of how you computed the parameters.
2. Verify the calculations for your parameters by using the Matlab *tf2latc* function. Turn in the parameters and the Matlab script you used to compute the parameters.
3. Write appropriate difference equations for the lattice implementation using your coefficients.

Problem 5.10. (System Transfer Function)

A discrete time system can be represented by the following difference equations:

$$
\begin{aligned}
r_1(n) &= -0.0072x(n) + 0.2642x(n-4) + 0.0562x(n-8), \\
s_1(n) &= x(n-1), \\
r_2(n) &= 0.0562s_1(n) + 0.2642s_1(n-4) - 0.0072s_1(n-8), \\
s_2(n) &= s_1(n-1), \\
r_3(n) &= -0.0201s_2(n) - 0.3008s_2(n-4), \\
s_3(n) &= s_2(n-1), \\
r_4(n) &= -0.3008s_3(n) - 0.0201s_3(n-4), \\
y(n) &= r_1(n) + r_2(n) + r_3(n) + r_4(n). \quad (5.281)
\end{aligned}
$$

Determine the system transfer function $H(z)$ for this discrete time system where $x(n)$ is the input and $y(n)$ is the output.

Problem 5.11. (Difference Equation and Frequency Response)

The system transfer function for a discrete time system is given by

$$
\begin{aligned}
H(z) = \ & 0.0123 + 0.1214z^{-1} + 0.3663z^{-2} + 0.3663z^{-3} \\
& + 0.1214z^{-4} + 0.0123z^{-5}. \quad (5.282)
\end{aligned}
$$

1. Derive a difference equation representation of the system that requires only 3 multiplications.
2. Derive an expression for $H(\omega)$ which is the frequency response of the system.
3. Derive an expression for $H(\omega)$ in the form

$$
H(\omega) = A_0 + e^{-jB\omega} \left[\sum_{k=1}^{3} A_k \cos(C_k\omega) \right] \quad (5.283)
$$

where all of A_0, A_k, B, and C_k are real constants. Show all of the details of your derivation.

Problem 5.12. (Parallel IIR Implementation)

The partial fraction expansion for an IIR filter is given by

$$
\begin{aligned}
H(z) = \ & \frac{(0.0413 + 0.0329j)z}{z + 0.4290 - 0.8482j} + \frac{(-0.3203 + 0.0197j)z}{z + 0.1519 - 0.7238j} \\
& + \frac{(0.0413 - 0.0329j)z}{z + 0.4290 + 0.8482j} + \frac{(-0.3203 - 0.0197j)z}{z + 0.1519 + 0.7238j} \\
& + \frac{1.9649z}{z - 0.2559} - 1.2489. \quad (5.284)
\end{aligned}
$$

Write a set of difference equations for the parallel implementation of $H(z)$ using parallel second or lower order sections with only real coefficients. Your final equation should be the sum of the outputs due to the parallel sections of the implementation.

Problem 5.13. (FIR Lattice Structure)
The system transfer function for an FIR filter is given by

$$H(z) = 2.1510 - 1.4318z^{-1} - 0.2428z^{-2} - 1.9937z^{-3}$$
$$+ 1.1593z^{-4} - 0.2163z^{-5}. \tag{5.285}$$

Determine the lattice parameters for an FIR lattice implementation of this filter. Show all of the details of your computations.

Problem 5.14. (IIR Lattice Structure)
The system transfer function for an all pole IIR filter is given by

$$H(z) = \frac{B(z)}{A(z)},$$
$$B(z) = 10.1270,$$
$$A(z) = 1.0000 - 1.9980z^{-1} + 2.9244z^{-2} - 2.4434z^{-3}$$
$$+ 1.3601z^{-4} - 0.4010z^{-5}. \tag{5.286}$$

Determine the K and V lattice parameters for an IIR lattice implementation of this filter. Show all of the details of your computations.

Problem 5.15. (FIR Lattice Structure)
The system transfer function for an FIR digital filter is given by

$$H(z) = 0.0211 + 0.1030z^{-1} + 0.3385z^{-2} + 0.2916z^{-3} + 0.2458z^{-4}. \tag{5.287}$$

Determine the K_m parameters for a lattice implementation of this FIR filter. Show the details of how you computed the parameters.

Problem 5.16. (State Space Representation)
The block diagram for a discrete time system is given in Fig. 5.35. Develop a state space representation for the system using the form

$$Q(n) = AQ(n-1) + Bx(n), \tag{5.288}$$

$$y(n) = CQ(n-1) + Dx(n). \tag{5.289}$$

Clearly identify the A, B, C, and D matrices for your implementation.

Problem 5.17. (IIR Lattice Implementation)
The system transfer function for a discrete time system is given by

$$H(z) = \frac{3.8613}{1.0 + 1.1347z^{-1} + 1.1860z^{-2} + 0.3409z^{-3} + 0.1997z^{-4}}. \tag{5.290}$$

Determine the lattice parameters for the IIR lattice implementation of this filter.

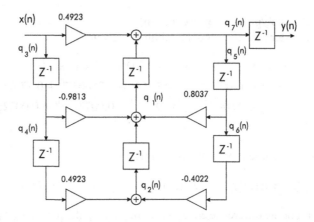

FIGURE 5.35

System block diagram for Problem 5.16

Problem 5.18. (Second Order Sections)
The system transfer function for an FIR digital filter is given by

$$
\begin{aligned}
H(z) \;=\; & 0.0052 - 0.0087z^{-1} - 0.0114z^{-2} + 0.1043z^{-3} \\
& - 0.2425z^{-4} + 0.3114z^{-5} - 0.2425z^{-6} + 0.1043z^{-7} \\
& - 0.0114z^{-8} - 0.0087z^{-9} + 0.0052z^{-10}.
\end{aligned}
\tag{5.291}
$$

1. Develop the difference equations for the implementation of this filter using a cascade of five second order sections. Show the details for deriving the difference equations.
2. Show which roots are combined to obtain the second order sections and explain why they were matched this way.

Problem 5.19. (State Space Representation)
The system transfer function for a discrete time digital system is given by

$$
H(z) = \frac{0.2871 - 0.5742z^{-2} + 0.2871z^{-4}}{1.0 - 0.2653z^{-1} - 0.0886z^{-2} - 0.0104z^{-3} + 0.0910z^{-4}}.
\tag{5.292}
$$

Develop a state space representation for the system using the form

$$
Q(n) = \mathbf{A}Q(n-1) + \mathbf{B}x(n),
\tag{5.293}
$$

$$
y(n) = \mathbf{C}Q(n-1) + \mathbf{D}x(n).
\tag{5.294}
$$

Clearly identify the **A**, **B**, **C**, and **D** matrices for your implementation.

Problem 5.20. (Polyphase Representation)
The system transfer function for a digital filter is given by

$$\begin{aligned}
H(z) \quad = \quad & 0.0022 - 0.0073z^{-1} - 0.0290z^{-2} + 0.0048z^{-3} \\
& + 0.1707z^{-4} + 0.3586z^{-5} + 0.3586z^{-6} + 0.1707z^{-7} \\
& + 0.0048z^{-8} - 0.0290z^{-9} - 0.0073z^{-10} + 0.0022z^{-11}.
\end{aligned}$$

$$(5.295)$$

1. Develop a polyphase representation of $H(z)$ in the form

$$H(z) = E_1(z^4) + z^{-1}E_2(z^4) + z^{-2}E_3(z^4) + z^{-3}E_4(z^4). \qquad (5.296)$$

2. Write appropriate difference equations to implement the polyphase version of the system.

Problem 5.21. (FIR System Implementation)
The system transfer function for a digital filter is given by

$$\begin{aligned}
H(z) \quad = \quad & 0.0022 - 0.0073z^{-1} - 0.0290z^{-2} + 0.0048z^{-3} \\
& + 0.1707z^{-4} + 0.3586z^{-5} + 0.3586z^{-6} + 0.1707z^{-7} \\
& + 0.0048z^{-8} - 0.0290z^{-9} - 0.0073z^{-10} + 0.0022z^{-11}.
\end{aligned}$$

$$(5.297)$$

1. Draw a block diagram for the implementation of $H(z)$ using only 6 multipliers.
2. Write one or more difference equations to implement $H(z)$ using only 6 multiplications.

Problem 5.22. (State Space Representation)
The block diagram for a discrete time digital system is given in Fig. 5.36. Develop a state space representation for the system using the form

$$Q(n) = \mathbf{A}Q(n-1) + \mathbf{B}x(n), \qquad (5.298)$$

$$y(n) = \mathbf{C}Q(n-1) + \mathbf{D}x(n). \qquad (5.299)$$

Clearly identify the $\mathbf{A}, \mathbf{B}, \mathbf{C}$, and $\mathbf{D}$ matrices for your implementation.

Problem 5.23. (Polyphase FIR Implementation)
The system transfer function for a digital filter is given by

$$\begin{aligned}
H(z) \quad = \quad & 0.0168 + 0.0264z^{-1} - 0.0409z^{-2} + 0.0334z^{-3} + 0.0287z^{-4} \\
- \quad & 0.1205z^{-5} + 0.2453z^{-6} + 0.6694z^{-7} + 0.2453z^{-8} - 0.1205z^{-9} \\
+ \quad & 0.0287z^{-10} + 0.0334z^{-11} - 0.0409z^{-12} + 0.0264z^{-13} + 0.0168z^{-14}.
\end{aligned}$$

$$(5.300)$$

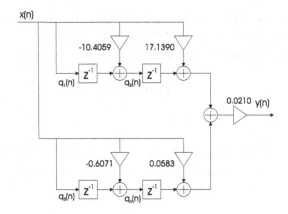

FIGURE 5.36

Block diagram for Problem 5.22.

1. Develop a polyphase representation of $H(z)$ in the form

$$H(z) = E_1(z^4) + z^{-1} E_2(z^4) + z^{-2} E_3(z^4) + z^{-3} E_4(z^4). \qquad (5.301)$$

Define $E_k(z) \ \forall \ 1 \leq k \leq 4$.

2. Write appropriate difference equations to implement the polyphase version of the system.

Problem 5.24. (Linear Phase FIR Filter)

The system transfer function for a digital filter is given by

$$
\begin{aligned}
H(z) \quad = \quad & 0.0062 + 0.0113z^{-1} - 0.0548z^{-2} - 0.2568z^{-3} + 0.6061z^{-4} \\
& - 0.2568z^{-5} - 0.0548z^{-6} + 0.0113z^{-7} + 0.0062z^{-8}. \qquad (5.302)
\end{aligned}
$$

1. Draw a block diagram for the implementation of $H(z)$ using only 5 multipliers.
2. Write one or more difference equations to implement $H(z)$ using only 5 multiplications.

5.11 MATLAB PROBLEMS FOR CHAPTER 5

This section provides homework problems that require the use of Matlab for their solution.

Problem 5.25. (Roundoff Errors)

This problem involves an investigation of the effects of roundoff on the location of the zeros and poles of the system transfer function and on the frequency response. The Matlab function *zplane* can be used to determine the location of the poles and zeros of the system transfer function and the Matlab function *freqz* can be used to determine the magnitude and phase of the frequency response.

1. The system transfer function for an IIR digital filter is given by

$$H(z) = \frac{B(z)}{A(z)} \tag{5.303}$$

where

$$
\begin{aligned}
B(z) \;=\; & 0.1006 - 0.4498z^{-1} + 1.1211z^{-2} - 1.8403z^{-3} \\
& + 2.1630z^{-4} - 1.8403z^{-5} + 1.1211z^{-6} \\
& - 0.4498z^{-7} + 0.1006z^{-8},
\end{aligned} \tag{5.304}
$$

$$
\begin{aligned}
A(z) \;=\; & 1.0 - 1.0130z^{-1} + 2.8890z^{-2} - 1.4483z^{-3} \\
& + 2.4901z^{-4} - 0.3959z^{-5} + 0.9169z^{-6} \\
& + 0.0324z^{-7} + 0.1867z^{-8}.
\end{aligned} \tag{5.305}
$$

2. Use the Matlab *zplane* function to compute a pole–zero plot for $H(z)$.
3. Use the Matlab *freqz* function to plot the magnitude and phase of the frequency response for $H(z)$.
4. The Matlab function *iirscale* has been provided in the course locker. Use this function to scale the filter coefficients to 10 bit integers.
5. Use the Matlab *zplane* function to compute a pole–zero plot for the scaled version of $H(z)$.
6. Use the Matlab *freqz* function to plot the magnitude and phase of the frequency response for scaled version of $H(z)$.
7. Compare the location of the poles and zeros and the frequency response of the resulting scaled filter to those for the original filter. Discuss any observations regarding the location of the poles and zeros as related to the frequency response and the stability of the filter.

Turn in:

1. The Matlab script for each part of the problem.
2. The plots for each part of the problem.

Problem 5.26. (Difference Equations)
The transfer function for a discrete time system is given by

$$H(z) = \frac{B(z)}{A(z)} \tag{5.306}$$

where

$$
\begin{aligned}
B(z) \;=\; & 0.0281 + 0.0790z^{-2} + 0.0170z^{-3} + 0.1000z^{-4} + 0.0170z^{-5} \\
& + 0.0790z^{-6} + 0.0281z^{-8}
\end{aligned} \tag{5.307}
$$

and

$$A(z) = 1.0 - 3.4739z^{-1} + 7.5998z^{-2} - 10.8672z^{-3} + 11.4306z^{-4}$$
$$- 8.6756z^{-5} + 4.7479z^{-6} - 1.7004z^{-7} + 0.3295z^{-8}. \quad (5.308)$$

1. Determine the coefficients for the use of second order sections to implement $H(z)$. **(Hint)** You can use the Matlab routine *tf2sos* to determine the coefficients.
2. Write an appropriate set of difference equations for the implementation of $H(z)$ using the second order section coefficients from part 1.

Problem 5.27. (Second Order Sections)

The system transfer function for an IIR digital filter is given by

$$H(z) = \frac{B(z)}{A(z)} \quad (5.309)$$

where

$$B(z) = 0.0609 - 0.3157z^{-1} + 0.8410z^{-2} - 1.4378z^{-3} + 1.7082z^{-4}$$
$$- 1.4378z^{-5} + 0.8410z^{-6} - 0.3157z^{-7} + 0.0609z^{-8},$$
$$A(z) = 1.0 - 0.8961z^{-1} + 2.6272z^{-2} - 0.9796z^{-3} + 2.1282z^{-4}$$
$$- 0.0781z^{-5} + 0.9172z^{-6} + 0.0502z^{-7} + 0.2602z^{-8}. \quad (5.310)$$

1. Use the Matlab routine *tf2sos* to obtain the coefficients for the cascade implementation of $H(z)$ using second order sections.
2. Determine the lattice parameters for each of the second order sections separately. Develop block diagrams for the IIR lattice implementation of each of the second order sections.
3. Develop appropriate difference equations for the implementation of the overall filter using the second order IIR lattice sections.

Problem 5.28. (IIR Lattice Implementation)

This problem involves the implementation of a sixth order IIR filter using second order lattice sections. The lattice parameters for section 1 are given by

```
K1 =
    0.4545
    0.5703

V1 =
    1.2459
    0.7137
   -1.0000
```

The lattice parameters for section 2 are

```
K2 =
   -0.1963
    0.8052
```

```
V2 =
    -0.0475
    -1.2347
     1.0000
```

The lattice parameters for section 3 are

```
K3 =
     0.8273
     0.8835
```

```
V3 =
    -0.1965
     0.3783
     1.0000
```

1. Write a Matlab function to implement a second order section IIR lattice filter.
2. The Matlab routine *sampdata* has been provided in the course locker under the *Matlab Introduction and Examples* item. Use the sequence generated by this routine as input for the filter simulations in this problem.
3. Write a Matlab routine to simulate the second order section implementation of $H(z)$ using second order IIR lattice sections and your function from part 1. Since the order of the filter is 6, your implementation should have 3 second order sections and you should call your routine 3 times.
4. Use the Matlab routine *latc2tf* to obtain the transfer function equivalent for each of the second order sections. Then write a Matlab routine to call the Matlab routine *filter* three times to filter the input sequence.
5. Compare the output from your routine that uses your second order lattice section filter with the output from your routine that uses the Matlab *filter* routine and note any differences.
6. Turn in your Matlab script, the output plots, and a summary of your observations.

Finite Word Length Effects

6

6.1 INTRODUCTION

The practical implementation of discrete time systems typically involves the representation of the system using digital circuits and systems. This requires the quantization of the coefficients for the discrete time system and the inputs to the system. Discrete time systems are typically represented by the use of finite difference equations and system functions with constant coefficients. The coefficients can take on any values as long as they are finite. In actual practice, the coefficients have to be represented in a form useful for digital circuits and systems and stored in registers with finite lengths. The arithmetic operations to implement the discrete time system need be performed through the use of finite precision arithmetic. The two most popular arithmetic approaches are floating point arithmetic and fixed point arithmetic. However, other approaches have been developed.

This chapter considers the representation of coefficients for discrete time systems with the use of finite precision arithmetic and the possible errors associated with the quantization of the input, quantization of the coefficients and quantization of the results from arithmetic operations so they can be stored in registers for later use or for use as the output samples. Possible errors include errors due to quantization of the coefficients for the discrete time system, errors due to rounding or truncation of intermediate results during the computations, errors due to overflow, and truncation and/or rounding of the output.

6.2 REPRESENTATION OF FLOATING POINT NUMBERS

A fixed point representation of numbers permits the representation of a finite range of numbers [4]. For example, if x_{max} is the maximum of the dynamic range and x_{min} is the minimum of the dynamic range, then the range of numbers with a resolution of

$$\Delta = \frac{x_{max} - x_{min}}{m - 1} \tag{6.1}$$

can be covered where $m = 2^b$ is the number of levels and b is the number of bits used to represent the numbers. The resolution for the fixed point representation is based upon the dynamic range to be represented and it does not change. The resolution

Digital Signal Processing. DOI: 10.1016/B978-0-12-804547-3.00006-1

351

Table 6.1 Examples of floating point representation of numbers

n	$x(n)$	$m(n)$	$e(n)$
1	53.719	$\frac{1719}{2048} = 0.8394$	6
2	−127.616	$\frac{-2041}{2048} = -0.9966$	7
3	0.04195	$\frac{1374}{2048} = 0.6709$	−4
4	−0.7914	$\frac{-1620}{2048} = -0.7910$	0

decreases as the dynamic range increases. A floating point representation can be used to cover a larger dynamic range. The binary representation of a number typically consist of a mantissa M, which represents the fractional part of the number, and an exponent E. The mantissa commonly falls within the range $0.5 \leq M < 1.0$. It is multiplied by the factor 2^E, where E can be positive or negative, to represent the number.

6.2.1 IEEE FLOATING POINT REPRESENTATION

The IEEE 754-1985 Standard for floating point arithmetic provides the commonly used format for the implementation of floating point arithmetic [14]. The Standard provides a 32 bit single format, a 64 bit double format, and a set of constraints for extended formats. Three fields are used to represent the numbers in both the single and double formats:

1. A 1 bit sign s,
2. A biased exponent $e = E +$ bias,
3. A fraction $f = .b_1 b_2 \cdots b_{p-1}$.

The fields for the 32 bit single format are as follows:

1. 1 bit is used for the sign (most significant bit),
2. 8 bits are used for the biased exponent, and
3. 23 bits are used for the fraction (least significant bits).

The exponent e is biased so that the actual exponent $E = e - 127$. The value of 1.0 must be added to the fractional part to obtain the actual mantissa for the number.

EXAMPLE 6.1
(Floating Point Representation of Numbers)

This example illustrates the concept used to represent floating point numbers. Assume that a 16 bit number is used to represent a floating point number with 12 bits used for the mantissa (11 bits for the magnitude and one bit for the sign) and 4 bits for the exponent (3 bits for the magnitude and 1 bit for the sign). Table 6.1 gives a tabulation of some decimal numbers and their floating point representations using this 16 bit representation.
End of the Example

6.3 COMPUTATIONAL ERRORS DUE TO ROUNDING

Rounding or truncation errors only affect the mantissa when floating point numbers are used to implement a discrete time system, provided the results of the computations involving the mantissa are within the available dynamic range. The mantissa and the exponent can be changed as appropriate to obtain the most accurate computation with floating point arithmetic. On the other hand, the exponent is predetermined for fixed point arithmetic. The exponents for all of the operands are typically the same or are predetermined for fixed point arithmetic, and no computations are done on the exponents. This gives floating point arithmetic an advantage over fixed point arithmetic for implementing discrete time systems for many applications. On the other hand, fixed point arithmetic has an advantage in that the system hardware required for implementation is usually less complicated than it is for floating point arithmetic. This can result in faster overall systems and systems with lower power dissipation. The use of fixed point arithmetic is more prevalent for application specific, special purpose systems and embedded systems for this reason. The problems to be solved in order to use fixed point arithmetic include scaling the input sequence, scaling the discrete time system parameters such as filter coefficients and scaling intermediate outputs for storage in memory or for use by subsequent stages of the system, etc.

6.3.1 ROUNDING ERRORS DURING MULTIPLICATION

One form of computational error result from the use of finite precision arithmetic for computations and then storing the results with finite precision. Example 6.2 demonstrates the effects of computational errors due to rounding with the use of an example.

EXAMPLE 6.2
(Multiplication Errors)

Consider two numbers

$$f_1 = 2.34692 \times 10^3,$$

$$f_2 = 8.76192 \times 10^5. \tag{6.2}$$

Consider the multiplication of these two numbers using floating point arithmetic by multiplying the two mantissas and adding the two exponents. It follows that

$$g_1 = f_1 \times f_2 = 2.3469 \times 8.7619 \times 10^{3+5},$$

$$g_1 = 2.0563 \times 10^9. \tag{6.3}$$

The mantissa has been represented using 5 decimal numbers in this example. The roundoff only occurs for the mantissa. It happens that the roundoff error for this example is

$$e_1 = 3.0311 \times 10^4. \tag{6.4}$$

The relative error is given by

$$re_2 = \frac{e_1}{g_1} = 1.474 \times 10^{-5}. \tag{6.5}$$

The relative error using floating point arithmetic is on the order of 10^{-5} which would be acceptable for many practical applications.

Now consider using fixed point arithmetic and also use 5 decimal numbers for the representation with the exponents the same for both numbers. No operations are done using the exponents when fixed point representation is used. The exponent at 4 for both numbers (a compromise) and 2 numbers can be used for the whole part and 3 numbers for the fractional part of the mantissas. Two numbers are needed for the whole part to avoid overflow. The representations then become

$$\widehat{f_1} = 0.235,$$
$$\widehat{f_2} = 87.619. \tag{6.6}$$

The corresponding output after adjusting the exponent is

$$\widehat{g_1} = 20.590 \times 10^8. \tag{6.7}$$

The corresponding error is

$$e_2 = -2.7162 \times 10^6. \tag{6.8}$$

The relative error is given by

$$re_2 = \frac{e_2}{g_1} = -0.0013. \tag{6.9}$$

The relative error using fixed point arithmetic is of the order of 10^{-3} or about 100 times the error using floating point arithmetic for this example. Of course, the relative error depends on the relative sizes of the two operands. The relative error would be less if both numbers were approximately the same size. However, this example illustrates the difference between using fixed point and floating point arithmetic for multiplication. Floating point arithmetic has a much greater advantage when the magnitudes of the two numbers are vastly different.
End of the Example

6.3.2 ROUNDING ERRORS DURING ADDITION

The exponents for two floating point numbers must be the same for floating point addition or subtraction. Typically, the number with the smaller exponent is shifted so that its exponent matches the exponent of the other number. If the required shift is too large, then the resulting mantissa will be zero and the sum is unaffected by the smaller number. The exponents are already equal for fixed point numbers so the fixed point addition only involves addition of the mantissas. The adjustment of the exponents so that they are equal adds a computational burden on floating point addition. Therefore, it is more computationally efficient to use fixed point addition than to use floating point addition. The errors in the results from the addition of the mantissas are comparable since the error is due to truncating the results for the mantissa. However, the exponent is set dynamically based upon the numbers involved for floating point numbers while the exponent is fixed for fixed point numbers. Therefore, floating point arithmetic will normally give better accuracy than fixed point arithmetic.

6.3.3 MULTIPLICATION OF FLOATING POINT NUMBERS

Multiplication of two floating point numbers requires the multiplication of the mantissas and adding the exponents [4]. The result of multiplying the two mantissas is then normalized so that the mantissas of the result falls within the range $0.5 \leq M < 1.0$ and the exponent is adjusted as needed to accommodate the normalization.

Rounding errors can occur if the number of bits used to store the result is the same as the number of bits used for the two input floating point numbers. However, the rounding is limited to the mantissa as long as the exponent falls within the appropriate dynamic range.

6.4 TWO'S COMPLEMENT REPRESENTATION OF NUMBERS

Most digital systems use two's complement arithmetic to represent numbers. A typical representation involves a sign, a whole number part, and a fractional part. The location of the decimal point is determined by the number of bits used for the whole number part and the number of bits used for the fractional part. The location of the decimal point is important for some operations such as addition. The decimal points for two or more operands must be aligned for addition. Thus, the exponents for the two operands must be the same for addition.

The most popular convention for representing fixed point numbers in two's complement arithmetic is to have the upper or left most bit as the sign bit and the other bits representing the fraction. Thus, a fixed point number with $m + 1$ bits would be represented as

$$f = \boxed{\begin{array}{|c|c|c|c|c|c|} s & b1 & b2 & b3 & \cdots & bm \end{array}} \qquad (6.10)$$

where bk indicates that if $bk = 1$, then 2^{-k} is added and if $bk = 0$ then zero is added. It is convenient to represent the number as an integer multiplied by 2^{-m}. For example,

$$f = \boxed{0\ |\ 1\ |\ 1\ |\ 1} \qquad (6.11)$$

represents

$$
\begin{aligned}
f &= 2^{-1} + 2^{-2} + 2^{-3} \\
&= 0.5 + 0.25 + 0.125 = 0.875, \\
f &= \frac{2^3 - 1}{2^3} = \frac{7}{8} \qquad (6.12)
\end{aligned}
$$

where $m = 3$ for this 4 bit number. We can also write this as

$$f = 7 \times 2^{-3}. \qquad (6.13)$$

In a similar way, negative numbers can be represented by using a 1 as the sign bit. For example, consider

$$f = \boxed{1\ |\ 0\ |\ 0\ |\ 0}. \qquad (6.14)$$

This number is negative due the 1 in the left most bit. Its magnitude can be determined by complementing each of the bits and adding 2^{-m}. Thus, the magnitude of f can be represented as

$$
\begin{aligned}
f_{mag} &= \boxed{0\ |\ 1\ |\ 1\ |\ 1} + 2^{-3} \\
&= 2^{-1} + 2^{-2} + 2^{-3} + 2^{-3} \\
&= 0.5 + 0.25 + 0.125 + 0.125 \\
&= 1.0. \qquad (6.15)
\end{aligned}
$$

Thus, it follows that

$$f = -1.0. \qquad (6.16)$$

6.5 ANALYTICAL BASIS FOR TWO'S COMPLEMENT NUMBERS

The representation of numbers using two's complement representation using an m bit word involves adding the value of 2^m to negative numbers to obtain a corresponding positive number. The range of the negative number is limited to

$$-2^{-(m-1)} \le f_k < 0 \qquad (6.17)$$

where f_k is the negative number to be represented. Thus, the range of negative numbers when represented using two's complement representation will be

$$2^m - 2^{-(m-1)} \le f_k < 2^m. \qquad (6.18)$$

For example, when $m = 16$ bit words are used, negative numbers will be in the range

$$32768 \le f_k < 65536, \qquad (6.19)$$

or

$$32768 \le f_k \le 65535. \qquad (6.20)$$

Two's complement numbers are convenient to use for addition and subtraction because the overflow above m bits can be ignored. This concept can be demonstrated as shown in Example 6.3.

EXAMPLE 6.3
(Two's Complement Arithmetic)

Consider the addition of the numbers

$$f_1 = -2.34692 \times 10^3,$$
$$f_2 = 8.76192 \times 10^5. \qquad (6.21)$$

The numbers should be divided by 2^{20} to convert them into fractional numbers in the desired range. The corresponding fractional numbers after performing the division are given by

$$r_1 = -0.002238 \times 2^{11},$$
$$r_2 = 0.835600 \times 2^{11}. \qquad (6.22)$$

Since $r_1 < 0$, 2^{12} needs to be added to make it a positive number in the desired range. Thus,

$$\widehat{r_1} = \text{round}\left(2^{12} + r_1\right) = 4091,$$
$$r_2 = 1711,$$
$$t = \widehat{r_1} + r_2 = 4091 + 1711 = 5802. \qquad (6.23)$$

Since $t > 2^{12} = 4096$, the correct output can be obtained by subtracting 4096. The subtraction is equivalent to ignoring the overflow bit that results from the addition. Thus,

$$s = t - 2^{12} = 5802 - 4096 = 1706. \qquad (6.24)$$

Finally, the results can be obtained by dividing s by 2^{20} to obtain

$$y_1 = 1076 \left(\frac{2^{20}}{2^{11}} \right) = 873472. \tag{6.25}$$

The results using floating point representation would be

$$y_2 = f_1 + f_2 = 873845. \tag{6.26}$$

The relative error is

$$re_4 = \frac{y_2 - y_1}{y_2} = 4.2694 \times 10^{-4}. \tag{6.27}$$

Thus, the relative error is of the order of 10^{-3}.
End of the Example

6.6 SCALING FIR DIGITAL FILTERS

The scaling of FIR digital filter coefficients for implementation using fixed point arithmetic will be considered in this section. The requirement is to convert the algorithm for the FIR digital filter so that fixed point arithmetic can be used for its implementation. The filter coefficients must be scaled and represented using fixed point arithmetic.

Consider the scaled FIR digital filter with transfer function given by

$$H(z) = \sum_{k=0}^{L} \widehat{b(k)} z^{-k} \tag{6.28}$$

where the $\widehat{b(k)}$ are scaled filter coefficients. The corresponding difference equation is given by

$$y(n) = \sum_{k=0}^{L} \widehat{b(k)} x(n-k). \tag{6.29}$$

Assume that the upper limit of the dynamic range for the output is a finite, real number R_1 such that

$$0 < R_1 < \infty. \tag{6.30}$$

Then the scaling requirements can be expressed as

$$|y(n)| = \left| \sum_{k=0}^{L} \widehat{b(k)} x(n-k) \right|,$$

$$|y(n)| \leq \sum_{k=0}^{L} \left| \widehat{b(k)} x(n-k) \right|,$$

$$|y(n)| \leq \sum_{k=0}^{L} \left| \widehat{b(k)} \right| |x(n-k)|. \tag{6.31}$$

The following inequalities have been used in Eq. (6.31):

$$|a+b| \leq |a| + |b|,$$
$$|ab| \leq |a| |b|. \tag{6.32}$$

Further specify that the upper limit of the dynamic range for the input is a finite, real number R_2 such that

$$0 < |x(n)| \leq R_2 < \infty. \tag{6.33}$$

Then, it follows that

$$|y(n)| \leq \sum_{k=0}^{L} \left| \widehat{b(k)} \right| R_2 < R_1. \tag{6.34}$$

The constraint on the filter coefficients to avoid overflow can therefore be expressed as

$$\sum_{k=0}^{L} |b(k)| \leq \frac{R_1}{R_2}. \tag{6.35}$$

Example 6.4 gives an example of the use of this constraint to avoid overflow for the scaling of the coefficients for an FIR filter.

EXAMPLE 6.4
(Overflow Constraint)

Assume that the input, $x(n)$, is scaled to 16 bit words and the output is limited to using 32 bit words. The word sizes, for this case, are limited as follows:

$$R_1 = 2^{31} - 1 \times 2^{-31} = 2,147,483,647 \times 2^{-31},$$
$$R_2 = 2^{15} - 1 = 32767 \times 2^{-15}. \tag{6.36}$$

The corresponding constraint to avoid overflow is

$$\sum_{k=0}^{L} \left| \widehat{b(k)} \right| \leq \frac{2,147,483,647 \times 2^{-31}}{32767 \times 2^{-15}}$$

$$= 65538 \times 2^{-16} = 1.0000. \tag{6.37}$$

This constraint can be used to avoid overflow for implementing an FIR filter under the stated conditions.
End of the Example

6.6.1 SCALING FIR FILTERS FOR A GIVEN WORD SIZE

It is often desirable to scale the filter coefficients to a particular word size to match the word size for a given system. For example, the system may have been designed to avoid overflow for the specified word sizes. For example, the coefficients and the input may be required to be scaled so that they can be represented by 16 bit two's complement fixed point numbers. Thus, the scale factors for the input and the coefficients, s_1 and s_2, need to be determined such that

1. The scaled input is $f(n) = s_1 x(n)$,
2. The coefficients are $\widehat{b(k)} = s_2 b(k)$,
3. The scaled output is $g(n) = s_1 s_2 y(n)$, and
4. The unscaled output is $y(n) = \frac{g(n)}{s_1 s_2}$.

This problem can be solved by using the following:

$$
\begin{aligned}
H(z) &= \frac{Y(z)}{X(z)} = \frac{s_1 Y(z)}{s_1 X(z)} = \frac{R(z)}{F(z)}, \\
F(z) &= s_1 X(z), \\
H_2(z) &= s_2 H(z) = \frac{G(z)}{F(z)}, \\
G(z) &= s_2 R(z), \\
Y(z) &= \frac{G(z)}{s_1 s_2}.
\end{aligned}
\tag{6.38}
$$

Thus, the scaled system transfer function is given by

$$
H_2(z) = s_2 H(z) = s_2 \sum_{k=0}^{L} b(k) z^{-k},
\tag{6.39}
$$

$$
\frac{G(z)}{F(z)} = \sum_{k=0}^{L} s_2 b(k) z^{-k}.
\tag{6.40}
$$

Since $\widehat{b(k)} = s_2 b(k)$, it follows that

$$
G(z) = \left[\sum_{k=0}^{L} \widehat{b(k)} z^{-k} \right] F(z),
\tag{6.41}
$$

or

$$
g(n) = \sum_{k=0}^{L} \widehat{b(k)} f(n-k).
\tag{6.42}
$$

The unscaled output can be obtained by using

$$y(n) = \frac{g(n)}{s_1 s_2}. \tag{6.43}$$

If 16 bits are used to represent the input and the coefficients, then s_1 and s_2 need to be computed such that

$$-32768 \times 2^{-15} \leq \widehat{b(k)} \leq 32767 \times 2^{-15},$$
$$-32768 \times 2^{-15} \leq f(n) \leq 32767 \times 2^{-15}. \tag{6.44}$$

EXAMPLE 6.5
(Scaling FIR Filters)

The following Matlab script can be used to design an FIR digital filter with 5 filter coefficients (order = 4) and relative cutoff frequency $\omega_c = 0.42$ and then scale it using the *fscale* function, which has been provided in Appendix D.1. This example uses 16 bits for the word size to represent the coefficients.

```
order = 4;
wc = 0.42;
bits = 16;
b = fir1(order, wc);
[bsc, scfac] = firscale(b, bits);
```

The filter coefficients are

$$b = [0.0080 \quad 0.2176 \quad 0.5488 \quad 0.2176 \quad 0.0080]. \tag{6.45}$$

The scaled output coefficients (ignoring the 2^{-15} exponent) are

$$bsc = [479 \quad 12989 \quad 32767 \quad 12989 \quad 479]. \tag{6.46}$$

The scale factor used in scaling the coefficients was

$$s_2 = scfac = 59702. \tag{6.47}$$

The original floating point difference equation was

$$\begin{aligned} y(n) &= 0.0080x(n) + 0.2176x(n-1) \\ &\quad + 0.5488x(n-2) + 0.2176x(n-3) \\ &\quad + 0.0080x(n-4). \end{aligned} \tag{6.48}$$

After scaling, the fixed point difference equation is

$$g(n) = 479f(n) + 12989f(n-1)$$

$$+ 32767 f(n-2) + 12989 f(n-3)$$
$$+ 479 f(n-4). \tag{6.49}$$

The exponent 2^{-15} was not explicitly given in Eq. (6.49) for this example. However, it should be remembered that the numbers are represented as fractions using two's complement fixed point numbers. In other words, the left most bit is the sign bit, and the remaining bits represent the fractional part of the number. Thus, all of the scaled coefficients have an exponent of 2^{-15}, and the output has an exponent of 2^{-30}.

End of the Example

6.6.2 SCALING FIR FILTERS USING THE OUTPUT OVERFLOW CRITERIA

One approach to scaling digital filters is to scale the coefficients such that overflow cannot occur. For example, the output may be stored in an accumulator, or the intermediate outputs need to be stored in registers of a particular size. Consider the FIR filter with transfer function

$$H(z) = \sum_{k=0}^{L} b(k)z^{-k} = \frac{Y(z)}{X(z)}. \tag{6.50}$$

The corresponding difference equation is

$$y(n) = \sum_{k=0}^{L} b(k)x(n-k). \tag{6.51}$$

Assume that the output is to be stored in a 40 bit word accumulator. This means that the maximum permitted value for the output is

$$y_{max} = \left[2^{39} - 1\right] \times 2^{-39}. \tag{6.52}$$

In addition, assume that the input is stored in 16 bit integer numbers such that

$$x_{max} = 2^{15} - 1 \times 2^{-15} \geq |x(n)| \tag{6.53}$$

where the $x(n)$ have also been scaled.

The coefficients can be scaled to avoid overflow by selecting s_2 such that

$$y_{max} \geq x_{max} \sum_{k=0}^{L} |b(k)| s_2. \tag{6.54}$$

The maximum permitted value of s_2 is therefore

$$s_2 = \frac{y_{max}}{x_{max} \sum_{k=0}^{L} |b(k)|}. \tag{6.55}$$

The Matlab script for a function to scale FIR filters using the output overflow criteria is provided in Appendix D.2 as the function *firscale2*. This function does not explicitly show the exponent because the exponent is known and it is based upon the chosen word sizes.

EXAMPLE 6.6

(An Output Overflow Scaling Example)

The same FIR filter that was used in Example 6.5 will be used in this example to illustrate scaling using the output overflow criteria. Thus, the filter coefficients are given by

```
b = [0.0080 0.2176 0.5488 0.2176 0.0080]
```

Assume that the input is to be scaled to 16 bits and the output is to be stored in a 32 bit word. The following Matlab script can be used to scale the given FIR filter:

```
ibits = 16;
obits = 32;
[bsc, scfac] = firscale2(b, ibits, obits);
```

The scaled coefficients are given by

```
bsc = [525 14259 35970 14259 525]
```

and

```
scfac = 65538
```

The difference equation to implement the scaled filter is

$$g(n) = 525x(n) + 14259x(n-1) + 35970x(n-2) + 14259x(n-3)$$
$$+ 525x(n-4). \tag{6.56}$$

Note that the scale factor is larger than it was for Example 6.5. Thus, this approach makes better use of the available dynamic range. Also, the exponents for the inputs and the outputs are not explicitly represented but the numbers will be represented using fixed point numbers as previously indicated.

End of the Example

Example 6.7 provides another example of scaling an FIR digital filter using the output overflow scaling criteria.

EXAMPLE 6.7

(Output Overflow Scaling Example Two)

Consider the FIR filter with transfer function given by

$$H(z) = -0.0023 - 0.0219z^{-1} + 0.0155z^{-2}$$
$$+ 0.2737z^{-3} + 0.4699z^{-4} + 0.2737z^{-5}$$
$$+ 0.0155z^{-6} - 0.0219z^{-7} - 0.0023z^{-8}. \tag{6.57}$$

Note that this filter was designed using the Matlab *fir1* function with the order $N = 8$ and the normalized cutoff frequency $\omega_c = 0.471$. The goal is to scale the coefficients for use of a 40 bit accumulator to implement the filter without overflow when the input is stored as 16 bit fixed point integers. The scale factor s_2 for this case can be determined as follows:

$$\sum_{k=0}^{8} |b(k)| = 1.0966,$$

$$s_2 = \frac{2^{39} - 1}{1.0966(2^{15} - 1)} = 15299688. \tag{6.58}$$

The scaled system transfer function is given by

$$H_2(z) = -34635 - 334875z^{-1} + 237724z^{-2} + 4186808z^{-3} + 7189645z^{-4}$$
$$+ 4186808z^{-5} + 237724z^{-6} - 334875z^{-7} - 34635z^{-8}. \tag{6.59}$$

End of the Example

6.7 SCALING IIR DIGITAL FILTERS

Scaling IIR filters is more complicated because of the use of feedback of previous output values to obtain the current output. The procedure for scaling IIR digital filters is complicated by the use of the denominator coefficients which are related to feedback. The scaling of both sets of coefficients can be done using a single scaling factor. However, improved performance can be obtained by using separate scaling factors for each set of coefficients.

The constraints for the scaling of IIR digital filters can be considered by starting with the difference equation

$$y(n) = \sum_{k=0}^{L} b(k)x(n-k) - \sum_{k=1}^{L} a(k)y(n-k). \tag{6.60}$$

The restriction for scaling the filter coefficients such that overflow is avoided is given by

$$|y(n)| \leq \left| \sum_{k=0}^{L} b(k)x(n-k) - \sum_{k=1}^{L} a(k)y(n-k) \right|, \qquad (6.61)$$

$$|y(n)| \leq \sum_{k=0}^{L} |b(k)x(n-k)| + \sum_{k=1}^{L} |a(k)y(n-k)|. \qquad (6.62)$$

Assume that

$$|y(n)| \leq y_{max} \qquad (6.63)$$

and

$$|x(n)| \leq x_{max} \qquad (6.64)$$

where y_{max} and x_{max} are positive, finite magnitude constants. Then, it follows that

$$|y(n)| \leq \sum_{k=0}^{L} |b(k)| x_{max} + \sum_{k=1}^{L} |a(k)| y_{max}. \qquad (6.65)$$

From a practical consideration, it is common to scale the output to the same dynamic range as the dynamic range of the input prior to using it as feedback. If this approach is used, then y_{max} can be replaced by x_{max} on the right side of the above equation to obtain

$$|y(n)| \leq \sum_{k=0}^{L} |b(k)| x_{max} + \sum_{k=1}^{L} |a(k)| x_{max}, \qquad (6.66)$$

or

$$|y(n)| \leq x_{max} \left[\sum_{k=0}^{L} |b(k)| + \sum_{k=1}^{L} |a(k)| \right]. \qquad (6.67)$$

If scaling factors are selected based on the right side of Eq. (6.67), then overflow will not occur. However, this is also based on the assumption that the output will be scaled back to the same dynamic range of the input (same number of bits) prior to using it as feedback. Example 6.8 illustrates this concept to scale the coefficients for an IIR filter.

EXAMPLE 6.8
(Scaling an IIR Digital Filter)

Consider the case where the goal is to scale the filter coefficients such that 16 bit integers can be used for the input and the output feedback and 32 bits can be used for the output. It follows, for this case, that

$$
\begin{aligned}
y_{max} &= 2^{31} - 1 = 2,147,483,647 \times 2^{-31}, \\
x_{max} &= 2^{15} - 1 = 32767 \times 2^{-15}.
\end{aligned}
\tag{6.68}
$$

The corresponding restriction on the scale factors is

$$
y_{max} = x_{max} \left[\sum_{k=0}^{L} s_2 |b(k)| + \sum_{k=1}^{L} s_3 |a(k)| \right],
\tag{6.69}
$$

or

$$
\begin{aligned}
\frac{y_{max}}{x_{max}} &= \frac{2,147,483,647}{32767} \\
&= 65,538 \times 2^{-16} = 1 \\
&= \left[\sum_{k=0}^{L} s_2 |b(k)| + \sum_{k=1}^{L} s_3 |a(k)| \right]
\end{aligned}
\tag{6.70}
$$

If s_3 is initially selected such that $s_2 = s_3$, then, s_2 can be determined as follows:

$$
s_2 = \frac{65,538}{\left[\sum_{k=0}^{L} |b(k)| + \sum_{k=1}^{L} |a(k)| \right]}.
\tag{6.71}
$$

It is practical for many applications to restrict s_3 to be a power of 2 so that scaling the output for feedback can be accomplished by shifting the output by the appropriate number of bit positions. This implies that the scale factor s_3 should be restricted such that

$$
s_3 = 2^k \leq s_2.
\tag{6.72}
$$

In a typical implementation of the IIR filter using fixed point arithmetic, it is necessary to divide the output by s_3 prior to using it as feedback for the next computation. If s_3 is a power of 2, then this division can be performed by shifting to the right by K bits where $s_3 = 2^K$. The scale factor s_3 can be computed as a power of 2 as follows:

$$
K = \lfloor \log_2(s_2) \rfloor
\tag{6.73}
$$

where $\lfloor \ \rfloor$ implies truncation. Then

$$
s_3 = 2^K.
\tag{6.74}
$$

The values of the exponents can be predicted for the computations of the various scale factors. Therefore, the exponents do not need to be explicitly represented in the computations.

6.7.1 SCALING IIR FILTER COEFFICIENTS

The implementation of IIR filters using fixed point arithmetic requires the determination of scale factors s_1, s_2, and s_3 such that

1. The input is scaled by s_1, i.e., $f(n) = s_1 x(n)$,
2. The numerator coefficients are scaled by s_2, i.e., $\widehat{b(k)} = s_2 b(k)$,
3. The denominator coefficients are scaled s_3, i.e., $\widehat{a(k)} = s_3 a(k)$,
4. The scaled output is given by $g(n) = \frac{s_1 s_2}{s_3} y(n)$, and
5. The unscaled output can be obtained by $y(n) = \frac{s_3 g(n)}{s_1 s_2}$.

The goal is to develop a procedure for scaling IIR filters by modifying the gain of the system and adjusting the feedback so that the overall transfer function is still valid except for the change in gain and the effects due to roundoff. The scaled transfer function is then given by

$$H_2(z) = \frac{s_2}{s_3} H(z) = \frac{s_2 s_1 H(z)}{s_3 s_1}. \tag{6.75}$$

If s_1 is used to scale the input sequence, then the output sequence will also be scaled by s_1 as a result. Further, if $H_2(z)$ is defined as

$$H_2(z) = \frac{G(z)}{F(z)} \tag{6.76}$$

then

$$f(n) = s_1 x(n), \tag{6.77}$$

$$g(n) = \frac{s_1 s_2 y(n)}{s_3}. \tag{6.78}$$

The scaled coefficients have been defined as follows:

$$\begin{aligned} \widehat{b(k)} &= s_2 b(k), \\ \widehat{a(k)} &= s_3 a(k). \end{aligned} \tag{6.79}$$

Thus,

$$H_2(z) = \frac{\displaystyle\sum_{k=0}^{L} \widehat{b(k)} z^{-k}}{s_3 + \displaystyle\sum_{k=1}^{L} \widehat{a(k)} z^{-k}} = \frac{G(z)}{F(z)}. \tag{6.80}$$

The following equations apply with these changes due to scaling:

$$
\begin{aligned}
G(z) &= \frac{s_1 s_2}{s_3} Y(z), \\
H_2(z) &= \frac{G(z)}{F(z)}, \\
s_3 G(z) &= \widehat{b(0)} F(z) \\
&+ \sum_{k=1}^{L} \left[\widehat{b(k)} F(z) - \widehat{a(k)} G(z) \right] z^{-k}.
\end{aligned}
\tag{6.81}
$$

Let $R(z) = s_3 G(z)$ for convenience in developing the difference equation. Then, it follows that

$$
\begin{aligned}
r(n) &= \widehat{b(0)} f(n) \\
&- \sum_{k=1}^{L} \left[\widehat{b(k)} f(n-k) - \widehat{a(k)} g(n-k) \right], \\
g(n) &= \frac{r(n)}{s_3}.
\end{aligned}
\tag{6.82}
$$

Eq. (6.82) shows that the output, $r(n)$, needs to be divided by s_3 in order to use it for feedback to compute the new values of $r(n)$. This division can be done by shifting if s_3 is constrained to be a power of 2 ($2, 4, 8, \ldots, 2^K$ where K is an integer). The unscaled output can be obtained by using

$$y(n) = \frac{s_3 g(n)}{s_1 s_2} = \frac{r(n)}{s_1 s_2}. \tag{6.83}$$

Example 6.9 illustrates the use of this procedure to scale an IIR filter.

EXAMPLE 6.9
(Scaling an IIR Digital filter)

A Matlab script for a function entitled *iirscale* to scale the coefficients for an IIR digital filter has been provided in Appendix D.3. The following Matlab script can be used to design and scale an IIR digital filter using the *iirscale* function.

Matlab Script 6.1.

```
wc = [0.12 0.68];
order = 2;
rp = 2;
rs = 45;
[b,a] = ellip(order, rp, rs, wc);
bits = 16;
[bsc, asc, scfac] = iirscale(b,a, bits);
```

End of the Script

The resulting b coefficients were

$$b = [0.3038 \ -0.0034 \ -0.5989 \ -0.0034 \ 0.3038].$$ (6.84)

The resulting a coefficients were

$$a = [1.0 \ -0.9061 \ 0.1651 \ -0.3150 \ 0.3903].$$ (6.85)

The scaled $\widehat{b}$ coefficients were

$$bsc = [16622 \ -188 \ -32767 \ -188 \ 16622].$$ (6.86)

The scaled $\widehat{a}$ coefficients were

$$asc = [16384 \ -14846 \ 2706 \ -5161 \ 6395].$$ (6.87)

Recall that each of the coefficients needs to be multiplied by 2^{-15} for representation using fixed point two's complement numbers.

The scale factors were

$$scfac = [54707 \ 16384].$$ (6.88)

The scale factors should also be multiplied by 2^{-15} for representation using two's complement numbers. Note that the scale factor for the denominator coefficients ($scfac(1, 2) = 16384 \times 2^{-15}$) is equal to $2^{14} \times 2^{-15} = 2^{-1}$ which is what was desired.

The floating point difference equation for the original system is

$$
\begin{aligned}
y(n) \ &= \ 0.3038x(n) - 0.0034x(n-1) \\
&\quad - 0.5989x(n-2) - 0.0034x(n-3) \\
&\quad + 0.3038x(n-4) + 0.9061y(n-1) \\
&\quad - 0.1651y(n-2) + 0.3150y(n-3) \\
&\quad - 0.3903y(n-4).
\end{aligned}
$$ (6.89)

The corresponding scaled difference equation is

$$
\begin{aligned}
r(n) \;=\;\; & 16622f(n) - 188f(n-1)\\
& - 32767f(n-2) - 188f(n-3)\\
& + 16622f(n-4) + 14846g(n-1)\\
& - 2706g(n-2) + 5161g(n-3)\\
& - 6395g(n-4),
\end{aligned}
$$

$$
g(n) \;=\; \frac{r(n)}{16384}. \tag{6.90}
$$

The above equation indicates that $r(n)$ needs to be divided by 16384 to obtain $g(n)$, which is needed for the next computation of $r(n)$. However,

$$
\log_2(16384) = 14. \tag{6.91}
$$

It follows that this division can be done for two's complement arithmetic by shifting $r(n)$ by to the left by 14 bit positions. This is much simpler than performing an actual division.

End of the Example

6.8 SCALING THE STATE SPACE REPRESENTATION

The implementation of the state space representation using fixed point arithmetic will be considered in this section. The scaling of the state space representation will be addressed in a general way and then an example will be given.

The state space representation for a digital filter can be written as

$$
\begin{aligned}
Q(n) &= AQ(n-1) + Bx(n),\\
y(n) &= CQ(n-1) + Dx(n).
\end{aligned} \tag{6.92}
$$

It will be assumed that the input has been scaled and rounded using a scale factor s_1. Thus,

$$
f(n) = \lfloor s_1 x(n) \rfloor \tag{6.93}
$$

where rounding by truncation has been used to show that $f(n)$ is represented using a fixed point number. It follows directly that the output will also be scaled such that

$$
g(n) = \lfloor s_1 y(n) \rfloor. \tag{6.94}
$$

The goal is to scale the coefficients of C and D by using a scale factor s_2 such that

$$
\widehat{C} \;=\; \lfloor s_2 C \rfloor,
$$

$$\widehat{D} = \lfloor s_2 D \rfloor. \tag{6.95}$$

After doing this, the output equation becomes

$$s_2 g(n) = s_2 C Q(n-1) + s_2 D f(n) = \widehat{C} Q(n-1) + \widehat{D} f(n). \tag{6.96}$$

Let

$$r(n) = s_2 g(n) = s_1 s_2 y(n) \tag{6.97}$$

where the above equation ignores roundoff errors. The output equation then becomes

$$r(n) = \widehat{C} Q(n-1) + \widehat{D} f(n). \tag{6.98}$$

The goal is also to scale the coefficients of A and B by using a scale factor s_3 such that

$$\begin{aligned} \widehat{A} &= \lfloor s_3 A \rfloor, \\ \widehat{B} &= \lfloor s_3 B \rfloor. \end{aligned} \tag{6.99}$$

After doing this, the state equation becomes

$$s_3 Q(n) = s_3 A Q(n-1) + s_3 B f(n) = \widehat{A} Q(n-1) + \widehat{B} f(n). \tag{6.100}$$

Let

$$P(n) = s_3 Q(n). \tag{6.101}$$

It then follows that

$$P(n) = \widehat{A} Q(n-1) + \widehat{B} f(n). \tag{6.102}$$

Note that $Q(n-1)$ needs to be used to compute $P(n)$ and $R(n)$ for the next iteration. Therefore, the output $P(n)$ needs to be divided by s_3 to obtain $Q(n)$. The modified state equations for the use of fixed point arithmetic is given by

$$\begin{aligned} P(n) &= \widehat{A} Q(n-1) + \widehat{B} f(n), \\ r(n) &= \widehat{C} Q(n-1) + \widehat{D} f(n), \\ Q(n) &= \frac{P(n)}{s_3}. \end{aligned} \tag{6.103}$$

It is appropriate to consider using constraints on s_2 and s_3 based upon the word sizes and implementation considerations. For example, it is practical to want $s_3 = 2^k$ where k is an integer so the division can be performed by shifting when two's complement numbers are used. The also consider using the output overflow criteria where we restrict the number of bits used to represent the elements of $P(n)$ and $r(n)$ for a given word size for $f(n)$ provide a restriction on the scale factors as well.

Example 6.10 provides an illustration on the use of this approach to scale the state space representation of a digital filter.

EXAMPLE 6.10
(Scaling a State Space Representation of a Digital Filter)

Consider the IIR filter with system transfer function

$$H(z) = \frac{B(z)}{A(z)} \tag{6.104}$$

where

$$B(z) = 0.0431 + 0.0816z^{-1} + 0.1103z^{-2} + 0.0816z^{-3} + 0.0431z^{-4} \tag{6.105}$$

and

$$A(z) = 1.0 - 1.7398z^{-1} + 1.8756z^{-2} - 1.0313z^{-3} + 0.2767z^{-4}. \tag{6.106}$$

The matrices for the transpose direct state space representation are

$$A = \begin{bmatrix} 1.7398 & 1.0 & 0 & 0 \\ -1.8756 & 0 & 1.0 & 0 \\ 1.0313 & 0 & 0 & 1.0 \\ -0.2767 & 0 & 0 & 0 \end{bmatrix}, \tag{6.107}$$

$$B = \begin{bmatrix} 0.1567 \\ 0.0294 \\ 0.1261 \\ 0.0312 \end{bmatrix}, \tag{6.108}$$

$$C = [1 \quad 0 \quad 0 \quad 0], \tag{6.109}$$

$$D = [0.0431]. \tag{6.110}$$

The infinity norm of a matrix is the largest row sum of the sum of the magnitude of the element of the rows of a matrix. Thus, if we restrict the elements of $P(n)$ to be represented by 32 bit numbers, we require that

$$\|P(n)\|_\infty \leq 2^{31} - 1 = p_{max}. \tag{6.111}$$

We can write

$$\|P(n)\|_\infty \leq \|\widehat{A}Q(n-1) + \widehat{B}f(n)\|_\infty. \tag{6.112}$$

We can form the matrix E using the columns of $\widehat{A}$ and the columns of $\widehat{B}$. Thus

$$E = \begin{bmatrix} \widehat{A} & \widehat{B} \end{bmatrix}. \tag{6.113}$$

If the input is represented using 16 bits, then

$$\|f(n)\|_\infty \le 2^{15} - 1 = f_{max} = 32767. \tag{6.114}$$

We also will specify that the values of the state variables used in the computations to be represented by 16 bits. Thus,

$$\|Q(n-1)\|_\infty \le f_{max}. \tag{6.115}$$

We can then write

$$\|P(n)\|_\infty \le \|E\|_\infty f_{max} \tag{6.116}$$

where we have eliminated the possibility of the input having a value of $f(n) = -32768$ for convenience. We need to choose s_3 such that

$$s_3 \le \frac{P_{max}}{f_{max}\|E\|_\infty}. \tag{6.117}$$

For our example, we have

$$\|E\|_\infty = 2.9051. \tag{6.118}$$

Thus, we have

$$s_3 \le \frac{2^{31} - 1}{32767(2.9051)} = t_1 = 22559.93. \tag{6.119}$$

We want to select s_3 such that $s_3 = 2^m$ where m is an integer:

$$m = \lfloor \log_2(t_1) \rfloor = 14. \tag{6.120}$$

Thus,

$$s_3 = 2^{14} = 16384. \tag{6.121}$$

In a similar way, we form the matrix

$$J = [C \quad D]. \tag{6.122}$$

Then

$$s_2 \le \frac{P_{max}}{f_{max}\|J\|_\infty}. \tag{6.123}$$

For our example,

$$\|J\|_\infty = 1.0431. \tag{6.124}$$

Thus,

$$s_2 \le \frac{2^{31} - 1}{32767(1.0431)} = 62828.22. \tag{6.125}$$

We can now determine the scaled matrices by multiplying A and B by s_3 and rounding the elements to integers and by multiplying C and D by s_2 and rounding the elements to integers. Thus, we have

$$\widehat{A} = \begin{bmatrix} 28506 & 16384 & 0 & 0 \\ -30730 & 0 & 16384 & 0 \\ 16897 & 0 & 0 & 16384 \\ -4534 & 0 & 0 & 0 \end{bmatrix}, \tag{6.126}$$

$$\widehat{B} = \begin{bmatrix} 2567 \\ 482 \\ 2066 \\ 511 \end{bmatrix}, \tag{6.127}$$

$$\widehat{C} = [62828 \ \ 0 \ \ 0 \ \ 0], \tag{6.128}$$

$$\widehat{D} = [2710]. \tag{6.129}$$

End of the Example

6.9 STATISTICAL ANALYSIS OF WORD LENGTH EFFECTS

Experimental methods can be used to analyze the effects of finite precision arithmetic for digital signal processing applications, but the effects are nonlinear and the accurate modeling is difficult. Statistical analysis is an alternate approach that often provides more representative results. This approach involves modeling the quantization errors as additive noise sequences at the point of the quantization [4]. We can use the simple example of a single pole IIR filter. The difference equation for implementing this filter is given by

$$y(n) = ay(n-1) + x(n). \tag{6.130}$$

We will analyze quantization errors due to rounding the results of the multiplication. We have

$$v(n) = Q\,[av(n-1)] + x(n) \tag{6.131}$$

where $Q[\ \ ]$ implies quantization. We can model this operation as

$$v(n) = av(n-1) + e(n) + x(n) \tag{6.132}$$

where $e(n)$ is a stationary, white noise sequence.

We have assumed that the system is linear and shift-invariant except for the quantization effects. Thus, we can separate the system response into one part due to the input $x(n)$ and one part due to the quantization error $e(n)$. Let

$$v(n) = y(n) + s(n). \tag{6.133}$$

Then

$$v(n) = ay(n-1) + as(n-1) + e(n) + x(n) \tag{6.134}$$

where

$$\begin{aligned} y(n) &= ay(n-1) + x(n), \\ s(n) &= as(n-1) + e(n). \end{aligned} \tag{6.135}$$

We can then determine the signal to noise ratio as the ratio of the power in the sequence $s(n)$ to the power in the sequence $y(n)$.

We can make the following assumptions to simplify the analysis [4]:

1. The quantization word size is m bits with a sign bit. Thus, for any n, the error sequence $e(n)$ is uniformly distributed over the range

$$-2^{-m} \leq e(n) \leq 2^{-m}. \tag{6.136}$$

   We have also assumed that the bits represent a fractional number with the second left most bit representing 2^{-1}, etc.

2. The error $e(n)$ is a stationary white noise sequence. In other words, $e(n)$ and the error $e(k)$ are uncorrelated for $n \neq k$.

3. The error sequence $e(n)$ is uncorrelated with the signal sequence $x(n)$.

EXAMPLE 6.11
(Roundoff Error Analysis for IIR SOS)

This example involves the analysis of the roundoff errors due to rounding the output from the multipliers in a second order section. Begin with the transposed direct form structure as given in Fig. 6.1. We can add an error input at the output of each multiplier to model the roundoff errors due to rounding the outputs from the multipliers. There are 5 multipliers, so we will have 5 different error inputs. Fig. 6.2 shows the block diagram used to derive the difference equations.

The output for each of the individual error inputs can be determined by making $x(n)$ equal to zero and by making the other error sequences equal to zero. This can be repeated for each error input. The system transfer function for the original system without any errors is given by

$$\frac{Y(z)}{X(z)} = \frac{b(0) + b(1)z^{-1} + b(2)z^{-2}}{1 + a(1)z^{-1} + a(2)z^{-2}}. \tag{6.137}$$

The system transfer functions and the corresponding difference equations for each of the error sequences are given by

1. Error due to e_0:

$$\frac{Y_0(z)}{E_0(z)} = \frac{1}{1 + a(1)z^{-1} + a(2)z^{-2}},$$
$$y_0(n) = e_0(n) - a(1)y_0(n-1) - a(2)y_0(n-2); \quad (6.138)$$

2. Error due to e_1:

$$\frac{Y_1(z)}{E_1(z)} = \frac{z^{-1}}{1 + a(1)z^{-1} + a(2)z^{-2}},$$
$$y_1(n) = e_1(n-1) - a(1)y_1(n-1) - a(2)y_1(n-2); \quad (6.139)$$

3. Error due to e_2:

$$\frac{Y_2(z)}{E_2(z)} = \frac{z^{-2}}{1 + a(1)z^{-1} + a(2)z^{-2}},$$
$$y_2(n) = e_2(n-2) - a(1)y_2(n-1) - a(2)y_2(n-2); \quad (6.140)$$

4. Error due to e_3:

$$\frac{Y_3(z)}{E_3(z)} = \frac{z^{-1}}{1 + a(1)z^{-1} + a(2)z^{-2}},$$
$$y_3(n) = e_3(n-1) - a(1)y_3(n-1) - a(2)y_3(n-2); \quad (6.141)$$

5. Error due to e_4:

$$\frac{Y_4(z)}{E_4(z)} = \frac{z^{-2}}{1 + a(1)z^{-1} + a(2)z^{-2}},$$
$$y_4(n) = e_4(n-2) - a(1)y_4(n-1) - a(2)y_4(n-2). \quad (6.142)$$

The principal of superposition can then be used to compute the overall error as

$$e(n) = e_0(n) + e_1(n) + e_2(n) + e_3(n) + e_4(n). \quad (6.143)$$

End of the Example

EXAMPLE 6.12
(Roundoff Error Analysis for FIR Lattice Filter)

This problem involves the statistical analysis of roundoff errors for an FIR second order lattice filter section. Fig. 6.3 gives a block diagram for a second order

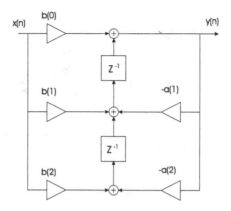

FIGURE 6.1

Block diagram for the transpose direct form for a second order IIR digital filter

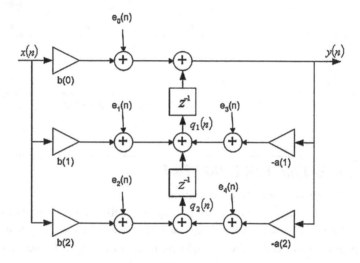

FIGURE 6.2

Block diagram of a second order IIR filter section with errors due to rounding the output of the multipliers

FIR filter that has been implemented as two FIR lattice sections. Fig. 6.4 gives a block diagram for the same filter showing the input errors $(e_0(n)$ to $e_3(n))$ associated with roundoff of the output from the adders. The outputs from the adders are rounded from 32 bits by multiplying the adder output by 2^{-16} and rounding the results. The error sequences represent the errors due to the rounding operations. **End of the Example**

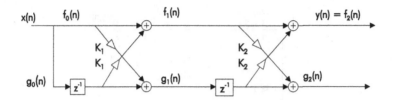

FIGURE 6.3

Block diagram for lattice filter for Example 6.12

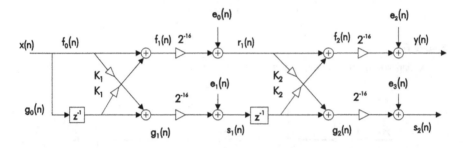

FIGURE 6.4

Block diagram for lattice filter for Example 6.12 showing roundoff errors

6.10 PROBLEMS FOR CHAPTER 6

Problem 6.1. (Fixed Point Representation)

Convert the following numbers to two's complement fixed point numbers assuming that a 10 bit word is used. Assume that one bit is used for the sign. Assume that the decimal point is located after the sign bit in each case. Specify the value of the exponent for the number using this assumption:

1. 0.0934
2. −0.00893
3. −0.552
4. 0.2565
5. 0.6071
6. 129.411
7. −75.192
8. −40.281
9. 5.898
10. 2.647

Problem 6.2. (Scaling an IIR Filter for Fixed Point Implementation)
The system transfer function for a discrete time system is given by

$$H(z) = \frac{B(z)}{A(z)},$$

$$B(z) = 0.1104 - 0.3838z^{-1} + 0.6582z^{-2} - 0.6582z^{-3}$$
$$+ 0.3838z^{-4} - 0.1104z^{-5},$$

$$A(z) = 1.0 - 0.1206z^{-1} + 1.2842z^{-2} + 0.2982z^{-3}$$
$$+ 0.4149z^{-4} + 0.2166z^{-5}. \tag{6.144}$$

1. Determine the scale factors for the numerator coefficients to scale the coefficients to 11 bit fixed point numbers.
2. Determine the scale factors for the denominator coefficients to scale the coefficients to 11 bit fixed point numbers with the restriction that the scale factor for the coefficients must be a power of 2 ($s_3 = 2^m$).
3. Determine the scaled numerator coefficients.
4. Determine the scaled denominator coefficients.

Problem 6.3. (Scaling an IIR Filter for Fixed Point Implementation)
The system transfer function for a discrete time system is given by

$$H(z) = \frac{B(z)}{A(z)},$$

$$B(z) = 0.3428 + 2.1417z^{-1} + 5.9713z^{-2} + 9.6003z^{-3}$$
$$+ 9.6003z^{-4} + 5.9713z^{-5} + 2.1417z^{-6} + 0.3428z^{-7},$$

$$A(z) = 1.0 + 4.3846z^{-1} + 8.9602z^{-2} + 10.5621z^{-3}$$
$$+ 7.5750z^{-4} + 3.1003z^{-5} + 0.5560z^{-6} - 0.0261z^{-7}. \tag{6.145}$$

1. Determine the scale factor for the numerator coefficients to scale the coefficients to 13 bit fixed point numbers with 1 bit representing the sign bit, 5 bits representing the whole number part, and 7 bits representing the fractional part. The scale factor for the coefficients is restricted to be a power of 2 ($s_2 = 2^m$). This is equivalent to the fixed point format $Q5.7$.
2. Determine the scale factor for the denominator coefficients to scale the coefficients to 13 bit fixed point numbers with 1 bit representing the sign bit, 5 bits representing the whole number part, and 7 bits representing the fractional part. The scale factor for the coefficients is restricted to be a power of 2 ($s_2 = 2^m$). This is equivalent to the fixed point format $Q5.7$.
3. Determine the fixed point numerator coefficients using the given format.
4. Determine the fixed point denominator coefficients using the given format.

Problem 6.4. (Two's Complement Numbers)
Convert the following numbers to two's complement fixed point numbers assuming that a 14 bit word is used. Specify the location of the decimal point and also the value of the exponent in each case:

1. 59.7407
2. −262.1167
3. −154.0914
4. 16.7390
5. 98.8602

Problem 6.5. (Two's Complement Numbers)
Convert the following numbers to two's complement fixed point numbers assuming that a 12 bit word is used.

1. 0.0062
2. −0.0111
3. −0.0552
4. 0.2565
5. 0.6071
6. 160.4711
7. −89.7895
8. −18.1259
9. 3.8986
10. 1.6470

Problem 6.6. (Scaling an IIR Filter for Fixed Point Implementation)
The system transfer function for a discrete time system is given by

$$H(z) = \frac{B(z)}{A(z)},$$
$$B(z) = 0.1296 - 0.5471z^{-1} + 1.1387z^{-2} - 1.4293z^{-3} + 1.1387z^{-4}$$
$$- 0.5471z^{-5} + 0.1296z^{-6},$$
$$A(z) = 1.0 - 1.0584z^{-1} + 2.0396z^{-2} - 0.8474z^{-3} + 1.0440z^{-4}$$
$$- 0.1153z^{-5} + 0.2654z^{-6}. \tag{6.146}$$

1. Determine the scale factor for the numerator coefficients to scale the coefficients to 14 bit fixed point numbers using the format Q5.8.
2. Determine the scale factor for the denominator coefficients to scale the coefficients to 14 bit fixed point numbers using the format Q5.8 with the restriction that the scale factor for the coefficients must be a power of 2 ($s_3 = 2^m$).
3. Determine the corresponding scaled numerator coefficients (format Q5.8).
4. Determine the corresponding scaled denominator coefficients (format Q5.8).

Problem 6.7. (Fixed Point Arithmetic)
The system transfer function for an FIR digital filter is given by

$$H(z) = -0.0087 - 0.1448z^{-1} + 0.0663z^{-2} + 0.6962z^{-3}$$
$$+ 0.0663z^{-4} - 0.1448z^{-5} - 0.0087z^{-6}. \tag{6.147}$$

Assume that you want to implement this filter using 14 bit fixed point arithmetic.

1. Determine the coefficients for the implementation of the filter using fixed point arithmetic. Represent the numbers in the form

$$\widehat{b(k)} = M \times 2^{-13} \qquad (6.148)$$

where M is an integer in the range $-2^{13} \le M \le 2^{13} - 1$.
2. Use the Matlab function *sampdata* provided in the course locker to obtain a sample sequence. Use the Matlab function *conv* to compute the convolution of the original filter with this sample sequence.
3. Use the Matlab function *fscale* to quantize the sample sequence for representation using 16 bit fixed point words. Use the Matlab function *conv* to compute the convolution of the quantized input and your fixed point representation of the filter.
4. Compare the two outputs:
   (a) Compute the maximum magnitude difference between samples of the two outputs.
   (b) Compute the average of the squared difference between the samples of the two outputs.
   (c) Plot the two output sequences on the same plot to compare them. Use a different plot symbol or line type for each sequence.

Problem 6.8. (Fixed Point Representation of an IIR Filter)
The system transfer function for a digital filter is given by

$$H(z) = \frac{B(z)}{A(z)} \qquad (6.149)$$

where

$$B(z) = 0.2324 + 1.6073z^{-1} + 5.0891z^{-2} + 9.5962z^{-3} + 11.7644z^{-4}$$
$$+ 9.5962z^{-5} + 5.0891z^{-6} + 1.6073z^{-7} + 0.2324z^{-8} \qquad (6.150)$$

and

$$A(z) = 1.0 + 4.4704z^{-1} + 9.9870z^{-2} + 13.7949z^{-3} + 12.9880z^{-4}$$
$$+ 8.5743z^{-5} + 4.0521z^{-6} + 1.3003z^{-7} + 0.2508z^{-8}. \qquad (6.151)$$

You are required to scale the filter coefficients for fixed point implementation using an input sequence with words scaled to 14 bits and with a requirement that the output can be represented using 32 bit words. Show all computations.

1. Determine the scale factor for the numerator coefficients.
2. Determine the corresponding scaled numerator coefficients.
3. Determine the scale factor for the denominator coefficients using the requirement that the scale factor must be a power of 2 (2^m where m is an integer).
4. Determine the corresponding scaled denominator coefficients.

Problem 6.9. (IEEE Format)

The system transfer function for an IIR digital filter is given by

$$H(z) = \frac{0.1345 - 0.4593z^{-1} + 0.6565z^{-2} - 0.4593z^{-3} + 0.1345z^{-4}}{1.0000 - 0.2188z^{-1} + 0.9438z^{-2} + 0.1229z^{-3} + 0.2819z^{-4}}. \quad (6.152)$$

The numbers in 32 bit IEEE format are represented in the form

$$p = (1.f)(-1)^s (2^{e-127}) \quad (6.153)$$

where p is the floating number to be represented, s is a sign bit, e is the biased exponent represented using 8 bits, and f is a positive fraction represented using 23 bits.

- s is the most significant bit in the 32 bit word (bit 32),
- e is the next most significant 8 bits (bits 24 through 31), and
- f is the 23 least significant bits (bits 1 through 23).

1. Determine the values for s, e, and f for each of the filter coefficients.
2. Determine the two's complement representation for each of the coefficients as represented using 32 bit IEEE floating point format.
3. Determine the hexadecimal representation for each of the coefficients as represented using 32 bit IEEE floating point format.

Problem 6.10. (Scaling the State Space Implementation)

The state space representation for a discrete time system is given by

$$\begin{aligned} q(n) &= Aq(n-1) + Bx(n), \\ y(n) &= Cq(n-1) + Dx(n) \end{aligned} \quad (6.154)$$

where

$$A = \begin{bmatrix} -1.7134 & 1.0000 & 0 & 0 \\ -1.4282 & 0 & 1.0000 & 0 \\ -0.5621 & 0 & 0 & 1.0000 \\ -0.0900 & 0 & 0 & 0 \end{bmatrix}, \quad (6.155)$$

$$B = \begin{bmatrix} 0.6851 \\ 1.3697 \\ 1.0300 \\ 0.2726 \end{bmatrix}, \quad (6.156)$$

$$C = [1 \quad 0 \quad 0 \quad 0], \quad (6.157)$$

$$D = [0.2996]. \quad (6.158)$$

Determine the scaled state space matrices to implement this system using fixed point arithmetic using the following specifications:

1. The coefficients of the scaled state space matrices should be represented using 12 bit fixed point numbers.
2. The values of the past state variables, $q_k(n - 1)$, used in the computations are represented using 12 bit fixed point numbers. The computed state variables will be truncated or rounded to 12 bits before they are used in later computations.
3. Assume that the input is represented using 12 bit fixed point numbers.
4. The outputs for the computations cannot exceed 24 bits.

Problem 6.11. (Finite Word Length Effects)
The polyphase representation for an FIR digital filter is given by

$$
\begin{aligned}
f_1(n) &= 0.0062x(n) - 0.0556x(n-2) + 0.6082x(n-4) \\
&\quad - 0.0556x(n-6) + 0.0062x(n-8), \\
s(n) &= x(n-1), \\
f_2(n) &= -0.0110s(n) + 0.2562s(n-2) + 0.2562s(n-4) \\
&\quad - 0.0110s(n-6), \\
y(n) &= f_1(n) + f_2(n).
\end{aligned}
\tag{6.159}
$$

You are required to scale the filter coefficients for fixed point implementation using an input sequence with words scaled to 15 bits and with a requirement that the output can be represented using 32 bit words.

1. Determine the scale factor for the coefficients.
2. Determine the corresponding scaled integer coefficients.

6.11 MATLAB PROBLEMS FOR CHAPTER 6

Problem 6.12. (Effects of Rounding)
This problem involves an investigation of the effects of roundoff on the location of the zeros and poles of the system transfer function and on the frequency response. The Matlab function *zplane* can be used to determine the location of the poles and zeros of the system transfer function and the Matlab function *freqz* can be used to determine the magnitude and phase of the frequency response.

1. Use the following Matlab script to design a low pass digital filter.
```
order = 8,
rp = 1.5;
rs = 40;
wc = 0.716;
[b, a] = ellip(order, rp, rs, wc);
```
2. Scale the filter coefficients to 10 bit fixed point numbers using the Matlab function *iirscale* provided in Appendix D.3.
```
[bsc, asc, scfac] = iirscale(b, a, bits);
```

3. Compare the location of the poles and zeros and the frequency response of the resulting filter to those for the original filter.

   (a) Use the Matlab function *zplane* to compute and plot the poles and zeros for the original floating point coefficients [b, a] and for the scaled coefficients [bsc, asc].

   (b) Use the Matlab function *freqz* to compute the magnitude and phase responses for the original floating point coefficients [b, a] and for the scaled coefficients [bsc, asc].

   (c) Summarize the effects of the scaling operation in terms of the location of the poles and zeros and the magnitudes and phases of the corresponding frequency responses.

Problem 6.13. (Effects of Rounding)
This problem involves an investigation of the effects of roundoff on the location of the zeros and poles of the system transfer function and on the frequency response when the system is implemented as a cascade of second order sections.

1. Use the following Matlab script to design a low pass digital filter.

```
order = 8,
rp = 1.5;
rs = 40;
wc = 0.716;
[b, a] = ellip(order, rp, rs, wc);
```

   Note that this is the same filter used in Problem 6.12.
2. Use the Matlab function *tf2sos* to develop a cascade implementation of the filter using second order sections.
3. Use the Matlab function *iirscale* to scale the coefficients for each of the individual second order sections to 10 bit fixed point numbers.
4. You will have 4 second order sections which we will call $H_1(z)$, $H_2(z)$, $H_3(z)$, and $H_4(z)$. Compute the equivalent overall system transfer function

$$H_2(z) = H_1(z)H_2(z)H_3(z)H_4(z) = \frac{\sum_{k=0}^{8} b2(k)z^{-k}}{1.0 + \sum_{k=1}^{8} a2(k)z^{-k}}. \qquad (6.160)$$

5. Compare the location of the poles and zeros and the frequency response of the resulting filter $H_2(z)$ to those for the original filter.

   (a) Use the Matlab function *zplane* to compute and plot the poles and zeros for the original floating point coefficients [b, a] and for the scaled coefficients [b2, a2].

   (b) Use the Matlab function *freqz* to compute the magnitude and phase responses for the original floating point coefficients [b, a] and for the scaled coefficients [b2, a2].

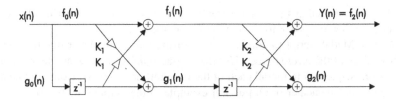

FIGURE 6.5

Block diagram for lattice filter for Problem 6.14

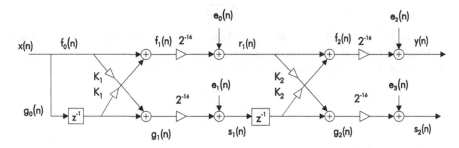

FIGURE 6.6

Block diagram for lattice filter for Problem 6.14 showing roundoff errors

    (c) Summarize the effects of the scaling operation in terms of the location of the poles and zeros and the corresponding frequency responses.

6. Give a short summary of your observations on the impact of rounding on the system transfer function and any benefits from using second order sections to implement the filter.

Problem 6.14. (Statistical Analysis of Roundoff Errors)
This problem involves the statistical analysis of roundoff errors for an IIR second order section. Fig. 6.5 gives a block diagram for a second order FIR filter that has been implemented as two FIR lattice sections. Fig. 6.6 gives a block diagram for the same filter showing the input errors ($e_0(n)$ to $e_3(n)$) associated with roundoff errors. The outputs from the adders are rounded from 32 bits by multiplying the adder output by 2^{-16} and truncating toward zero. The scaled coefficients are given by

$$K_1 = 22921,$$
$$K_2 = -7504. \qquad (6.161)$$

1. Use the Matlab script *sampdata*, provided in Appendix A.1, to obtain an input sample. Use the Matlab script *fscale*, provided in Appendix D.1, to scale the input sequence to 16 bits.

2. Write a Matlab script to obtain the output sequence with the output from the adders multiplied by 2^{-16} but the results left in a floating point number (no rounding).

This will be used as the output without error. Note that this experiment does not consider any errors due to scaling the K parameters.

3. Write a Matlab script to compute the contributions to the error due to rounding from each of the error sources. Use the Matlab *rand* script to generate appropriate random sequences to model each of the roundoff error inputs. Compute the average power for each error output. For example, if the output due to $e_0(n)$ is $d_0(n)$, then the average power for $d_0(n)$ can be computed as

$$P_0 = \frac{1}{N} \sum_{k=0}^{N-1} d_0^2(n) \qquad (6.162)$$

where N is the number of samples in $d_0(n)$. Add all of the average power values to obtain the total average noise power.

4. Compute the average power in the error free output from part 2 as

$$PY = \frac{1}{N} \sum_{k=0}^{N-1} y^2(n) \qquad (6.163)$$

where $y(n)$ is the error free output.

5. Compute the roundoff error signal to noise power as the ratio of the average output power for the error free output to the total average noise power.

Problem 6.15. (FIR Filter Design and Implementation)
Assume that a signal is sampled at 8000 samples per second.

1. Design an FIR band stop filter using the Matlab script *firls* using the following parameters:

- The pass band is 1000 to 2500 Hz.
- The stop bands are
  (a) 0 to 750 Hz,
  (b) 2750 to 4000 Hz.
- The maximum desired pass band ripple is ± 2.0 dB.
- The maximum desired stop band ripple is -35 dB.
- The order of the desired filter is 32.

Use the Matlab help utility to obtain more information on the *firls* script.

2. Obtain the sample input using the *sampdata* function provided in Appendix A.1. Use the Matlab *conv* function to filter the input signal using your FIR band pass filter.

3. Use the Matlab function *firscale2* in Appendix D.2 to scale the filter coefficients to avoid overflow if a 32 bit word is used for the output and a 13 bit word is used for the input.

4. Use the Matlab function *fscale* in Appendix D.1 to scale the sample sequence to be represented using 13 bit words.

5. Use the Matlab function *conv* function to filter the scaled input sequence using your the scaled band pass filter.
6. Compare the output from the fixed point implementation with the floating point output:
    (a) Compute the maximum magnitude difference between samples of the two outputs.
    (b) Compute the average of the squared difference between the samples of the two outputs.
    (c) Plot the two output sequences on the same plot to compare them. Use a different plot symbol or line type for each sequence.
7. Observe that the number of bits needed to represent the scaled filter coefficients obtained in part 3 is greater than 16 bits.
    (a) Determine the maximum word size we can use for scaling the input sequence and represent the filter coefficients using 16 bit words without getting overflow when a 32 bit word is used for the output.
    (b) Use the Matlab function *firscale2* to scale the filter coefficients using this word size for the input sequence.
    (c) Repeat part 5 using the revised scaled filter coefficients. The input for this part of the problem should be scaled to the new word size using *fscale*.
    (d) Repeat part 6 to compare the output for the convolution of the revised fixed point coefficients and the input scaled to the new word size with the output from the floating point implementation.
8. **To be turned in:**
    (a) The design parameters for your FIR filter. This includes the input arrays to *firls* for specifying the frequencies, the frequency response, the weights, etc.
    (b) The Matlab script for designing and implementing the floating point and both versions of the fixed point implementations of the filter.
    (c) The magnitude and phase plots for the filter.
    (d) The stem plots for all of the output sequences.
    (e) The parameters for comparing the outputs to the floating point output from *conv*.

Problem 6.16. (Effects of Coefficient Scaling)
The system transfer function for an IIR digital filter is given by

$$H(z) = \frac{B(z)}{A(z)} \tag{6.164}$$

where

$$\begin{aligned} B(z) \;=\; & 0.1006 - 0.4498z^{-1} + 1.1211z^{-2} - 1.8403z^{-3} + 2.1630z^{-4} \\ & - 1.8403z^{-5} + 1.1211z^{-6} - 0.4498z^{-7} + 0.1006z^{-8}, \tag{6.165} \end{aligned}$$

$$\begin{aligned} A(z) \;=\; & 1.0 - 1.0130z^{-1} + 2.8890z^{-2} - 1.4483z^{-3} + 2.4901z^{-4} \\ & - 0.3959z^{-5} + 0.9169z^{-6} + 0.0324z^{-7} + 0.1867z^{-8}. \tag{6.166} \end{aligned}$$

1. Develop a cascade implementation of the filter using second order sections.
2. Scale the filter coefficients for each individual second order section to 10 bits.
3. Compute the overall system transfer function for the cascade of the scaled second order sections with no additional rounding beyond that used for the scaling.
4. Use the Matlab *zplane* function to compute the pole–zero plot for the cascade of the scaled second order sections version of $H(z)$.
5. Use the Matlab *freqz* function to plot the magnitude and phase of the frequency response for the cascade of the scaled second order sections version of $H(z)$.
6. Compare the location of the poles and zeros and the frequency response of the resulting cascade of the scaled second order sections version of $H(z)$ to those for the original filter.
7. Give a short summary of your observations on the impact of rounding on the system transfer function and any benefits from using second order sections to implement the filter.

To be turned in:

1. The Matlab script for each part of the problem.
2. The plots for each part of the problem.

Problem 6.17. (Coefficient Scaling)
Use the following Matlab script to compute the coefficients for an FIR digital filter.

```
order = 40;
wc = 0.491;
type = 'high';
b1 = fir1(order, wc, type);
```

1. Use the Matlab function *freqz* to plot the magnitude and phase plots for the filter designed using this script which we will call $H(z)$ for this problem.
2. Use the Matlab function *zplane* to plot the location of the zeros for $H(z)$.
3. The Matlab function *fscale* has been provided in Appendix A.1. Use this function to scale the filter coefficients to 6 bits.
4. Use the Matlab function *freqz* to plot the magnitude and phase plots for the scaled version of $H(z)$.
5. Use the Matlab function *zplane* to plot the location of the zeros for the scaled version of $H(z)$.
6. Compare the location of the poles and zeros and the frequency response of the scaled version of $H(z)$ to those for the original filter.
7. Give a short summary of your observations on the impact of rounding on the system transfer function.

To be turned in:

1. The Matlab script for each part of the problem.
2. The plots for each part of the problem.

Multirate Digital Signal Processing

7

Many digital systems perform all of the required operations at a single sampling rate. However, there are many practical digital systems that require the sampling rate to be changed one or more times during normal operations. For example, telecommunication systems transmit and receive different types of signals (e.g., teletype, facsimile, speech, and video) and require the processing of the signal at different rates commensurate with the corresponding bandwidths of the signals [4]. In general, manipulation of the sampling rate of a digital signal is accomplished by increasing or decreasing the sampling rate by some factor. This process is generally called *sampling rate conversion*. Systems that use digital signals at different sampling rates are called *multirate digital signal processing systems*.

There are two general methods for performing sampling rate conversions [4]. One method is to pass the digital signal through a D/A converter and then sample the reconstructed, continuous time signal at the new sampling rate. This method often leads to signal distortion due to signal reconstruction and quantization effects in the D/A converters and/or in the A/D converter. The second method involves performing the sampling rate conversion directly in the digital domain. Performing the sample rate conversion directly in the digital domain can often result in less distortion due to the sampling rate conversion process.

This chapter gives a conceptual description of the impact that manipulating the sample rate has on the resulting Nyquist frequency and the resulting frequency content contained within the Nyquist range of the new resampled signal. This chapter describes the time domain and frequency domain representations of two basic sample rate alteration devices, namely *upsamplers* and *downsamplers*. Further details on how these operations are used to perform sample rate conversions are explored. This often requires low pass filtering to avoid distortions in the newly resampled signal, resulting in an interpolated (upsampling then filtering) or a decimated (filtering then downsampling) signal. The sample rate changing devices, along with the polyphase filtering structure, can be used to implement computationally efficient ways to perform interpolation and decimation. Multistage implementations of decimators are also discussed, which are particularly efficient when modifying the sample rate by large factors. Later sections in this chapter focus on how these sample alteration devices combined with specially designed filters can be used to implement *multirate filter banks*. Multirate filter banks allow for the analysis of signals over specific spectral ranges.

Digital Signal Processing. DOI: 10.1016/B978-0-12-804547-3.00007-3

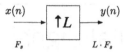

FIGURE 7.1

Block diagram of the upsampling process

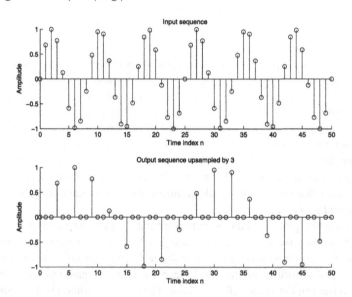

FIGURE 7.2

Original signal (top) and the signal upsampled by a factor of 3 (bottom)

7.1 CONCEPTUAL VIEW OF MODIFYING SAMPLING RATE

Fig. 7.1 shows a block diagram of the upsampling process. The output $y(n)$ and the input $x(n)$ are related to each other as shown in Eq. (7.1):

$$y(n) = \begin{cases} x(n/L) & n = 0, \pm L, \pm 2L, \pm 3L, \ldots, \\ 0, & \text{otherwise.} \end{cases} \tag{7.1}$$

The upsampling process increases the number of samples by inserting $L - 1$ zeroes between the original samples of $x(n)$. This operation results in the increase of the sampling rate by a factor of L. Fig. 7.2 shows an original signal and the resulting upsampled signal where $L = 3$. Consider the implications of increasing this sampling rate on the effective Nyquist rate of the signal $y(n)$ and subsequent change in the frequency spectrum of $y(n)$. Let $x(n)$ be a signal whose frequency spectrum is shown in Fig. 7.3. Here, the Nyquist frequency $\omega = \pm\pi$ corresponds to the analog frequencies $F = \pm F_s/2$. Here, F_s is the respective sampling rate of the signal $x(n)$.

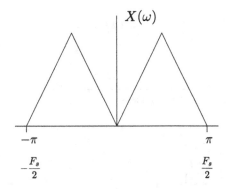

FIGURE 7.3

Spectrum of some signal $x(n)$

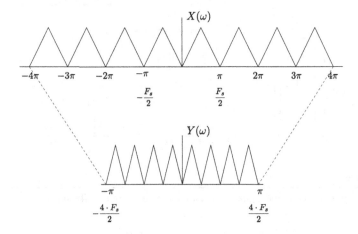

FIGURE 7.4

Spectrum of the upsampled signal $y(n)$, $L = 4$

As shown in Fig. 7.1, it is clear that sampling rate of the up sampled signal $y(n)$ is $L \cdot F_s$ or L-times the original sampling rate. This results in a Nyquist frequency for $y(n)$ of $L \cdot F_s/2$. Thus, the range of analog frequencies contained between $\pm\pi$ in the frequency spectrum of $y(n)$ represents analog frequency content between $\pm L \cdot F_s/2$. Without loss of generality, consider an upsampling factor $L = 4$. Looking back at the spectral content of $x(n)$, it is clear that the spectral content corresponding to the range analog frequency $\pm 4 \cdot F_s/2$ contains portions of the periodic repeats associated with the DTFT. All of the content of the spectrum of $x(n)$ between the range of $\pm 4 \cdot F_s/2$ will be squeezed into the range $\pm\pi$ in the spectrum of $y(n)$ (shown in Fig. 7.4). Thus, one can expect up sampling process to introduce additional frequency images or spectral content, where the additional content corresponds to the

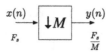

FIGURE 7.5

Block diagram of the downsampling process

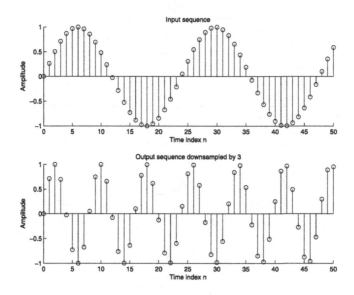

FIGURE 7.6

Original signal (top) and the signal downsampled by a factor of 3 (bottom)

periodic spectra associated with a sampled signal. We will see in later sections that we can identify ways to effectively eliminate this additional content using adequately designed filters.

A similar assessment of how downsampling an input signal influences the resulting spectral content of a signal can be made. Fig. 7.5 shows a block diagram of the downsampling process. We can relate the output $y(n)$ of a downsampling device to the input $x(n)$ using Eq. (7.2):

$$y(n) = x(nM). \tag{7.2}$$

The downsampling process decreases the number of samples by obtaining every Mth sample of $x(n)$ and throwing away every other sample. This operation results in an overall decrease in the sampling rate of the output signal $y(n)$ by a factor of M. Fig. 7.6 shows a sampled signal and the resulting downsampled signal for $M = 3$. We again consider the implications that modifying the sample rate has on the spectral content of the downsampled signal. Consider the situation where the signal $x(n)$ has

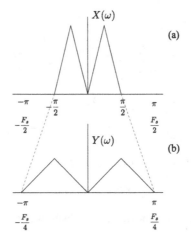

FIGURE 7.7

(a) Sample spectrum of signal $x(n)$; (b) spectrum of downsampled signal, $M = 2$

the frequency spectrum shown in Fig. 7.7(a), where the Nyquist frequency $\omega = \pm\pi$ again corresponds to the analog frequencies $F = \pm F_s/2$. Downsampling of the signal results in the reduction of the sampling rate to F_s/M. Thus, the Nyquist frequencies $\pm\pi$ for the spectrum of the downsampled signal $y(n)$ corresponds to the analog frequency range $\pm F_s/(2M)$. Without loss of generality, consider a downsampling factor $M = 2$. Looking back at the spectral content of $x(n)$, we see that the spectral content corresponding to the range $\pm F_s/4$ is a subset of the Nyquist range of the original signal corresponding to the content over the range $\pm\pi/2$. Thus the spectral content of $x(n)$ over the range $\pm\pi/2$ is spread of the range $\pm\pi$ in the spectral content of $y(n)$ (shown in Fig. 7.7(b)). In this case, the lack of spectral content over the range $\pi/2$ to π and $-\pi/2$ to $-\pi$ ensures that downsampling by a factor of $M = 2$ does not introduce any level of aliasing. However, in general, spectral content outside of the range of $\pm\pi/M$ ($|\omega| \leq \pi$) will cause aliasing if the signal is downsampled by a factor of M. As shown later in the chapter, this problem can be alleviated if the signal is adequately filtered prior to down sampling.

7.2 FREQUENCY REPRESENTATION OF THE UPSAMPLER

A mathematical description of the effects of the upsampler on the frequency domain characteristics of the original signal can be derived. This is first done by deriving the DTFT of the upsampled signal $y(n)$ in terms of the DTFT of $x(n)$ for an upsampling factor of $L = 2$. From (7.1), the Z Transform of the upsampled signal can be written as

$$
\begin{aligned}
Y(z) &= \sum_{n=-\infty}^{\infty} y(n)z^{-n} = \sum_{n=-\infty}^{\infty} x(n/2)z^{-n} \\
&= \sum_{m=-\infty}^{\infty} x(m)z^{-2m} \\
&= X(z^2).
\end{aligned}
\tag{7.3}
$$

Similarly for a general L, where $y(n) = x(n/L)$,

$$
Y(z) = X(z^L).
\tag{7.4}
$$

The frequency response can be obtained by substituting $z = e^{j\omega}$ as

$$
Y(e^{j\omega}) = X(e^{jL\omega}),
\tag{7.5}
$$

or

$$
Y(\omega) = X(L\omega).
\tag{7.6}
$$

We see from the above equation that the spectrum $Y(\omega)$ is a frequency scaled or compressed version of the spectrum of $X(\omega)$ since $L \geq 1$. This results in the inclusion of $L - 1$ additional images in the Nyquist range of $Y(\omega)$. These additional images result from the periodic repeats in the frequency spectrum of sampled signals. These additional images are a manifestation of the discontinuities introduced when the additional zeroes that were incorporated in the upsampled signal. The original spectral content can be recovered by low pass filtering $Y(\omega)$, removing the $L - 1$ additional images. This, in effect, averages the signal filling in the inserted zeroes with interpolated values. The process of upsampling and then filtering is known as *interpolation*.

7.3 FREQUENCY INTERPRETATION OF THE DOWNSAMPLER

The frequency domain representation a downsampler can be derived in a similar fashion. For some down sampling factor M, the Z Transform of the downsampled signal in (7.2) can be written as

$$
\begin{aligned}
Y(z) &= \sum_{n=-\infty}^{\infty} y(n)z^{-n} \\
&= \sum_{n=-\infty}^{\infty} x(nM)z^{-n}.
\end{aligned}
\tag{7.7}
$$

Unfortunately, this representation does not allow $Y(z)$ to be written directly as a function of $X(z)$. An intermediate signal can be generated, which can link $Y(z)$ and

$X(z)$. This provides the needed link between the Z Transforms of the input and output of the downsampler. Define a new signal $x_{int}(n)$

$$x_{int}(n) = \begin{cases} x(n), & n = 0, \pm M, \pm 2M, \ldots, \\ 0, & \text{otherwise.} \end{cases} \tag{7.8}$$

Since every Mth sample of $x_{int}(n)$ is equal to every Mth sample of $x(n)$ (i.e., $x_{int}(nM) = x(nM)$), we proceed from (7.7) yielding

$$\begin{aligned} Y(z) &= \sum_{n=\infty}^{\infty} x_{int}(nM)z^{-n} \\ &= \sum_{k=\infty}^{\infty} x_{int}(k)z^{-k/M} \\ &= X_{int}(z^{1/M}). \end{aligned} \tag{7.9}$$

The signal $X_{int}(z^{1/M})$ can be obtained by deriving $X_{int}(z)$ in terms of $X(z)$. It is clear that $x_{int}(n)$ can be written as

$$x_{int}(n) = c(n)x(n) \tag{7.10}$$

where

$$c(n) = \begin{cases} 1, & n = 0, \pm M, \pm 2M, \ldots, \\ 0, & \text{otherwise.} \end{cases} \tag{7.11}$$

The signal $c(n)$ can be rewritten as

$$c(n) = \frac{1}{M} \sum_{k=0}^{M-1} e^{j\frac{2\pi kn}{M}} \tag{7.12}$$

where this equality can be proven using the geometric series. Thus

$$\begin{aligned} X_{int}(z) &= \sum_{n=-\infty}^{\infty} c(n)x(n)z^{-n} \\ &= \frac{1}{M} \sum_{n=-\infty}^{\infty} \left(x(n) \sum_{k=0}^{M-1} e^{j\frac{2\pi kn}{M}} \right) z^{-n} \\ &= \frac{1}{M} \sum_{k=0}^{M-1} \sum_{n=-\infty}^{\infty} x(n) \left(ze^{-j\frac{2\pi k}{M}} \right)^{-n} \\ &= \frac{1}{M} \sum_{k=0}^{M-1} X\left(ze^{-j\frac{2\pi kn}{M}} \right). \end{aligned} \tag{7.13}$$

Substituting $X_{int}(z)$ into Eq. (7.9), we get

$$Y(z) = \frac{1}{M} \sum_{k=0}^{M-1} X\left(z^{\frac{1}{M}} e^{-j\frac{2\pi k}{M}}\right) \tag{7.14}$$

with a frequency response of

$$\begin{aligned}Y(e^{j\omega}) &= \frac{1}{M} \sum_{k=0}^{M-1} X\left(e^{j\frac{\omega}{M}} e^{-j\frac{2\pi k}{M}}\right) \\ &= \frac{1}{M} \sum_{k=0}^{M-1} X\left(e^{j\frac{\omega-2\pi k}{M}}\right),\end{aligned} \tag{7.15}$$

or

$$Y(\omega) = \frac{1}{M} \sum_{k=0}^{M-1} X\left(\frac{\omega - 2\pi k}{M}\right). \tag{7.16}$$

To get an idea of what is going on conceptually, let us look at the case when $M = 2$. This gives a frequency response of

$$Y(\omega) = \frac{1}{2}\left(X\left(\frac{\omega}{2}\right) + X\left(\frac{\omega - 2\pi}{2}\right)\right). \tag{7.17}$$

$X(\omega/2)$ is just an expanded version of $X(\omega)$ and $X((\omega - 2\pi)/2)$ is the expanded version shifted by 2π. These two signals will not overlap (alias) only if $X(\omega)$ is bandlimited to $\pm\pi/2$ or $X(\omega) = 0$, $|\omega| \geq \pi/2$. In general, aliasing due to a factor of M downsampling is absent if and only if the signal $x(n)$ is bandlimited to $\pm\pi/M$. As was demonstrated conceptually above, if the original signal $x(n)$ possessed frequency content outside of $\pi/2$, aliasing would occur. The process of filtering a signal to bandlimit it over the range $\pm\pi/M$ and then downsampling by a factor of M is known as *decimation*. In the following section, we discuss the process of *interpolation* and *decimation* in greater detail.

7.4 MULTIRATE STRUCTURES FOR SAMPLING RATE CONVERSION

As previously stated, any signal that is decimated by M must be bandlimited to the range $|\omega| \leq \pi/M$ before the signal is downsampled to prevent aliasing. Thus, the signal must be filtered by a low pass filter with a cutoff frequency of π/M. The process of interpolation also requires filtering. In this case, the signal is filtered after the upsampling process by a low pass filter with a cutoff frequency of π/L to enforce interpolation between the samples. This eliminates the additional images that are

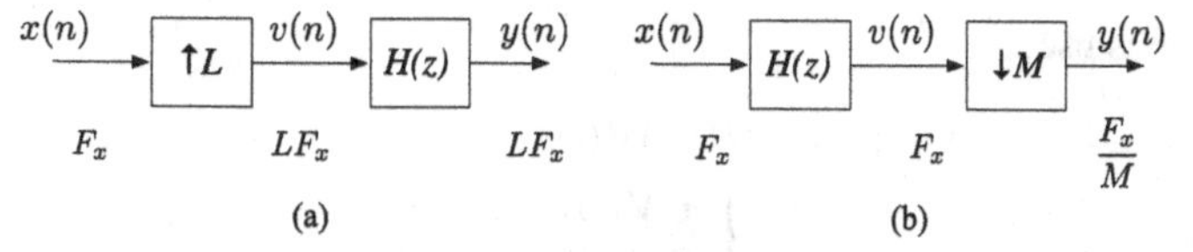

$x(n) \rightarrow \boxed{\uparrow L} \xrightarrow{v(n)} \boxed{H(z)} \xrightarrow{y(n)}$ $x(n) \rightarrow \boxed{H(z)} \xrightarrow{v(n)} \boxed{\downarrow M} \xrightarrow{y(n)}$

$F_x \qquad\qquad LF_x \qquad\qquad LF_x$ $F_x \qquad\qquad F_x \qquad\qquad \dfrac{F_x}{M}$

(a)   (b)

FIGURE 7.8

Interpolator (a) and decimator (b)

shifted into the baseband due to the insertion of zeros in the time signal. Fig. 7.8 illustrates the block diagrams for a decimator and an interpolator.

7.4.1 IDEAL CHARACTERISTICS OF DECIMATION/INTERPOLATION FILTERS

7.4.1.1 *Interpolator*

Interpolation requires filtering after the upsampling process to get rid of the additional images in the baseband. This is shown in Fig. 7.8(a). The question is what are the ideal characteristics of such a filter. Specifically, the interest is in identifying the frequency range and filter magnitude needed to ensure that the values of the samples in $y(n)$ that correspond to the original samples in $x(n)$ are conserved (i.e., $y(n) = x(n/L)$ for $0, \pm L, \pm 2L, \ldots$). In this section, these ideal characteristics will be derived.

Based on the interpolator shown in Fig. 7.8(a), we know that the discrete frequency relationship between the output of the upsampler $v(n)$ and the input $x(n)$ is

$$\omega_v = \frac{\omega_x}{L}. \tag{7.18}$$

The idea is to design a filter that attenuates any additional images due to upsampling that occur above $\omega_v \geq \pi/L$. Thus, we can design an ideal filter with an arbitrary gain C as follows:

$$H(\omega_v) = \begin{cases} C & 0 \leq |\omega_v| \leq \frac{\pi}{L}, \\ 0 & \text{otherwise.} \end{cases} \tag{7.19}$$

Using the following Fourier relationships:

$$x(n) \overset{\mathcal{F}}{\Longleftrightarrow} X(\omega), \tag{7.20}$$

$$v(n) \overset{\mathcal{F}}{\Longleftrightarrow} V(\omega) = X(L\omega), \tag{7.21}$$

we can derived the functional form of $Y(\omega_y)$. Note that $\omega_y = \omega_v$ because these signals operate at the same sampling rate $L \cdot F_x$. As such, the subscript is dropped and only ω is used:

$$Y(\omega) \;=\; H(\omega)V(\omega) \tag{7.22}$$

$$= \begin{cases} CV(\omega) & 0 \le |\omega| \le \frac{\pi}{L}, \\ 0 & \text{otherwise}, \end{cases} \tag{7.23}$$

$$= \begin{cases} CX(L\omega) & 0 \le |\omega| \le \frac{\pi}{L}, \\ 0 & \text{otherwise}. \end{cases} \tag{7.24}$$

The goal is to select C such that $y(n) = x(n/L)$ for $n = 0, \pm L, \pm 2L, \ldots$. We can write the inverse Fourier relationship of $y(n)$ and $Y(\omega)$ as

$$y(n) \;=\; \frac{1}{2\pi} \int_{-\pi}^{\pi} Y(\omega)e^{j\omega n}\,d\omega \tag{7.25}$$

$$= \frac{1}{2\pi} \int_{-\pi}^{\pi} H(\omega)V(\omega)e^{j\omega n}\,d\omega. \tag{7.26}$$

Without loss of generality, selecting $n = 0$, we get

$$y(0) \;=\; \frac{1}{2\pi} \int_{-\pi}^{\pi} H(\omega)V(\omega)\,d\omega \tag{7.27}$$

$$= \frac{C}{2\pi} \int_{-\frac{\pi}{L}}^{\frac{\pi}{L}} X(L\omega)\,d\omega. \tag{7.28}$$

Performing a change of variable where $\hat{\omega} = L\omega$, we get

$$y(0) \;=\; \frac{C}{L} \cdot \frac{1}{2\pi} \int_{-\pi}^{\pi} X(\hat{\omega})\,d\hat{\omega} \tag{7.29}$$

$$= \frac{C}{L}x(0). \tag{7.30}$$

Thus, for $y(0) = x(0)$ a filter with an amplitude of $C = L$ in the pass band is needed. This in fact applies to all $y(n) = x(n/L)$, $n = 0, \pm L, \pm L, \ldots$. This follows because the process of upsampling compresses multiple images in the baseband range without changing the amplitude or increasing the energy in the signal. Removing the additional images thus decreases the energy in the signal. Amplifying the signal by a factor of L recovers the energy lost due to removal of the additional images.

7.4.1.2 Decimator

Decimation requires filtering prior to the downsampling process to ensure that the output signal of the decimator is not distorted due to aliasing. This is shown in Fig. 7.8(b). The question once again is what are the ideal characteristics of this filter. As stated previously, the discrete frequency relationship between the output of the downsampler $y(n)$ and the input $v(n)$ is

$$\omega_y = M\omega_v. \tag{7.31}$$

The idea is to design a filter that bandlimits the input signal of the decimator, $x(n)$, within the range $0 \le |\omega| \le \pi/M$ to prevent aliasing when the signal is downsampled. We can design an ideal filter with an arbitrary gain B as follows:

$$H(\omega) = \begin{cases} B & 0 \le |\omega| \le \frac{\pi}{M}, \\ 0 & \text{otherwise.} \end{cases} \tag{7.32}$$

Recall the following Fourier relationships:

$$x(n) \overset{\mathcal{F}}{\Longleftrightarrow} X(\omega), \tag{7.33}$$

$$v(n) \overset{\mathcal{F}}{\Longleftrightarrow} V(\omega), \tag{7.34}$$

$$y(n) \overset{\mathcal{F}}{\Longleftrightarrow} Y(\omega) = \frac{1}{M} \sum_{k=0}^{M-1} V\left(\frac{\omega - 2\pi k}{M}\right). \tag{7.35}$$

We can write the frequency response of $v(n)$ as

$$V(\omega) = H(\omega)X(\omega) \tag{7.36}$$

$$= \begin{cases} BX(\omega) & 0 \le |\omega| \le \frac{\pi}{M}, \\ 0 & \text{otherwise.} \end{cases} \tag{7.37}$$

The idea is to set the gain of the decimation filter, B, to preserve the values of the samples that will be kept during the downsampling process. This is equivalent to preserving the shape and energy of the original signal $x(n)$ between $-\pi/M \le \omega \le \pi/M$. We again go back to the inverse Fourier relationship between $y(n)$ and $Y(\omega)$,

$$y(n) = \frac{1}{2\pi} \int_{-\pi}^{\pi} Y(\omega)e^{j\omega n} d\omega. \tag{7.38}$$

Without loss of generality, selecting $n = 0$, we get

$$y(0) = \frac{1}{2\pi} \int_{-\pi}^{\pi} Y(\omega)d\omega \tag{7.39}$$

$$= \frac{1}{2\pi} \cdot \frac{1}{M} \int_{-\pi}^{\pi} \sum_{k=0}^{M-1} V\left(\frac{\omega - 2\pi k}{M}\right) d\omega. \tag{7.40}$$

Since the filter $H(z)$ attenuates all frequencies corresponding to $|\omega| > \pi/M$, all terms in the summation for $k > 0$ are eliminated. This results in

$$y(0) = \frac{1}{2\pi} \cdot \frac{1}{M} \int_{-\pi}^{\pi} V\left(\frac{\omega}{M}\right) d\omega. \tag{7.41}$$

Performing a change of variable from ω to $\hat{\omega}$ where $\hat{\omega} = \omega/M$, we get

$$y(0) \;=\; \frac{1}{M} \cdot M \cdot \frac{1}{2\pi} \int_{-\frac{\pi}{M}}^{\frac{\pi}{M}} V(\hat{\omega}) d\hat{\omega} \tag{7.42}$$

$$=\; B \cdot \frac{1}{2\pi} \int_{-\frac{\pi}{M}}^{\frac{\pi}{M}} X(\hat{\omega}) d\hat{\omega} \tag{7.43}$$

$$=\; Bx(0). \tag{7.44}$$

Thus, for $y(0) = x(0)$, the magnitude of the decimation filter should be $B = 1$. This follows because the process of downsampling expands the spectral content *and* reduces the magnitude of the spectral content by a factor of M. Thus, no change in magnitude due to the decimation filter is needed to preserve the original spectral content $X(\omega)$ for $|\omega| \leq \pi/M$.

7.5 FREQUENCY-BASED INTERPOLATION USING THE FFT

The process of interpolation involves the steps of upsampling then filtering to remove the additional images that are introduced into the new Nyquist region from $-\pi$ to π. The same process can be accomplished by performing specific operations on the FFT of the original signal. This translates to the implementation of time interpolation of a signal by performing operations entirely in the frequency domain.

Consider an N point signal $x(n)$ whose N-point FFT is given as $X(k)$. Suppose we generate an $\hat{N}$ point FFT, $\hat{X}(k)$, composed of $\hat{N}$ samples of $X(k)$ where $\hat{N}$ is an integer multiple of N (i.e., $\hat{N} = M \cdot N$, M an integer). Due to the N-periodic nature of $X(k)$, we can generate the $\hat{N}$ point signal $\hat{X}(k)$ by repeating $X(k)$ M-times. Taking an $\hat{N}$-point IFFT of $\hat{X}(k)$ results in an $\hat{N}$ length time signal $\hat{x}(n)$ where

$$\hat{x}(n) = \begin{cases} x\left(\frac{n}{M}\right), & n = 0, \pm M, \pm 2M, \ldots, \\ 0, & \text{otherwise.} \end{cases} \tag{7.45}$$

In other words, taking an $\hat{N}$-point IFFT of an $\hat{N}$-point signal $\hat{X}(k)$ ($\hat{N} = M \cdot N$) containing M repeats of the sampled spectrum $X(k)$ results in an upsampled version of the original signal upsampled by a factor of M. Fig. 7.9 shows an $N = 11$ point magnitude response $|X(k)|$ and its corresponding time signal $x(n)$. Fig. 7.10 shows the magnitude response of an $\hat{N} = 3 \cdot N$ extension of this signal, $\hat{X}(k)$, and the corresponding time signal $\hat{x}(n)$. As before, we can conceptually interpret the additional images as resulting from the additional high frequency content introduced by the abrupt transitions between the signal and the inserted zeroes in time. Thus, we can achieve an interpolated signal by replacing the repeated spectral content by zeroes. Specifically, for an $\hat{N} = M \cdot N$ signal, we can achieve interpolation by inserting $(M - 1)N$ zeroes in the center of the original FFT. The center corresponds to the value of k at or around the normalized frequency π. Fig. 7.11 shows the frequency

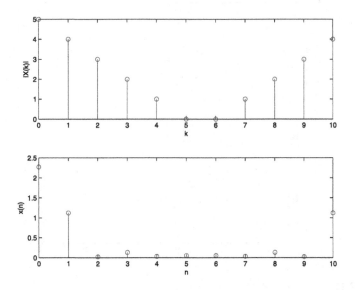

FIGURE 7.9

Sample $N = 11$ point magnitude response $X(k)$ and corresponding time signal $x(n)$

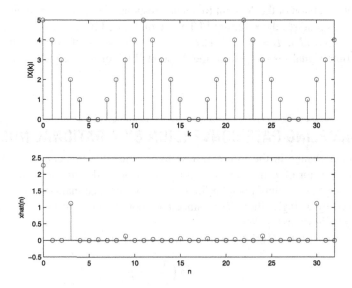

FIGURE 7.10

Extended $\hat{N} = 3 \cdot N$ point magnitude response $\hat{X}(k)$ and corresponding time signal $\hat{x}(n)$

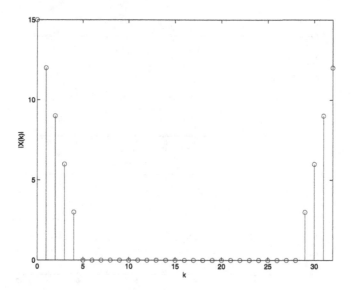

FIGURE 7.11

Frequency interpolation version of the extended signal $\hat{X}(k)$ where the spectral repeats are replaced with zeroes

interpolated version of the $\hat{N}$ point frequency response. Note that removing the additional zeroes requires the resulting FFT to be multiplied by a factor of M to ensure that $y(n) = x(n/L)$, $L = 0, \pm L, \pm 2L, \ldots$. Fig. 7.12 shows the original time signal and the time signal resulting from the $\hat{N}$-point IFFT of the frequency interpolated signal.

7.6 SAMPLING RATE CONVERSION BY A RATIONAL NUMBER

A fractional change in the sampling rate by a rational factor can be achieved by cascading a factor of L interpolator with a factor of M decimator, L and M being positive integers. The interpolation filter $H_u(z)$ and the decimation filter $H_d(z)$ can be replaced by a single filter, $H(z)$, since they both function at the same sampling rate. This is shown in Fig. 7.13.

The low pass filter $H(z)$ has a stop band cutoff frequency of

$$\omega_s = \min\left(\frac{\pi}{L}, \frac{\pi}{M}\right). \tag{7.46}$$

This filter can be designed as an FIR or IIR filter.

For the case where $F_y > F_x$ ($\frac{L}{M} > 1$), the overall effect of the sampling conversion is an upsampling by a rational factor [4]. The low pass filter acts like an

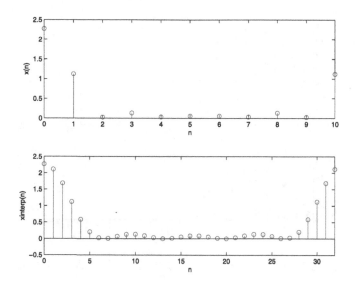

FIGURE 7.12

Comparison between the original signal $x(n)$ and the resulting $\hat{N}$-point IFFT of the frequency interpolated frequency response in Fig. 7.11

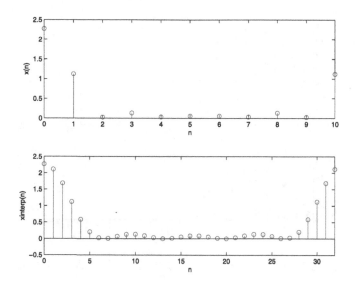

FIGURE 7.13

Equivalent filter representation

anti-imaging post filter that removes the spectral replicas at multiples of F_x, but not at multiples of LF_x. For the case where $F_y < F_x$ ($\frac{L}{M} < 1$), the overall effect is a downsampling at a rational factor. The low pass filter thus acts as an antialiasing prefilter that limits spectral content so that down shifted spectral replicas at multiples of F_y do not cause aliasing in the baseband.

EXAMPLE 7.1
(Sample Rate Change)

We consider changing the sampling rate of a discrete signal by a ratio of two positive numbers. We use the MATLAB function *resample* to perform a fractional change in the sampling rate by a factor of L/M where L is the upsampling factor

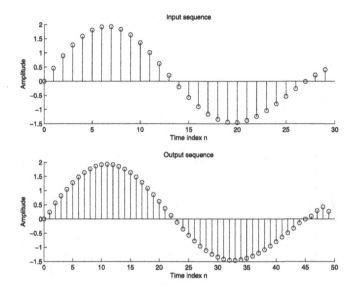

FIGURE 7.14

Illustration of sampling rate increase by a rational number 5/3

and M is the downsampling factor. The signal of interest is a sum of two sinusoids where the discrete frequency of the first sinusoid is $f_1 = 0.043$ and the discrete frequency of the second sinusoid is $f_2 = 0.031$. The code to implement this is illustrated below along with resulting plots given in Fig. 7.14.

Matlab Script 7.1.

```
% This routine using the MATLAB function resample.  The sampling rate
% will be changed by the fractional factor L/M where L is the
% upsampling factor and M is the downsampling factor.  The signal
% of interest is a sum of two sinusoid signals, one with discrete
% frequency f1 and the other with discrete frequency f2.

clear

SavePlots = 1;

L = 5; M = 3;    % set the up- and downsampling factors
N = 30;       % length of the signal
f1 = 0.043; % frequency of 1st sinusoid
f2 = 0.031; % frequency of 2nd sinusoid
n = 0:N-1;
xn = sin(2*pi*f1*n) + sin(2*pi*f2*n);

ym = resample(xn, L, M);
```

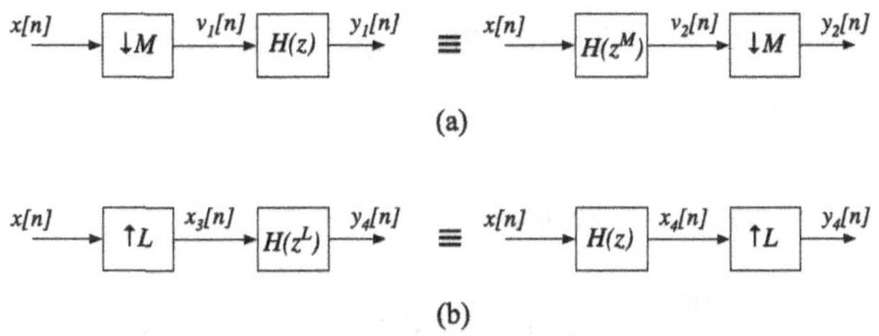

$x[n] \rightarrow \downarrow M \rightarrow v_1[n] \rightarrow H(z) \rightarrow y_1[n] \quad \equiv \quad x[n] \rightarrow H(z^M) \rightarrow v_2[n] \rightarrow \downarrow M \rightarrow y_2[n]$

(a)

$x[n] \rightarrow \uparrow L \rightarrow x_3[n] \rightarrow H(z^L) \rightarrow y_4[n] \quad \equiv \quad x[n] \rightarrow H(z) \rightarrow x_4[n] \rightarrow \uparrow L \rightarrow y_4[n]$

(b)

FIGURE 7.15

Noble identity equivalent representations of interpolator (a) and decimator (b)

```
m = 0:length(ym)-1;  % create a new index

figure(1)
subplot(2, 1, 1)
stem(n, xn)
xlabel('Time index n'); ylabel('Amplitude')
title('Input sequence')

subplot(2, 1, 2)
stem(m, ym)
xlabel('Time index n'); ylabel('Amplitude')
title('Output sequence')
```

End of the Script

End of the Example

7.7 NOBLE IDENTITIES

In general, the order of the filter and the rate changing device cannot be changed unless proper modifications to the filters are made. Fig. 7.15 shows the filter modifications that are needed to change the order of the filter and the sample rate changing device. As shown in later sections, the Noble identities are particularly useful when constructing polyphase implementation of interpolators and decimators.

7.8 INTERPOLATION USING A POLYPHASE FILTER

We can perform interpolation by upsampling and then filtering to remove the signal images introduced by upsampling. Fig. 7.16 gives a conceptual block diagram of this interpolation procedure. We can use a polyphase decomposition of the interpolation filter to more efficiently perform the interpolation as shown in Fig. 7.17. Note that

FIGURE 7.16

Conceptual block diagram for interpolation

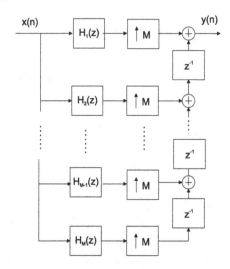

FIGURE 7.17

Conceptual block diagram for interpolation using a polyphase decomposition

the order of the filter and the upsampler were modified using the equivalences shown in the Noble identities (Fig. 7.15). We illustrate the efficiency of the polyphase implementation using the following example.

EXAMPLE 7.2
(Polyphase Interpolation)

We can use an example of interpolation by a factor of 3 to demonstrate the efficiency of using the polyphase interpolation procedure. Fig. 7.18 gives the block diagram for interpolating the input sequence by a factor of 3 using the polyphase decomposition.

We can use the following Matlab script to design a low pass filter to remove the unwanted high frequencies introduced by upsampling. This is just one example of many possible interpolation filter implementations that would be appropriate for this problem. Note that for all implementations, the stop band of the filter must be no greater that $\omega = \pi/3$. In Matlab, this corresponds to a normalized frequency (ω/π) of $1/3$.

Matlab Script 7.2.

```
order = 20;
wp = 0.3;
ws = 0.33;
delp = 1.0 - 10^(-1/20);
dels = 10^(-35/20);
F1 = [0.0 wp ws 1.0];
A1 = [1.0 (1.0-delp) dels 0.0];
ratio = delp/dels;
W1 = [1.0 ratio];
b = firls(order, F1, A1, W1);
```

End of the Script

The system transfer function for the resulting FIR filter is given by

$$
\begin{aligned}
H(z) = {} & 0.0110 + 0.0286z^{-1} + 0.0321z^{-2} + 0.0110z^{-3} \\
& - 0.0269z^{-4} - 0.0533z^{-5} - 0.0373z^{-6} + 0.0353z^{-7} \\
& + 0.2449z^{-11} + 0.1425z^{-8} + 0.2449z^{-9} + 0.2918z^{-10} \\
& + 0.1425z^{-12} + 0.0353z^{-13} - 0.0373z^{-14} - 0.0533z^{-15} \\
& - 0.0269z^{-16} + 0.0110z^{-17} + 0.0321z^{-18} + 0.0286z^{-19} \\
& + 0.0110z^{-20}.
\end{aligned} \tag{7.47}
$$

Fig. 7.19 gives the magnitude and phase plots for this filter. For a 3 level polyphase decomposition, the decomposed filters are given by

$$
\begin{aligned}
H_1(z) = {} & 0.0110 + 0.0110z^{-1} - 0.0373z^{-2} + 0.2449z^{-3} + 0.1425z^{-4} \\
& - 0.0533z^{-5} + 0.0321z^{-6}, \\
H_2(z) = {} & 0.0286 - 0.0269z^{-1} + 0.0353z^{-2} + 0.2918z^{-3} + 0.0353z^{-4} \\
& - 0.0269z^{-5} + 0.0286z^{-6}, \\
H_3(z) = {} & 0.0321 - 0.0533z^{-1} + 0.1425z^{-2} + 0.2449z^{-3} - 0.0373z^{-4} \\
& + 0.0110z^{-5} + 0.0110z^{-6}.
\end{aligned} \tag{7.48}
$$

We can use the Matlab *sampdata* function to obtain an input sequence. Fig. 7.20 gives the stem plot of the input sample sequence. The following Matlab script can be used to implement the interpolation procedure by upsampling followed by filtering by the original filter to remove the high frequencies due to upsampling.

Matlab Script 7.3.

```
x = sampdata;
n2 = 3*length(x);
v1 = zeros(1, n2);
v1(1, 1:3:end) = x;
y1 = conv(b, v1);
```

End of the Script

The Matlab routine `sampdata.m` produces a sample input signal (see Appendix). We can use the following Matlab script to implement the polyphase decomposition interpolation procedure.

Matlab Script 7.4.

```
b1 = b(1, 1:3:end);
b2 = b(1, 2:3:end);
b3 = b(1, 3:3:end);
r1 = conv(b1, x);
r2 = conv(b2, x);
r3 = conv(b3, x);
n3 = length(r1);
n4 = length(y1);
s1 = zeros(1,n4);
s2 = s1;
s3 = s1;
n5 = 3*length(r1);
s1(1,1:3:n5) = r1;   % upsample by 3
s2(1,2:3:n5) = r2;   % delay by 1, upsample by 3
s3(1,3:3:n5) = r3;   % delay by 2, upsample by 3
y2 = s1+ s2+ s3;
```

End of the Script

Fig. 7.21 gives the stem plots of the output using upsampling followed by filtering and the output using the polyphase decomposition with filtering followed by upsampling. Fig. 7.22 gives plots of the two output sequences on the same plot. The normalized peak error was zero, and the average mean squared error was zero. Thus, the outputs are the same.

We can determine the number of multiplications and additions required for each implementation to compare computational complexity of the two approaches. The number of multiplications required for a convolution operation is equal to the product of the length of the two sequences. Thus, if the filter length is L_1 and the length of the sample sequence is L_2, then the number of multiplications required is

$$n_1 = L_1 L_2. \tag{7.49}$$

The number of additions is

$$n_2 = (L_1 - 1)L_2. \tag{7.50}$$

Table 7.1 gives a comparison of the number of additions and multiplications required for the two approaches. Note that for the conventional method $L_1 = 21$ for a 20 order filter and $L_2 = 256 \cdot 3$, which is the length of the up sampled signal $x(n)$. For the polyphase implementation 3 convolutions are implemented,

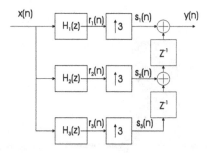

FIGURE 7.18

Block diagram of the use of the polyphase decomposition for interpolation by a factor of 3

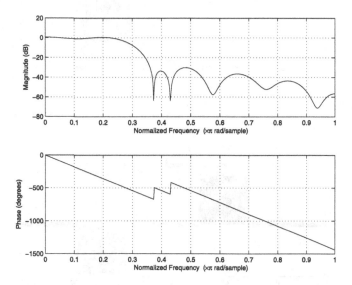

FIGURE 7.19

Magnitude and phase plots for the low pass filter

corresponding to the filters $H_1(z)$, $H_2(z)$, and $H_3(z)$. Each filter is of length $L_1 = 21/3 = 7$. These filters are each convolved with a sequence $x(n)$ that is of length $L_2 = 256$ because the filtering occurs before the upsampling. Adding up the number of additions and multiplications for the convolution at these three levels yields the values shown in Table 7.1. Note the savings of a factor of 3 for the multiplications and a factor of 3.3333 for additions for this example.

End of the Example

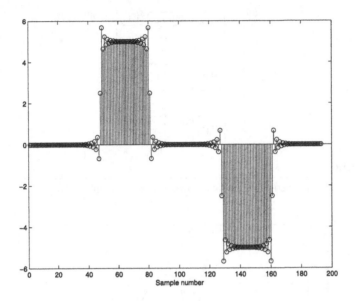

FIGURE 7.20

An input sample sequence

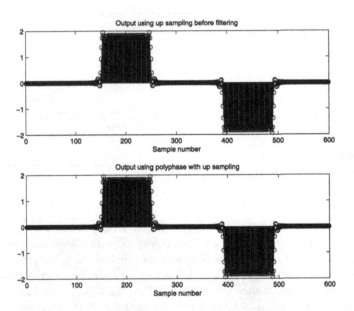

FIGURE 7.21

Stem plots of the two output sequences

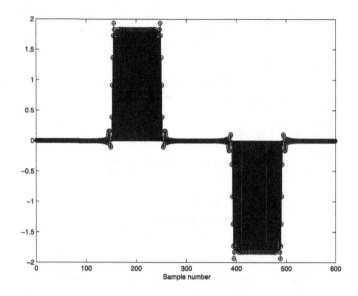

FIGURE 7.22

Plots of the two output sequences on the same plot

Table 7.1 Comparison of the number of additions and multiplications for the two interpolation methods

Method	Additions	Multiplications
Filtering and downsampling	11580	12159
Downsampling with polyphase	3474	4053

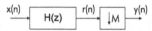

FIGURE 7.23

Conceptual Block Diagram for Decimation

7.9 DECIMATION USING A POLYPHASE FILTER

We can use a low pass filter to avoid aliasing and downsample by a factor M to implement decimation as shown in Fig. 7.23. We can use polyphase filtering to perform this operation more efficiently. The idea is to only compute the outputs that are saved. This can lead to considerable savings in computations. Fig. 7.24 gives a conceptual diagram of decimation by a factor of M using a polyphase decomposition. The efficiency of the polyphase implementation is emphasized with the following example.

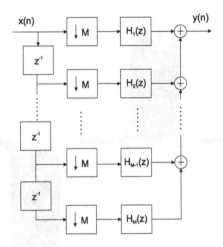

FIGURE 7.24

Conceptual block diagram for decimation using the polyphase implementation

EXAMPLE 7.3
(Polyphase Downsampling)

The following Matlab script can be used to design a decimator that downsamples a signal by a factor of 3. Note that the stop band edge should be no greater than $\pi/3$ corresponding to a Matlab normalized frequency (ω/π) of 1/3 to prevent aliasing.

Matlab Script 7.5.

```
order = 14;
wp = 0.3;
ws = 0.33;
delp = 1.0 - 10^(-1/20);
dels = 10^(-30/20);
F1 = [0.0 wp ws 1.0];
A1 = [1.0 (1.0-delp) dels 0.0];
ratio = delp/dels;
W1 = [1.0 ratio];
b = firls(order, F1, A1, W1);
```

End of the Script

The transfer function for the resulting antialiasing filter is given by

$$
\begin{aligned}
H(z) = \quad & -0.0003 - 0.0315z^{-1} - 0.0526z^{-2} - 0.0350z^{-3} \\
& + 0.0366z^{-4} + 0.1401z^{-5} + 0.2438z^{-6} + 0.2956z^{-7} \\
& + 0.2438z^{-8} + 0.1401z^{-9} + 0.0366z^{-10} - 0.0350z^{-11} \\
& - 0.0526z^{-12} - 0.0315z^{-13} - 0.0003z^{-14}.
\end{aligned} \qquad (7.51)
$$

Fig. 7.25 gives magnitude and phase plots for the antialiasing low pass filter.
   The following Matlab script is used to obtain the polyphase filters.

Matlab Script 7.6.

```
b1 = b(1,1:3:end);
b2 = b(1,2:3:end);
b3 = b(1,3:3:end);
```

End of the Script

   The corresponding polyphase filters are given by

$$
\begin{aligned}
H_1(z) &= -0.0003 - 0.0350z^{-1} + 0.2438z^{-2} + 0.1401z^{-3} \\
&\quad - 0.0526z^{-4}, \\
H_2(z) &= -0.0315 + 0.0366z^{-1} + 0.2956z^{-2} + 0.0366z^{-3} \\
&\quad - 0.0315z^{-4}, \\
H_3(z) &= -0.0526 + 0.1401z^{-1} + 0.2438z^{-2} - 0.0350z^{-3} \\
&\quad - 0.0003z^{-4}.
\end{aligned}
\tag{7.52}
$$

A sample input sequence can be obtained using the *sampdata* Matlab function
(see Appendix). Fig. 7.26 gives the stem plot of the input sequence. We can use
the following Matlab script to compute the output using filtering following by
downsampling.

Matlab Script 7.7.

```
x = sampdata;
v1 = conv(b, x);
y1 = v1(1,1:3:end);
```

End of the Script

   The following Matlab script can be used to compute the output using the
polyphase implementation with downsampling before filtering.

Matlab Script 7.8.

```
t1 = [x 0 0];
r1 = t1(1,1:3:end);
t2 = [0 x 0];
r2 = t2(1,1:3:end);
t3 = [0 0 x];
r3 = t3(1,1:3:end);
s1 = conv(b1, r1);
s2 = conv(b2, r2);
s3 = conv(b3, r3);
y2 = s1 + s2 + s3;
```

End of the Script

Fig. 7.27 gives the stem plots of the output using filtering followed by down-sampling as shown in Fig. 7.23 and the output using the polyphase decomposition with downsampling before filtering as shown in Fig. 7.27. Fig. 7.28 gives the two stem plots of the two outputs on the same plot. The normalized peak error was 3.514168×10^{-16}. The average mean squared error was 2.351461×10^{-31}. These errors are within the range expected for floating point computations using Matlab.

We can again determine the number of multiplications and additions required for each implementation to compare computational complexity of the two approaches. The number of multiplications required for a convolution operation is equal to the product of the length of the two sequences. Thus, if the filter length is L_1 and the length of the sample sequence is L_2, then the number of multiplications required is

$$n_1 = L_1 L_2. \tag{7.53}$$

The number of additions is

$$n_2 = (L_1 - 1)L_2. \tag{7.54}$$

Table 7.2 gives a comparison of the number of additions and multiplications required for the two approaches. The number of additions and multiplications for each approach were computed in a manner similar to that in the previous example. For the conventional implementation of the decimator, $L_1 = 15$ (length of the filter) and $L_2 = 256$ (length of the input signal). For the polyphase implementation, filter outputs are computed at three levels for a 3 level polyphase decomposition implementation. At each level, $L_1 = 5$ and $L_2 \approx 256/3$. Note the savings of a factor of 3 for the multiplications and a factor of 3.5 for additions for this example. **End of the Example**

7.10 MULTISTAGE DECIMATOR

Consider the decimation of a signal $x(n)$ by a factor of $M \gg 1$. This requires a filter capable of passing frequency content for frequencies $|\omega| \leq \pi/M$. Given that M is large, we would need a filter that has a very sharp transition band. From filter design, we know that a sharp transition band translates to filters with a large number of coefficients.

Assume that the decimation factor M can be factored into a product of J positive integers

$$M = \prod_{i=1}^{J} M_i. \tag{7.55}$$

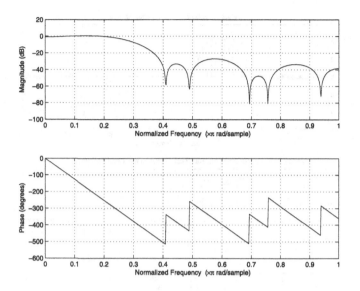

FIGURE 7.25

Magnitude and phase plots for the antialiasing filter

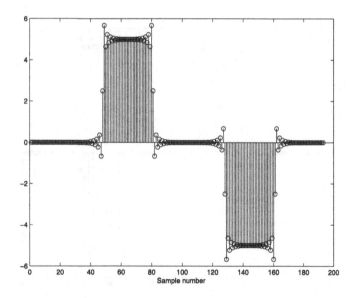

FIGURE 7.26

Stem plot of the sample input sequence

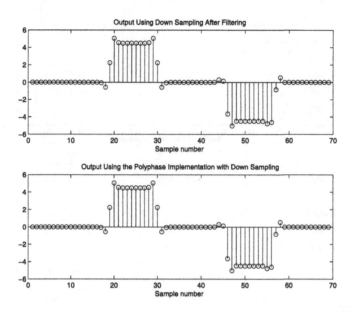

FIGURE 7.27

Stem plots of the two output sequences

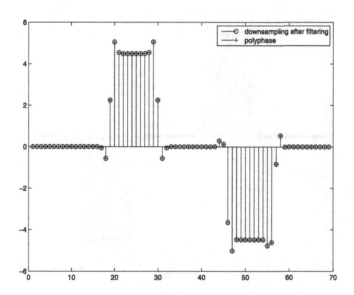

FIGURE 7.28

Stem plots of the two output sequences on the same plot

Table 7.2 Comparison of the number of additions and multiplications for the two decimation methods

Method	Additions	Multiplications
Filtering and downsampling	2702	2895
Downsampling with polyphase	773	965

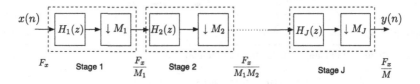

FIGURE 7.29

Multistage decimator

A decimation factor M can be implemented by cascading J decimators, with the ith stage having a decimation factor of M_i. This is illustrated in Fig. 7.29. In this multistage decimator, the sampling rate at the output of the ith stage is given as

$$F_i = \frac{F_{i-1}}{M_i}, \qquad i = 1, 2, \ldots, J. \tag{7.56}$$

Correct implementation of the overall cascaded decimator requires the design of filters at each of the J stages such that aliasing in the overall decimator is eliminated. Consider the decimation of a signal by a factor of M. We have learned previously that this means that the signal must be ideally filtered only keeping spectral content within the frequency range (radians) $|\omega| \leq \pi/M$. This translates to a frequency range (Hz) of $|F| \leq F_x/2M$. Considering a non-ideal filter with a pass band, a transition band, and a stop band, the overall filter should have a response as follows:

$$\text{(Pass Band)} \qquad 0 \leq F \leq F_{pass}, \tag{7.57}$$

$$\text{(Transition Band)} \qquad F_{pass} \leq F \leq F_{stop}, \tag{7.58}$$

$$\text{(Stop Band)} \qquad F_{stop} \leq F \leq \frac{F_x}{2}, \tag{7.59}$$

where $F_{stop} \leq F_x/2M$. An approximation of this response is illustrated in Fig. 7.30.

The approach for designing the filters for the multistage decimator is done by closely examining the first stage. The first stage of the multistage decimator is illustrated in Fig. 7.31. The frequency response relationships between the signals are as

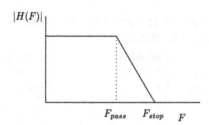

FIGURE 7.30

Approximate overall filter response of a decimation filter

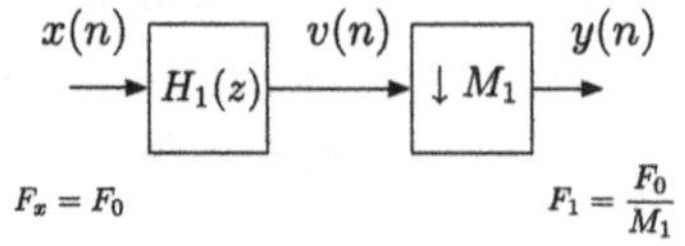

$$x(n) \xrightarrow{\quad} \boxed{H_1(z)} \xrightarrow{v(n)} \boxed{\downarrow M_1} \xrightarrow{y(n)}$$

$$F_x = F_0 \qquad\qquad F_1 = \frac{F_0}{M_1}$$

FIGURE 7.31

Stage 1 of multistage decimator

follows:

$$Y(\omega) \;=\; \frac{1}{M_1} \sum_{k=0}^{M_1-1} V\left(\frac{\omega - 2\pi k}{M_1}\right) \tag{7.60}$$

$$=\; \frac{1}{M_1} \sum_{k=0}^{M_1-1} H_1\left(\frac{\omega - 2\pi k}{M_1}\right) X\left(\frac{\omega - 2\pi k}{M_1}\right), \tag{7.61}$$

This can be rewritten in terms of the analog frequency F as

$$Y(F) = \frac{1}{M_1} \sum_{k=0}^{M_1-1} H_1\left(\frac{F - kF_0}{M_1}\right) X\left(\frac{F - kF_0}{M_1}\right). \tag{7.62}$$

The spectral shape of the signal $Y(F)$ is completely dependent on the shape of $H_1(z)$. Due to the downsampling, we also know that the spectral characteristics of $Y(F)$ has the shape of $H_1(x)$ and is shifted and expanded every kF_0/M_1.

The filter $H_1(F)$ can be designed by taking the following things into consideration. Recall that we want to ensure that all content within the pass band of $H(F)$ is undisturbed. The transition band, however, can be a little more relaxed. We can explore just how relaxed by looking at the representation of $H_1(F)$ and the subsequent repeats that occur every kF_0/M_1 shown in Fig. 7.32. We can identify what the end of the transition band of $H_1(F)$ should be by examining the first shifted image at F_0/M_1. We can allow some aliasing or overlap between the image centered at zero and the signal centered at F_0/M_1 as long as we do not permit overlap for any frequency less

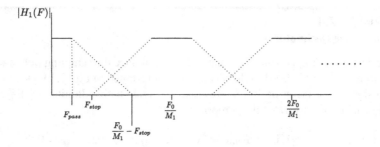

FIGURE 7.32

Magnitude response of $H_1(z)$

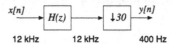

FIGURE 7.33

Block-diagram representation of a single state factor-of-30 decimator

than the stop band edge of $H(F)$, or $F < F_{stop}$. This is because we expect any aliasing that does take place over this region will be eliminated in subsequent stages of the decimator. Thus the distance between the center of the image at F_0/M_1 and the transition band/stop band edge of the overall filter is $F_0/M_1 - F_{stop}$. This means that for the image centered at zero, the transition band ends at $F_0/M_1 - F_{stop}$. All other content greater than $F_0/M_1 - F_{stop}$ should be eliminated in order to get rid of the additional images. Using this information, we can derive the filter characteristics of the $H_1(z)$. They are as follows:

$$\text{(Pass Band)} \qquad 0 \leq F \leq F_{pass}, \tag{7.63}$$

$$\text{(Transition Band)} \qquad F_{pass} \leq F \leq F_0/M_1 - F_{stop}, \tag{7.64}$$

$$\text{(Stop Band)} \qquad F_0/M_1 - F_{stop} \leq F \leq \frac{F_0}{2}. \tag{7.65}$$

We can generalize this for the ith stage as

$$\text{(Pass Band)} \qquad 0 \leq F \leq F_{pass}, \tag{7.66}$$

$$\text{(Transition Band)} \qquad F_{pass} \leq F \leq F_i - F_{stop}, \tag{7.67}$$

$$\text{(Stop Band)} \qquad F_i - F_{stop} \leq F \leq \frac{F_{i-1}}{2}. \tag{7.68}$$

We use this formulation to design the filter specifications for a multistage decimator in the following example.

EXAMPLE 7.4
(Multistage Decimator)

The desire is to implement a decimator that will reduce the sampling rate of a signal from 12 kHz to 400 Hz. This decimator is illustrated in Fig. 7.33. Decimation of a signal by a factor of 30 requires that the signal be bandlimited to $\pm\frac{\pi}{30}$. The following design for $H(z)$ is appropriate

$$H(z): \quad F_p = 180\,\text{Hz}, \quad F_{stop} = 200\,\text{Hz}, \quad \delta_p = 0.002, \quad \delta_s = 0.001.$$
$$(7.69)$$

The MATLAB functions *firpmord* and *firpm* can be used to create an optimal equiripple FIR filter with these specifications. The following MATLAB code is used

Matlab Script 7.9.

```
M = 30;
Fpass = 180;
Fstop = 200;
delta_p = 0.002;
delta_s = 0.001;
Fs = 12000;

[N_H,Fo,Ao,w] = firpmord([Fpass Fstop], [1 0], [delta_p delta_s],
                Fs);
h = firpm(N_H,Fo,Ao,w);
```

End of the Script

The function *firpmord* returns a filter order of $N_H = 1827$. We can calculate the number of multiplications per second of a decimator with an input sample rate of F_{i-1}, a decimation factor of M_i, and an output sampling rate of F_i as $(N+1)F_{i-1}/M_i$. The variable N corresponds to the number of filter coefficients. Thus, we have the following characteristics for this signal stage decimator:

```
--------------------Single Stage Implementation----------------
Number of filter coefficients: 1828
Number of samples generated per second: 12000
Number of samples processed at a decimation of 30: 400
Number of multiplications per second: 731200
--------------------------------------------------------------
```

The downsampling factor M can be factored into $M = M_1 M_2$ where $M_1 = 15$ and $M_2 = 2$. A block diagram representation of the 2-stage decimator is shown in Fig. 7.34.

The filters $H_1(z)$ and $H_2(z)$ must be designed such that they eliminate aliasing in the overall decimation. From the previous section, the filter specifications for $H_i(z), i = 1, 2$, are

$$\text{(Pass Band)} \qquad 0 \le F \le F_{pass}^i, \qquad (7.70)$$

$$\text{(Transition Band)} \qquad F_{pass} \le F \le F_{stop}^i, \qquad (7.71)$$

$$\text{(Stop Band)} \qquad F_{stop}^i \le F \le \frac{F_{i-1}}{2} \qquad (7.72)$$

where

$$F_{pass}^i = F_{pass}, \qquad F_{stop}^i = \frac{F_{i-1}}{M_i} - F_{stop}. \qquad (7.73)$$

This yields the following filter specifications for $H_1(z)$ and $H_2(z)$:

$$H_1(z): \quad F_{pass}^1 = F_{pass} = 180\,\text{Hz}, \quad F_{stop}^1 = \frac{F_0}{M1} - F_{stop}$$

$$= \frac{12000}{15} - 200 = 600\,\text{Hz},$$

$$H_2(z): \quad F_{pass}^2 = F_{pass} = 180\,\text{Hz}, \quad F_{stop}^2 = \frac{F_1}{M2} - F_{stop}$$

$$= \frac{800}{2} - 200 = 200\,\text{Hz}. \qquad (7.74)$$

Even though the width of the transition region for $H_2(z)$ is the same as for $H(z)$, we see that the normalized width of the region $\Delta f = (F_{stop} - F_{pass})/Fs$ is very different. It is clear that

$$\Delta f^H = \frac{F_{stop} - F_{pass}}{F_s} = \frac{200 - 180}{12000} = 0.0025, \qquad (7.75)$$

$$\Delta f^2 = \frac{F_{stop}^2 - F_{pass}^2}{F_1} = \frac{200 - 180}{800} = 0.025. \qquad (7.76)$$

Thus $\Delta f^H < \Delta f^2$, effectively making the transition region for $H(z)$ smaller, leading to a filter with more coefficients. Since the overall filter has a pass band ripple of δ_p, we can split this ripple across the two stages. We then have

$$\delta_p^1 = \delta_p^2 = \frac{\delta_p}{2} = 0.001. \qquad (7.77)$$

The stop band ripple of the overall filter is, at worst, the highest stop band ripple of any filter within the cascade. Thus we can set

$$\delta_s^1 = \delta_s^2 = \delta_s. \qquad (7.78)$$

The following code is used to obtain the minimum filter order of $H_1(z)$ and $H_2(z)$ given the specifications designed above.

Matlab Script 7.10.

```
M1 = 15; M2 = 2;
F0 = Fs;
F1 = F0/M1;
F2 = F1/M2;

% First Filter
Fpass1 = Fpass;
Fstop1 = F1 - Fstop;
delta_p1 = delta_p/2;
delta_s1 = delta_s;

[N_H1,Fo1,Ao1,w1] = firpmord([Fpass1 Fstop1], [1 0],
                     [delta_p1 delta_s1], F0);
h1 = firpm(N_H1,Fo1,Ao1,w1);

% Second Filter
Fpass2 = Fpass;
Fstop2 = F2 - Fstop;
delta_p2 = delta_p/2;
delta_s2 = delta_s;

[N_H2,Fo2,Ao2,w2] = firpmord([Fpass2 Fstop2], [1 0],
                     [delta_p2 delta_s2], F1);
h2 = firpm(N_H2,Fo2,Ao2,w2);
```

End of the Script

This results in the following characteristics for the two-stage decimator:

```
---------------------2 Stage Implementation---------------
Number of filter coefficients for first stage: 94
Number of filter coefficients for second stage: 131
Number of samples per second processed at first stage decimation
of 15: 800
Number of multiplications per second at first stage: 75200
Number of samples per second processed at second stage decimation
of 2: 400
Number of multiplications per second at second stage: 52400
Total number of multiplications per second: 127600
-----------------------------------------------------------------
```

This results in a savings of

$$ratio = \frac{731200}{127600} = 5.73. \tag{7.79}$$

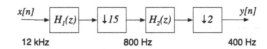

FIGURE 7.34

Block diagram of a two-stage factor-of-30 decimator

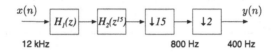

FIGURE 7.35

Equivalent representation of a multistage decimator

The overall frequency response of the multistage implementation can be compared to the single state implementation in the following way. The position of the first sample rate modification device and the second filter are first rearranged using the Noble identities in Fig. 7.15(a). The resulting structure is shown in Fig. 7.35.

The frequency response of both implementations can be calculated and illustrated using the following MATLAB code.

Matlab Script 7.11.

```
% Frequency response of single stage implementation
figure(1)
freqz(h, 1)

% Frequency response of multistage implementation
figure(2)
h2_up = upsamp(h2, M1);
h_casc = conv(h1, h2_up);
freqz(h_casc, 1)
```

End of the Script

The resulting responses are shown in Figs. 7.36 and 7.37. By visual inspection, we see that the filter specifications have been met.
End of the Example

Fig. 7.38 shows a general multistage implementation of a decimator and interpolator.

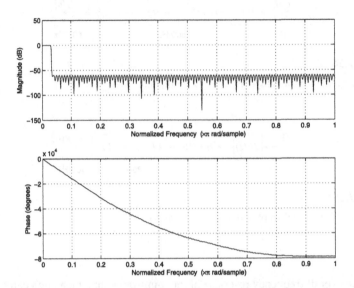

FIGURE 7.36

Frequency response of a single stage implementation

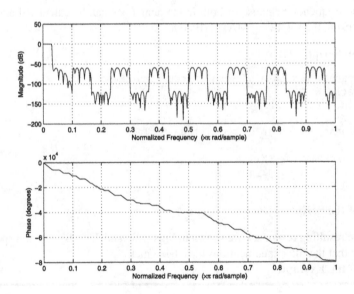

FIGURE 7.37

Frequency response of a multistage implementation

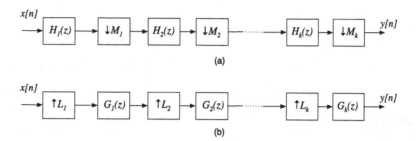

FIGURE 7.38

General multistage structure for decimation and interpolation

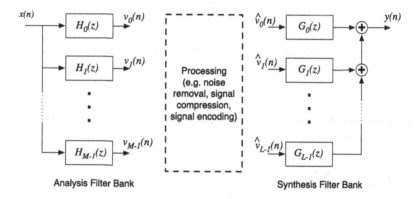

FIGURE 7.39

Analysis and synthesis filter bank [8]

7.11 MULTIRATE FILTER BANKS

7.11.1 UNIFORM FILTER BANKS

A digital filter bank is a set of digital band pass filters that is used to analyze a given input signal by separating it into multiple signals with non-overlapping frequency content. They can also be used to synthesize or construct multiple input signals of non-overlapping frequency content into a single output. Filter banks are often used for performing spectrum analysis and signal synthesis [4]. Fig. 7.39 shows an example of an analysis filter bank and a synthesis filter bank.

It is common for these filter banks to be generated using filters with non-overlapping magnitude responses of equal pass band widths. This class of banks is called a *Uniform DFT filter bank* [8]. Consider a causal low pass filter with the

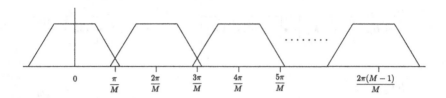

FIGURE 7.40

Magnitude response of a set of M filter banks

impulse response

$$H_0(z) = \sum_{n=0}^{\infty} h_0(n)z^{-n} \tag{7.80}$$

with pass band and stop band around π/M for some integer M. Now consider another filter

$$h_k(n) = h_0(n)e^{\frac{j2\pi kn}{M}}. \tag{7.81}$$

This phase shift in the time domain modulates the original filter $h_0(n)$. The resulting Z Transform gives

$$
\begin{aligned}
H_k(z) &= \sum_{n=0}^{\infty} h_k(n)z^{-n} = \sum_{n=0}^{\infty} h_0(n)e^{\frac{j2\pi kn}{M}} z^{-n} \\
&= \sum_{n=0}^{\infty} h_0(n)\left(ze^{-\frac{j2\pi k}{M}}\right)^{-n} = H_0\left(ze^{-\frac{j2\pi k}{M}}\right), \quad 0 \le k \le M-1.
\end{aligned}
\tag{7.82}
$$

The frequency response is then

$$H_k(\omega) = H_0\left(\omega - \frac{2\pi k}{M}\right). \tag{7.83}$$

Fig. 7.40 illustrates a conceptual view of the frequency response of the M filter banks.

7.11.2 POLYPHASE IMPLEMENTATION OF UNIFORM FILTER BANKS

The M-band polyphase decomposition of the original filter $h_0(n)$ can be written as

$$H_0(z) = \sum_{l=0}^{M-1} z^{-l} E_l(z^M) \tag{7.84}$$

where $E_l(z)$ is the lth polyphase component of $H_0(z)$ (adopted from [8]),

$$E_l(z) = \sum_{n=0}^{\infty} e_l(n)z^{-n} = \sum_{n=0}^{\infty} h[l+nM]z^{-n}, \quad 0 \le l \le M-1. \tag{7.85}$$

We can use (7.84) and (7.82) to formulate the polyphase decomposition of the kth analysis filter as

$$\begin{aligned}
H_k(z) &= H_0\left(ze^{-\frac{j2\pi kn}{M}}\right) = \sum_{l=0}^{M-1} \left(ze^{-\frac{j2\pi k}{M}}\right)^{-l} E_l\left(\left(ze^{-\frac{j2\pi k}{M}}\right)^M\right) \\
&= \sum_{l=0}^{M-1} z^{-l} E_l\left(z^M\right) e^{\frac{j2\pi kl}{M}}, \quad 0 \le k \le M-1.
\end{aligned} \tag{7.86}$$

with the last equality resulting because $e^{-\frac{j2\pi kM}{M}} = 1$. It is apparent that (7.86) resembles the equation for the inverse Fourier transform, thus it can be rewritten as

$$H_k(z) = M\mathcal{F}^{-1}\left\{z^{-l}E_l(z^M)\right\}. \tag{7.87}$$

Equation (7.87) shows that each filter $H_k(z)$ can be written as a linear combination of the filters $E_l(z^M)$, $l = 0, \ldots, M-1$, where the coefficients of the linear combination are determined by the inverse Fourier transform. The following example will illustrate this.

EXAMPLE 7.5
(Uniform Filter Banks)

Say we have an $L = 4$ band polyphase decomposition of a filter $H_0(z)$ where

$$H_0(z) = E_0(z^4) + z^{-1}E_1(z^4) + z^{-2}E_2(z^4) + z^{-3}E_3(z^4). \tag{7.88}$$

From (7.86) we know

$$H_k(z) = \sum_{l=0}^{3} z^{-l} E_l\left(z^4\right) e^{\frac{j2\pi kl}{4}}. \tag{7.89}$$

This results in the following equations:

$$H_0(z) = E_0(z^4) + z^{-1}E_1(z^4) + z^{-2}E_2(z^4) + z^{-3}E_3(z^4), \quad k = 0, \tag{7.90}$$

$$H_1(z) = E_0(z^4) + jz^{-1}E_1(z^4) + (-1)z^{-2}E_2(z^4) + (-j)z^{-3}E_3(z^4), \quad k = 1, \tag{7.91}$$

$$H_2(z) = E_0(z^4) + (-1)z^{-1}E_1(z^4) + z^{-2}E_2(z^4) + (-1)z^{-3}E_3(z^4), \quad k = 2, \tag{7.92}$$

$$H_3(z) = E_0(z^4) + (-j)z^{-1}E_1(z^4) + (-1)z^{-2}E_2(z^4) + jz^{-3}E_3(z^4), \quad k = 3. \tag{7.93}$$

These equations can be written in matrix form

$$\begin{bmatrix} H_0(z) \\ H_1(z) \\ H_2(z) \\ H_3(z) \end{bmatrix} = \begin{bmatrix} 1 & 1 & 1 & 1 \\ 1 & j & -1 & -j \\ 1 & -1 & 1 & -1 \\ 1 & -j & -1 & j \end{bmatrix} \begin{bmatrix} E_0(z^4) \\ z^{-1}E_1(z^4) \\ z^{-2}E_2(z^4) \\ z^{-3}E_3(z^4) \end{bmatrix}. \tag{7.94}$$

Define the Fourier transform matrix for $M = 4$ as

$$\mathbf{D} = \begin{bmatrix} 1 & 1 & 1 & 1 \\ 1 & -j & -1 & j \\ 1 & -1 & 1 & -1 \\ 1 & j & -1 & -j \end{bmatrix}. \tag{7.95}$$

The inverse Fourier transform operator is defined as

$$\mathbf{D}_{inv} = \mathbf{D}^{-1} = \frac{1}{4}\mathbf{D}^H$$

where the operator H is the Hermitian transpose. It is clear that Eq. (7.94) can be written as

$$\begin{bmatrix} H_0(z) \\ H_1(z) \\ H_2(z) \\ H_3(z) \end{bmatrix} = 4 \cdot \mathbf{D}_{inv} \begin{bmatrix} E_0(z^4) \\ z^{-1}E_1(z^4) \\ z^{-2}E_2(z^4) \\ z^{-3}E_3(z^4) \end{bmatrix}. \tag{7.96}$$

End of the Example

A block diagram interpretation of the polyphase implementation of the analysis filter is shown in Fig. 7.41. In this case, for each value of n, the sequence $[v_0(n), v_1(n), \dots, v_{M-1}(n)]$ is generated by taking the inverse DFT (e.g., Matlab function *ifft*) of the sequence $[u_0(n), u_1(n), \dots, u_{M-1}(n)]$.

7.11.3 TWO CHANNEL QUADRATURE MIRROR FILTER BANKS

It may be necessary in many applications to decompose an input signal $x(n)$ into a number of subband signals $\{v_k(n)\}$ by means of the analysis filters shown in the analysis section of Fig. 7.39. These subband signals are then processed and recon-

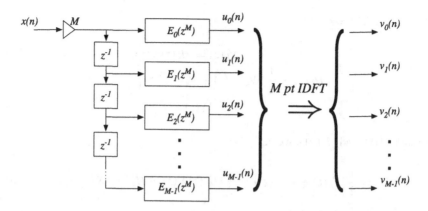

FIGURE 7.41

Polyphase implementation of the analysis filters of a uniform filter bank (adopted from [8])

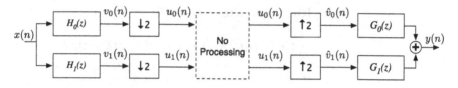

FIGURE 7.42

A two channel quadrature mirror filter (QMF) bank (adopted from [8])

structed by a synthesis filter bank, as shown in the synthesis section of Fig. 7.39, to produce the output signal $y(n)$. We learned in the previous chapter that signals bandlimited to only portions of the frequency space can be downsampled in order to make processing more efficient. Once processed, these signals are upsampled before being reconstructed by the synthesis filter banks. The combined structure used for decomposition and reconstruction of the signals are called *Quadrature Mirror filter Banks* (QMF). Fig. 7.42 illustrates a two channel quadrature mirror filter bank. Each set of filters, $H_i(z)$ and $G_i(z)$, $i = 0, 1$, must be constructed such that the constructed output $y(n)$ is as close to the input $x(n)$ as possible. Naturally, errors are bound to occur due to the sample rate alterations and imperfect filters. Approaches for generating appropriate filters for implementing these filter banks will be discussed below. The frequency relationship between the input and output of the sample rate alteration devices will be used in the analysis and are restated here

$$X_u(z) = X(z^L), \quad L \text{ upsampler;} \tag{7.97}$$

$$X_d(z) = \frac{1}{M} \sum_{k=0}^{M-1} X\left(z^{1/M} e^{\frac{j2\pi k}{M}}\right), \quad M \text{ downsampler.} \tag{7.98}$$

Using Fig. 7.42, we get [8]

$$V_k(z) = H_k(z)X(z), \qquad (7.99)$$

$$U_k(z) = \frac{1}{2}\left\{V_k(z^{1/2}) + V_k(-z^{1/2})\right\}, \qquad (7.100)$$

$$\hat{V}_k(z) = U_k(z^2). \qquad (7.101)$$

From (7.100) and (7.101), we get

$$\hat{V}_k(z) = \frac{1}{2}\{V_k(z) + V_k(-z)\} = \frac{1}{2}\{H_k(z)X(z) + H_k(-z)X(-z)\}. \qquad (7.102)$$

The output can be written as

$$
\begin{aligned}
Y(z) &= G_0(z)\hat{V}_0(z) + G_1(z)\hat{V}_1(z) \\
&= \frac{1}{2}\{G_0(z)H_0(z)X(z) + G_0(z)H_0(-z)X(-z)\} \\
&\quad + \frac{1}{2}\{G_1(z)H_1(z)X(z) + G_1(z)H_1(-z)X(-z)\} \\
&= \frac{1}{2}\{G_0(z)H_0(z) + G_1(z)H_1(z)\}\,X(z) \\
&\quad + \frac{1}{2}\{G_0(z)H_0(-z) + G_1(z)H_1(-z)\}\,X(-z) \qquad (7.103) \\
&= T(z)X(z) + A(z)X(-z) \qquad (7.104)
\end{aligned}
$$

where

$$T(z) = \frac{1}{2}\{G_0(z)H_0(z) + G_1(z)H_1(z)\} \qquad (7.105)$$

is called the *distortion transfer function* and

$$A(z) = \frac{1}{2}\{G_0(z)H_0(-z) + G_1(z)H_1(-z)\} \qquad (7.106)$$

is the *aliasing term* [8].

As stated in the previous section, upsamplers and downsamplers are linear time-varying (LTV) systems. This means that the QMF in Fig. 7.42 is also an LTV system [8]. This, in general, means that even with no processing, the output of the two channel QMF, $y(n)$, may not closely resemble the input $x(n)$. There does exist, however, a class of filters $H_i(z)$ and $G_i(z)$, $i = 0, 1$, that eliminates the aliasing term ($A(z) = 0$) and minimizes the distortion term $T(z)$. This class of filters generates what are known as perfect reconstruction QMF filter banks. The following section first looks at the conditions needed to eliminate the aliasing term $A(z)$.

7.11.4 ELIMINATING ALIASING

The aliasing term $A(z)$ should be 0 (i.e., $A(z) = 0$) to completely eliminate aliasing. In other words, it is required that

$$H_0(-z)G_0(z) = -H_1(-z)G_1(z). \tag{7.107}$$

The equality is derived directly from (7.106). The frequency response relationship can be written as

$$H_0(e^{j(\omega-\pi)})G_0(e^{j\omega}) = -H_1(e^{j(\omega-\pi)})G_1(e^{j\omega}). \tag{7.108}$$

The relationship in (7.108) can be satisfied by setting

$$
\begin{aligned}
G_0(e^{j\omega}) &= H_1(e^{j(\omega-\pi)}), \\
G_1(e^{j\omega}) &= -H_0(e^{j(\omega-\pi)}).
\end{aligned}
\tag{7.109}
$$

The Z Transform relationship can be written as

$$
\begin{aligned}
G_0(z) &= H_1(-z), \\
G_1(z) &= -H_0(-z).
\end{aligned}
\tag{7.110}
$$

Designing analysis and synthesis filters that satisfy (7.110) will completely eliminate the aliasing term $A(z)$.

7.11.5 CONDITIONS FOR PERFECT RECONSTRUCTION

With $A(z) = 0$, the representation of $Y(z)$ becomes

$$
\begin{aligned}
Y(z) &= \frac{1}{2}\{G_0(z)H_0(z) + G_1(z)H_1(z)\}\, X(z) \\
&= T(z)X(z).
\end{aligned}
\tag{7.111}
$$

With the aliasing term eliminated, identifying the additional conditions on the analysis and synthesis filters, $H_i(z)$ and $G_i(z)$, $i = 0, 1$, that result in perfect reconstruction becomes the next step. Start first with the simplifying assumption that the analysis filters $H_0(z)$ and $H_1(z)$ are mirrored versions of each other. More specifically, let us assume that $H_0(e^{j*\omega})$ is a low pass filter $H(e^{j\omega})$ and that $H_1(e^{j\omega})$ is the mirrored high pass filter. Thus

$$
\begin{aligned}
H_0(e^{j\omega}) &= H(e^{j\omega}), \\
H_1(e^{j\omega}) &= H(e^{j(\omega-\pi)})
\end{aligned}
\tag{7.112}
$$

with the Z Transform equivalent represented as

$$
\begin{aligned}
H_0(z) &= H(z), \\
H_1(z) &= H(-z).
\end{aligned}
\tag{7.113}
$$

Utilizing the relationship in (7.113) and the relationship necessary for eliminating aliasing, the distortion function, $T(z)$, can be rewritten as

$$T(z) = \frac{1}{2}\left\{H^2(z) - H^2(-z)\right\},\qquad(7.114)$$

or equivalently,

$$T(e^{j\omega}) = \frac{1}{2}\left\{H^2(e^{j\omega}) - H^2(e^{j(\omega-\pi)})\right\}.\qquad(7.115)$$

For the overall response of the QMF, we want the output $y(n)$ to be equal to $x(n)$ if no processing is performed, except for some arbitrary delay and constant magnitude. When this is satisfied, the filter bank is called a perfect reconstruction QMF [8]. This condition is satisfied when $T(e^{j\omega})$ has constant magnitude and linear phase or

$$T(e^{j\omega}) = Ce^{-jk\omega}.\qquad(7.116)$$

We will now look at cases when we can design an FIR filters $H_i(z)$ and $G_i(z)$ ($i = 0, 1$), which are both linear phase or nonlinear phase that satisfy this perfect reconstruction condition.

7.11.6 LINEAR PHASE QMF BANKS

This section explores the conditions under which we can design perfect reconstruction versions of $H_i(z)$ and $G_i(z)$ ($i = 0, 1$) where these analysis and synthesis filters are individually linear phase. First, continue with the assumption that $H_0(z)$ and $H_1(z)$ are mirrored filters based on a prototype filter $H(z)$. Assume this prototype filter, $H(z)$, to be an N-tap linear phase filter with frequency response

$$H(e^{j\omega}) = e^{\frac{j\omega(N-1)}{2}}|H(e^{j\omega})|\qquad(7.117)$$

and

$$H^2(e^{j\omega}) = |H(e^{j\omega})|^2 e^{j\omega(N-1)}.\qquad(7.118)$$

The overall frequency response of $T(z)$, and thus overall response of the QMF, is then written as

$$
\begin{aligned}
T(e^{j\omega}) &= \frac{1}{2}\left\{|H(e^{j\omega})|^2 e^{j\omega(N-1)} - |H(e^{j(\omega-\pi)})|^2 e^{j(\omega-\pi)(N-1)}\right\}\\
&= \frac{1}{2}\left\{|H(e^{j\omega})|^2 e^{j\omega(N-1)} - (-1)^{N-1}|H(e^{j(\omega-\pi)})|^2 e^{j\omega(N-1)}\right\}\\
&= \frac{e^{j\omega(N-1)}}{2}\left\{|H(e^{j\omega})|^2 - (-1)^{N-1}|H(e^{j(\omega-\pi)})|^2\right\}.
\end{aligned}
$$

Thus

$$|T(e^{j\omega})| = \frac{1}{2}\left\{|H(e^{j\omega})|^2 - (-1)^{N-1}|H(e^{j(\omega-\pi)})|^2\right\},\qquad(7.119)$$

$$\angle T(e^{j\omega}) \;=\; (N-1)\omega. \tag{7.120}$$

The arbitrary phase delay criteria for perfect reconstruction are met. We now examine the constant magnitude criteria. We see that for N odd at $\omega = \pi/2$, we have $|T(e^{j\pi/2})| = 0$. This is undesirable for a QMF, so we restrict N to be even. With N even, we have

$$|T(e^{j\omega})| = \frac{1}{2}\left\{|H(e^{j\omega})|^2 + |H(e^{j(\omega-\pi)})|^2\right\} = C. \tag{7.121}$$

Without loss of generality, assume $2C = 1$. This yields

$$|T(e^{j\omega})| = |H(e^{j\omega})|^2 + |H(e^{j(\omega-\pi)})|^2 = 1. \tag{7.122}$$

This is only possible in the most trivial cases when $H(e^{j\omega})$ is restricted to a linear phase filter. Thus, given that $H_0(z)$ and $H_1(z)$ are mirrored filters based on some linear phase prototype $H(z)$, any nontrivial $H(e^{j\omega})$ will introduce some amplitude distortion. The amount of distortion can be minimized, however, by optimizing for the smallest amount of amplitude distortion with respect to the filter coefficients.

Perfect reconstruction linear phase filters in the QMF can be generated if we relax the mirrored condition on $H_0(e^{j\omega})$ and $H_1(e^{j\omega})$. In other words, $H_1(e^{j\omega})$ is no longer the mirror high pass filter of $H_0(e^{j\omega})$. Still using the relationship needed to eliminate aliasing in (7.110), we rewrite $T(z)$ as

$$T(z) \;=\; \frac{1}{2}\{H_0(z)G_0(z) + H_1(z)G_1(z)\} \tag{7.123}$$

$$\;=\; \frac{1}{2}\{H_0(z)H_1(-z) - H_0(-z)H_1(z)\}. \tag{7.124}$$

For perfect reconstruction, we require

$$T(z) = \frac{1}{2}\{H_0(z)H_1(-z) - H_0(-z)H_1(z)\} = Cz^{-k}. \tag{7.125}$$

The delay k must be odd. This is easily shown by defining the following relationships:

$$P_0(z) \;=\; H_0(z)H_1(-z), \tag{7.126}$$
$$P_1(z) \;=\; H_0(-z)H_1(z). \tag{7.127}$$

Since $P_1(z) = -P_0(z)$, we have

$$T(z) = \frac{1}{2}\{P_0(z) + P_1(z)\} = \frac{1}{2}\{P_0(z) - P_0(-z)\} = Cz^{-k}. \tag{7.128}$$

We see that k must be odd because only odd powers of z will remain. We can rewrite (7.125) as

$$\frac{1}{C}z^k H_0(z)H_1(-z) - \frac{1}{C}z^k H_0(-z)H_1(z) = 2. \tag{7.129}$$

Define

$$P(z) = \frac{1}{C} z^k H_0(z) H_1(-z). \qquad (7.130)$$

Because k is odd,

$$P(-z) = -\frac{1}{C} z^k H_0(-z) H_1(z) \qquad (7.131)$$

Substituting into (7.129), we get

$$P(z) + P(-z) = 2. \qquad (7.132)$$

This means that $P(z)$ is a *half-band zero phase* filter [8]. The equality can be satisfied if all even power coefficients of $P(z)$ are zero. A zero phase filter can take the form

$$Q(z) = \sum_{l=0}^{N} q_l(z^l + z^{-l}). $$

These restrictions lead us to the following form of $P(z)$:

$$P(z) = 1 + p_1(z + z^{-1}) + p_3(z^3 + z^{-3}) + p_5(z^5 + z^{-5}) + \cdots. \qquad (7.133)$$

The general form of $P(z)$ is given in [15].
    $P(z)$ can be factored into the form

$$P(z) = (1 + z^{-1})^m (1 + z)^m R(z) \qquad (7.134)$$

where $R(z)$ can be written as [8]

$$R(z) = r_0 + \sum_{s=1}^{m-1} r_s(z^s + z^{-s}). \qquad (7.135)$$

$P(z)$ can be used to obtain $H_0(z)$ and $G_0(z)$ by factoring $P(z)$ appropriately. It is important to note that if $H_0(z)$ and $G_0(z)$ are linear filters, $H_1(z)$ and $G_1(z)$ will also be linear filters. Example 7.6 will illustrate one way to factor this filter [8].

EXAMPLE 7.6
(Linear Phase QMF Filter Bank)

We want to create a set of linear phase filters by factoring the polynomial $P(z)$. Setting $m = 2$, we get

$$
\begin{aligned}
P(z) &= (1 + z^{-1})^2 (1 + z)^2 \cdot (r_1 z + r_0 + r_1 z^{-1}) \\
&= (1 + 2z^{-1} + z^{-2})(1 + 2z + z^2)(r_1 z + r_0 + r_1 z^{-1}) \\
&= r_1 z^3 + (4r_1 + r_0)z^2 + (7r_1 + 4r_0)z + (8r_1 + 6r_0) \\
&\quad + (7r_1 + 4r_0)z^{-1} + (4r_1 + r_0)z^{-2} + r_1 z^{-3}.
\end{aligned}
\tag{7.136}
$$

The even coefficients must be zero and $p_0 = 1$ so

$$
4r_1 + r_0 = 0, \quad 8r_1 + 6r_0 = 1.
\tag{7.137}
$$

This gives $r_0 = 1/4$ and $r_1 = -1/16$. Substituting back into $P(z)$ gives

$$
P(z) = \frac{1}{16} z^3 (1 + 2z^{-1} + z^{-2})^2 (-1 + 4z^{-1} - z^{-2}).
\tag{7.138}
$$

Recall that

$$
\begin{aligned}
P(z) &= \frac{z^k}{C} H_0(z) H_1(-z)
\tag{7.139} \\
&= \frac{z^k}{C} H_0(z) G_0(z).
\tag{7.140}
\end{aligned}
$$

We let $k = 3$ and $C = 1$. Thus, one possible factorization of $P(z)$ is to define

$$
H_0(z) = \frac{1}{2}(1 + 2z^{-1} + z^{-2})
\tag{7.141}
$$

$$
\begin{aligned}
G_0(z) &= H_1(-z) = \frac{1}{8}(1 + 2z^{-1} + z^{-2})(-1 + 4z^{-1} + z^{-2})
\tag{7.142} \\
&= \frac{1}{8}(-1 + 2z^{-1} + 6z^{-2} + 2z^{-3} - z^{-4}).
\tag{7.143}
\end{aligned}
$$

From (7.110), we can determine the remaining filters:

$$
H_1(z) = G_0(-z) = \frac{1}{8}(-1 - 2z^{-1} + 6z^{-2} - 2z^{-3} - z^{-4}),
\tag{7.144}
$$

$$
G_1(z) = -H_0(-z) = -\frac{1}{2}(1 - 2z^{-1} + z^{-2}).
\tag{7.145}
$$

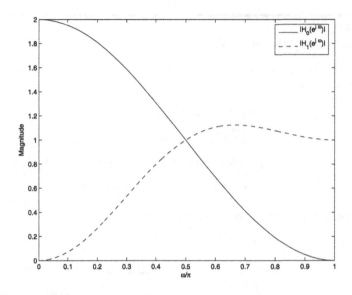

FIGURE 7.43

Magnitude Response of $H_0(z)$ and $H_1(z)$

This particular factorization is known as the LeGall 3/5 tap filter [16]. We can verify that $T(z)$ meets the perfect reconstruction criteria:

$$
\begin{aligned}
T(z) &= \frac{1}{2}\{H_0(z)H_1(-z) - H_0(-z)H_1(z)\} \\[4pt]
&= \frac{1}{2}\left\{\frac{1}{2}(1 + 2z^{-1} + z^{-2}) \cdot \frac{1}{8}(-1 + 2z^{-1} + 6z^{-2} + 2z^{-3} - z^{-4})\right. \\[4pt]
&\quad \left. -\frac{1}{2}(1 - 2z^{-1} + z^{-2}) \cdot \frac{1}{8}(-1 - 2z^{-1} + 6z^{-2} - 2z^{-3} - z^{-4})\right\} \\[4pt]
&= \frac{1}{32}(-1 + 9z^{-2} + 16z^{-3} + 9z^{-4} - z^{-6}) \\[4pt]
&\quad -\frac{1}{32}(-1 + 9z^{-2} - 16z^{-3} + 9z^{-4} - z^{-6}) \\[4pt]
&= \frac{32}{32}z^{-3} \\[4pt]
&= z^{-3}.
\end{aligned}
\tag{7.146}
$$

Fig. 7.43 shows the magnitude responses of the linear filters $H_0(z)$ and $H_1(z)$. It is apparent that these are not mirrored filters. QMF banks made up of perfect reconstruction linear phase filters are often referred to as *biorthogonal filter banks* [8].
End of the Example

7.11.7 NONLINEAR PHASE PERFECT RECONSTRUCTION FILTER BANKS

Perfect reconstruction filter banks composed of nonlinear phase filters are designed using a different scheme than that used for filter banks composed of linear filters. In these cases, the factorization of $P(z)$ is different and the resulting filters will not have a linear phase but they will be mirrored filters. Recall that $P(z)$ can be written in the following form:

$$P(z) = (1 + z^{-1})^m (1 + z)^m \left(r_0 + \sum_{s=1}^{m-1} r_0(z^s + z^{-s}) \right). \qquad (7.147)$$

This filter has $2m$ zeros on the unit circle, $m - 1$ zeros inside the unit circle, and $m - 1$ zeros outside the unit circle. Let $H(z)$ be a filter corresponding to half of the zeros on the unit circle and all of the zeros inside the unit circle. Thus $H(z)$ is a filter of length $N = 2m$. Using this definition, we can factor $P(z)$ into the following form:

$$P(z) = H(z)H(z^{-1}). \qquad (7.148)$$

Since $P(z)$ is a zero phase half-band filter,

$$P(z) + P(-z) = H(z)H(z^{-1}) + H(-z)H(-z^{-1}) = \alpha. \qquad (7.149)$$

Since $P(z)$ and $P(-z)$ are not causal filters, we can introduce a delay of $2m - 1$ or $N - 1$. We can do this by multiplying both sides by $z^{-(N-1)}$:

$$H(z)H(z^{-1})z^{-(N-1)} + H(-z)H(-z^{-1})z^{-(N-1)} = \alpha z^{-(N-1)}. \qquad (7.150)$$

Recall that for perfect reconstruction with no aliasing

$$T(z) = \frac{1}{2}\{H_0(z)G_0(z) + H_1(z)G_1(z)\} = Cz^{-k} \qquad (7.151)$$

with

$$G_0(z) = H_1(-z), \quad G_1(z) = -H_0(-z). \qquad (7.152)$$

Using these relationships and (7.150), we can make the following assignments assuming $\alpha = 2C$:

$$
\begin{aligned}
H_0(z) &= H(z), \\
H_1(z) &= -z^{-(N-1)}H(-z^{-1}) = -z^{-(N-1)}H_0(-z^{-1}), \\
G_0(z) &= z^{-(N-1)}H(z^{-1}) = z^{-(N-1)}H_0(z^{-1}), \\
G_1(z) &= -H(-z) = -H_0(-z).
\end{aligned}
\qquad (7.153)
$$

We also see that

$$H_1(e^{j\omega}) = -e^{-j\omega(N-1)} H_0(-e^{-j\omega}) \qquad (7.154)$$

$$= -e^{-j\omega(N-1)} H_0(e^{j\pi} e^{-j\omega}) \qquad (7.155)$$

$$= -e^{-j\omega(N-1)} H_0(-e^{-j(\omega-\pi)}). \qquad (7.156)$$

From this we observe that

$$|H_1(e^{j\omega})| = |H_0(e^{-j(\omega-\pi)})| \qquad (7.157)$$

$$= |H_0(e^{j(\omega-\pi)})|. \qquad (7.158)$$

Thus, we see that $H_1(z)$ is the mirrored high pass filter of the low pass filter $H_0(z)$. Such filters are perfect reconstruction power symmetric filter banks and are often called orthogonal filter banks. Thus, the filter bank design problem reduces to designing a power symmetric low pass filter $H(z)$.

Given the zero-phase representation of $P(z)$ in (7.134), its factors are of the form $(\alpha z + 1)(1 + \alpha z^{-1})$. We can assign the factors $(1 + \alpha z^{-1})$ to $H_0(z)$ and the factors $(\alpha z + 1)$ to $G_0(z)$. Then we have

$$G_0(z) = z^{-N} H_0(z^{-1}). \qquad (7.159)$$

EXAMPLE 7.7
(Perfect Reconstruction Filter Bank)

Using the $P(z)$ from Example 7.6, we have

$$P(z) = -\frac{1}{16} z^3 (1 + z^{-1})^4 (1 - (\sqrt{3} + 2)z^{-1})(1 - (2 - \sqrt{3})z^{-1}).$$

Let $H_0(z)$ be composed of half the zeros on the unit circle and the zeros inside the unit circle [8], i.e.,

$$H_0(z) = G(1 + z^{-1})^2 (1 - (2 - \sqrt{3})z^{-1})$$

where G is adjusted to ensure unit gain. Expanding we get

$$H_0(z) = G(1 + 1.732z^{-1} + 0.4641z^{-2} - 0.26795z^{-3}).$$

For unit gain, we get $G = 1/(4\sqrt{3} - 4)$, thus

$$H_0(z) = 0.3415 + 0.5915z^{-1} + 0.1585z^{-2} - 0.0915z^{-3}.$$

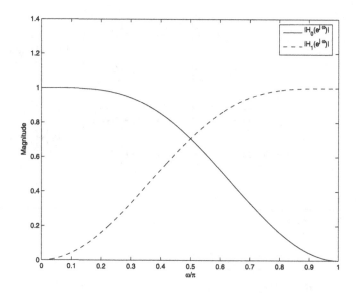

FIGURE 7.44

Magnitude response of $H_0(z)$ and $H_1(z)$

Using Eq. (7.159),

$$
\begin{aligned}
G_0(z) &= z^{-3}H_0(z^{-1}) = G \cdot z^{-3}(1+z)^2(1-(2-\sqrt{3})z) \\
&= -G \cdot (1+z^{-1})^2((2-\sqrt{3})-z^{-1}) \\
&= -\frac{1}{(\sqrt{3}+2)(4\sqrt{3}-4)}(1+z^{-1})^2(1-(\sqrt{3}+2)z^{-1}) \\
&= -0.0915 + 0.1585z^{-1} + 0.5915z^{-2} + 0.3415z^{-3}. \quad (7.160)
\end{aligned}
$$

We can see that $G_0(z)$ is composed of the remaining terms of $P(z)$ multiplied by a constant to ensure unit gain. Again, choosing the remaining filters according to the antialiasing property, the filters $H_1(z)$ and $G_1(z)$ are obtained as follows:

$$
\begin{aligned}
H_1(z) &= G_0(-z) = -0.0915 - 0.1585z^{-1} + 0.5915z^{-2} - 0.3415z^{-3}, \\
& \quad (7.161)
\end{aligned}
$$

$$
\begin{aligned}
G_1(z) &= -H_0(-z) = -0.3415 + 0.5915z^{-1} - 0.1585z^{-2} - 0.0915z^{-3}. \\
& \quad (7.162)
\end{aligned}
$$

Fig. 7.44 shows plots of the magnitude responses of $H_0(z)$ and $H_1(z)$.
End of the Example

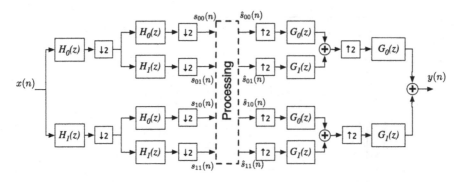

FIGURE 7.45

A two level four channel QMF

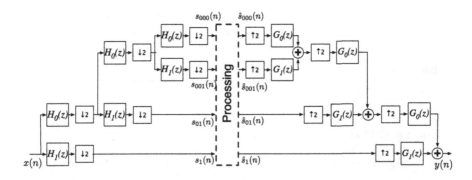

FIGURE 7.46

A four channel QMF with unequal pass band widths

7.12 MULTILEVEL FILTER BANKS

We can add more levels of filter banks to Fig. 7.42 to form a multilevel filter bank. Fig. 7.45 shows an example of a four channels filter bank. This overall system is sometimes called a *tree structured filter bank* [8]. Using this type of structure, we can form four filter banks, each with a pass band of $\pi/4$. Notice that an M channel filter bank, where M is a power of 2, will form banks with a pass band width of π/M for each filter.

Another useful structure can be used to form filter banks of unequal pass band widths. Fig. 7.46 illustrates such a structure for a four-channel QMF bank. A typical magnitude response of the analysis (synthesis) filter banks is shown in Fig 7.47. Here, the response H_{000} corresponds to the overall filter that generates the output $s_{000}(n)$. These filter banks are often referred to as *octave band QMF banks* [8]. There is a distinct relationship between this class of filter banks and the discrete wavelet

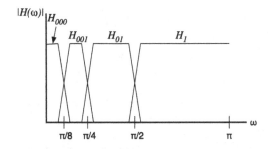

FIGURE 7.47

Magnitude response of a four channel QMF with unequal pass band widths

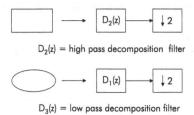

$D_2(z)$ = high pass decomposition filter

$D_3(z)$ = low pass decomposition filter

FIGURE 7.48

Decomposition filters for wavelet decomposition

transform. The following section discusses in greater detail how the individual paths through a multilevel filter bank are analyzed.

7.12.1 MULTILEVEL DECOMPOSITION OF SIGNALS

Many applications require the multilevel decomposition of a sampled sequence using decomposition filters. The wavelet decomposition scheme is an example of a multilevel QMF. Such a structure can be used to perform processing operations such as subband coding of speech signals for compression and signal denoising.

Consider the two level decomposition as given in Figs. 7.48 and 7.49. Fig. 7.48 shows the symbols used for the low pass (approximation) and high pass (detail) analysis filters.

Consider the computation of $s_{21}(n)$ as shown in Fig. 7.50. We can interchange the order of filtering and down sampling using the Noble identities as shown in Fig. 7.51 with a corresponding change to replace $H_1(z)$ by $H_1(z^2)$.

We can proceed in the same manner to obtain the equivalent representation for the three level decomposition of $x(n)$. Consider the computation of $s_{31}(n)$ as shown in Fig. 7.52. We can again interchange the order of filtering and downsampling as shown in Fig. 7.53.

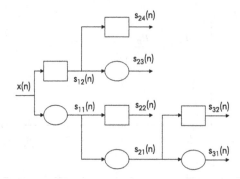

FIGURE 7.49

Multilevel wavelet decomposition of a sample sequence

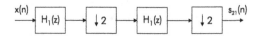

FIGURE 7.50

Two level decomposition of the approximation signal

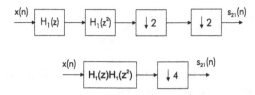

FIGURE 7.51

Two level decomposition of the approximation signal with a change in the order of filtering and downsampling

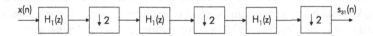

FIGURE 7.52

Three Level Decomposition of the Approximation Signal

The polyphase implementation can be used to efficiently implement a digital filter followed by downsampling as discussed earlier in Sect. 7.9. This concept is shown in Fig. 7.54 for the three level decomposition of the approximation signal as shown in Fig. 7.53.

This concept can be illustrated with an example.

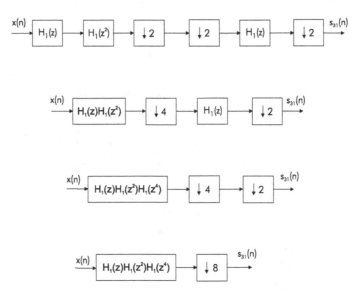

FIGURE 7.53

Three level decomposition of the approximation signal with a change in the order of filtering and downsampling

EXAMPLE 7.8
(Multilevel Decomposition)

We can use the Daubechies wavelet filters *db4* to obtain appropriate subband decomposition filters. The following Matlab script can be used to compute the required filter coefficients.

Matlab Script 7.12.

```
[b1, b2, b3, b4] = wfilters('db4');
```

End of the Script

The filter with coefficients $b1$ is the low pass decomposition filter, and the filter with the coefficients $b2$ is the high pass decomposition filter. The filter with coefficients $b3$ is the low pass reconstruction filter, and the filter with the coefficients $b4$ is the high pass reconstruction filter. The system transfer function for the low pass decomposition filter $H_1(z)$ with coefficients $b1$ is given by

$$H_1(z) = -0.0106 + 0.0329z^{-1} + 0.0308z^{-2} - 0.1870z^{-3}$$
$$- 0.0280z^{-4} + 0.6309z^{-5} + 0.7148z^{-6} + 0.2304z^{-7}.$$

$$(7.163)$$

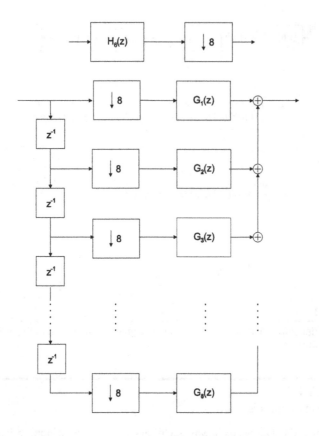

FIGURE 7.54

Three level decomposition of the approximation signal using a polyphase filter implementation

The system transfer function for the high pass decomposition filter $H_2(z)$ with coefficients $b2$ is given by

$$H_2(z) = -0.2304 + 0.7148z^{-1} - 0.6309z^{-2} - 0.0280z^{-3}$$
$$+ 0.1870z^{-4} + 0.0308z^{-5} - 0.0329z^{-6} - 0.0106z^{-7}.$$

$$(7.164)$$

Fig. 7.51 gives the operations for the multilevel subband decomposition of an input sequence $x(n)$. Fig. 7.50 gives the filtering and downsampling operations to obtain the sequence $s_{21}(n)$. Fig. 7.51 shows the modification to change the order of filtering and downsampling for the first downsampling operation and the second filtering operation.

We can obtain the filter $H_1(z^2)$ by replacing z by z^2 in $H_1(z)$. It follows that

$$
\begin{aligned}
H_1(z^2) = \ & -0.0106 + 0.0329z^{-2} + 0.0308z^{-4} - 0.1870z^{-6} \\
& - 0.0280z^{-8} + 0.6309z^{-10} + 0.7148z^{-12} + 0.2304z^{-14}.
\end{aligned}
$$

(7.165)

This is equivalent to placing a zero between each filter coefficient in $b1$. We can define $H_5(z) = H_1(z)H_1(z^2)$ with corresponding filter coefficients $b5$. It follows that $b5$ is the convolution of $b1$ and $b1$ with a zero between each coefficient. We can use the following Matlab script to compute $b5$ from $b1$.

Matlab Script 7.13.

```
n1 = length(b1);
n2 = 2*n1 - 1; % We do not need to use the zero at the end.
c1 = zeros(1, n2);
c1(1, 1:2:end) = b1;
b5 = conv(b1, c1);
```

End of the Script

It follows that the system transfer function with filter coefficients $b5$ is given by

$$
\begin{aligned}
H_5(z) = \ & 0.0001 - 0.0003z^{-1} - 0.0007z^{-2} + 0.0031z^{-3} + 0.0010z^{-4} \\
& - 0.0118z^{-5} - 0.0056z^{-6} + 0.0064z^{-7} + 0.0172z^{-8} \\
& + 0.0611z^{-9} + 0.0197z^{-10} - 0.0849z^{-11} - 0.1210z^{-12} \\
& - 0.1552z^{-13} - 0.0181z^{-14} + 0.2654z^{-15} + 0.4381z^{-16} \\
& + 0.5532z^{-17} + 0.5046z^{-18} + 0.3100z^{-19} + 0.1647z^{-20} \\
& + 0.0531z^{-21}.
\end{aligned}
$$

(7.166)

End of the Example

Consider the three level decomposition as shown in Fig. 7.53. We can define

$$
H_6(z) = H_1(z)H_1(z^2)H_2(z^4).
$$

(7.167)

The signal $s_{32}(n)$ can be obtained by filtering $x(n)$ using $H_6(z)$ followed by downsampling by a factor of 8. This can be done by using a polyphase FIR filter implementation similar to the illustration for computing $s_{31}(n)$ in Fig. 7.54. We can use an example to illustrate this concept.

Table 7.3 Table of coefficients for $b6$

n	0	1	2	3	4	5	6
$b6(n)$	−0.0000	0.0001	0.0002	−0.0007	−0.0001	0.0025	0.0008
n	7	8	9	10	11	12	13
$b6(n)$	0.0007	−0.0033	−0.0223	−0.0081	0.0222	0.0395	0.0869
n	14	15	16	17	18	19	20
$b6(n)$	0.0218	−0.1260	−0.1983	−0.2767	−0.1416	0.1723	0.3513
n	21	22	23	24	25	26	27
$b6(n)$	0.4773	0.3705	0.0578	−0.1520	−0.2957	−0.3143	−0.2188
n	28	29	30	31	32	33	34
$b6(n)$	−0.1383	−0.0757	−0.0167	0.0381	0.0730	0.0953	0.0932
n	35	36	37	38	39	40	41
$b6(n)$	0.0689	0.0481	0.0314	0.0159	0.0017	−0.0080	−0.0149
n	42	43	44	45	46	47	48
$b6(n)$	−0.0164	−0.0130	−0.0101	−0.0076	−0.0053	−0.0033	−0.0017
n	49						
$b6(n)$	−0.0006						

EXAMPLE 7.9
(A Three Level Polyphase Decomposition)

We will use the same wavelet filters used for Example 7.8 for this example. The system transfer function for $H_1(z)H_1(z^2) = H_5(z)$ is given in Eq. (7.166). It follows that

$$H_6(z) = H_5(z)H_2(z^4). \tag{7.168}$$

We have

$$
\begin{aligned}
H_2(z^4) = \quad &-0.2304 + 0.7148z^{-4} - 0.6309z^{-8} - 0.0280z^{-12} \\
&+ 0.1870z^{-16} + 0.0308z^{-20} - 0.0329z^{-24} - 0.0106z^{-28}.
\end{aligned}
\tag{7.169}
$$

This is equivalent to inserting three zeros between each of the coefficients in the coefficient sequence $b2$. We can use the following Matlab script to obtain the coefficients for $H_6(z)$.

Matlab Script 7.14.

```
n1 = length(b1);
n2 = 2*n1 - 1; % We do not need to use the zero at the end.
c1 = zeros(1, n2);
c1(1, 1:2:end) = b1;
b5 = conv(b1, c1);
```

```
% conventional approach
n3 = 4*n1;
c2 = zeros(1, n3);
c2(1, 1:4:end) = b2;
b6 = conv(b5, c2);
```

End of the Script

The resulting filter coefficients are given in Table 7.3.

This approach has the disadvantage that the resulting overall filter $H_6(z)$ with filter coefficients $b6$ has a long length. If the length of the original filter $H_1(z) = m_1 = 8$, then the length of $H_5(z)$ can be determined as

$$m_2 = m_1 + 2m_1 - 2 = 22 \qquad (7.170)$$

where the trailing zeros have been eliminated from $H_1(z^2)$. The length of $H_6(z)$ can be determined as

$$m_3 = m_1 + 2m_1 - 2 + 4m_1 - 4 = 7m_1 - 6 = 50 \qquad (7.171)$$

where the trailing zeros have been eliminated from both $H_1(z^2)$ and $H_2(z^4)$. We need to implement $H_6(z)$ followed by downsampling by a ratio of 8. We can use a polyphase filter to do this more efficiently as shown in Fig. 7.54.

Fig. 7.55 gives a sample input sequence for this example.

We can use the Matlab *dwt* function to compute $s_{32}(n)$ using the following Matlab script.

Matlab Script 7.15.

```
[s11, s12] = dwt(x, b1, b2);
[s21, s22] = dwt(s11, b1, b2);
[s31, s32] = dwt(s21, b1, b2);
```

End of the Script

Fig. 7.56 gives a stem plot of the output $s_{32}(n)$. We can also compute $s_{32}(n)$ by filtering the input sequence $x(n)$ using $H_6(z)$ followed by down sampling by a factor of 8. The following Matlab script can be used to compute $s_{32}(n)$ using this approach.

Matlab Script 7.16.

```
n4 = length(s32);
tmp3 = conv(b6, x);
s3 = tmp3(1, 8:8:end);
y3 = s3(1, 1:n4);
```

End of the Script

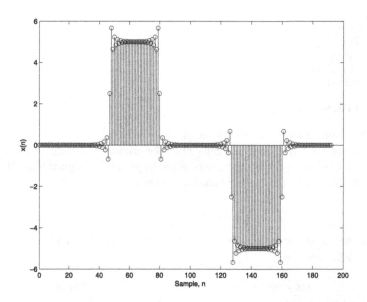

FIGURE 7.55

Stem plot of a sample input sequence for Example 7.9

Fig. 7.57 gives a stem plot of the output with $s_{32}(n)$ and $y3(n)$ on the same plot to compare the two outputs. This plot verifies that the outputs are the same.

We can more efficiently compute $s_{32}(n)$ using a polyphase FIR filter implementation with downsampling by a factor of 8 as shown in Fig. 7.54.

Appendix E.1 gives a Matlab function *polydownfil* that can be used to implement the FIR polyphase FIR filter with downsampling by a factor of m. The following Matlab script can be used to compute $s_{32}(n)$ using *polydownfil*.

Matlab Script 7.17.

```
tmp4 = polydownfil(b6, 8, x);
y4 = tmp4(1, 1:n4);
```

End of the Script

Fig. 7.58 gives a stem plot of the output with $s_{32}(n)$ and $y4(n)$ on the same plot to compare the two outputs. This plot verifies that the outputs are the same.

End of the Example

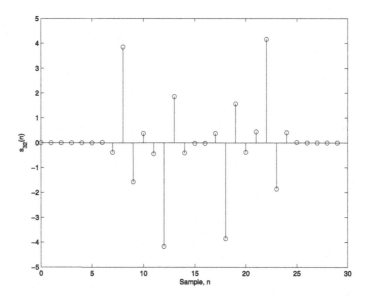

FIGURE 7.56

Stem plot of the output $s_{32}(n)$

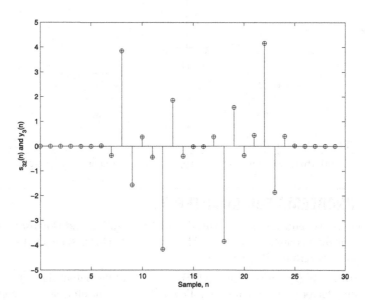

FIGURE 7.57

Stem plots of the outputs $s_{32}(n)$ and $y3(n)$ on the same plot

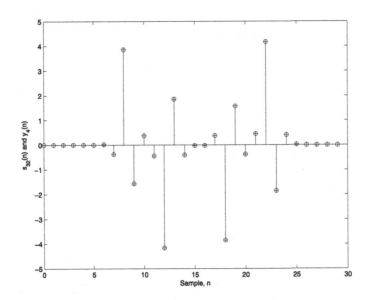

FIGURE 7.58

Stem plots of the outputs $s_{32}(n)$ and $y4(n)$ on the same plot

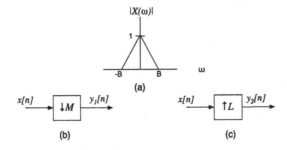

FIGURE 7.59

(a) DTFT of $x[n]$, (b) downsampling block diagram, (c) upsampling block diagram

7.13 PROBLEMS FOR CHAPTER 7

Problem 7.1. (Frequency Representation of Upsampling and Downsampling)
The discrete time Fourier transform (DTFT) of a signal $x[n]$ is shown in Fig. 7.59(a).
The response is zero for $B \leq |\omega| \leq \pi$.

1. Referring to Fig. 7.59(b), draw the DTFT $|Y_1(\omega)|$ of the output signal for a down-sampling factor of $M = 2$ for the following cases. Your plots should range from $-\pi \leq \omega \leq \pi$. Label all relevant frequencies.
   (a) $B = \frac{\pi}{4}$
   (b) $B = \frac{\pi}{2}$

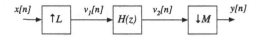

FIGURE 7.60

Rational sample rate conversion

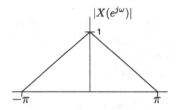

FIGURE 7.61

Magnitude spectrum of $X(e^{j\omega})$

    (c) $B = \frac{3\pi}{4}$

    (d) $B = \pi$

2. Referring to Fig. 7.59(c), draw the DTFT $|Y_2(\omega)|$ of the output signal for the follow cases. Your plots should range from $-\pi \leq \omega \leq \pi$. Label all relevant frequencies.

    (a) $B = \frac{\pi}{2}; L = 2$

    (b) $B = \frac{\pi}{2}; L = 3$

    (c) $B = \frac{\pi}{2}; L = 4$

    (d) $B = \frac{\pi}{2}; L = 5$

3. Use the MATLAB function *fir2* to produce a 100-sample sequence $x[n]$ whose magnitude response matches the response in Fig. 7.59(a). Using *your own* upsampling and downsampling routines, test the results above. Did your drawings match the results MATLAB gave? If they didn't, explain why not. Turn in your drawings and the MATLAB plots.

Problem 7.2. (Rational Sample Rate Conversion)

Using the block diagram in Fig. 7.60 and the spectrum of $x[n]$ in Fig. 7.61, draw the spectra corresponding to a rational sampling rate conversion by (a) $L/M = 7/4$ and (b) $L/M = 4/7$. Draw the response of all intermediate signals and label all relevant frequencies. What is the overall effect for each case?

Problem 7.3. (Multistage Decimators)

1. Refer to the decimator structure in Fig. 7.62. Use the MATLAB routine *firpmord* to design a filter that meets the following specifications:

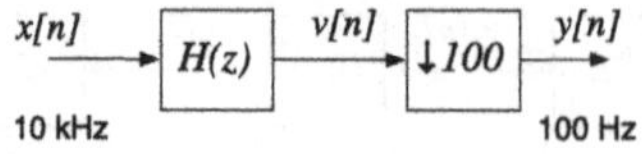

FIGURE 7.62

Decimation by a factor of 100

(Pass Band)	$0 \leq F \leq 45$ Hz,
(Transition Band)	45 Hz $\leq F \leq 50$ Hz,
(Pass Band Ripple)	0.1,
(Stop Band Ripple)	60 dB.

Calculate the number of multiplications per second needed to implement this single stage decimator.

2. Use the method discussed in class to implement a 2-stage decimator. Set $M_1 = 10$ and $M_2 = 10$ for the corresponding stages. Calculate the number of multiplications per second needed to implement this filter. Comment on the efficiency of this method versus the single stage implementation.

3. Implement a 3-stage decimator. Set $M_1 = 25$, $M_2 = 5$, and $M_3 = 5$ for the corresponding stages. Again, calculate the number of multiplications per second needed to implement this decimator. Comment on the efficiency of this method versus the previous two.

4. Use the MATLAB routine *firpm* to obtain the filter coefficients of the filters used for each implementation. For each decimator implementation, plot the overall frequency response. You should have one magnitude/phase plot for each of the three implementations. How are the frequency responses similar? How do they differ? Comment on your results.

Problem 7.4. (Perfect Reconstruction)

The prototype low pass decomposition filter, $H(z)$, for a two channel, quadrature mirror filter (QMF) bank is given by

$$H(z) = -0.0758 - 0.0296z^{-1} + 0.4976z^{-2} + 0.8037z^{-3}$$
$$+ 0.2979z^{-4} - 0.0992z^{-5} - 0.0126z^{-6} + 0.0322z^{-7}. \quad (7.172)$$

The requirements for a perfect reconstruction QMF bank are given by:

$$\begin{aligned}
H_0(z) &= H(z), \\
H_1(z) &= -z^{-(N-1)}H(-z^{-1}) = -z^{-(N-1)}H_0(-z^{-1}), \\
G_0(z) &= z^{-(N-1)}H(z^{-1}) = z^{-(N-1)}H_0(z^{-1}), \quad (7.173) \\
G_1(z) &= -H(-z) = -H_0(-z). \quad (7.174)
\end{aligned}$$

1. Determine the other three filters, $H_1(z)$, $G_0(z)$, and $G_1(z)$.

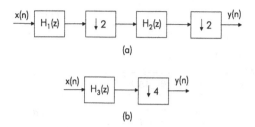

FIGURE 7.63

System block diagram

2. Verify that the alias term

$$A(z) = \frac{1}{2}\{G_0(z)H_0(-z) + G_1(z)H_1(-z)\} \qquad (7.175)$$

is equal to zero for your two channel QMF.
3. Verify that the distortion transfer function

$$T(z) = \frac{1}{2}\{G_0(z)H_0(z) + G_1(z)H_1(z)\} = Cz^{-k} \qquad (7.176)$$

where C is a constant and k is an integer for your two channel QMF.

Problem 7.5. (Polyphase Representation)
Use the following Matlab script to design a 38th order FIR filter.

```
order = 38;
F1 = [0.0  0.0625  0.1250  0.6250  0.6875  1.0];
A1 = [0.0  0.0  1.0  1.0  0.0  0.0];
W1 = [3.4389  1.0000  3.4389];
b = firls(order, F1, A1, W1);
```

1. Define $H_k(z)$ for $1 \le k \le 3$ to implement the filter using a polyphase representation of the form

$$H(z) = H_1(z^3) + z^{-1}H_2(z^3) + z^{-2}H_3(z^3). \qquad (7.177)$$

There will be 13 coefficients in each filter since the original filter has 39 coefficients (order = 38).
2. Write an appropriate set of difference equations to implement the filter $H(z)$ using this polyphase form.

Problem 7.6. (Multirate Signal Processing)
Fig. 7.63(a) shows a part of the two level decomposition of a sequence using subband decomposition filters. The system transfer functions for the filters are given by

$$\begin{aligned}
H_1(z) &= -0.1768z^{-1} + 0.3536z^{-2} + 1.0607z^{-3} + 0.3536z^{-4} - 0.1768z^{-5}, \\
H_2(z) &= 0.3536z^{-1} - 0.7071z^{-2} + 0.3536z^{-3}.
\end{aligned} \qquad (7.178)$$

FIGURE 7.64

Block diagram of a system to interpolate by a factor of four

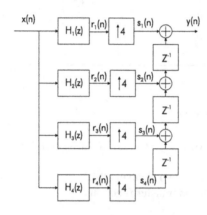

FIGURE 7.65

Block diagram of a system to interpolate by a factor of four using a polyphase filter

Fig. 7.63(b) gives another implementation using $H_3(z)$. Determine the filter coefficients for $H_3(z)$.

Problem 7.7. (Polyphase Implementation with Upsampling)
Fig. 7.64 shows the block diagram for the implementation of upsampling a sequence by a factor of 4 and followed by filtering the output using a low pass filter. The system transfer function for the low pass filter is given by

$$
\begin{aligned}
H_L(z) \quad = \quad & -0.0238 - 0.0271z^{-1} - 0.0148z^{-2} + 0.0165z^{-3} + 0.0637z^{-4} \\
& + 0.1175z^{-5} + 0.1649z^{-6} + 0.2007z^{-7} + 0.2007z^{-8} + 0.1649z^{-9} \\
& + 0.1175z^{-10} + 0.0637z^{-11} + 0.0165z^{-12} - 0.0148z^{-13} \\
& - 0.0271z^{-14} - 0.0238z^{-15}.
\end{aligned} \tag{7.179}
$$

Fig. 7.65 shows the implementation of this interpolation procedure using a polyphase filter.

Required:

1. Determine the system transfer functions for $H_k(z)$ $\forall\, 1 \leq k \leq 4$.
2. Write appropriate difference equations to implement the polyphase filter as shown in Fig. 7.65.

CHAPTER

Digital Signal Processing Systems Design

8

8.1 INTRODUCTION

Digital signal processing has emerged as the core for many electronic products in an information oriented society. Examples include telecommunications, wireless smart phones, digital television, multimedia information storage, retrieval and transmission, digital cameras, automobile subsystems operation and control, global positioning system (GPS) and digital audio systems. These products have evolved due to advances in the design and implementation of digital electronic systems. The design of high performance systems for handling the extremely stringent real time processing requirements for digital signal processing systems associated with these electronic products is unique in that it involves a cross-disciplinary study of application requirements, algorithms, architectures, and the technological aspects of digital system design.

While there are many reasons for this rapid development of digital systems for information technology, some of the reasons include:

1. Advances in computer systems technology make it possible to design systems with the capability to perform real time processing of many of the signals of interest for applications in communications, control, image processing, multimedia, etc.
2. Digital systems are inherently more accurate than corresponding analog systems because they are insensitive to changes in temperature, aging, or component variations. Accuracy is determined by the number of bits used in the system. Therefore, as long as the digital system operates properly, any uncertainty in the results can be predicted.
3. Advances in computer systems have made it practical to add more functions to hand held computers and wireless phones. Thus, there is increased emphasis on designing more sophisticated computer systems for wireless smartphones, handheld tablet computers, and systems that combine all of these functions in a single, handheld device.
4. Wireless communication networks have become widely available and together with the availability of home computers and handheld devices capable of assessing the internet, they have facilitated the development of commerce, email, image transmission, etc., over the internet.

Digital systems have replaced or will soon replace analog systems for many applications such as communications, television, and radio.

Digital Signal Processing. DOI: 10.1016/B978-0-12-804547-3.00008-5

455

Technological improvements have made it possible to implement several million devices on a single chip. In addition, software tools are becoming available to enable designers to take advantage of this capability. Thus, the design of application specific and special purpose digital systems for high performance applications has become practical.

Recently, there has been a lot of emphasis on designing high performance computer systems and computer systems at the functional or behavioral level. However, the development of design tools to design systems automatically from the algorithm specification are still lacking. Typically, ad hoc methods are used to translate the algorithm specification of an application to the register transfer level (RTL) to implement the algorithms. This chapter addresses this problem and presents some systematic procedures for making this task easier.

The traditional approach to developing high performance algorithms for digital signal processing, image processing, and scientific computations was to reduce the computational complexity (for example, the number of complex multiplications). Memory and processing power are relatively cheap in modern digital systems, and the main emphasis of the design of high performance digital systems has been shifted to reducing the overall interconnection complexity, minimizing the data communication requirements, and keeping the overall architecture highly regular, parallel, and/or pipelined [17].

8.2 GENERAL PURPOSE PROCESSORS

General purpose processors can be used to implement digital signal processing algorithms. A general purpose processor has the advantage that it is programmable, and therefore it can be programmed to solve many different types of problems without modifying the hardware. However, a general purpose processor is often too slow to meet the requirements of many compute intensive signal processing applications. The most common general purpose computer architecture is a bus oriented von Neumann Architecture [18]. A general purpose processor generally includes a data path and a control unit. An arithmetic logic unit (ALU) is a basic logic unit, with two inputs, that is used in the data path. It can perform arithmetical functions such as add, subtract, multiply or divide and logical functions such as AND, OR, XOR, etc. The ALU includes the adder, multiplier, comparator, data shifters, logical operators, registers, etc. Registers are used to hold intermediate input and output data for the ALU, and for coprocessors. Registers are also used to facilitate the movement of data from place to place. A bus is typically used as a communication path between parts of the system. Fig. 8.1 gives a block diagram for a general purpose processor.

Several approaches have been used to speed up this simple bus oriented system. Some of the ways to accomplish this are:

1. Use one or more data caches to speed up data communications.
2. Use interleaving and/or overlapping to speed up data access from memory or disk.

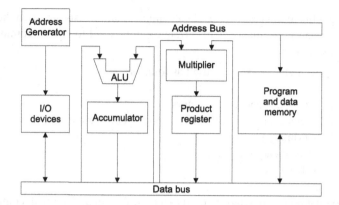

FIGURE 8.1

A simplified architecture for a general purpose processor (the control unit is not shown [18])

3. Use pipelining for instruction fetch, instruction decode, operand fetch, operate, write back, etc.
4. Use multiple functional units such as two or more ALUs in a processor. The functional units can then be individually controlled by using a long instructional word. Processors designed using this approach are commonly called very long instruction word (VLIW) processors.
5. Use a high speed local bus for data transfer between main memory and the processor(s).
6. Use several processors operating in parallel. This includes the use of attached coprocessors such as a floating point processor, a graphics accelerator or an attached field programmable resource such as a field programmable gate array (FPGA).

A communications protocol must be established for bus communications. General considerations include:

1. How does the sender know that it is okay to send a message?
2. How does the receiver know that a message is available?
3. How does the sender know that the message has been received?
4. How is a response message associated with the request that initiated it?

As a general rule, at most only one read/write request will be in progress at any one time and no other request can be honored until that request has been serviced. A bus may be dedicated (always used for data communications between the same two devices) or it may be shared. An arbitration scheme must be used for the control of a shared bus. A bus control system controls which subsystem is allowed to secure use of the bus at any given time. Common bus arbitration schemes include multi-slave, master-slave, time division multiplexing, subsystems given addresses, etc.

Software considerations are also a part of the design. Often there is a tradeoff between flexibility by implementing a function in software and performance by implementing a function in hardware. Data communication costs are often the limiting factor for system performance. Concurrence can be used to reduce communication costs. For example,

1. Hierarchical memories may be used as follows:

   - level 1 – registers and accumulators,
   - level 2 – cache memory (often two levels of cache are used),
   - level 3 – primary memory,
   - level 4 – secondary memory,
   - level 5 – disk drive

2. Input/output processors may be used to handle data in parallel with computations,
3. Attached coprocessors (i.e., floating point processor) may be used to speed up complicated operations,
4. Additional buses (multi-port memory, local bus for the CPU, etc.) may be used to permit data communications in parallel.

8.3 PROGRAMMABLE DIGITAL SIGNAL PROCESSOR

The development of the programmable digital signal processor (PDSP) accelerated the development of digital signal processing technology [19]. The PDSP could compute a fixed point multiply–accumulate in only one clock cycle. Later improvements included floating point multiply–accumulate, barrel shifters, memory banks, and special interfaces to an ADC and a DAC. The PDSP can be considered to be a special purpose computing system. A special purpose computing system is a hardware and/or software system designed to give high performance for a group of applications with similar computational requirements. The PDSP was designed to provide high performance for regular computational algorithms such as convolution and filtering. The architecture of a digital signal processor is optimized for implementing these DSP functions for real time applications. It is characterized by the following [18]:

1. Multiple buses with separate memory space for data and program. Typically, the data memories holds input data, intermediate data values, output values, and algorithm coefficients or parameters such as filter coefficients.
2. The input/output (I/O) ports provide a means for passing data to and from external devices such as the ADC, the DAC, or other processors. Direct memory access (DMA) provides the capability for rapid transfer of blocks of data directly to or from data memories.
3. Arithmetic units for logical and arithmetic operations. This normally includes an ALU, a single cycle hardware multiplier, shifters to perform bit-shift operations and a multiplier–accumulator.

Fig. 8.2 gives a basic hardware architecture for a digital signal processor [18].

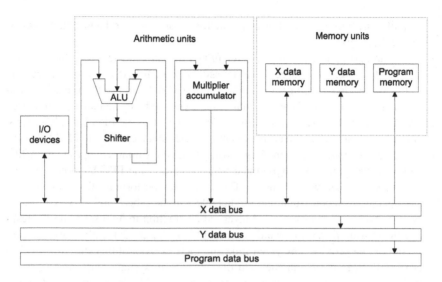

FIGURE 8.2

A basic generic architecture for a digital signal processor (the control unit is not shown [18])

8.4 FIELD PROGRAMMABLE GATE ARRAY

An FPGA is a programmable device that uses prebuilt logic blocks and programmable routing resources. An FPGA can be configured to implement the functionality comparable to custom hardware without requiring additional fabrication steps, printed circuit board fabrication, etc. A user can develop computing tasks and compile them into a bitstream to be downloaded to the FPGA using software. The downloaded bitstream can then be used to configure the FPGA to implement the computing tasks. The FPGA is very versatile because it can be completely reconfigured to implement a completely different group of computing tasks by repeating the process for the new tasks. The FPGA facilitates parallel computing because each different computing task can be assigned to a dedicated group of logic blocks.

FPGA technology has several benefits [20]:

1. FPGAs can be used to transform a sequential executing paradigm to a parallel computing paradigm. Thus, FPGAs can deliver many times the processing power per dollar for DSP solutions in some areas through the implementation of parallel or pipeline computing.
2. FPGA technology offers the capability for rapid prototyping which can lead to an advantage in competitive time-to-market scenarios. Ideas can be tested or concepts can be verified without going through the long fabrication process required for ASIC designs. The increased availability of high-level software tools for FPGA design decreases the learning curve required to implement a design. IP cores or

prebuilt functions are also often available for advanced control and signal processing applications.

3. The nonrecurring engineering cost for FPGA design is far less than that for custom ASIC design. The large initial investment cost for ASIC designs is easy to justify for original equipment manufacturers (OEMs) that use thousands of chips based upon the design. However, many end users need custom hardware functionality for tens to hundreds of systems in development so the FPGA is a lower cost alternative. In addition, the costs of making incremental changes to FPGA designs are negligible compared to the large costs of making similar changes in an ASIC.

4. Although the FPGA is programmed using software, the FPGA circuitry implementation is a hardware solution. The timing for performing the computations and operations is consistent and predictable. On the other hand, processor based systems typically involve several layers of abstraction to help schedule the tasks and the sharing of resources among multiple processes. Typically, only one processor can be active at any given time and there is a possibility that time-critical tasks will preempt one another. On the other hand, systems implemented using FPGAs do not require an operating system and reliability concerns related to timing can be minimized through the use of true parallel execution and the use of deterministic hardware dedicated to each task.

5. FPGAs can be reconfigured to keep up with modifications that may be necessary over time due to changes in standards, system requirements, or changes in specifications. Thus, systems designed using FPGAs can be upgraded to meet new system requirements which reduces the cost of long term maintenance.

The resources available on FPGA chips have improved considerably with advances in technology. The chips are often divided into configurable logic cells, block RAM, DSP slices, and transceivers. The DSP slices may include a multiplier, and accumulator and a pre-adder–adder for high performance filtering. The FPGA fabric may also include fixed logic blocks oriented toward, internet access and one or more programmable processors [21].

8.5 APPLICATION SPECIFIC COMPUTING SYSTEMS

An application specific computer system (ASCS) is a hardware/software system designed specifically to address a particular application or a family of related applications. In contrast to developing an algorithm to run on a general purpose computer, an ASCS matches the computing structure to the problem. ASCS structures can offer 100 to 1000 fold improvements in cost/performance over general purpose computers on applications such as interactive graphics, signal processing, heuristic search, numerical simulation, etc.

For some applications, the entire system must be considered when designing the ASCS. In other cases, the design of an ASCS may begin with an algorithm or a small set of algorithms that form the heart of the application. The system designer must transform this application into forms suitable for different hardware/software architectures. For example, a designer may devise a variant of an algorithm that is suitable for a pipeline processor array and another variant that requires parallel processors that share memory. Exploring each variation requires the system designer to model the system performance and tune the system parameters such as memory speed or data path width to reduce bottlenecks. When a structure is finally chosen, the design must be mapped onto physical hardware, with appropriate designs of subsystems. Software, firmware, and configuration data (e.g., for electrically programmable chips) must be created. The physical hardware must be fabricated, tested and interfaced to the host computer or the input/output devices. Finally, the system must be evaluated and measured against its design goals. Although this is a long and cumbersome process, the cost reduction and performance advantages achievable from ASCS solutions far exceed what can be obtained from software running on general purpose computer systems.

8.6 SYSTEM ON A CHIP

The development of semiconductor technology along with the corresponding development of electronic computer aided design tools have made it practical to design highly sophisticated digital systems on a single chip. However, the emergence of the system-on-chip (SOC) era has created many new challenges at all stages of the design process. The traditional approach to the development of a system for digital signal processing applications involves the development of the algorithm(s) using Matlab or C/C++. However, system hardware designers typically use a hardware description language (HDL) such as Verilog or VHDL to design the system at the register transfer level (RTL). There is a gap in the design process because the Matlab or C/C++ model must be translated to the HDL level manually. In addition, there are usually many different ways to perform this translation and each way has advantages and disadvantages related to the overall performance of the system. Thus, the motivation is very high to develop an approach to exploring the advantages and disadvantages of various potential implementations at a high level of abstraction such as the functional level or the system level.

SystemC is a digital system modeling language based upon C++. It is intended to enable the system level design of sophisticated systems and permit intellectual property (IP) exchange at various levels of abstraction. SystemC is entirely based upon C++. SystemC permits system designers to construct structural design using modules, ports and signals similar to the way it is done using an HDL. It supports commonly used data types such as single bits, bit vectors, characters, integers, floating point numbers, etc., and it also includes support for four-state logic signals (i.e., signals that model 0, 1, X, and Z). SystemC also supports the modeling of both the hardware and software parts of a system in a single simulation model.

Many of the high performance multiprocessor systems designed for demanding digital signal and image processing applications have been special purpose single instruction, multiple data (SIMD) systems [22,23]. Many of these have used bit serial arithmetic. These systems have proven to give very high performance for many applications. However, they have not totally solved the problem of providing high performance for many digital signal and image processing applications because of input/output problems, synchronization difficulty as the number of processors is increased, and the lack of flexibility in adapting to different applications.

The most successful multiprocessor systems for digital signal and image processing applications either have mesh architectures or are pipeline. Mesh architectures often provide very large speedup after the input data has been loaded but overall performance often suffers from input/output (I/O) limitations. Pipeline machines can accept data at a fast rate and often they can accept it at real time rates. However, pipeline multiprocessor systems have historically been difficult to program or reconfigure for different tasks.

There are several reasons why a multiprocessor system may not achieve advertised performance on a given application. The most important of these include:

- There is a mismatch between the parallelism in the algorithm and the multiprocessor architecture.
- A synchronization bottleneck occurs because some of the processors must wait for results from other processors or for input from the input device.
- A resource contention problem occurs because two or more processors need to simultaneously use the same system resource. For example, two processors may need to read data from a shared memory at the same time.
- Timing problems due to clock skew, clock distribution, etc., limit the performance of the system as the number of processors increases.
- Programmers find that high performance can only be achieved by careful consideration of the target architecture, and laborious manual optimization of the application software. Thus, they are reluctant to use the system.

The traditional approach to developing high performance algorithms for digital signal processing, image processing, and scientific computations was to reduce the computational complexity (for example, the number of complex multiplications). Memory and processing power are relatively cheap in modern digital systems, and the main emphasis of the design of high performance computing systems has been shifted to reducing the overall interconnection complexity and keeping the overall architecture highly regular, using parallel operations, and using a pipeline architecture as appropriate [17].

8.7 COMPUTATIONAL STRUCTURES

The computational structure used to implement a particular DSP algorithm impacts the requirement for data communications as well as the computational complexity. Several methods to map DSP algorithms to computational structures to be implemented on special purpose or application specific processors are discussed in the following sections of this chapter. These methods are applicable to mapping algorithms to embedded processors, to FPGAs or to ASIC systems. Many algorithms used in scientific and engineering applications such as matrix operations, signal processing, image processing and communications require extensive computations but the operations are regular and localized. The regular features of these algorithms can be exploited to facilitate the use of concurrent computing to design high performance systems. The methods to develop computational structures in the following sections are intended to be applied to this special class of algorithms.

The early work in developing algorithms for digital signal and image processing was oriented toward single processor von Neumann computer systems. Emphasis was placed on minimizing data storage, minimizing the number of multiplications and divisions, etc. However, the capability of digital circuits and systems has advanced tremendously in the last few years. Therefore, the parameters associated with the design of high performance algorithms and systems have been modified. It is practical to use more hardware in the form of larger memories, multiple functional units, etc., while the cost of transferring data from one part of the system to another has not decreased nearly as fast. Therefore, the cost of data communications has become more important than computational complexity relative to the overall hardware cost and performance for many digital systems used for DSP applications.

The algorithm chosen to implement an application can have a dramatic effect on the computational efficiency on a given system. Important issues include the order and frequency of data transfers to and/or from memory, the number and type of computational operations involved, the overhead associated with computing memory addresses for storage or retrieval of data, etc. The matrix addition operation can be used as an example to illustrate the role of the algorithm on the performance of a computer system. Operations performed on vectors or matrices can typically be performed in iterative loops. For example, consider the following C language algorithm for matrix addition $C = A + B$ [17]:

```
for (i=0; i<IMAX; i++)
    {
    for(j=0; j<JMAX; j++)
        {
        c[i][j] = a[i][j] + b[i][j];
        }
    }
```

Here, the elements of **A** and **B** are addressed in column major order, which is the order which they are typically stored in memory. Thus, the address for the next data value is just the incremented address of the current data value. If this order is reversed, then the algorithm will not execute as efficiently because of the additional overhead associated with computing the addresses of the input data. There may also be a penalty associated with obtaining the data as well since the next data value to be used is not adjacent to the data currently being used. There may also be a penalty due to the retrieval of the next data value due to the inherent inefficiency of computer systems in randomly retrieving data. On the other hand, changing to addressing the matrices in row major order does not affect the number or complexity of the computations. This simple example reinforces the point that data communications are just as important as computational complexity in determining the efficiency of a given algorithm.

Sequential programming languages, such as C and Fortran, imply an order in which computations are performed. However, one must be careful when specifying the ordering. In some cases, the algorithm can be described without specifying the ordering. The compiler can then choose the order which will optimize the efficiency of the computations.

There has been a great amount of work done on mapping digital signal processing algorithms to systolic arrays [24]. However, the practical application of systolic arrays has been limited because of timing problems with large systolic arrays. These timing problems can be overcome for pipeline arrays where the data flow is in one direction. Therefore, developing computational structures that can be mapped into pipeline architectures should be given priority. On the other hand, systolic type computational structures can provide important advantages as long as the number of cells in the array is small.

8.8 PARALLEL ALGORITHM EXPRESSIONS

Parallel algorithms may be derived by

1. Modification of sequential algorithm expressions.
2. Direct parallel algorithm expressions, such as snapshots, recursive equations, parallel codes, single assignment codes, dependence graphs, etc.

Much of the work in developing parallel algorithms has been ad hoc. The search for more general techniques that can be applied to a wide variety of applications is still active.

8.8.1 SINGLE ASSIGNMENT CODE

A *single assignment code* is a form where every single variable is assigned one value only during the execution of the algorithm. A single assignment code can be readily mapped unto an array structure once it has been developed. Consider the following algorithm for matrix–vector multiplication [17]:

```
for (i=0; i<IMAX; i++)
    {
    c[i] = 0.0;
    for(j=0; j<JMAX; j++)
        {
        c[i] = c[i] + a[i][j]*b[j];
        }
    }
```

Note that the storage location indicated by $c[i]$ is used $\mathsf{JMAX} + 1$ times in computing each element of the output vector. This code can be modified to make it a single assignment code by making **c** a matrix as follows:

```
for (i=0; i<IMAX; i++)
    {
    c[i,0] = 0.0;
    for(j=0; j<JMAX; j++)
        {
        c[i][j+1] = c[i][j] + a[i][j]*b[j];
        }
    }
```

In this case, each storage location is used only once and the corresponding output is $\mathsf{IMAX} + 1$ rows $\times$ JMAX columns (includes the initialization row $c[i, 0]$). The resulting algorithm can more easily be mapped onto an array.

8.8.2 RECURSIVE ALGORITHMS

A convenient and concise way to represent algorithms is to use recursive equations. Z Transforms and/or transfer functions for DSP applications can be manipulated to derive recursive difference equations. However, recursive difference equations can be used to solve many other types of problems. Consider the matrix–vector multiplication problem **c** = **Ab**. A recursive equation for this algorithm using a single assignment code is given by [17]:

$$c[i][j + 1] = c[i][j] + a[i][j] * b[j], \quad 0 \le j < \mathsf{JMAX} \qquad (8.1)$$

where j is the recursion index, and

$$
\begin{aligned}
c[0] &= 0, \\
a[i][j] &= \mathbf{A}(i, j), \\
b[j] &= \mathbf{B}(j).
\end{aligned}
\qquad (8.2)
$$

Often a recursive equation uses one index for time and the other index for space (memory location). The activities of a parallel algorithm can be adequately expressed by doing so. These roles could just as well be reversed. Thus, from a mathematical point of view, the indices in Eq. (8.1) could just as well be reversed. However, the order does affect the communication overhead as was previously discussed.

8.9 PIPELINE IMPLEMENTATION OF DSP SYSTEMS

Current digital system design technology makes it possible to realize many computational units on a single chip. In particular, an implementation of digital filters can be used to achieve a high processing rate by dividing the computational load among many computational units working concurrently.

Parallelism and pipelining are two forms of concurrent computation for high speed processing. Parallelism is defined as the case where the same or different inputs are applied to different computational units and the outputs are used independently or are collected in some way to form the total output. Alternately, pipelining is defined as the case where the output from one computational unit is the input to another computational unit with the final computational unit in the pipeline producing the final output. Generally, pipelining is more suitable for the design of high performance systems because of the higher likelihood of data communication bottlenecks with the use of parallelism. This is especially true when a parallel system involves broadcasting the same data to a large number of computational units in parallel. Broadcasting becomes a problem when large numbers of connections and cross connections, which are expensive and difficult to implement, are required to deliver the data to the computational units as needed on the chip. In addition, inter-chip communications must be minimized to keep pin count reasonable. For example, the implementation of a large tree adder on a single chip is not a good idea since it has a large number of inputs. However, the addition can be distributed over a pipeline of adders for a practical design. Therefore, pipeline implementations of algorithms for DSP are generally more desirable because a pipeline implementation only involves a few local communications. However, there are tradeoffs for all design concepts to implement DSP applications such as a tradeoff of size or complexity versus throughput. The final design decision is often a compromise designed to meet given performance requirements.

Digital signal processing applications typically involve a stream of data to be processed in the same way. Thus, a pipeline architecture is a good choice for many DSP applications. Once the pipeline has been filled, a new operation can be started at each computational cycle. The following generality may be applied to pipeline systems:

- Given K processors in a pipeline with each processor having a unit processing interval of T.
- Assume that a given process takes KT time on a single processor.
- Given N such task $(N > K)$, it would take $KT + (N - 1)T$ time.
- Also, after $(K - 1)T$ time, a new task would be completed after each interval T.

As an example, consider a filtering operation that requires 5 cycles on a single processor. If the 5 cycle operation can be broken up to be performed by 5 processors such that a new operation can be started each cycle, then after 4 cycles, a new result will be available each cycle. Thus, on cycles 5 through 14, a new result will be available for a total of 10 results in 14 computational cycles. If the processing for a new sample is started each cycle without any interruption, then a new output will be available from one of the processors each cycle once the pipeline fill up time of 4 cycles has passed. Several pipelined structures of digital filters have been proposed in the literature [24–26]. However, they have been mainly derived with the use of ad hoc or intuitive approaches [24,26].

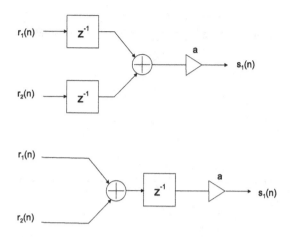

FIGURE 8.3

Moving delays in a block diagram involving an adder and a multiplier

Other desirable features for the implementation of DSP algorithms include modularity, regularity, and few and simple computational primitives. These properties are considered in the approaches presented below for mapping DSP algorithms to computational structures.

8.10 MOVING DELAYS AROUND

There are occasions when delays need to be moved locally to achieve a desired computational structure. Fig. 8.3 gives a part of a block diagram that shows an example of moving delays involving an adder and a multiplier. The following equation applies both before and after the move:

$$s_1(n) = a \left[r_1(n-1) + r_2(n-1) \right]. \tag{8.3}$$

Thus, moving the delay did not affect the computational accuracy of the result.

Fig. 8.4 gives a part of a block diagram that shows an example of moving delays involving a branch and a multiplier. The following equation applies both before and after the move:

$$
\begin{aligned}
s_1(n) &= ar_1(n-1), \\
s_2(n) &= r_1(n-2).
\end{aligned}
\tag{8.4}
$$

Thus, moving the delay did not affect the computational accuracy of the result. It is a good practice to verify that the block diagram, resulting from moving delays around, is still valid by verifying that the corresponding equations are consistent. In

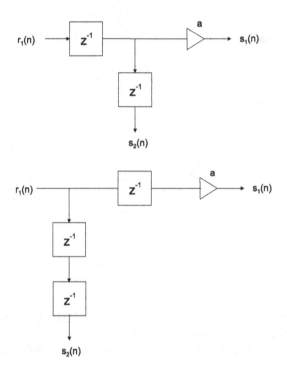

FIGURE 8.4

Moving delays in a block diagram involving a branch and a multiplier

addition, moving delays to achieve desirable computational results is typically not as easy as indicated in Figs. 8.3 and 8.4. Thus, a more general approach for moving delays around is presented in Sect. 8.11.

8.11 RETIMING OF CELLULAR ARRAYS

Broadcasting is a major problem that should be avoided in developing computational cells for cellular arrays. A procedure called **cut set retiming** can be used to implement interleaving or delay in a computational structure [17].

If a cut is made completely through the signal flow diagram (or block diagram) for an algorithm, then delays can be added and/or removed as needed at the cut without altering the overall timing for the computations. These delays can then be moved as needed to solve broadcasting or direct path problems. Since delays are typically implemented in registers or memory, the added delays permit the saving of intermediate results in registers or memory associated with the delays and then these results can be passed on to the next node in the signal flow diagram on the next computational cycle.

The rules for adding and removing delays along the cut are as follows:

1. If a delay is added to an arc with direction to the left, then a delay must be added to all arcs with direction to the left and a delay must be removed from all arcs with direction to the right.

2. If a delay is added to an arc with direction to the right, then a delay must be added to all arcs with direction to the right and a delay must be removed from all arcs with direction to the left.

Often, after adding a desired delay in one direction, there is no corresponding delay in a given arc in the opposite direction to remove. This problem can be usually be solved by replacing all delays in the computational cell by an integer multiple of delays prior to using cutset retiming. The input and output rates must also be scaled by the same integer multiple. This is equivalent to implementing the interleaving procedure described in Sect. 8.13.1.

8.11.1 RETIMING OF COMPUTATIONAL CELLS FOR ARRAYS

This section covers the partitioning of algorithms such that they can be implemented using array structures. There are several options to efficiently implement an algorithm once it has been mapped into a computational array that supports pipelining. These options include the efficient implementation on general purpose computers using iterative loops as well as hardware implementation on a pipeline processor. The efficiency is due to the use of only local data communications without increasing computational requirements. Generally, the requirements for a linear computational array that supports pipelining are as follows:

1. One computational cell should be developed to be used as a regular cell that can be replicated as many times as needed to form the interior cells of the array.

2. One or two input/output cells should be developed, as needed, to handle the special requirements of the interface with the input device and/or the output device.

3. None of the cells should have a path without a delay between a primary input and a primary output.

4. There should be a delay between any of the computational units in the cell and a primary output.

Thus, the goal is to define at most three computational cells that can be used as many times as needed to form the desired computational array. This approach has the following additional advantages.

1. The design of the hardware or software to implement the regular computational cells has to be done only once for all of the interior cells of the array.

2. The design of the hardware or software for only one or two computational input/output cells has to be done to complete the system design.

8.11.2 TIMING EXAMPLES

This subsection provides 3 cutset timing examples to illustrate the cutset timing concept.

EXAMPLE 8.1

(Timing Example 1)

Consider the block diagram of the computational cell given in Fig. 8.5. A set of difference equations relating the inputs to the outputs is given by

$$
\begin{aligned}
q_{k+1}(n) &= q_k(n-1), \\
y_k(n) &= y_{k+1}(n-1), \\
r_k(n) &= b(k)[q_k(n-1) + y_{k+1}(n-1)] + r_{k+1}(n). \quad (8.5)
\end{aligned}
$$

There is a problem because there is no delay between the input $r_{k+1}(n)$ and the output $r_k(n)$. If this cell is used in a cellular array by interconnecting several cells, then the computations for the bottom path would have to be done for all of the cells in a single computational cycle. This problem can be solved by using the following steps:

1. Double the delays in the cell (change z^{-1} to z^{-2}). Fig. 8.6 shows the results of this step.
2. Move one of the delays in the top arc (delay of $q_k(n)$) past the branch. Fig. 8.7 shows the results of this step.
3. Use cut set retiming in the right side of the cell for all inputs and outputs. Add a delay in all arcs directed to left and subtract a delay from all arcs directed to the right. Fig. 8.8 shows the results of this step.

  A set of difference equations for the modified cell is given by:

$$
\begin{aligned}
q_{k+1}(n) &= q_k(n-1), \\
y_k(n) &= y_{k+1}(n-3), \\
r_k(n) &= b(k)[q_k(n-2) + y_{k+1}(n-3)] + r_{k+1}(n-1). \quad (8.6)
\end{aligned}
$$

Note that there is at least one delay between all paths between a primary input and a primary output for the modified circuit.

End of the Example

EXAMPLE 8.2

(Timing Example 2)

Fig. 8.9 gives a representation of the block diagram for a computational cell to be implemented in a cellular array. The following set of difference equations can be

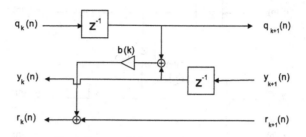

FIGURE 8.5

Block diagram of a computational cell

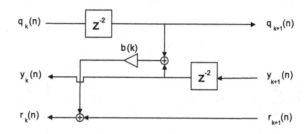

FIGURE 8.6

Cell after doubling delays

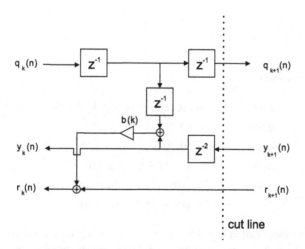

FIGURE 8.7

Cell after doubling delays and moving one delay

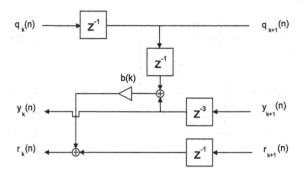

FIGURE 8.8

Modified computational cell without broadcasting

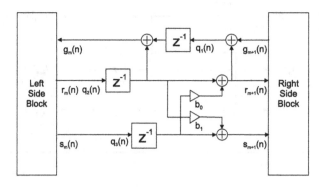

FIGURE 8.9

Block diagram for cutset Example 8.2

used to implement the cell:

$$
\begin{aligned}
q_1(n) &= g_{m+1}(n) + b_0 q_3(n-1) + q_2(n-1), \\
q_2(n) &= r_m(n), \\
q_3(n) &= s_m(n), \\
g_m(n) &= q_1(n-1) + q_2(n-1), \\
r_{m+1}(n) &= b_0 q_3(n-1) + q_2(n-1), \\
s_{m+1}(n) &= b_1 q_2(n-1) + q_3(n-1).
\end{aligned} \tag{8.7}
$$

The computational cell in Fig. 8.9 does not meet the desired requirements because there is no delay between a computational unit and the primary outputs $r_{m+1}(n)$ and $s_{m+1}(n)$. The computational cell cannot be modified as desired directly by

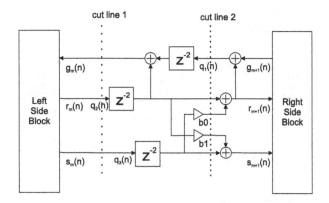

FIGURE 8.10

Computational cell after doubling delays for cutset Example 8.2

only using cutset retiming. However, the desired cell can be obtained by doubling the delays and then applying cutset retiming as appropriate. Fig. 8.10 shows the cell after doubling the delays. This figure also shows two cutsets.

Fig. 8.11 shows the cell after doubling the delays and implementing cutset retiming. The delays can then be moved past the two adders to obtain the final computational cell as shown in Fig. 8.12.

The following set of difference equations can be used to implement the computational cell:

$$
\begin{aligned}
q_1(n) &= g_{m+1}(n) + q_5(n-1), \\
q_2(n) &= r_m(n), \\
q_3(n) &= s_m(n), \\
q_4(n) &= q_1(n-1) + q_2(n-1), \\
q_5(n) &= b_0 q_3(n-1) + q_2(n-1), \\
q_6(n) &= b_0 q_3(n-1) + q_2(n-1), \\
q_7(n) &= b_1 q_2(n-1) + q_3(n-1), \\
g_m(n) &= q_4(n-1), \\
r_{m+1}(n) &= q_6(n-1), \\
s_{m+1}(n) &= q_7(n-1). \quad\quad (8.8)
\end{aligned}
$$

End of the Example

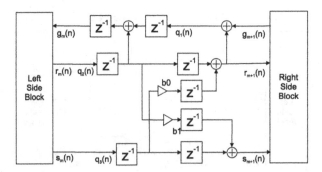

FIGURE 8.11

Computational cell after doubling delays and cutset retiming for cutset Example 8.2

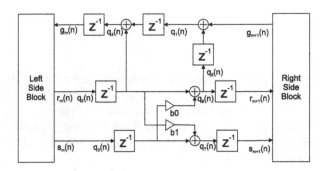

FIGURE 8.12

Final computational cell for cutset Example 8.2

EXAMPLE 8.3
(Timing Example 3)

Fig. 8.13 gives a computational cell that can be used to implement a digital filter. This computational cell does not meet the requirements for a computational cell for a computational array because there is no delay between the primary input $r_{m+1}(n)$ and the primary output $r_m(n)$. In addition, there is no delay between one of the adders and the primary output $s_m(n)$. The goal is to modify this cell to solve these problems.

1. The requirement is to add a delay in the path between the input $r_{m+1}(n)$ and the output $r_m(n)$ which goes from right to left.

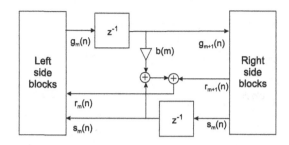

FIGURE 8.13

Computational cell for cutset Example 8.3

2. However, if a delay is added in that path, then it will be necessary to remove the delay between input $g_m(n)$ and the output $g_{m+1}(n)$ which leaves a path between a primary input and a primary output without a delay.
3. There is also the need to add a delay on the arc for the output $s_m(n)$.

These problems cannot be solved by directly using cutset retiming. However, these problems can be solved by doubling the delays first and then using cutset retiming.

Fig. 8.14 gives a block diagram to show the doubling of the delays and a cut line where delays can be added and/or deleted as necessary to solve the problems.

Fig. 8.15 gives the block diagram for the modified cell after adding and removing delays along the cut line.

Fig. 8.16 gives the block diagram for the modified cell after moving delays to achieve the final desired computational structure. The difference equations for the modified computational cell are given below:

$$
\begin{aligned}
g_{m+1}(n) &= g_m(n-1), \\
r_m(n) &= b(m)g_m(n-2) + s_{m+1}(n-3), \\
&\quad + r_{m+1}(n-1), \\
s_m(n) &= s_{m+1}(n-3).
\end{aligned}
\tag{8.9}
$$

End of the Example

8.12 COMPUTATIONAL CELLS FOR ARRAYS

The approach, in this chapter, to the design of computational cells for the implementation of DSP applications using linear arrays has been oriented toward the hardware implementation of the applications. However, the concepts used lead to the development of computational cells with only local data communication requirements which can be used to efficiently implement the applications using software as well as hardware. The cells in a typical linear array are identical except for the cells at the edges

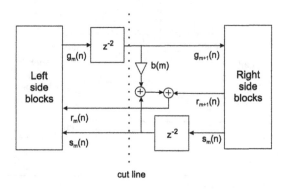

FIGURE 8.14

Computational cell for cutset Example 8.3 after doubling delays

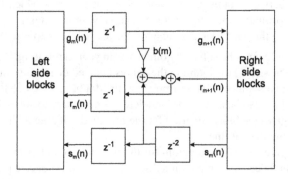

FIGURE 8.15

Computational cell for cutset Example 8.3 after doubling the delays and using cutset retiming

of the array which may involve initialization, data input or data output. One standard approach to designing such a cell is to develop an iterative expression for the algorithm of interest and use this iterative expression to design the computational cell. Some suggestions for the development of computational cells for the implementation of a pipeline computational using hardware follow:

1. The cells of a pipeline computational array should be identical. This will provide for good utilization of the hardware/software and simplify synchronization.
2. Broadcasting of data to several cells is not a good practice. Broadcasting of data often requires long wire lengths and large capacitance loads which will limit the performance of the system because of the corresponding reduction in the maximum clock rates.
3. Data flow in only one direction is very desirable. If data flows in more than one direction simultaneously, then tighter constraints must be placed on circuit timing.

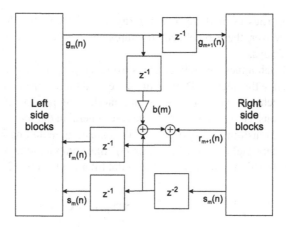

FIGURE 8.16

Final computational cell for cutset Example 8.3

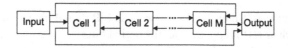

FIGURE 8.17

General representation of a linear array

This may result in a requirement to reduce the clock rate to ensure synchronization, which in turn will reduce overall performance.

4. Data communication protocols must be as simple as possible. Complex communication protocols typically require additional time, additional hardware or both to implement.

Fig. 8.17 gives a diagram of a general linear array.

Each input in Fig. 8.17 may represent multiple inputs, and each output may represent multiple outputs. The interior cells are identical while the input is inserted at the left edge and the output is collected at the right edge. However, these could be reversed or both input and output can be on the same side.

8.13 PIPELINING TECHNIQUES

There are two techniques that can be used to develop algorithms for a pipeline implementation of a DSP application: the *interleaving* and the *delay* techniques [27]. These techniques can be applied to a particular DSP application to yield two pipelined structures: the *interleaved-data* and the *delayed-data* structures. Both of these structures feature the properties of modularity, regularity, and few and simple computational primitives. Moreover, these pipelined structures can easily be mapped to practical computational arrays.

Two formal techniques for adding delays using the properties of the Z Transform will be discussed in this section. Digital filters are used for the discussion of the application of these techniques to obtain delay and interleaving pipeline implementations of DSP applications. However, the techniques are general in the sense that they can be applied to the Z Transform of any digital DSP application, including 2-D algorithms, in well defined steps yielding corresponding pipeline computational structures.

The transfer function of a *causal*, 1-D, digital filter is given by [28]

$$H(z) = \frac{G(z)}{F(z)} = \frac{\displaystyle\sum_{k=0}^{L} b(k)z^{-k}}{1 + \displaystyle\sum_{k=1}^{L} a(k)z^{-k}}. \tag{8.10}$$

If all of the $a(k) = 0.0$, then $H(z)$ represents an FIR digital filter. However, if any of the $a(k) \neq 0$, then $H(k)$ represents an IIR digital filter.

The transfer function in Eq. (8.10) has a variety of realization structures such as *direct I* and *II forms, cascade, parallel, lattice,* etc. [28]. However, these realization structures are not suitable for direct implementation in a pipeline computational array because they involve data-broadcasting and global interconnections. Therefore, they need to be transformed to pipelined structures that avoid these problems and at the same time have a high degree of modularity. The *interleaving technique* and the *delay technique* can be used to develop pipeline computational structures for digital filters [27]. These techniques are presented in the next two subsections.

8.13.1 INTERLEAVING TECHNIQUE

This technique is based upon replacing the unit-time delay z^{-1} by z^{-K}, where K is a positive integer ($K \geq 2$). Fig. 8.18(a) shows a block diagram of a digital filter where the input and output are related by

$$G(z) = H(z)F(z). \tag{8.11}$$

This relationship is still valid after the replacement as shown in Fig. 8.18(b) whose input and output are related by [27]

$$G(z^K) = H(z^K) F(z^K). \tag{8.12}$$

However, this replacement has an effect on both the input and output sequences and the digital filter transfer function.

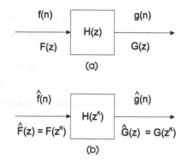

FIGURE 8.18

Block diagram for interleaving

The effect of the replacement on the input and output sequences can be demonstrated in the Z Transform domain. Consider a causal sequence $\{x(n)\}$ whose Z Transform is given by [28]

$$X(z) = \sum_{n=0}^{\infty} x(n)\, z^{-n}. \tag{8.13}$$

Assume $\{\hat{x}(n)\}$ is a new sequence that corresponds to $X(z^K)$. Then its Z Transform is given by

$$\hat{X}(z) = X(z^K) = \sum_{n=0}^{\infty} x(n)\, z^{-nK},$$

$$\hat{X}(z) = \sum_{n=0,K,2K,\dots} x\left(\frac{n}{K}\right) z^{-n} = \sum_{n=0}^{\infty} \hat{x}(n)\, z^{-n},$$

$$\hat{X}(z) = \sum_{n=0}^{\infty}\sum_{m=0}^{\infty} x\left(\frac{n}{K}\right) \delta(n - Km) z^{-n}. \tag{8.14}$$

Note that $x\left(\frac{n}{K}\right)$ is nonzero only when $\frac{n}{K}$ is an integer. It follows from the last equality in Eq. (8.14) that

$$\hat{x}(n) = \begin{cases} x\left(\frac{n}{K}\right) & \text{for } n = 0,\ K,\ 2K,\dots, \\ 0 & \text{elsewhere.} \end{cases} \tag{8.15}$$

Therefore, the new input and output sequences after the replacement, $\{\hat{f}(n)\}$ and $\{\hat{g}(n)\}$, are *zero-interleaved* sequences of $\{f(n)\}$ and $\{g(n)\}$ before the replacement, respectively. In addition, the number of interleaving zeros is equal to $(K-1)$. As an example, for $K = 3$

$$\{\hat{f}(n)\} = \{\, f(0),\ 0,\ 0,\ f(1),\ 0,\ 0,\ f(2),\ \dots\, \}. \tag{8.16}$$

The effect of this replacement on the digital filter is the introduction of extra delays in the transfer function that can be used for pipelining the computational structure. This can be achieved by moving the extra delays in the block diagram to branches where data-broadcasting and global connections need to be eliminated. The following examples illustrate the pipelining procedure by using the interleaving technique.

EXAMPLE 8.4
(Interleaving Example)

Fig. 8.19(a) shows a block diagram of a second-order FIR digital filter in the *transposed direct* realization [28]. This filter realization will be used in this example to illustrate the interleaving technique. This filter is widely used in many applications of signal processing. In addition, it is suitable for multiprocessor implementation since it involves inner-product computations that can be implemented efficiently by processors of the multiplier–adder type. However, the structure in Fig. 8.19(a) has the problem of broadcasting the input $f(n)$ to all of the multipliers. The pipelined structure can be obtained by first replacing z^{-1} by z^{-2}, $f(n)$ by $\hat{f}(n)$, and $g(n)$ by $\hat{g}(n)$ as shown in Fig. 8.19(b). Then, the extra delays can be relocated to eliminate the broadcasting and obtain the pipelined structure shown in Fig. 8.19(c). Notice that, except for the first stage, the pipelined structure is modular and can be extended to higher-order filters. In addition, the computational primitives are simple and only require local data communications.

End of the Example

8.13.2 DELAY TECHNIQUE

The delay technique is based on delaying the output (or input) of the filter by integer multiples of the unit-time delay. Then, the extra delays are distributed uniformly throughout the structure of the filter. Usually, the number of delays introduced is equal to the order of the filter. This procedure can be used to eliminate data-broadcasting and global connections resulting in a pipelined structure. The procedure is illustrated by the following examples.

EXAMPLE 8.5
(An FIR Filter Delay Sample)

Fig. 8.20(a) shows a block diagram of a second-order FIR filter in the *direct form*, also called a *transversal filter* [28].

The output has been delayed by introducing two delays, which is the same as the order of the filter. These extra delays will be used to eliminate the global

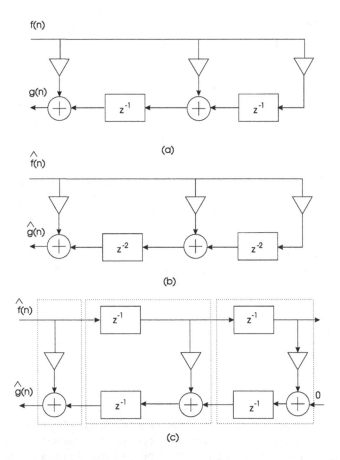

FIGURE 8.19

Transposed direct realization of an FIR digital filter

connections of the adder at the output. These delays can be moved backward and distributed to obtain the pipelined filter shown in Fig. 8.20(b).

End of the Example

EXAMPLE 8.6
(An IIR Filter Delay Sample)

Fig. 8.21(a) shows a second-order IIR filter in the *parallel form* [28]. This form is widely used because of its good roundoff noise performance and inherent par-

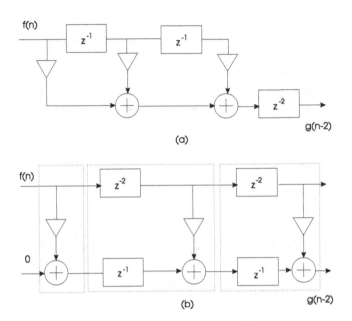

FIGURE 8.20

Direct realization of an FIR (transversal) filter

allelism. The output is delayed by introducing two delays (the number of delays is the same as the order of the filter). This example is an interesting one because it involves both data-broadcasting at the input and global interconnection to the adder at the output. Both of these can be eliminated simultaneously by distributing the additions and the extra delays throughout the structure. This results in the pipelined structure as shown in Fig. 8.21(b). The structure is modular and has local interconnections.

End of the Example

The delay technique can achieve speed-up that is proportional to the number of processors used, in contrast to the interleaving technique, since it does not involve zero-interleaving. However, it introduces a latency equal to the order of the filter. This is not a problem for most digital signal processing applications. However, the delay technique cannot be applied to all filter structures. In particular, it cannot eliminate data-broadcasting and global interconnections if they are involved in a feedback loop. Therefore, most IIR filters cannot be pipelined by using this technique.

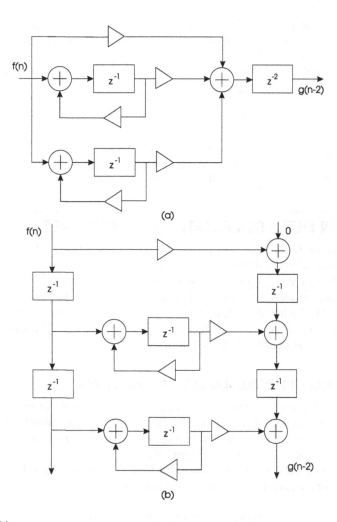

FIGURE 8.21

Parallel form for a second order IIR digital filter

The interleaving technique is applicable to all filter structures. However, it decreases the speed-up of the processing rate due to zero-interleaving of the input and output sequences. The speed-up can be improved by interleaving several input and output sequences. The delay technique has linear increase in speed-up. However, it introduces latency and is not applicable to all filter structures, especially IIR filters. Both techniques yield pipeline computational structure that can be implemented on processor arrays, such as systolic or wave-front arrays, or in software using iterative loops. In addition, they can be extended to 2-D digital filters and other 2-D DSP applications.

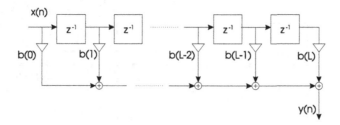

FIGURE 8.22

Block diagram for direct implementation of an FIR digital filter

8.14 FIR FILTER COMPUTATIONAL STRUCTURES

The application of the techniques presented in this chapter will now be considered for the implementation of FIR filters using computational arrays. The concepts presented for the FIR filter also apply to 1-D convolution and moving average (MA) filters. All initial conditions are set to zero for 1-D convolution but they may or may not be zero for a filter. In addition, the length of the output is the same as the length of the input for a filter which is different than the length of the output for 1-D convolution. FIR filters are sometimes called MA filters.

8.14.1 MULTIPLIER/ACCUMULATOR – FIR FILTER

The multiplier/accumulator computational primitive is very popular for digital signal processing applications. It can be derived from the iterative equation required to implement one of the direct forms for the FIR filter. Fig. 8.22 gives the block diagram for the direct realization of the FIR digital filter.

An iterative equation can be developed for this implementation by writing the equations for the output of the adder in the kth stage as follows:

$$s_k(n) = b(k)r_{k-1}(n) + s_{k-1}(n),$$
$$r_k(n) = r_{k-1}(n-1). \tag{8.17}$$

The inputs and outputs can then be defined as

$$s_{-1}(n) = 0.0,$$
$$r_{-1}(n) = x(n),$$
$$y(n) = s_L(n). \tag{8.18}$$

Fig. 8.23 gives a regular computational cell based upon the recursive algorithm in Eq. (8.17). A delay in a block diagram can be equated to a register or other memory in hardware. The assumption that clocked circuits will be used for all operations and that all operations being performed by combinational logic between registers will be

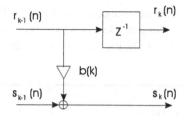

FIGURE 8.23

Computational cell for an FIR digital filter

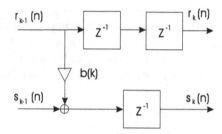

FIGURE 8.24

Modified computational cell for FIR filter

performed during a single clock cycle apply to this example. Since there is no delay between the input $s_{k-1}(n)$ and $s_k(n)$, the output is considered to be generated in less than one clock cycle. This causes a problem in implementing the cell in hardware if it is to be used as a regular cell in a computational array. This problem can fixed by using cutset retiming to add a delay in both branches of this computational cell. The resulting computational cell is given in Fig. 8.24 and the resulting recursive equations are given by

$$
\begin{align}
s_k(n) &= b(k)r_{k-1}(n-1) + s_{k-1}(n-1), & (8.19)\\
r_k(n) &= r_{k-1}(n-2),\\
s_{-1}(n) &= 0.0,\\
r_{-1}(n) &= x(n),\\
y(n) &= s_L(n). & (8.20)
\end{align}
$$

The initialization and data insertion equations have not been modified by this process. However, it takes longer to generate the output for a given input using the modified cells as compared to using the original cells. This means that the latency has been increased by this procedure. Thus, the broadcasting problem has been eliminated with the cost of additional latency.

A Matlab function to implement the equations for this cell and a function to implement an arbitrary FIR filter using this function have been provided in Appendix F.1. Example 8.7 uses these functions to illustrate the use of the pipelining techniques to implement the MA filter.

EXAMPLE 8.7
(Moving Average – FIR Filter)

A Matlab script called *firmac_cell.m* to implement the equations in Eq. (8.19) and a Matlab script for a function called *firmac_pipe.m* to use the *firmac_cell.m* to implement an arbitrary order FIR filter have been provided in Appendix F.1. The Matlab script for a function called *sampdata.m* to generate a sample sequence is given in Appendix A.1. This function can be used to generate a test input sequence to verify that the computational array implementation. The following Matlab script can be used to design a sixth order high pass FIR filter with a normalized pass band frequency of 0.341π and verify the operation of the computational array.

Matlab Script 8.1.

```
clear
order = 6;
wc = 0.341;
b = fir1(order, wc, 'high');
x = sampdata;
y = firmac_pipe(b, x);
```

End of the Script

Fig. 8.25 gives a stem plot of the input sequence generated by the *sampdata* function. Fig. 8.26 gives a stem plot of the output sequence from the *firmac_pipe* function. The pipeline fill up time for the array is equal to the number of filter coefficients.

The Matlab function *filter* can also be used to filter the input sequence with the same filter using the following Matlab script.

Matlab Script 8.2.

```
a = 1.0;
y2 = filter(b, a, x);
```

End of the Script

Fig. 8.27 gives a stem plot of the output from the *firmac_pipe* function after adjusting for the pipeline fill up time and from the Matlab *filter* function on the same plot. The outputs are the same except for the delay in the output from the *firmac_pipe* function associated with the pipeline fill up time.

End of the Example

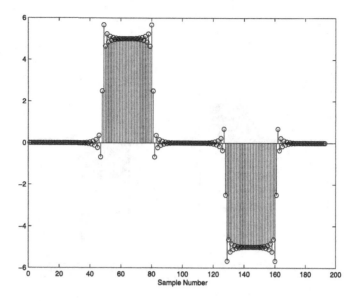

FIGURE 8.25

Stem plot of an input sample sequence

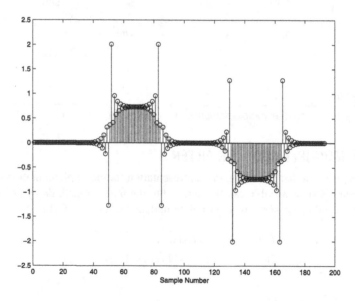

FIGURE 8.26

Stem plot of an output sequence for the cellular array for Example 8.7

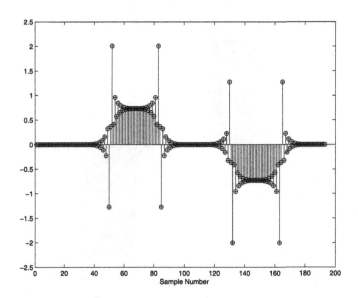

FIGURE 8.27

Stem plot of both output sample sequences

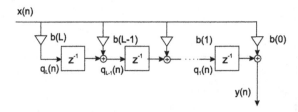

FIGURE 8.28

Block diagram for the direct implementation of an FIR digital filter

8.14.2 MULTIPLY/ADD – FIR FILTER

Fig. 8.28 gives another block diagram representation for the implementation of a 1-D FIR system. A state variable can be assigned to the input of each delay as indicated in Fig. 8.28 and equations can be written to update the states as follows:

$$
\begin{aligned}
q_L(n) &= b(L)x(n), \\
q_{L-1}(n) &= b(L-1)x(n) + q_L(n-1), \\
&\vdots \\
q_1(n) &= b(1)x(n) + q_2(n-1), \\
y(n) &= b(0)x(n) + q_1(n-1).
\end{aligned}
$$

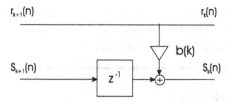

FIGURE 8.29

Computational cell for direct implementation of the FIR digital filter

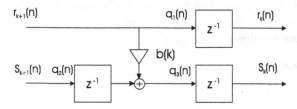

FIGURE 8.30

Modified cell for direct implementation of the FIR digital filter

It follows that a recursive algorithm can be derived to implement the filter as follows:

$$
\begin{aligned}
s_k(n) &= b(k)r_{k+1}(n) + s_{k+1}(n-1), \quad 0 \le k \le L, \\
r_k(n) &= r_{k+1}(n), \\
s_{L+1}(n) &= 0.0, \\
r_{L+1}(n) &= x(n), \\
y(n) &= s_0(n).
\end{aligned}
\tag{8.21}
$$

Fig. 8.29 gives a computational cell based upon Eq. (8.21). Note that there is no delay between the input $r_k(n+1)$ and the output $r_k(n)$. As previously mentioned, this will cause problems with hardware implementation if the cell is to be used in a computational array. This problem can be fixed by adding a delay in this output and doubling the delay for the input of $q_{k+1}(n)$. This does not affect the throughput but it does increase the latency. The modified cell is shown in Fig. 8.30.

State variables can be assigned to the inputs of the delays in the modified computational cell to obtain the following equations:

$$
\begin{aligned}
q_{1,k}(n) &= r_{k+1}(n), \\
q_{2,k}(n) &= s_{k+1}(n), \\
q_{3,k}(n) &= b(k)r_{k+1}(n) + q_{2,k}(n-1), \\
r_k(n) &= q_{1,k}(n-1), \\
s_k(n) &= q_{3,k}(n-1).
\end{aligned}
\tag{8.22}
$$

A Matlab script *firmadd_cell.m* for a function to implement the cell and a Matlab script *firmadd_pipe.m* to use the *firmadd_cell* function to implement an arbitrary order FIR filter have been provided in Appendix F.4.

EXAMPLE 8.8
(Multiply/Add – FIR Filter)

Example 8.8 illustrates the use of the *firmadd_cell.m* and the *firmadd_pipe.m* functions to filter an input sequence to verify that their performance. The same FIR filter and sample input sequence that we used in Example 8.7 will be used for this example as well. The following Matlab script can be used to verify the performance of these functions.

Matlab Script 8.3.

```
clear
order = 6;
wc = 0.341;
b = fir1(order, wc, 'high');
x = sampdata;
y = firmac_pipe(b, x);
```

End of the Script

Fig. 8.31 gives a stem plot of the output sequence from the *firmadd_pipe* function. The pipeline fill up time for the array is equal to the number of filter coefficients as was observed for Example 8.7. Fig. 8.32 gives a stem plot of the output from the *firmadd_pipe* function after adjusting for the pipeline fill up time and from the Matlab *filter* function on the same plot. Thus, the outputs are the same except for the delay associated with the pipeline fill up time.

End of the Example

8.15 INTERLEAVE FIR DIGITAL FILTER

Many applications require filtering of multiple sequences using the same filter. The interleaving technique as discussed in Sect. 8.13.1 can be used to develop a structure to filter more than one sequence at the same time using a single system. This concept will be illustrated for two different sequences in this section. Consider the direct form for the FIR filter as shown in Fig. 8.33. This is the same form that was used to develop the multiplier/accumulator computational structure except the input arc has been rotated 180 degrees. This filter will be used to illustrate a 2:1 interleave structure to filter two sequences simultaneously using one system.

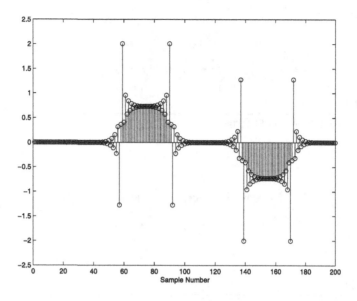

FIGURE 8.31

Stem plot of an output sequence for the cellular array for Example 8.8

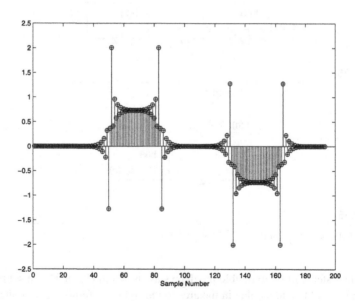

FIGURE 8.32

Stem plot of both output sample sequences

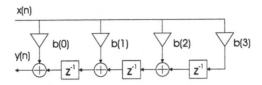

FIGURE 8.33

Direct form for a third order FIR filter

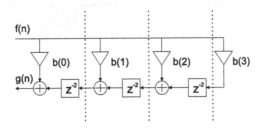

FIGURE 8.34

Direct form after doubling the delays

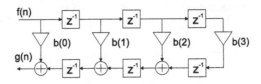

FIGURE 8.35

Block diagram after use of cutset retiming

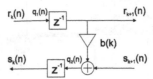

FIGURE 8.36

Computational cell for interleave FIR digital filter

The first step will be to double the delays as shown in Fig. 8.34. Cutset retiming can then be used to add a delay in the arcs to the right and remove one in the arcs to the left. The result is shown in Fig. 8.35. Fig. 8.36 shows a computational cell that can be used to implement an arbitrary order FIR digital filter using the result shown in Fig. 8.36.

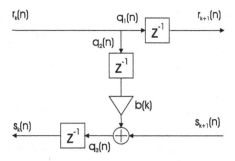

FIGURE 8.37

Computational cell for interleave FIR digital filter with assigned state variable

The multiplication operation takes longer than the addition operation if fixed point arithmetic is used to implement the filter. The delay at the top of the cell can be moved past the branch without affecting the equations for the cell. This delay can then be incorporated into the multiplier to permit two cycles rather than one to implement the multiplication. A multiplier that requires more than one cycle to implement the multiplication is typically called a pipeline multiplier. Providing the capability to use a pipeline multiplier to implement the computational cell can increase throughput compared to using a single cycle multiplier to implement the cell because a longer cycle time would be needed to accommodate the single cycle multiplier. The resulting computational cell is shown in Fig. 8.37.

A state variable can be assigned to the input of each delay in Fig. 8.37. The corresponding difference equations for the implementation of the modified cell follow:

$$
\begin{aligned}
q_1(n) &= r_k(n), \\
q_2(n) &= r_k(n), \\
q_3(n) &= b(k)q_2(n-1) + s_{k+1}(n), \\
r_{k+1}(n) &= q_1(n-1), \\
s_k(n) &= q_3(n-1).
\end{aligned}
\tag{8.23}
$$

The delays were doubled to obtain this computational cell. Thus, the typical procedure to use this cell would be to interleave the input sequence with zeros with the result that the output sequence would also be interleaved with zeros. However, a second input can be interleaved with the first input instead of interleaving with zeros. The output will then be two independent interleaved outputs corresponding to the two independent, interleaved inputs.

The same cell, modified with one additional delay, can be used for the input cell. Fig. 8.38 shows how to connect three cells to implement the third order filter in Fig. 8.33. The input $x(n)$ is connected to the input $r_1(n)$ for cell 0. The output $y(n)$ is the output $s_1(n)$ for cell 0. The input $s_4(n)$ for cell 3 is connected to ground and the

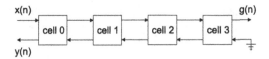

FIGURE 8.38

Implementation of a third order FIR filter using the FIR interleave array cell

output $g(n) = r_4(n)$ is not needed. This cell could be used in a computational array to implement any arbitrary order, FIR digital filter.

A Matlab script *firintlv_cell.m* for a function to implement the cell and a Matlab script *firintlv_pipe.m* to use the *firintlv_cell* function to implement an arbitrary order FIR filter have been provided in Appendix F.6. Example 8.9 provides an example of the use of these functions to filter two independent input sequences using one computational array through the use of the interleaving capability of the cell.

EXAMPLE 8.9

(An Interleave FIR Filter)

The same FIR filter used in Example 8.7 was used for this example. This example involves interleaving two independent inputs for the input sequence for the computational array. The input used for Example 8.7 was used for one of the input sequences. A second Matlab function to generate an input sequence has been provided in Appendix F.3. This function was used to generate the second input sequence, which was interleaved with the first sequence, to form the input sequence for the computational array. Fig. 8.39 gives stem plots for the two input sequences. Fig. 8.40 gives stem plots for the input sequences formed by interleaving the two independent input sequences.

The filtered output sequences can be obtained as every other sample of the output sequence from the *firintlv_pipe* function. This is shown in the following Matlab script.

Matlab Script 8.4.

```
f1 = sampdata;
f2 = sampdata2;
% Interleave f1 and f2 to obtain x
n1 = length(f1);
n2 = length(f2);
n3 = 2*max(n1, n2);
x = zeros(1, n3);
x(1, 1:2:end) = f1;
x(1, 2:2:end) = f2;
g1 = firintlv_pipe(b, x);
% Separate the two independent output sequences
```

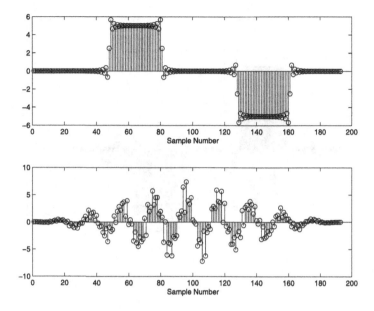

FIGURE 8.39

Stem plot of the two input sequences for the interleave cellular array for Example 8.9

```
y1 = g1(1, 2:2:end);
y2 = g1(1, 3:2:end);
```

End of the Script

Fig. 8.41 gives stem plot of the output sequence from the *firintlv_pipe* function.

Fig. 8.42 gives stem plots of the two individual output sequences after they have been separated.

Fig. 8.43 gives stem plots of the output sequences from the *firintlv_pipe* function and from the Matlab *filter* function on the same plot. It follows that the two independent input sequences were filtered independently through the use of the interleaving capability of this computational array. This concept can be extended to accommodate additional independent sequences by tripling, quadrupling, etc., the delays in the computational cell.

This structure has the advantage that there is no additional latency associated with using cutset retiming. However, the input sequences have to be interleaved. It there were only one input sequence, then it would have to be interleaved with zeros. However, this is not a disadvantage if two sequences need to filtered.

End of the Example

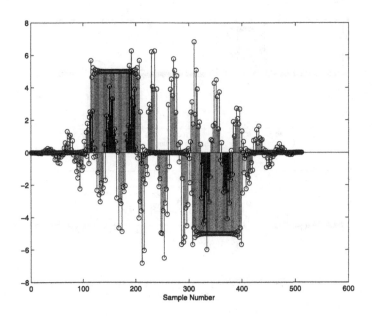

FIGURE 8.40

Stem plot of the input sequence formed by interleaving the two sequences

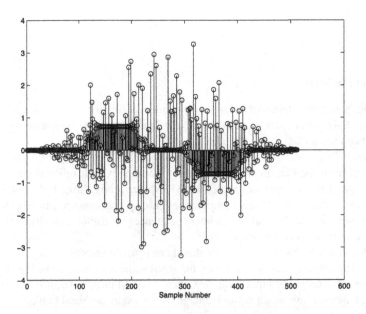

FIGURE 8.41

Stem plot of the output sequence for the interleave cellular array for Example 8.9

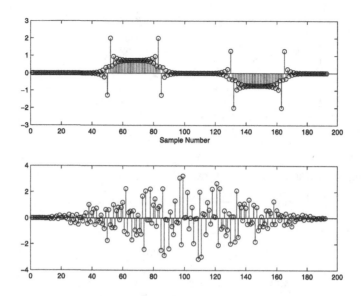

FIGURE 8.42

Stem plot of the individual output sequences for the interleave cellular array for Example 8.9

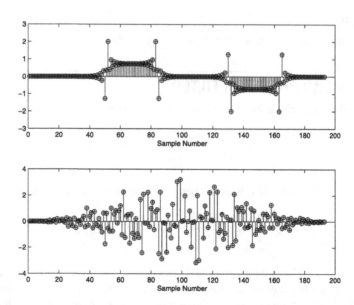

FIGURE 8.43

Stem plot of both output sample sequences

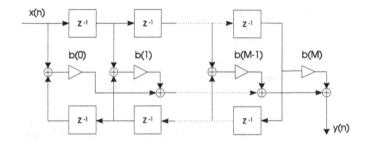

FIGURE 8.44

Block diagram for a linear phase FIR digital filter

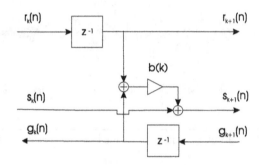

FIGURE 8.45

Regular computational cell for a linear phase FIR digital filter

8.16 LINEAR PHASE FIR FILTER

Consider the linear phase FIR filter with system transfer function given by

$$H(z) = \sum_{n=0}^{L} h(n)z^{-n} \qquad (8.24)$$

where $h(k) = h(L - k)$. If L is even, then $H(z)$ can be written in the form

$$H(z) = \sum_{n=0}^{M-1} h(n)\left[z^{-n} + z^{-(L-n)}\right] + h(M)z^{-M} \qquad (8.25)$$

where $M = L/2$. Fig. 8.44 gives a block diagram of this filter where the symmetry has been taken advantage of to reduce the number of multiplications as indicated in Eq. (8.25). Fig. 8.45 gives a block diagram of a regular computational cell for this filter. The following equations relate the inputs to outputs for this cell:

$$r_{k+1}(n) \quad = \quad r_k(n-1),$$

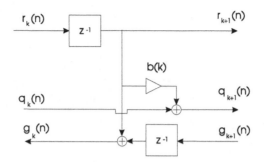

FIGURE 8.46

Output cell for the linear phase FIR filter

$$
\begin{aligned}
s_{k+1}(n) &= b(k)[r_k(n-1) + g_{k+1}(n-1)] + s_k(n), \\
g_k(n) &= g_{k+1}(n-1) .
\end{aligned}
\tag{8.26}
$$

Since coefficient $b(M)$ only multiplies one sample, an output cell that is different from the regular cell as shown in Fig. 8.46 can be used. The equations for this cell are given by:

$$
\begin{aligned}
r_{k+1}(n) &= r_k(n-1), \\
s_{k+1}(n) &= b(k)r_k(n-1) + s_k(n), \\
g_k(n) &= g_{k+1}(n-1) + r_k(n-1) .
\end{aligned}
\tag{8.27}
$$

Observe that there is a path between the input $q_k(n)$ and the output $q_{k+1}(n)$. This broadcast problem can solved through the use of the following steps:

1. Double all delays in the cell.
2. Use cut set retiming in the right side and add a delay in the arcs to the right while subtracting a delay from the arcs to the left.
3. Move one of the delays in the input arc for $r_k(n)$ past the branch.
4. Move the delay in the input arc for $g_{k+1}(n)$ past the branch.
5. Move the two delays as appropriate past the adder to the arc with the multiplier.

The modified cell is shown in Fig. 8.47.

The equations for the modified regular computational cell are given by:

$$
\begin{aligned}
r_{k+1}(n) &= r_k(n-3), \\
s_{k+1}(n) &= b(k)[r_k(n-3) + g_{k+1}(n-2)] + s_k(n-1), \\
g_k(n) &= g_{k+1}(n-1) .
\end{aligned}
\tag{8.28}
$$

State variables can be assigned to the individual delays as shown in Fig. 8.48. The corresponding equations are given by

$$
q_1(n) = r_k(n),
$$

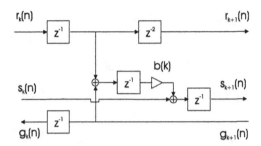

FIGURE 8.47

Modified regular computational cell for a linear phase FIR digital filter

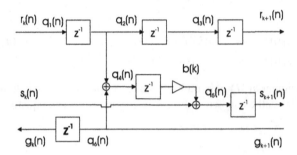

FIGURE 8.48

Modified computational cell for a linear phase FIR digital filter showing assignment of state variables

$$
\begin{aligned}
q_2(n) &= q_1(n-1), \\
q_3(n) &= q_2(n-1), \\
q_4(n) &= q_1(n-1) + g_{k+1}(n), \\
q_5(n) &= b(k)q_4(n-1) + q_5(n), \\
r_{k+1}(n) &= q_3(n-1), \\
s_{k+1}(n) &= q_5(n-1), \\
g_k(n) &= q_5(n-1).
\end{aligned}
\tag{8.29}
$$

The same steps can be used to modify the output cell. The modified output cell is shown in Fig. 8.49.

The equations for the modified output computational cell are given by:

$$
\begin{aligned}
r_{k+1}(n) &= r_k(n-3), \\
s_{k+1}(n) &= b(k)r_k(n-3) + s_k(n-1), \\
g_k(n) &= g_{k+1}(n-1) + r_k(n-2).
\end{aligned}
\tag{8.30}
$$

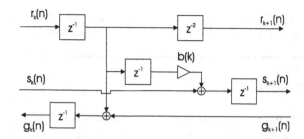

FIGURE 8.49

Modified output cell for the linear phase FIR digital filter

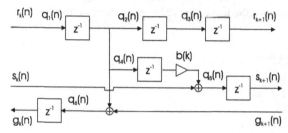

FIGURE 8.50

Modified output cell for the linear phase FIR digital filter showing the assigned state variables

State variables can be assigned to the individual delays in the modified output cell as shown in Fig. 8.50.

The corresponding equations are given by

$$
\begin{aligned}
q_1(n) &= r_k(n), \\
q_2(n) &= q_1(n-1), \\
q_3(n) &= q_2(n-1), \\
q_4(n) &= q_1(n-1), \\
q_5(n) &= b(k)q_4(n-1) + s_k(n), \\
q_6(n) &= q_1(n-1) + g_{k+1}(n), \\
r_{k+1}(n) &= q_3(n-1), \\
s_{k+1}(n) &= q_5(n-1), \\
g_k(n) &= q_6(n-1).
\end{aligned} \tag{8.31}
$$

A Matlab script *firlin_rcell.m* for a function to implement the regular cell, a Matlab script *firlin_ecell.m* for a function to implement the end cell, and a Matlab script *firln_pipe.m* to use the *firlin_rcell* function and the *firlin_ecell* function to implement an arbitrary order FIR filter have been provided in Appendix F.8. Example 8.10 provides an example of the use of these functions to filter an input sequence.

EXAMPLE 8.10
(A Linear Phase FIR Filter)

The same FIR filter and sample input sequence that was used in Example 8.7 can be used for this example as well. However, the input sequence needs to be interleaved with zeros, or another independent sequence, because the delays were doubled to develop the computational cells. The output can be retrieved as every other sample of the output sequence from the *firlin_pipe* function. The following Matlab script shows the use of the *firlin_pipe* function to filter the input sequence as described.

Matlab Script 8.5.

```
f = sampdata;
% Interleave f to obtain x
n1 = length(f);
n2 = 2*n1;
x = zeros(1, n2);
x(1, 1:2:end) = f;
g1 = firlin_pipe(b, x);
y1 = g1(1, 1:2:end);
```

End of the Script

Fig. 8.51 gives a stem plot of the output sequence from the *firlin_pipe* function. The pipeline fill-up time for the array is equal to one half of the number of filter coefficients plus the two additional delays used to obtain regular computational cells. Fig. 8.52 gives a stem plot of the output from the *firlin_pipe* function after adjusting for the pipeline fill up time and from the Matlab *filter* function on the same plot. Observe that the outputs are the same except for the delay associated with the pipeline fill-up time.

End of the Example

8.17 SECOND ORDER IIR FILTER SECTION

IIR filters are typically implemented using second order sections to reduce error due to rounding, speed up the processing, and improve stability. Therefore, the concepts presented in this section are based upon the assumption that an IIR filter will be implemented using a cascade of second order sections. Second order sections are realized by factoring the transfer function for the IIR filter and combining poles and zeros to obtain second order sections. Since the transfer function for a linear shift

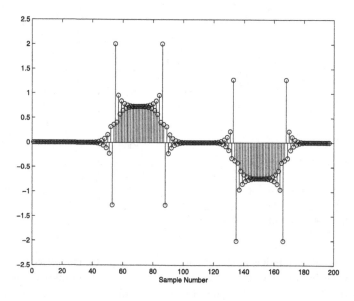

FIGURE 8.51

Stem plot of the output sequence for the cellular array for Example 8.10

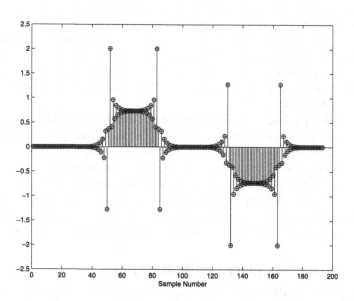

FIGURE 8.52

Stem plot of both output sample sequences

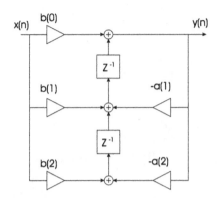

FIGURE 8.53

Second order transpose direct IIR filter section

invariant digital system has constant coefficients, the poles and zeros are either real or they occur in complex conjugate pairs. Real roots and complex conjugate pairs of roots should be combined in pairs for the numerator and for the denominator to obtain second order filter sections as appropriate. On the other hand, if the poles and zeros are real, the filter can be implemented using first order filter sections. It is common to realize a higher-order IIR filter by using a cascade of second order sections. However, the higher precision of modern digital signal processors makes it practical to implement fourth-order direct form sections for many applications [29].

This section covers the use of interleaving to obtain a pipeline structure for the transposed direct form and for the direct form II second order sections. Typically, a delay is used between each second order section in an actual implementation to limit data broadcasting between the sections. Thus, both interleaving and delay may be used in actual practice.

In general, a pipeline implementation can achieve $O(P)$ speed-up in the processing rate compared to a single processor implementation, where P is the number of processors in the pipeline. However, the interleaving technique yields P/K speed-up due to zero-interleaving of the input and output sequences by $(K - 1)$ zeros. The speed-up can be brought back to P if K independent sequences need to be processed by the same filter. This can be achieved by interleaving the sequences instead of zero-interleaving each individual sequence [17]. It should be mentioned that array structures of other filter realizations can be obtained in a similar manner [17,30,31].

8.17.1 TRANSPOSE DIRECT IIR REALIZATION

The transpose direct IIR realization for a second order IIR filter is shown in Fig. 8.53. A second order section for use in a pipeline implementation of an arbitrary order IIR filter, based upon this second order realization, can be developed by using interleaving and cutset retiming. The procedure to develop this realization should begin

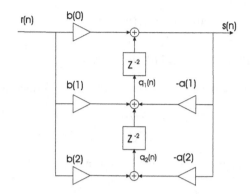

FIGURE 8.54

Second order transpose direct IIR filter stage with delays doubled

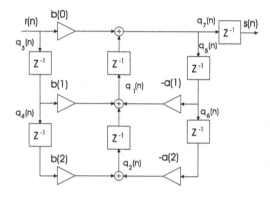

FIGURE 8.55

Modified computational cell for an IIR second order stage

by doubling the delays in the second order section to obtain $H_i(z)$ as shown in Fig. 8.54.

Cutset retiming can be used to obtain the interleaved computational structure given in Fig. 8.55. A delay has been at the output to avoid having a path between a primary input and a primary output without a delay. This will cause a delay of one sampling interval for each second order section.

State variables can be assigned to the input of the delay elements to obtain

$$
\begin{aligned}
s(n) &= q_7(n-1), \\
q_1(n) &= q_2(n-1) + b(1)q_3(n-1) - a(1)q_5(n-1), \\
q_2(n) &= b(2)q_4(n-1) - a(2)q_6(n-1), \\
q_3(n) &= r(n), \\
q_4(n) &= q_3(n-1),
\end{aligned}
$$

$$
\begin{aligned}
q_5(n) &= b(0)r(n) + q_1(n-1), \\
q_6(n) &= q_5(n-1), \\
q_7(n) &= q_5(n) = b(0)r(n) + q_1(n-1).
\end{aligned}
\tag{8.32}
$$

These equations can be written in the state space matrix form as follows:

$$
\begin{aligned}
Q(n) &= \mathbf{A}Q(n-1) + \mathbf{B}r(n), \\
s(n) &= \mathbf{C}Q(n-1) + \mathbf{D}r(n).
\end{aligned}
\tag{8.33}
$$

The corresponding matrices and vectors are given by

$$
Q(n) = \begin{bmatrix} q_1(n) \\ q_2(n) \\ q_3(n) \\ q_4(n) \\ q_5(n) \\ q_6(n) \\ q_7(n) \end{bmatrix},
$$

$$
\mathbf{A} = \begin{bmatrix}
0 & 1 & b(1) & 0 & -a(1) & 0 & 0 \\
0 & 0 & 0 & b(2) & 0 & -a(2) & 0 \\
0 & 0 & 0 & 0 & 0 & 0 & 0 \\
0 & 0 & 1 & 0 & 0 & 0 & 0 \\
1 & 0 & 0 & 0 & 0 & 0 & 0 \\
0 & 0 & 0 & 0 & 1 & 0 & 0 \\
1 & 0 & 0 & 0 & 0 & 0 & 0
\end{bmatrix},
$$

$$
\begin{aligned}
\mathbf{B}^{\mathrm{T}} &= [0 \ \ 0 \ \ 1 \ \ 0 \ \ b(0) \ \ 0 \ \ b(0)], \\
\mathbf{C} &= [0 \ \ 0 \ \ 0 \ \ 0 \ \ 0 \ \ 0 \ \ 1], \\
\mathbf{D} &= [0].
\end{aligned}
\tag{8.34}
$$

8.17.2 DIRECT FORM II REALIZATION

Fig. 8.56 shows a block diagram of a typical second order section of an IIR filter using a direct form II realization. The computational structure can be modified by doubling the delays and using cut set retiming to redistribute the delays. Fig. 8.57 shows a block diagram of this second order section after replacing z^{-1} by z^{-2}. Fig. 8.58 shows the result after using cut set retiming to redistribute the delays. A delay has also been added at the output so there would not be a path between a primary input and a primary output without a delay. This will cause a delay of one sampling interval for each second order section.

State variables can be assigned to the inputs of the delays and equations can be written for the second order section in state space form. The state equations are

$$
q_1(n) = r(n) + q_3(n-1),
$$

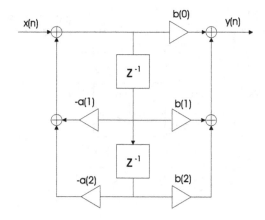

FIGURE 8.56

Direct form II realization of a second order IIR digital filter

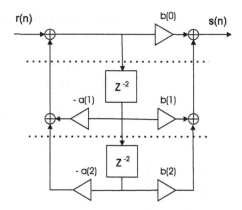

FIGURE 8.57

Direct form II realization of a second order IIR digital filter with delays doubled

$$
\begin{aligned}
q_2(n) &= q_1(n-1), \\
q_3(n) &= -a(1)q_1(n-1) + q_4(n-1), \\
q_4(n) &= -a(2)q_2(n-1), \\
q_5(n) &= b(1)q_1(n-1) + q_6(n-1), \\
q_6(n) &= b(2)q_2(n-1), \\
q_7(n) &= b(0)q_3(n-1) + q_5(n-1) + b(0)r(n). \qquad (8.35)
\end{aligned}
$$

The output equation is

$$
s(n) = q_7(n-1). \qquad (8.36)
$$

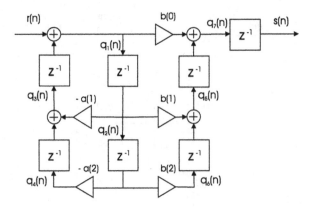

FIGURE 8.58

Modified direct form II realization of a second order IIR digital

The equations for the computational cell can be written in the matrix form of the state space representation as follows:

$$\begin{aligned}
Q(n) &= \mathbf{A}Q(n-1) + \mathbf{B}r(n), \\
s(n) &= \mathbf{C}Q(n-1) + \mathbf{D}r(n).
\end{aligned} \quad (8.37)$$

The definitions for the corresponding matrices and vectors for this representation follow:

$$Q(n) = \begin{bmatrix} q_1(n) \\ q_2(n) \\ q_3(n) \\ q_4(n) \\ q_5(n) \\ q_6(n) \\ q_7(n) \end{bmatrix}, \quad (8.38)$$

$$\mathbf{A} = \begin{bmatrix}
0 & 0 & 1 & 0 & 0 & 0 & 0 \\
1 & 0 & 0 & 0 & 0 & 0 & 0 \\
-a(1) & 0 & 0 & 1 & 0 & 0 & 0 \\
0 & -a(2) & 0 & 0 & 0 & 0 & 0 \\
b(1) & 0 & 0 & 0 & 0 & 1 & 0 \\
0 & b(2) & 0 & 0 & 0 & 0 & 0 \\
0 & 0 & b(0) & 0 & 1 & 0 & 0
\end{bmatrix},$$

$$\begin{aligned}
\mathbf{B}^T &= [1 \ 0 \ 0 \ 0 \ 0 \ 0 \ b(0)], \\
\mathbf{C} &= [0 \ 0 \ 0 \ 0 \ 0 \ 0 \ 1], \\
\mathbf{D} &= [0].
\end{aligned} \quad (8.39)$$

8.18 BLOCK IMPLEMENTATION OF IIR DIGITAL FILTERS

The IIR digital filter involves the use of previous output values as well as the current and previous inputs to compute a new output value. This requirement limits the use of parallelism to increase the throughput of the hardware used to implement the filter. In addition, many hardware systems can practically be implemented using a 32 or 64 bit data path. On the other hand, many IIR filter applications can be implemented with sufficiently adequate precision using a 16 bit word for the filter coefficients and the input values. Fortunately, the IIR filter computational structure can be modified to overcome the limitation imposed by using previous values of the output and simultaneously permit the use of multiple input values in parallel. The block implementation of the IIR filter is discussed in this section as an approach to increase the potential use of parallelism for the implementation of IIR digital filters.

The state space representation of a generic discrete time system will be used to develop the procedure for block implementation of IIR digital filters. Specific examples on how to use this approach to implement IIR digital filters will then be given. Consider the general state space model as given by

$$\begin{aligned} Q(n) &= AQ(n-1) + Bx(n), \\ y(n) &= CQ(n-1) + Dx(n). \end{aligned} \tag{8.40}$$

This equation represents a system with a single input data stream $x(n)$ and a single output data stream $y(n)$. The vector $Q(n-1)$ includes the most recent values of the states which is related to the feedback of previous values of the output for an IIR digital filter.

Equation (8.40) can be implemented recursively with different values of the sample index n to develop a block implementation of the filter as follows:

$$Q(n-1) = AQ(n-2) + Bx(n-1). \tag{8.41}$$

Substituting this into Eq. (8.40) for $Q(n)$ results in the following equations:

$$\begin{aligned} Q(n) &= A[AQ(n-2) + Bx(n-1)] + Bx(n), \\ y(n) &= C[AQ(n-2) + Bx(n-1)] + Dx(n). \end{aligned} \tag{8.42}$$

The output equation for index $n-1$ can be written as

$$y(n-1) = CQ(n-2) + Dx(n-1). \tag{8.43}$$

The combined equations can be written as follows:

$$\begin{aligned} Q(n) &= A^2 Q(n-2) + Bx(n) + ABx(n-1), \\ y(n) &= CAQ(n-2) + Dx(n) + CBx(n-1), \\ y(n-1) &= CQ(n-2) + Dx(n-1). \end{aligned} \tag{8.44}$$

Note that Eq. (8.44) now represents a system with two inputs, $x(n)$ and $x(n-1)$, and two corresponding outputs, $y(n)$ and $y(n-1)$. These equations can also be written as

$$Q(n) = A^2 Q(n-2) + \begin{bmatrix} B & AB \end{bmatrix} \begin{bmatrix} x(n) \\ x(n-1) \end{bmatrix},$$

$$\begin{bmatrix} y(n) \\ y(n-1) \end{bmatrix} = \begin{bmatrix} CA \\ C \end{bmatrix} Q(n-2) + \begin{bmatrix} D & CB \\ 0 & D \end{bmatrix} \begin{bmatrix} x(n) \\ x(n-1) \end{bmatrix}.$$

$$(8.45)$$

The following matrices can be defined to simplify the representation of Eq. (8.45).

$$\widehat{A} = A^2,$$
$$\widehat{B} = \begin{bmatrix} B & AB \end{bmatrix},$$
$$\widehat{C} = \begin{bmatrix} CA \\ C \end{bmatrix},$$
$$\widehat{D} = \begin{bmatrix} D & CB \\ 0 & D \end{bmatrix},$$
$$X(n) = \begin{bmatrix} x(n) \\ x(n-1) \end{bmatrix},$$
$$Y(n) = \begin{bmatrix} y(n) \\ y(n-1) \end{bmatrix}. \qquad (8.46)$$

The corresponding block state representation follows:

$$Q(n) = \widehat{A} Q(n-2) + \widehat{B} X(n),$$
$$Y(n) = \widehat{C} Q(n-2) + \widehat{D} X(n). \qquad (8.47)$$

This system uses two inputs in the column vector $X(n)$ to produce two outputs in the column vector $Y(n)$. In addition, the inputs are needed at one half the original rate.

This process can be continued to obtain a system with any arbitrary number of inputs and corresponding outputs. The computations for each set of inputs can be computed in parallel because the data dependency problem has been modified by using multiple inputs to compute multiple outputs.

The block state matrices for a given number of inputs, M, can be computed as

$$Q(n) = \widehat{A} Q(n-M) + \widehat{B} X(n),$$
$$Y(n) = \widehat{C} Q(n-M) + \widehat{D} X(n). \qquad (8.48)$$

The corresponding matrices and vectors for this representation are given by

$$X(n) = \begin{bmatrix} x(n) \\ x(n-1) \\ \vdots \\ x(n-M-1) \end{bmatrix}, \tag{8.49}$$

$$Y(n) = \begin{bmatrix} y(n) \\ y(n-1) \\ \vdots \\ y(n-M-1) \end{bmatrix}, \tag{8.50}$$

$$\begin{aligned} \widehat{A} &= A^M, \\ \widehat{B} &= \sum_{m=0}^{M-1} A^m B, \\ \widehat{C} &= \sum_{m=0}^{M-1} CA^{M-1-m}, \end{aligned} \tag{8.51}$$

$$\widehat{D} = \begin{bmatrix} D & CB & CAB & \cdots & CA^{M-2}B \\ 0 & D & CB & \cdots & CA^{M-3}B \\ \vdots & \vdots & \vdots & \ddots & \vdots \\ 0 & 0 & \cdots & D & CB \\ 0 & 0 & \cdots & 0 & D \end{bmatrix}. \tag{8.52}$$

Note that $\widehat{D}$ is an upper triangular matrix. This can be shown through the development of the corresponding matrices and vectors for a four input, four output block state implementation as follows:

$$\begin{aligned} \widehat{A} &= A^4, \\ \widehat{B} &= \begin{bmatrix} B & AB & A^2B & A^3B \end{bmatrix}, \\ \widehat{C} &= \begin{bmatrix} CA^3 \\ CA^2 \\ CA \\ C \end{bmatrix}, \end{aligned}$$

$$\hat{D} = \begin{bmatrix} D & CB & CAB & CA^2B \\ 0 & D & CB & CAB \\ 0 & 0 & D & CB \\ 0 & 0 & 0 & D \end{bmatrix},$$

$$X(n) = \begin{bmatrix} x(n) \\ x(n-1) \\ x(n-2) \\ x(n-3) \end{bmatrix},$$

$$Y(n) = \begin{bmatrix} y(n) \\ y(n-1) \\ y(n-2) \\ y(n-3) \end{bmatrix}. \tag{8.53}$$

The block state model, for implementing this four input, four output system, is the same as given in Eq. (8.48) for $M = 4$. Four inputs are used simultaneously and four outputs are produced for each iteration. It follows that a hardware implementation of this modified system can be developed by designing a system to implement these equations in parallel. The system design can also be parameterized based upon the sizes of the corresponding matrices and vectors in the equations. The block state implementation also has the advantage that the data requirements have been localized to only include the current inputs and the most recent values of the state variables.

8.19 MATRIX–VECTOR MULTIPLICATION

The concepts and techniques discussed in this chapter for the development of computational arrays can be extended to more complex problems. Consider the single assignment code algorithm for matrix–vector multiplication as follows [17]:

```
for (i=0; i<IMAX; i++)
    {
    c[i][0] = 0.0;
    for(j=0; j<JMAX; j++)
        {
        b[i+1][j] = b[i][j];
        c[i][j+1] = c[i][j] + a[i][j]*b[i][j];
        }
    }
```

The index i represents the column address and the index j represents the row address. Note that the input, $b[i][j]$, is broadcast to the next column and the output, $c[i][j + 1]$, is computed for the next row. Thus, this representation of the matrix–vector multiplication can be represented as a two-dimensional cellular array. Fig. 8.59 gives a typical computational cell for this array. There are no delays in

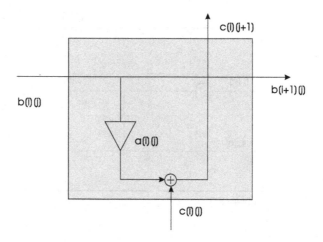

FIGURE 8.59

Typical cell for matrix–vector multiplication

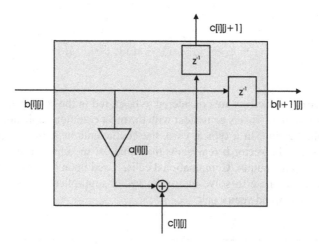

FIGURE 8.60

Modified cell for the matrix–vector multiplication

this cell as given in Fig. 8.59. Cutset retiming can be used to eliminate broadcasting so the cell can be implemented in a computational cellular array. This can be accomplished by adding a delay in either both the input branches or both of the output branches. Fig. 8.60 gives the modified computational cell with delays added in the output branches.

The modified algorithm for this computational cell is given by

```
b[0][0][0] = 0.0;
c[0][0][0] = 0.0;
```

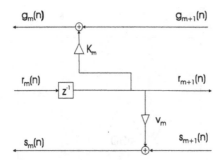

FIGURE 8.61

Computational cell for Problem 8.1

```
n = 1;
for (i=0; i<IMAX; i++)
    {
    c[i][0][n] = 0.0;
    for(j=0; j<JMAX; j++)
        {
        b[i+1][j][n] = b[i][j][n-1];
        c[i][j+1][n] = c[i][j][n-1] + a[i][j]*b[i][j][n-1];
        }
    }
```

The matrix coefficients are considered to be stored in the computational cells for this implementation. This is consistent with the most common application of matrix–vector multiplication. In a typical case, the coefficients in **A** represent the system coefficients and the vector **b** represents the values of the input. Therefore, the vector **c** represents the output. Computational cells, based upon Fig. 8.60 can be sacked together as appropriate to solve matrix by vector multiplication problems for any compatible matrix and vector pair.

8.20 PROBLEMS FOR CHAPTER 8

Problem 8.1. (Cutset Retiming)
Fig. 8.61 gives a computational cell that can be used to implement a digital filter.

1. Use cutset retiming and/or other procedures as appropriate:
   (a) To add or remove delays as appropriate to modify your computational cell to ensure that there are no paths from a primary input to a primary output.
   (b) To have a delay associated with (next to) each of the multipliers. The combined delay and multiplier can be implemented as a pipeline multiplier, and
   (c) To have a delay at each primary output ($g_m(n)$, $r_{m+1}(n)$, and $s_m(n)$).
2. Write an appropriate set of equations to implement your computational cell.

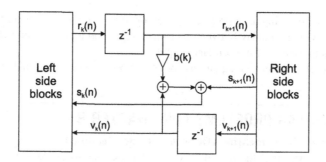

FIGURE 8.62

Computational cell for Problem 8.2

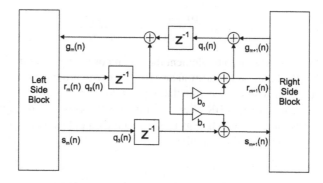

FIGURE 8.63

Block Diagram for Problem 8.3

Problem 8.2. (Cutset Retiming)

Fig. 8.62 gives a computational cell for a computational array.

1. Use cutset retiming and/or other procedures as appropriate to add or remove delays as appropriate:
   (a) To modify this computational cell to ensure that there are no paths from a primary input to a primary output without a delay.
   (b) To have a delay associated with (next to) the multiplier. The combined delay and multiplier can be implemented as a pipeline multiplier, and
   (c) To have a delay at each primary output ($r_{k+1}(n)$, $s_k(n)$, and $v_k(n)$).
2. Write an appropriate set of equations to implement your computational cell.

Problem 8.3. (Cutset Retiming)

Fig. 8.63 gives a representation of the block diagram for a computational cell to be implemented in a cellular array.

1. Write a set of difference equations for the cell as given in Fig. 8.63.

516 **CHAPTER 8** Digital Signal Processing Systems Design

2. Use cutset retiming as appropriate to add or remove delays to ensure that there is no path without a delay between a primary input and a primary output and that a delay is in each arc for a primary output.
3. Write a set of difference equations for your modified computational cell.

8.21 MATLAB PROBLEMS FOR CHAPTER 8

Problem 8.4. (Implementing IIR Second Order Sections)
This problem involves the use of second order sections to implement an IIR filter. The idea is to implement any arbitrary order IIR filter using a cascade of second order sections with the output of one section being the input to the next section. The system transfer function for a second order IIR filter is given by

$$H(z) = \frac{b(0) + b(1)z^{-1} + b(2)z^{-2}}{1.0 + a(1)z^{-1} + a(2)z^{-2}}. \tag{8.54}$$

1. Write a Matlab function to implement the second order IIR filter in Eq. (8.54).
2. Write a Matlab function to partition any IIR filter with arbitrary order into second order sections and call the function that you wrote for part 1 of Problem 8.4 to implement an arbitrary order IIR filter. The filter coefficients and the input sequence should be passed to the function as inputs and the output should be returned. The call to the function should be similar to

```
y = your_IIRSOS(b, a, x);
```

Use the following specifications to design a 13th order IIR filter using the Matlab *ellip* function.

```
order = 13;
passband ripple = 1 dB
stopband ripple = 40 dB
relative cutoff frequency = 0.618
filter type = lowpass
```

Write a Matlab function to use the functions from parts 1 and 2 of Problem 8.4 to implement this IIR filter using second order sections. Use the sample input sequence from the *sampdata* function in Appendix A.1 to verify your function. Compare your results with the use of the Matlab *filter* function.
To be turned in:
(a) The Matlab code for your function to implement a second order IIR filter section.
(b) The Matlab code for your function to partition an arbitrary order IIR filter into second order sections and implement the filter using the second order sections.
(c) Stem plots of the output from your function and the output from the Matlab *filter* function.
(d) A comparison of the two outputs.

Problem 8.5. (Implementing IIR Filters using a Cellular Array)
This problem involves the implementation of an arbitrary order IIR filter using an array of computational cells for second order sections.

1. Write a Matlab function to implement the second order IIR filter section using the transposed direct form as shown in Fig. 8.55. Write the function so that it will take one input sample $r(n)$ as input and generate the corresponding output $s(n)$ for that input sample.

2. Write a Matlab function to implement a general order IIR filter using your function from part 1 of Problem 8.5 to implement the filter using second order sections. For example, if the order of the filer is N, then your filter function should call your function $\lceil \frac{N}{2} \rceil$ times per input sample for the implementation. Use an outer *for* loop to access the input sequence one sample at a time.

3. Use the following specifications to design an 11th order IIR filter using the Matlab *ellip* function.

```
order = 11;
passband ripple = 1 dB
stopband ripple = 40 dB
relative cutoff frequency = 0.418
filter type = lowpass
```

Use this filter to test your Matlab functions. Create an interleaved input using input sample sequences from the Matlab functions *sampdata* in Appendix A.1 and *sampdata2* in Appendix F.3 as the sample inputs.

4. Adjust your outputs to account for any delays introduced by the retiming procedure and compare your outputs to the corresponding output from the Matlab *filter* function.

Hardware Implementation

9

9.1 INTRODUCTION

The speed and complexity of digital circuits has increased rapidly. As electronic design automation (EDA) tools have been developed to take care of the low level implementation details, designers have responded by designing at higher levels of abstraction. Electronic applications demand ever increasing flexibility and performance while at the same time demands are increasing for a decrease in power consumption. Parallel efforts to reduce design cost and increase the life of products have placed emphasis on configurability and programmability [32]. Thus, pressure has increased to integrate analog circuits, memory, digital logic, and soft programmable units onto a single system chip.

The need for EDA software tools to design at higher levels of abstraction has clearly been recognized as the complexity of digital systems has continued to increase. System level EDA tools such as System Verilog and SystemC have been developed with cooperation by several companies to meet the need for system level EDA tools.

It is also possible to design at the register transfer level (RTL) using the SystemC language, and many systems have been designed using this approach. This approach has the advantage of making it easier to mix RTL descriptions of a part of a complex system with high level behavioral level descriptions of other parts of the system so that modeling and simulation can be done using a mixed level model. System Verilog and VHDL also permit modeling at the behavioral level. With either approach, the mapping of a given behavioral level model to an equivalent RTL level model using a hardware description language (HDL) is not trivial, and usually there are many different solutions to this problem for a particular application. The design process usually involves evaluating the performance of these possible solutions. This chapter presents a conceptual approach to this problem using examples of the modeling and performance evaluation of DSP applications.

9.2 PIPELINE FIR FILTER IMPLEMENTATION

The interleave procedure to develop a computational cell for a transposed direct realization of an FIR digital filter to be used in a cellular array was presented in Chap. 8.

Digital Signal Processing. DOI: 10.1016/B978-0-12-804547-3.00009-7

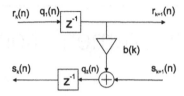

FIGURE 9.1

Computational cell for a transposed direct FIR digital filter

The computational cell for this filter is represented in Fig. 8.36 and repeated here as Fig. 9.1 for convenience.

The equations for this cell follow:

$$
\begin{aligned}
q_1(n) &= r_k, \\
q_2(n) &= b(k)q_1(n-1) + v_k, \\
u_k(n) &= q_1(n-1), \\
s_k(n) &= q_2(n-1).
\end{aligned}
\tag{9.1}
$$

This cell will be used as an example to discuss the procedures for implementing DSP applications.

The goal is to develop a design for the implementation of FIR filters using this cell and then develop a Verilog RTL description for the hardware implementation of the filter. This hardware implementation will use fixed point arithmetic. Thus, roundoff errors need to be considered as a part of the design process. It is also important to verify the correctness of the design and estimate the expected performance of the hardware implementation. The following procedure will be used to accomplish this goal:

1. Develop a floating point model for the algorithm using Matlab (or SystemC). This model should use the same architecture to be used for the hardware implementation.

2. Develop a fixed point arithmetic model for the algorithm using Matlab (or SystemC). This model should give the same results as expected from the hardware implementation. Thus, it can be used to generate test sequences to test the hardware.

3. Develop a SystemC or Verilog behavioral model for the hardware implementation at the RTL level. This model will use two's complement fixed point arithmetic and provide realistic modeling for the timing of the hardware implementation as well.

9.2.1 THE MATLAB FLOATING POINT FIR FILTER MODEL

The Matlab floating point model for the transposed direct FIR digital filter along with the floating point model for the computational cell is given in Appendix G.1.

The current values (output state variables) of the state variables for the cell are stored in the array *qn* for the design of the function to implement the computational cell. The previous values of the state variables are stored in the array *qnm*1. The primary inputs are *rk* and *skp*1. The primary outputs are *sk* and *rkp*1. The filter coefficient for the cell is *bk*.

An arbitrary order FIR filter with interleaving of the inputs can be implemented by connecting these cells in a cellular array. The input to the array will be the *r*1 input to the first (left most) cell. The output will be the *s*1 for the first (left most) cell. The number of cells to be used is the same as the number of coefficients for the filter. Thus, if the number of filter coefficients is *m*2, then the *skp*1 input for the last (right most) cell should be set to zero. The *rkp*1 output is not needed. The cellular array can be used to filter two inputs by interleaving the inputs to form a single input to the array. The corresponding outputs can be obtained by reversing the interleaving. The Matlab function *fir3array* to use the *fir3cell* function to implement an arbitrary FIR digital filter with interleaved inputs is given in Appendix G.2.

The number of cells needed for the implementation has been determined to be the same as the number of filter coefficients. The state variables are stored in a register array *reg* for convenience, and the connections to the cells are made by assigning the inputs and outputs from the *fir3cell* function to the appropriate elements of the *reg* array. This is equivalent to assigning variable names to make the connections in Verilog. The output from this model should be essentially the same as the output from the Matlab *filter* function with an assignment of $a = 1$ for the denominator coefficients.

The Matlab function *sampdata*, which has been provided in Appendix A.1, can be used to obtain one of the inputs and the Matlab function *sampdata2*, which has been provided in Appendix F.3, can be used to obtain the other input. Fig. 9.2 gives a stem plot of sample sequence 1, and Fig. 9.3 gives a stem plot of sample sequence 2.

The Matlab script to interleave the two inputs and test the computational array using a high pass FIR filter with order equal to 24 and pass band frequency equal to 0.317π is given is Appendix G.9. Fig. 9.4 gives plots of the output for sample sequence 1 from the Matlab *filter* function and for the *fir3array* function on the same plot. Fig. 9.5 gives plots of the output for sample sequence 2 from the Matlab *filter* function and for the *fir3array* function on the same plot.

The outputs from the Matlab *filter* function were compared with those from the *fir3array* function. The peak value of the absolute value of the difference between the two outputs for sample sequence 1 was 2.573198×10^{-18}. The average mean squared error was 5.496290×10^{-38}. The peak value of the absolute value of the difference between the two outputs for sample sequence 2 was 8.840598×10^{-17}. The average mean squared error was 3.483791×10^{-34}. These errors confirm that the results obtained using the *fir3array* are correct.

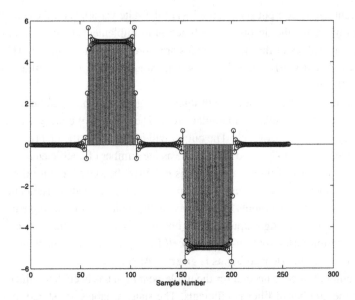

FIGURE 9.2

Stem plot for sample sequence 1

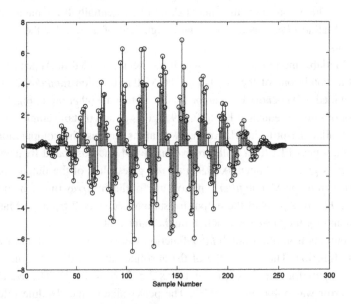

FIGURE 9.3

Stem plot for sample sequence 2

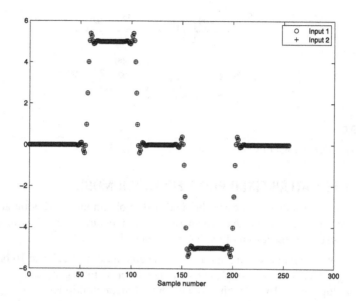

FIGURE 9.4

Plots of the output for sample sequence 1 from the Matlab *filter* function and for the *fir3array* function on the same plot

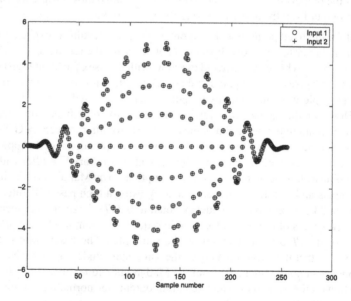

FIGURE 9.5

Plots of the output for sample sequence 2 from the Matlab *filter* function and for the *fir3array* function on the same plot

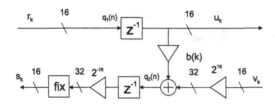

FIGURE 9.6

Modified interleave cell for fixed point arithmetic

9.2.2 THE MATLAB FIXED POINT FIR FILTER MODEL

The floating point model can easily be modified to obtain the fixed point arithmetic model since both models use the same hardware architecture. The following assumptions were used for the design of the hardware model:

1. The inputs and outputs to the module for the computational cell use 16 bit ports.
2. The registers inside the module can be either 16 or 32 bit registers.
3. The multiplier is a 16×16 bit, fixed point arithmetic multiplier.
4. The adder adds two 32 bit, fixed point numbers and produces a 32 bit sum and a carry bit. However, the inputs have been scaled such that the output can be represented using a 32 bit number.
5. All of the outputs use 16 bit fixed point words. The 16 most significant bits of the 32 bit output from the adder is used as the $s_k(n)$ output.

The goal for the fixed point model is to obtain the same results as expected from the hardware and to ensure that overflow does not occur for the registers. Thus, the filter coefficients need to be scaled to avoid overflow with the use of a 32 bit register for the output. Fig. 9.6 gives the computational cell as modified for fixed point arithmetic.

This example uses the same two input sequences as used for the floating point model. However, the inputs were individually scaled to 16 bit fixed point numbers to test the fixed point implementation of the FIR interleave filter. Appendix G.4 gives the *dfir3* Matlab function to implement the fixed point computational cell. Appendix G.5 gives the *dfir3array* Matlab function to implement an arbitrary FIR filter with the use of fixed point arithmetic using the above assumptions. The Matlab script to interleave the two inputs and test the computational array using a high pass FIR filter with order equal to 24 and pass band frequency equal to 0.317π is given in Appendix G.6. Fig. 9.7 gives plots of the output for sample sequence 1 from the Matlab *filter* function and for the *dfir3array* function on the same plot. The fixed point output was normalized so that the two outputs have the same magnitude. Fig. 9.8 gives plots of the output for sample sequence 2 from the Matlab *filter* function and for the *dfir3array* function on the same plot. The fixed point output was normalized so that the two outputs have the same magnitude.

The outputs from the Matlab *filter* function were compared with the normalized outputs from the *dfir3array* function. The peak value of the absolute value of the difference between the two outputs for sample sequence 1 was 6.344178×10^{-4}.

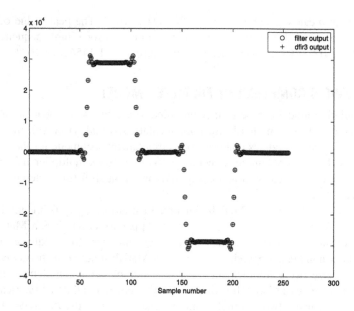

FIGURE 9.7

Plots of the output for sample sequence 1 from the Matlab *filter* function and for the *fir3array* function on the same plot

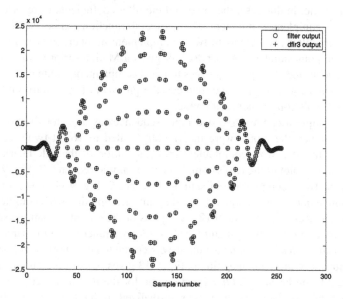

FIGURE 9.8

Plots of the output for sample sequence 2 from the Matlab *filter* function and for the *fir3array* function on the same plot

The average mean squared error was 2.209894×10^{-8}. The peak value of the absolute value of the difference between the two outputs for sample sequence 2 was 6.860017×10^{-4}. The average mean squared error was 1.145518×10^{-7}.

9.2.3 TWO'S COMPLEMENT FIR FILTER MODEL

The simulation model can be further modified to use two's complement arithmetic in the same way that the hardware implementation would. However, this requires the development of modules that use two's complement arithmetic for the arithmetic operations. The required modules include a 16 bit by 16 bit multiplier, a 32 bit adder with a carry-out (the carry-in is not required) and a bit shift (to implement division by powers of 2).

Appendix G.12 gives a Matlab function to multiply two 16 bit numbers and return the 32 bit two's complement product. Appendix G.10 gives a Matlab function *twoscomp* to convert a sequence of integer numbers to a sequence of two's complement numbers. Appendix G.11 gives a Matlab function *invtwoscomp* to obtain the inverse two's complement for a sequence of two's complement numbers. Appendix G.7 gives a Matlab function *tfir3cell* to implement the FIR interleave computational cell using two's complement arithmetic. It uses the *twoscomp* function, the *invtwoscomp* function and the *multi16* function to implement the arithmetic operations for the two's complement numbers. Appendix G.8 gives a Matlab function to call the *tfir3cell* function to implement an arbitrary FIR filter using two's complement arithmetic. In this case, the input and the filter coefficients were assumed to be 16 bit two's complement numbers. The *twoscomp* function can be used to covert the input and the filter coefficients to two's complement numbers after scaling them to 16 bit fixed point numbers. Appendix G.9 gives a Matlab script *testtfir3* to convert the coefficients and the input sequences to two's complement numbers, interleave the two's complement sequences and then call the *tfir3array* function to implement the filter using two's complement arithmetic.

This example uses the same two input sequences as used for the floating point model. However, the inputs were individually scaled to 16 bit two's complement numbers to test the two's complement implementation of the FIR interleave filter. Fig. 9.9 shows a plot of the two's complement of sample sequence 1. Fig. 9.10 shows a plot of the two's complement of sample sequence 2. Fig. 9.11 shows a plot of the two's complement of input signal formed by interleaving the two's complements for sample sequences 1 and 2. Fig. 9.12 gives plots of the output for sample sequence 1 from the Matlab *filter* function and for the *tfir3array* function on the same plot. The inverse of the two's complement output was computed and normalized so that the two outputs have the same magnitude. Fig. 9.13 gives plots of the output for sample sequence 2 from the Matlab *filter* function and for the *tfir3array* script on the same plot. The two's complement output was normalized so that the two outputs have the same magnitude.

The outputs from the Matlab *filter* function were compared with the normalized outputs from the *tfir3array* function. The peak value of the absolute value of the

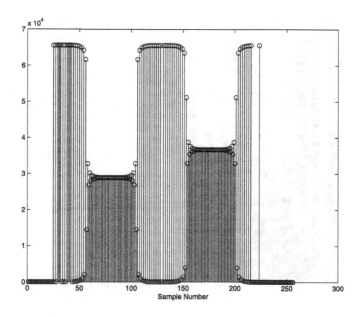

FIGURE 9.9

Plot of the two's complement of sample sequence 1

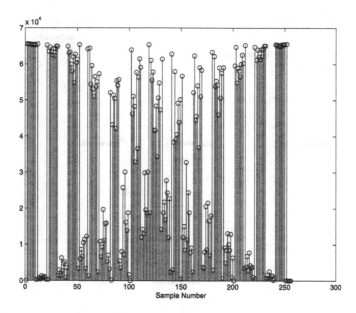

FIGURE 9.10

Plot of the two's complement of sample sequence 2

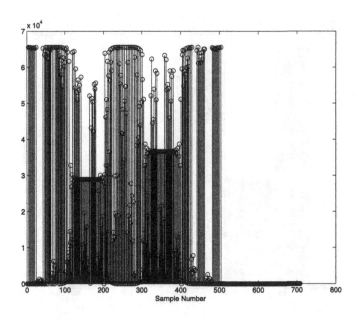

FIGURE 9.11

Plot of the interleaved two's complements of the two sample sequences

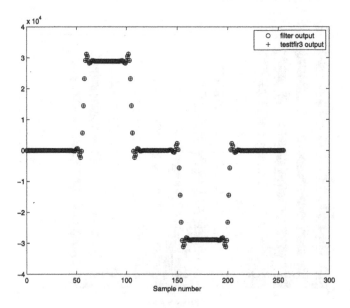

FIGURE 9.12

Plots of the output for sample sequence 1 from the Matlab *filter* function and for the *tfir3array* function on the same plot

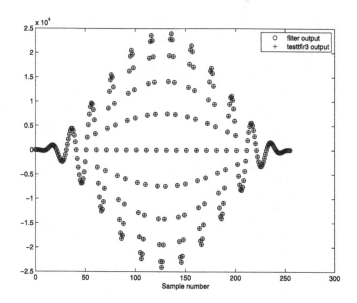

FIGURE 9.13

Plots of the output for sample sequence 2 from the Matlab *filter* function and for the *tfir3array* function on the same plot

difference between the two outputs for sample sequence 1 was 1.150817×10^{-3}. The average mean squared error was 3.150224×10^{-7}. The peak value of the absolute value of the difference between the two outputs for sample sequence 2 was 1.661650×10^{-3}. The average mean squared error was 6.264718×10^{-7}.

9.2.4 THE VERILOG FIR FILTER MODEL

The development of the Verilog hardware description model will now be considered. Several modules that will be needed for the Verilog model are as follows:

1. A delay module for a 16 bit word size. Two of these modules will be used.
2. A 16 bit $\times$ 16 bit multiplier for two's complement numbers.
3. A 32 bit $\times$ 32 bit adder with a 32 bit output and a carry out.

9.2.4.1 *The Multiplier*

The following Verilog module can be used to multiply two 16 bit two's complement numbers.

```
module mult16(c, a, b);

output [31:0] c;
reg [31:0] c;
input [15:0] a, b;
```

```
wire [15:0] a, b;
reg [15:0] reg1, reg2;
reg [31:0] reg3;
reg s1, s2, s3;

always @(a or b)
        begin
            s1 = a[15];
            s2 = b[15];
            s3 = s1^s2;
            if (s1 == 1)
                reg1 = ~a + 16'h0001;
            else if (s1 == 0)
                reg1 = a;
            if (s2 == 1)
                reg2 = ~b + 16'h0001;
            else if (s2 == 0)
                reg2 = b;
            reg3 = reg1*reg2;
            if (s3 == 1)
                c = ~reg3 + 32'h00000001;
            else if (s3 == 0)
                c = reg3;
            end
    endmodule
```

Each word is checked in this module to determine if it is negative. If it is negative, then the inverse two's complement is obtained and the sign bit is stored in a one bit register. The magnitudes of the two numbers are then multiplied to obtain the product. The exclusive *OR* of the two sign bits is obtained to determine the sign of the output. If it is negative, then the two's complement of the product is obtained to form the output. Otherwise, the product is the output.

9.2.4.2 *The Delay*

The following Verilog module can be used to implement a 16 bit delay register.

```
module dreg16(clk, out1, in1);
input [15:0] in1;
input clk;
wire clk;
wire [15:0] in1;
output [15:0] out1;
reg [15:0] out1;

reg [15:0] reg1, reg2;

initial begin
    reg1 = 16'h0000;
    reg2 = 16'h0000;
    end
```

```
always @(posedge clk) begin
    out1 <= reg1;
    # 4 reg1 <= in1;
    end
endmodule
```

This module uses a 16 bit register to store the previous input. On the positive clock edge, the contents of the register are copied to the output. After a delay of 4 ticks, the input is stored into the register for the next clock period.

9.2.4.3 *The Adder*

The following Verilog module can be used to add two 32 bit two's complement numbers.

```
module add32(cout, sum, a, b);

output cout;
output [31:0] sum;
input a, b;
wire [31:0] a, b;
reg [31:0] sum;
reg cout, tmp;
reg [32:0] reg1, reg2, reg3;

initial begin
    reg1 = 33'h000000000;
    reg2 = 33'h000000000;
    end

always @(a or b)
    begin
        reg1 = {a[31],a[31:0]};
        reg2 = {b[31], b[31:0]};
        {tmp, reg3} = reg1 + reg2;
        sum = reg3[31:0];
        cout = reg3[32];
    end
endmodule
```

This modules copies the 32 bit words into the lower 32 bits of 33 bit registers and copies the sign bit into bit 32. They are then added together. This ensures that the carry out is correct. The contents of the 33 bit registers are added. The lower 32 bits are sent to the output as the sum and bit 32 is sent to the output as the carry out.

9.3 IIR FILTER IMPLEMENTATION

A computational structure for the direct form 1 IIR second order filter was developed in Chap. 8. This structure is shown here is Fig. 9.14. A delay can also be added at the input to correct the problem of having a direct path between the primary input and

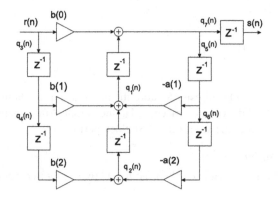

FIGURE 9.14

Computational cell for an IIR second order stage

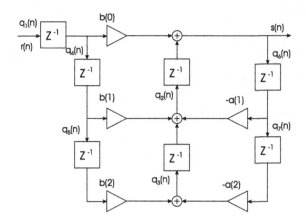

FIGURE 9.15

Computational cell with an input delay

an input to a mathematical unit (multiplier) as shown in Fig. 9.15. The implementation and simulation of the computational structure shown in Fig. 9.15 will now be considered.

The inputs to the delays can be considered to be state variables for this model. This is also convenient for the implementation because the inputs to the delays must be computed so that the corresponding values can be stored in registers to be used on the next clock cycle. The state space model is consistent with this requirement. The state equations to implement this computational cell are given by

$$q_1(n) = r(n),$$
$$q_2n) = b(1)q_4(n-1) - a(1)q_6(n-1) + q_3(n-1),$$

$$
\begin{aligned}
q_3(n) &= b(2)q_3(n-1) - a(2)q_7(n-1), \\
q_4(n) &= q_1(n-1), \\
q_5(n) &= q_4(n-1), \\
q_6(n) &= q_2(n-1) + b(0)q_1(n-1), \\
q_7(n) &= q_6(n-1), \\
s(n) &= q_6(n) = q_2(n-1) + b(0)q_1(n-1).
\end{aligned}
\tag{9.2}
$$

9.3.1 THE MATLAB FLOATING POINT IIR FILTER MODEL

The Matlab function *iir1cell* below can be used to use a single sample as the input and update each of the state variables and the output. This operation would be done on each computational cycle.

Matlab Script 9.1.

```
function [sk, rout] = iir1cell(rk, bk, ak, rin)
% [sk, rout] = iir1cell(rk, bk, ak, rin);
%
% This function simulates a computational cell for the
%     implementation of the transpose direct form
%     realization of a second order section for an
%     IIR digital filter.  Extra delays have been added to
%     pipeline the filter using these computational cells.
%
% Inputs:
%   rk - input to the cell (passed through from previous
%          sos or the input)
%   bk - 1x3 array of numerator filter coefficients
%   ak - 1x3 array of denominator filter coefficients
%   rin - a 7x1 array for the register values which are
%          the previous values of the state variables.
%
% Outputs:
%   sk - the output from the cell
%   rout - a 7x1 array of returned values for the registers
%          state in the cell. These are the updated values for
%          the variables.
%
rout(1,1) = rk;
rout(2,1) = bk(1,2)*rin(4,1) - ak(1,2)*rin(6,1) + rin(3,1);
rout(3,1) = bk(1,3)*rin(5,1) - ak(1,3)*rin(7,1);
rout(4,1) = rin(1,1);
rout(5,1) = rin(4,1);
rout(6,1) = rin(2,1) + bk(1,1)*rin(1,1);
rout(7,1) = rin(6,1);
sk = rin(2,1) + bk(1,1)*rin(1,1);
return
```

End of the Script

The parameter name *rin* ($q_k(n-1)$) was used in the Matlab function *iir1cell* for the array of previous state variables and the parameter name *rout* was used for the array of current state values ($q_k(n)$). The following Matlab function can be used to call the above function once per sample to filter the input sequence.

The following Matlab function can be used to call the cell as required to implement an IIR filter with coefficients in Matlab SOS format.

Matlab Script 9.2.

```
function y = iirlarray(sos, x);
%  y = iirlarray(sos, x);
%
% This function is used to implement a pipeline IIR filter
%    using the iir1cell computational cell.
%
%  Inputs:
%       sos - Filter coefficients in Matlab sos format
%       x - interleaved input sequence
%

n1 = length(x);
[m1, m2] = size(sos);
cellreg = zeros(7,m1);     % array for state variables
cellrk = zeros(m1,1);      % array for cell inputs
cellsk = zeros(m1,1);      % array for cell outputs
for k3=1:n1
    for k2 = 1:m1          % Simulate all cells
        if(k2==1)
            cellrk(k2,1) = x(1,k3);  % in\put to the first cell
        else
            cellrk(k2,1) = cellsk(k2-1,1);
        end
        bk = sos(k2,1:3);
        ak = sos(k2,4:6);
        rk = cellrk(k2,1);
        regin = cellreg(:,k2);
        [sk, regout] = iir1cell(rk, bk, ak, regin);
        cellsk(k2,1) = sk;
        cellreg(:,k2) = regout;
    end
    y(1,k3) = cellsk(m1,1);
end
return
```

End of the Script

The function *iir1array* uses the parameter name *cellreg* to hold the state variables for each SOS. It uses the parameter name *cellrk* for the inputs to each SOS and the parameter *cellsk* for the outputs from each SOS.

The input must be interleaved, either with zeros or with another sample sequence, prior to using the modified computational cell for the IIR transpose direct form because the delays were doubled in developing the cell. An additional delay was also added at the input of each cell to avoid having a path from a primary input to a primary output without a delay. A Matlab script can be written to perform the following operations in a procedure to validate the performance of the cell for an arbitrary IIR filter:

1. Obtain an input sequence using the Matlab *sampdata* function, which has been provided in Appendix A.1;
2. Pad the end of the sequence with zeros to accommodate the delay in the filter due to the added delays;
3. Interleave the input with zeros or with a second input sequence;
4. Extract the output from the interleaved output from the *iir1array* script;
5. Compute the desired output, for comparison, using the Matlab *sosfilt* function;
6. Compare the two outputs using performance measures and stem plots.

The following Matlab script can be used to test *iir1cell* and *iir1array* using the above procedure.

Matlab Script 9.3.

```
% testiirlarray
%
% This routine is used to test the iirlarray and
%   iirlcell functions.  The inputs are obtained
%   for sampdata and sampdata2 and interleaved to
%   form the test inputs.
%
clear
close all
ptime = 2;
order = 8;
wc = 0.75;
rp = 1;
rs = 40;
[b, a] = ellip(order, rp, rs, wc);
n1 = length(b);
f1 = sampdata;
num = 1;
figure(num);
stem(f1);
xlabel('Sample Number')
print -deps '../../eps/iirla.eps';
pause(ptime);

f2 = sampdata2;
num = gcf + 1;
figure(num);
stem(f2);
xlabel('Sample Number')
print -deps '../../eps/iirlb.eps';
pause(ptime);

%  Interleave sequences for the input
n2 = length(f1);
```

```
n3 = length(f2);
n4 = 2*max(n2, n3) + ceil(0.5*n1);
x = zeros(1, n4);
x(1, 1:2:2*n2-1) = f1;
x(1, 2:2:2*n3) = f2;
num = gcf + 1;
figure(num);
stem(x);
xlabel('Sample Number');
print -deps '../../eps/iir1c.eps';
pause(ptime);

[m1, m2] = size(b);
[m3, m4] = size(a);
if (m1 ~= 1)
    display(m1);
    return
end
if (m3~=1)
    display(m3);
    return
end
[sos, gain] = tf2sos(b,a);  % get sos filter coefficients
[m5, m6] = size(sos);
y1 = gain*sosfilt(sos, f1);
y2 = gain*sosfilt(sos, f2);
y3 = iir1array(sos, x);
delay = m5;

g1strt = delay + 1;
g1end = g1strt + 2*n2 - 1;
g1 = y3(1, g1strt:2:g1end);
g2strt = delay + 2;
g2end = g2strt + 2*n3 - 1;
g2 = y3(1, g2strt:2:g2end);

[err1, mea1, mea2] = ncompare(y1, g1);
legend('sosfilt', 'iir1array');
[err2, mea3, mea4] = ncompare(y2, g2);
legend('sosfilt', 'iir1array');
```

End of the Script

Fig. 9.16 gives the stem plots of the output for the filter using the Matlab *sosfilt* function and from the *iir1array* function. Fig. 9.17 gives the stem plots of the output for the filter using the Matlab *sosfilt* function and from the *iir1array* function on the same plot for comparison.

9.3.2 THE MATLAB FIXED POINT IIR FILTER MODEL

The fixed point model requires for the input and the filter coefficients to be quantized for using fixed point arithmetic. In addition, the functions need to accommodate fixed

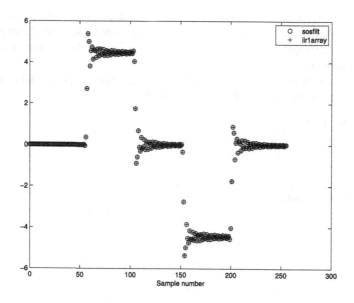

FIGURE 9.16

Stem plots of the output from *sosfilt* and from *iir1array*

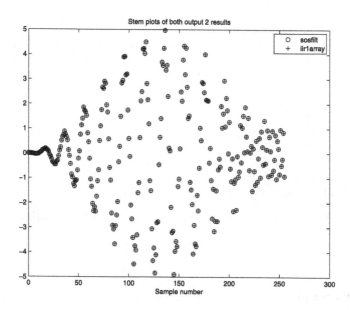

FIGURE 9.17

Stem plots of the output from *sosfilt* and from *iir1array* on the same plot

point arithmetic, including rounding or truncation as appropriate, after each operation that may generate a fractional result. Rounding has been implemented using truncation for this example. The approach to rounding is less likely to result in overflow due to rounding.

The following function can be used to implement the computational cell for the transpose direct IIR filter using fixed point arithmetic.

Matlab Script 9.4.

```
function [sk, regout] = diir1cell(rk, bk, ak, afac, regin)
% [sk, regout] = diir1cell(rk, bk, ak, regin, clk);
%
% This function simulates a computational cell for the implementation
%       of the transpose direct form realization of a second order
%       section for an IIR digital filter.  Extra delays have been added
%       to pipeline the filter using these computational cells.
%
% Inputs:
%    rk - input to the cell (passed through from previous sos or the
%         input)
%    bk - array of numerator filter coefficient for this cell
%    ak - array of denominator filter coefficients for this cell
%    afac - log2 of scale factor for denominator coefficients
%    regin - a 7x7 array for the register values which are the previous
%            values of the state variables.
%
% Outputs:
%    sk - the output from the cell
%    regout - a 7x7 array for the returned values for the registers in
%         the cell. These are the updated values for the state variables.
%
iscale = 2^(-2);
fdbk = 2^(-afac);
rk = fix(rk*iscale);
sk = regin(7,1);
regout(1,1) = bk(1,2)*regin(3,1) - ak(1,2)*regin(5,1) + regin(2,1);
regout(2,1) = bk(1,3)*regin(4,1) - ak(1,3)*regin(6,1);
regout(3,1) = rk;
regout(4,1) = regin(3,1);
regout(5,1) = fix(fdbk*(bk(1,1)*rk + regin(1,1)));
if(abs(regin(5,1)) > 32767)
     fprintf(1, 'overflow occurred. regin(5,1) = %d\n', regin(5,1));
end
regout(6,1) = regin(5,1);
regout(7,1) = regin(5,1);
return
```

End of the Script

The following Matlab function can be used to call the *diir1cell* function above as needed to implement an arbitrary order IIR filter with scaled coefficients in the Matlab SOS format.

Matlab Script 9.5.

```
%  y = diir1_pipeline(scsos, scfac, x);
%
%  This function is used to implement a pipeline IIR filter using
%       the iir1cell pipeline computational cell.
%
%  Inputs:
%       scsos - Filter coefficients in Matlab sos format
%       scfac - array of scale factors
%       x - scaled input sequence
%
function y = diir1_pipeline(scsos, scfac, x);

n1 = length(x);
[m1, m2] = size(scsos);
m3 = length(x);
m1p1 = m1 + 1;
cellreg = zeros(7,m1);
cellrk = zeros(m1p1,1);       % identify extra set of inputs to simplify
                              % the code
cellsk = zeros(m1p1,1);       %
clk = 0;
stop = 0;
for k3=1:n1
    for k2 = 1:m1          % Simulate all cells
        if(k2==1)
            cellrk(k2,1) = x(1,k3);  % initialize the first cell
        else
            cellrk(k2,1) = cellsk(k2-1,1);
        end
        bk = scsos(k2,1:3);
        ak = scsos(k2,4:6);
        afac = scfac(k2, 3);
        rk = cellrk(k2,1);
        regin = cellreg(:,k2);
        [sk, regout] = diir1cell(rk, bk, ak, afac, regin);
        cellsk(k2,1) = sk;
        cellreg(:,k2) = regout;
    end
    y(1,k3) = cellsk(m1,1);
end
return
```

End of the Script

The following Matlab function can be used to obtain two sample sequences, interleave them into a single sample sequence, design a digital filter, convert the coefficients to SOS format, scale the filter coefficients, and call the *diir1_pipeline* function to implement the IIR filter using fixed point arithmetic.

Matlab Script 9.6.

```
% test_iir1cell
%
% This script is used to test the iir1cell function
%
% Design the filter
%
clear
close all
order = 8;
wc = 0.75;
rp = 1;
rs = 40;
[b, a] = ellip(order, rp, rs, wc);
n1 = length(b);
%
% obtain the two inputs
%
x1 = sampdata;
x2 = sampdata2;
x1p = [x1 zeros(1,2*n1+1)];     %increase length of x1 for extra delays
x2p = [x2 zeros(1,2*n1+1)];     %increase length of x1 for extra delays

%   Scale and Interleave the inputs

ibits = 16;
x1sc = fscale(x1p, ibits);
x2sc = fscale(x2p, ibits);
len = length(x1sc);
for k1=1:len
    k2 = 2*k1 - 1;
    k3 = k2 + 1;
    dx(1,k2) = x1sc(1,k1);
    dx(1,k3) = x2sc(1,k1);
end
p1 = length(x1p)-1;
order = 8;
wc = 0.75;
rp = 1;
rs = 40;
[b, a] = ellip(order, rp, rs, wc);
n1 = length(b);
[m1, m2] = size(b);
[m3, m4] = size(a);
if (m1 ~= 1)
    display(m1);
    return
end
if (m3~=1)
    display(m3);
    return
end
```

```
[sos, gain] = tf2sos(b,a);   % get sos filter coefficients
[m5, m6] = size(sos);
pwr = 1/m5;
gfac = (gain)^pwr;
%sos(:,1:3) = gfac*sos(:,1:3);          % distribute gain to all sections
ibits = 16;
obits = 32;
cbits = 16;
%[sos, gain, scsos, scfac] = iircoef2(b, a, ibits, obits);
[sos, gain, scsos, scfac] = iircoef(b, a, cbits);
y1 = sosfilt(sos, x1p);
my1 = sosfilt(sos, x2p);
n3 = length(y1) - 5;
out1 = [0 0 0 0 y1(1,1:n3)];
mout1 = [0 0 0 0 my1(1,1:n3)];
y2 = diir1_pipeline(scsos, scfac, dx);
m7 = length(y2);
m8 = fix(0.5*(length(y2)-m5+2));
for k1=1:m8          % Extract output
    k2 = 2*k1 + m5 - 5;
    y3(1,k1) = y2(1,k2);
    y4(1,k1) = y2(1,k2+1);
end
out2 = y3(1,1:p1);
out3 = y4(1,1:p1);

% Compare the outputs

[nerr, mea1, mea2] = ncompare(out1, out2);
legend('sosfilt', 'diir1');
print -deps '../../eps/diir1a.eps';
[nerr, mea1, mea2] = ncompare(mout1, out3);
print -deps '../../eps/diir1a.eps';
legend('sosfilt', 'diir1');
```

End of the Script

Fig. 9.18 gives a plot of the output for the first sample sequence using the Matlab *sosfilt* function and the normalized output from *diir1_pipeline* on the same plot. The normalized peak error was $1.186354e\text{-}02$. The normalized average mean squared error was $2.024579e\text{-}05$.

Fig. 9.19 gives a plot of the output for the second sample sequence using the Matlab *sosfilt* function and the normalized output from *diir1_pipeline* on the same plot. The normalized peak error was $3.282445e\text{-}02$. The normalized average mean squared error was $4.901706e\text{-}05$.

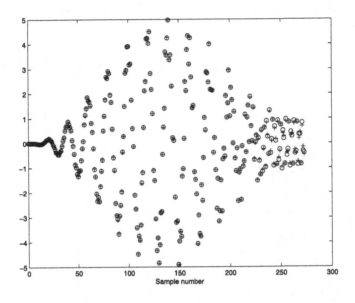

FIGURE 9.18

Stem plot of the two outputs for the first sample sequence on the same plot

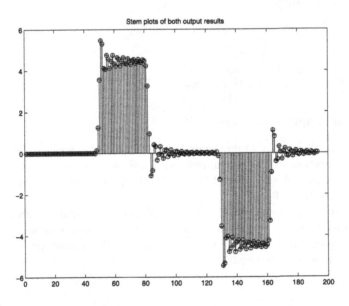

FIGURE 9.19

Stem plot of the two outputs for the second sample sequence on the same plot

Multidimensional Discrete Time Systems

10.1 INTRODUCTION

Extensive research and development have been devoted to multidimensional (M-D) digital signal processing [33]. Recent advances in the performance of digital systems has made it practical to implement many of the M-D digital signal processing applications in real-time. Practical applications of M-D digital signal processing include remote sensing, industrial inspection, computer vision, data compression for communications, processing biomedical images for diagnosis, character recognition, finger print recognition, weather forecasting, beam-forming, etc. In general, these applications are computationally intensive and require substantial data communications. This chapter covers some of the fundamentals of M-D digital signal processing and presents some algorithms for the implementation of M-D digital signal processing algorithms.

10.2 MULTIDIMENSIONAL SEQUENCES

A two-dimensional (2-D) discrete time signal (also referred to as a sequence or an array) may be defined over a set of ordered integers such as

$$[x(n_1, n_2)] = x(n_1, n_2), \quad -\infty \le n_1, n_2 \le \infty. \tag{10.1}$$

A single element of $[x(n_1, n_2)]$ may be defined as a sample (a sample of an image may be called a pixel or a pel).

An M-D discrete time signal may be defined over a set of ordered integers such as

$$[x(n_1, n_2, \ldots, n_N)] = x(n_1, n_2, \ldots, n_N), \quad -\infty \le n_1, n_2, \ldots, n_N \le \infty. \tag{10.2}$$

The array $[x(n_1, n_2, \ldots, n_N)]$ may be a function derived from samples of a continuous function or it may be a collection of samples with no relationship to any continuous function. For this reason, it is advantageous to develop the theory for 2-D and M-D systems in terms of a normalized sample space where the interval between samples in each dimension is unity. Appropriate scaling in each dimension can then be used to relate the derived system equations to their real-world counterparts.

Digital Signal Processing. DOI: 10.1016/B978-0-12-804547-3.00010-3

In practice, most 2-D and M-D discrete time signals are defined only over a finite region. In such cases, it is often convenient to assume the values of the function outside the defined region are equal to zero. However, it is sometimes important to define initial conditions at the boundaries of the data set. This problem will be addressed later in the text.

10.2.1 SPECIAL SEQUENCES

There are several special sequences that will in the analysis of M-D signals and systems. These special sequences are introduced below:

1. M-D unit impulse

$$\delta(n_1, n_2, \dots, n_M) \;=\; 1, \quad \prod_{k=1}^{M} n_k = 0,$$

$$\delta(n_1, n_2, \dots, n_M) \;=\; 0; \quad \text{otherwise.} \tag{10.3}$$

Note that this is equivalent to

$$\delta(n_1, n_2, \dots, n_M) = \delta(n_1)\delta(n_2)\cdots\delta(n_M). \tag{10.4}$$

2. Line impulse [33]

$$h(n_1, n_2) \;=\; \delta_x(n_1) \quad \text{(column),} \tag{10.5}$$
$$h(n_1, n_2) \;=\; \delta_y(n_2) \quad \text{(row).} \tag{10.6}$$

3. Unit step

$$u(n_1, n_2, \dots, n_M) = 1, \quad \prod_{k=1}^{M} n_k \geq 0. \tag{10.7}$$

Note that this is equivalent to

$$u(n_1, n_2, \dots, n_M) = u(n_1)u(n_2)\cdots u(n_M). \tag{10.8}$$

4. Exponential sequences

$$x(n_1, n_2, \dots, n_M) = a_1^{n_1} a_2^{n_2} \cdots a_M^{n_M}. \tag{10.9}$$

If

$$|a_k| = 1, \quad a_k = e^{j\omega_k} \; \forall \; a_k \tag{10.10}$$

then

$$
\begin{aligned}
x(n_1, n_2, \dots, n_M) \;=\;& e^{j\omega_1 n_1} e^{j\omega_2 n_2} \cdots e^{j\omega_M n_M} \\
=\;& \cos(\omega_1 n_1 + \omega_2 n_2 + \cdots + \omega_M n_M) \\
& + j \sin(\omega_1 n_1 + \omega_2 n_2 + \cdots + \omega_M n_M). \tag{10.11}
\end{aligned}
$$

10.2.2 SEPARABLE SEQUENCES

Under certain circumstances, an M-D sequence can be written as a product of separable sequences

$$f(n_1, n_2, \ldots, n_M) = f_1(n_1) f_2(n_2) \cdots f_M(n_M), \tag{10.12}$$

or a sum of separable sequences

$$f(n_1, n_2, \ldots, n_M) = f_1(n_1) + f_2(n_2) + \cdots + f_M(n_M). \tag{10.13}$$

If a linear, shift-invariant system can be represented in either of these forms, then its evaluation may be simplified. In particular, if the impulse response of a linear, shift-invariant system is product or sum separable, then the output for any arbitrary input can be computed as a set of one-dimensional convolutions. If a linear, shift-invariant system is product or sum separable, then the basic theorems for analysis and design of one-dimensional systems can be used for analysis and design.

10.2.3 FINITE EXTENT SEQUENCES

Many M-D sequences are considered to have finite extent such that

$$0 \leq n_1 \leq N_1, \quad 0 \leq n_2 \leq N_2, \quad \ldots, \quad 0 \leq n_M \leq N_M. \tag{10.14}$$

However, M-D sequences can have other regions of support. Such irregular regions of support can often be enclosed in a larger regular region of support to simplify the representation of the sequence and the operations performed on it. Thus, for example, finite extent 2-D sequences can be generalized to have rectangular regions of support.

10.2.4 PERIODIC SEQUENCES

Periodic discrete time signals form another important class of M-D sequences. This section only considers 2-D periodic sequences. However, the concepts can be extended to the M-D case. A periodic 2-D sequence can be considered to be a waveform that repeats itself at regularly spaced intervals. A periodic 2-D signal must repeat itself in two different directions. Thus, the representations of 2-D periodic sequences are generally more complicated.

The vector between the periodic elements that repeat must be determined in order to determine periodicity for M-D discrete signals. Thus, the general periodicity constraints for a 2-D sequence are given by

$$\begin{aligned}
x(n_1 + N_{11}, n_2 + N_{21}) &= x(n_1, n_2), \\
x(n_1 + N_{12}, n_2 + N_{22}) &= x(n_1, n_2).
\end{aligned} \tag{10.15}$$

The corresponding periodicity matrix P can be formed as

$$P = \begin{bmatrix} N_{11} & N_{12} \\ N_{21} & N_{22} \end{bmatrix}. \tag{10.16}$$

$$
\begin{array}{cccccccccccccccc}
1 & 1 & 1 & 0 & 0 & 0 & 0 & 0 & 1 & 1 & 1 & 1 & 0 & 0 \\
0 & 0 & 0 & 0 & 0 & 0 & 0 & 0 & 1 & 1 & 1 & 1 & 0 & 0 \\
0 & 1 & 1 & 1 & 1 & 0 & 0 & 0 & 1 & 1 & 1 & 1 & 0 & 0 \\
0 & 1 & 1 & 1 & 1 & 0 & 0 & 0 & 0 & 0 & 0 & 0 & 0 & 0 \\
0 & 1 & 1 & 1 & 1 & 0 & 0 & 0 & 0 & 0 & 1 & 1 & 1 & 1 \\
0 & 0 & 0 & 0 & 0 & 0 & 0 & 0 & 0 & 0 & 1 & 1 & 1 & 1 \\
0 & 0 & 0 & 1 & 1 & 1 & 1 & 0 & 0 & 0 & 1 & 1 & 1 & 1 \\
0 & 0 & 0 & 1 & 1 & 1 & 1 & 0 & 0 & 0 & 0 & 0 & 0 & 0 \\
0 & 0 & 0 & 1 & 1 & 1 & 1 & 0 & 0 & 0 & 0 & 0 & 1 & 1 \\
0 & 0 & 0 & 0 & 0 & 0 & 0 & 0 & 0 & 0 & 0 & 0 & 1 & 1 \\
\end{array}
$$

FIGURE 10.1

An example of a periodic 2-D sequence

The number of points in the periodic parallelogram is given by the magnitude of the determinant of P as follows:

$$|\det(P)| = |N_{11}N_{22} - N_{12}N_{21}| \neq 0. \tag{10.17}$$

For example, consider the sequence given in Fig. 10.1. The vector between periodic elements for this example is given by

$$
\begin{aligned}
x(n_1 - 2, n_2 + 4) &= x(n_1, n_2), \\
x(n_1 + 7, n_2 + 2) &= x(n_1, n_2).
\end{aligned} \tag{10.18}
$$

Thus, the periodicity matrix is given by

$$P = \begin{bmatrix} -2 & 7 \\ 4 & 2 \end{bmatrix}. \tag{10.19}$$

The magnitude of the determinant of P gives the number of elements in the periodic parallelogram. The number of elements in the periodic parallelogram for this case is given by

$$|\det(P)| = |(-2)(2) - (4)(7)| = 32. \tag{10.20}$$

Note if P is diagonal, then the sequence is rectangularly periodic (vectors form a 90 degree angle). On the other hand, if $\det(P) = 0$, the vectors are parallel to each other. If P is diagonal, then the sequence follows [7]

$$x(n_1, n_2) = x(n_1 + N_{11}, n_2) = x(n_1, n_2 + N_{22}) \tag{10.21}$$

where the period is $N_{11} \times N_{22}$. The periodicity relationship in vector form for the M-D equivalent of this case can be written as

$$x(\mathbf{n} + \mathbf{P}_m) = x(\mathbf{n}), \quad m = 1, 2, \ldots, M. \tag{10.22}$$

The vectors, $\mathbf{P}_m$, are the periodicity vectors and form the columns of the $M \times M$ periodicity matrix P.

10.3 MULTIDIMENSIONAL SYSTEMS

A general operator can be represented using the notation

$$y = T[x]. \tag{10.23}$$

The output sequence y is obtained by performing some operation on the sequence x. This section considers those operations which can be related to M-D digital signal processing.

In general, the operator may be nonlinear and shift variant. For example, the operator

$$y(n_1, n_2) = [x(n_1, n_2)]^2 \tag{10.24}$$

is a memoryless nonlinear operation. The operation

$$y(n_1, n_2) = b(n_1, n_2)x(n_1, n_2) \tag{10.25}$$

is a shift variant operation if $b(n_1, n_2)$ is not a constant. An image obtained by a charged coupled device array is nonlinear because each device in the array has its own response characteristics which varies over its surface. This must be considered during the processing of the resultant image.

10.3.1 LINEAR SYSTEMS

If $L[\ ]$ represents a linear system and if the input/output relationships for two signals, x_1 and x_2, are given by

$$L[x_1] = y_1, \ \ L[x_2] = y_2 \tag{10.26}$$

then

$$L[ax_1 = bx_2] = ay_1 + by_2. \tag{10.27}$$

This general statement for a linear system, which is called the principle of superposition, is appropriate regardless of the dimensions of x_1, x_2, y_1, and y_2 [33]. If a system is linear, then the principle of superposition can be used to obtain the output for given input sequences.

10.3.2 SHIFT INVARIANT SYSTEMS

A shift invariant system is a system for which a shift in the input sequence implies a corresponding shift in the output sequence. Thus, if an operator on a 2-D system with input $x(n_1, n_2)$ yields the output $y(n_1, n_2)$ as follows:

$$T[x(n_1, n_2)] = y(n_1, n_2). \tag{10.28}$$

The system is shift invariant if and only if

$$T[x(n_1 - m_1, n_2 - m_2)] = y(n_1 - m_1, n_2 - m_2) \tag{10.29}$$

for all n_1, n_2, m_1, and m_2 for which the system is defined.

10.4 TWO-DIMENSIONAL CONVOLUTION

The 2-D convolution sum is given by

$$y(n_1, n_2) = \sum_{k_1=-\infty}^{\infty} \sum_{k_2=-\infty}^{\infty} x(k_1, k_2)h(n_1 - k_1, n_2 - k_2). \tag{10.30}$$

The 2-D convolution sum can be shown to be a commutative operation by letting

$$m_1 = n_1 - k_1, \quad m_2 = n_2 - k_2. \tag{10.31}$$

Then

$$k_1 = n_1 - m_1, \quad k_2 = n_2 - m_2. \tag{10.32}$$

The ranges of the variables are unchanged since they are infinite. It follows that

$$y(n_1, n_2) = \sum_{m_1=-\infty}^{\infty} \sum_{m_2=-\infty}^{\infty} x(n_1 - m_1, n_2 - m_2)h(m_1, m_2). \tag{10.33}$$

Thus, the 2-D convolution summation is a commutative operation. Note that 2-D convolution is similar to its 1-D counterpart.

This result can be extended to M-D convolution. Vector notation can be used to express the M-D convolution as

$$y(\mathbf{n}) = \sum_{\mathbf{k}} x(\mathbf{k})h(\mathbf{n} - \mathbf{k}) \tag{10.34}$$

where $\mathbf{n}$ and $\mathbf{k}$ are vectors of appropriate length.

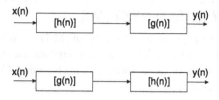

FIGURE 10.2

Commutative property for convolution of LSI systems

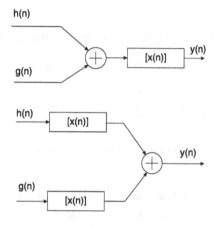

FIGURE 10.3

Distributive property for convolution of LSI systems

10.5 CASCADE AND PARALLEL CONNECTION OF SYSTEMS

The order of transfer functions in a cascade connected LSI system is immaterial because convolution is commutative for LSI systems. Thus

$$y(\mathbf{n}) = x(\mathbf{n}) **h(\mathbf{n}) **g(\mathbf{n}) = x(\mathbf{n}) **g(\mathbf{n}) **h(\mathbf{n})$$

$$** \rightarrow \text{ convolution.} \tag{10.35}$$

Fig. 10.2 gives a block diagram representation of the commutative property for convolution of LSI systems.

Convolution also obeys the distributive law with respect to addition. Thus, we can write

$$x(\mathbf{n}) ** \big[h(\mathbf{n}) + g(\mathbf{n})\big] = x(\mathbf{n}) **h(\mathbf{n}) + x(\mathbf{n}) **g(\mathbf{n}). \tag{10.36}$$

Fig. 10.3 gives a block diagram representation of the distributive property for the convolution of LSI systems.

10.6 STABILITY OF M-D SYSTEMS

The most often used criterion for stability of M-D systems is bounded-input, bounded-output (BIBO) stability. This criterion requires the output for a stable system to remain bounded as long as the input remains bounded. Consider a discrete LSI system with impulse response given by

$$[h(n_1, n_2, \ldots, n_M)], \quad -\infty \le n_k \le \infty \ \forall n_k, 1 \le k \le M. \tag{10.37}$$

The convolution summation can be used to obtain the output for a given bounded input sequence

$$[x(n_1, n_2, \ldots, n_M)], \quad |x(n_1, n_2, \ldots, n_M)| \le R < \infty \tag{10.38}$$

where R is a positive real number. This can be written using vector notation as

$$\lim_{\mathbf{n} \to \infty} y(\mathbf{n}) = \sum_{k_1=-\infty}^{\infty} \cdots \sum_{k_M=-\infty}^{\infty} h(\mathbf{k}) x(\mathbf{n} - \mathbf{k}). \tag{10.39}$$

The magnitude of the product of two or more numbers is less than the product of their magnitudes. Thus,

$$\begin{aligned} |h(\mathbf{k}) x(\mathbf{n} - \mathbf{k})| &\le |h(\mathbf{k})||x(\mathbf{n} - \mathbf{k})|, \\ |h(\mathbf{k})||x(\mathbf{n} - \mathbf{k})| &\le |h(\mathbf{k})| R. \end{aligned} \tag{10.40}$$

It follows that

$$\lim_{\mathbf{n} \to \infty} y(n_1, \ldots, n_M) = \sum_{k_1=-\infty}^{\infty} \cdots \sum_{k_M=-\infty}^{\infty} |h(k_1, \ldots, k_M)| R. \tag{10.41}$$

Since R is a positive, real constant, the discrete LSI system is stable only if

$$\sum_{k_1=-\infty}^{\infty} \cdots \sum_{k_M=-\infty}^{\infty} |h(k_1, \ldots, k_M)| < \infty. \tag{10.42}$$

For the necessary condition for BIBO stability, consider

$$y(n_1, \ldots, n_M) = \sum_{k_1=-\infty}^{\infty} \cdots \sum_{k_M=-\infty}^{\infty} h(k_1, \ldots, k_M) x(n_1 - k_1, \ldots, n_M - k_M). \tag{10.43}$$

Select $x(n - k)$ such that

$$h(\mathbf{k}) x(\mathbf{n} - \mathbf{k}) = |h(\mathbf{k})||x(\mathbf{n} - \mathbf{k})|, \tag{10.44}$$

or

$$x(\mathbf{k}) = \frac{|h(\mathbf{k})||x(\mathbf{n} - \mathbf{k})|}{h(\mathbf{k})} \tag{10.45}$$

when $h(\mathbf{k}) \neq 0$. Otherwise, $x(\mathbf{n} - \mathbf{k}) = 0$ when $h(\mathbf{k}) = 0$. It follows that

$$\lim_{\mathbf{n} \to \infty} y(n_1, \ldots, n_M) = \sum_{k_1=-\infty}^{\infty} \cdots \sum_{k_M=-\infty}^{\infty} |h(k_1, \ldots, k_M)| = \infty \tag{10.46}$$

unless

$$\sum_{k_1=-\infty}^{\infty} \cdots \sum_{k_M=-\infty}^{\infty} |h(k_1, \ldots, k_M)| < \infty. \tag{10.47}$$

10.7 REGIONS OF SUPPORT

Physically realizable one-dimensional (1-D) systems can be characterized as being casual if their outputs cannot precede their inputs. This is a very valid constraint for systems with time as the independent variable. There are two independent variables for 2-D systems and often, neither of the independent variable is time. Therefore, the principle of causality must be generalized. One generalization of the concept of causality is to require that the impulse response be zero outside some region of support.

A 1-D causal system has the restriction that $h(n) = 0$ for $n < 0$. If this concept is extended to 2-D, the requirement becomes

$$h(n_1, n_2) = 0 \quad \forall \, n_1 < 0, \, n_2 < 0. \tag{10.48}$$

This region is said to be first quadrant support (sometimes called quarter plane support). This concept can be extended to define a region or sector which is nonzero only at points on or inside two rays emanating from the origin, provided that the angle between the two rays is strictly less than 180 degrees. This is called wedge support. The asymmetric half-plane region of support is an example [34].

Any sequence with support on a wedge can be mapped into a sequence with support on the first quadrant through a linear transformation. The transformation matrix may be defined as

$$S = \begin{bmatrix} s_{11} & s_{12} \\ s_{21} & s_{22} \end{bmatrix}. \tag{10.49}$$

The determinant of S must be nonzero. The new indices can be determined from

$$\begin{aligned} m_1 &= s_{22}n_1 - s_{12}n_2, \\ m_2 &= -s_{21}n_1 + s_{11}n_2. \end{aligned} \tag{10.50}$$

If there are no common factors between s_{11} and s_{21} or between s_{12} and s_{22}, then the requirement is that

$$\det(S) = s_{11}s_{22} - s_{12}s_{21} \neq 0. \tag{10.51}$$

The change of variables will map the wedge into the first quadrant. However, not every point in the transformed region will lie in the range space of the linear transformation. This is because of the use of discrete time signals. However, the mapping can be restricted in order for every integer valued order pair to lie in the range space of the linear transformation by requiring $\det(P) = 1$.

10.8 FREQUENCY DOMAIN CHARACTERISTICS

Consider a 2-D LSI system for which the input is a general sinusoid of unit magnitude as follows:

$$x(n_1, n_2) = e^{(j\omega_1 n_1 + j\omega_2 n_2)}. \tag{10.52}$$

The output for such a system with impulse response $[h(n_1, n_2)]$ is given by

$$
\begin{aligned}
y(n_1, n_2) &= \sum_{k_1=-\infty}^{\infty} \sum_{k_2=-\infty}^{\infty} h(k_1, k_2) e^{[j\omega_1(n_1-k_1)+j\omega_2(n_2-k_2)]} \\
&= e^{j(\omega_1 n_1 + \omega_2 n_2)} \sum_{k_1=-\infty}^{\infty} \sum_{k_2=-\infty}^{\infty} h(k_1, k_2) e^{-j(\omega_1 k_1 + \omega_2 k_2)}.
\end{aligned}
\tag{10.53}
$$

The frequency response of a linear system is typically considered to be the output, $y(n_1, n_2)$, divided by the input $x(n_1, n_2)$, which is a general sinusoid of unit magnitude. Thus, the frequency response is defined as

$$H(\omega_1, \omega_2) = \sum_{k_1=-\infty}^{\infty} \sum_{k_2=-\infty}^{\infty} h(k_1, k_2) e^{-j(\omega_1 k_1 + \omega_2 k_2)}. \tag{10.54}$$

Note that $e^{-j(\omega_1 k_1 + \omega_2 k_2)}$ is periodic in ω_1 with period 2π and in ω_2 with period 2π. Thus,

$$
\begin{aligned}
H(\omega_1 + 2k_1\pi, \omega_2) &= H(\omega_1, \omega_2), \\
H(\omega_1, \omega_2 + 2k_2\pi) &= H(\omega_1, \omega_2).
\end{aligned}
\tag{10.55}
$$

EXAMPLE 10.1
(Frequency Response Example 1)

Consider the impulse response given by

$$h(n_1, n_2) = \delta(n_1+1, n_2)+\delta(n_1-1, n_2)+\delta(n_1, n_2+1)+\delta(n_1, n_2-1). \quad (10.56)$$

Using the above development, it follows that

$$H(\omega_1, \omega_2) = e^{j\omega_1} + e^{-j\omega_1} + e^{j\omega_2} + e^{-j\omega_2}. \quad (10.57)$$

This can be simplified to obtain

$$H(\omega_1, \omega_2) = 2\cos(\omega_1) + 2\cos(\omega_2). \quad (10.58)$$

The frequency response for any arbitrary impulse response (corresponding to an LSI system) can be determined in a similar manner [33].

End of the Example

EXAMPLE 10.2
(Frequency Response Example 2)

Consider the impulse response given by

$$\begin{aligned} h(n_1, n_2) &= 0.125[\delta(n_1+1, n_2+1) + \delta(n_1+1, n_2-1) \\ &+ \delta(n_1-1, n_2+1) + \delta(n_1-1, n_2-1)] \\ &+ 0.25[\delta(n_1+1, n_2) + \delta(n_1-1, n_2) \\ &+ \delta(n_1, n_2+1) + \delta(n_1, n_2-1)] \\ &+ 0.5[\delta(n_1, n_2)]. \end{aligned} \quad (10.59)$$

Recall that

$$\begin{aligned} \delta(n_1, n_2) &= \delta_1(n_1)\delta_2(n_2), \\ \delta(n_1+1, n_2+1) &= \delta_1(n_1+1)\delta_2(n_2+1), \text{ etc.} \end{aligned} \quad (10.60)$$

Thus, $h(n_1, n_2)$ can be factored to obtain

$$\begin{aligned} h(n_1, n_2) &= 0.125\delta_1(n_1+1)[\delta_2(n_2+1) + 2\delta_2(n_2) + \delta_2(n_2-1)] \\ &+ 0.25\delta_1(n_1)[\delta_2(n_2+1) + 2\delta_2(n_2) + \delta_2(n_2-1)] \\ &+ 0.125\delta_1(n_1-1)[\delta_2(n_2+1) + 2\delta_2(n_2) + \delta_2(n_2-1)]. \end{aligned} \quad (10.61)$$

It follows, that with further simplification,

$$h(n_1, n_2) = 0.125 h_1(n_1) h_2(n_2) \tag{10.62}$$

where

$$
\begin{aligned}
h_1(n_1) &= [\delta_1(n_1+1) + 2\delta_1(n_1) + \delta_1(n_1-1)], \\
h_2(n_2) &= [\delta_2(n_2+1) + 2\delta_2(n_2) + \delta_2(n_2-1)].
\end{aligned} \tag{10.63}
$$

It follows that the frequency response can be simplified to obtain

$$H(\omega_1, \omega_2) = 0.125[e^{j\omega_1} + 2 + e^{-j\omega_1}][e^{j\omega_2} + 2 + e^{-j\omega_2}]. \tag{10.64}$$

$H(\omega_1, \omega_2)$ can be simplified further to obtain [33]

$$H(\omega_1, \omega_2) = 0.5[1 + \cos(\omega_1)][1 + \cos(\omega_2)]. \tag{10.65}$$

End of the Example

10.9 THE 2-D IMPULSE RESPONSE

In general, the frequency response of a discrete time 2-D LSI system is continuous and periodic over the normalized radial frequency range $-\pi \le \omega_1, \omega_2 \le \pi$. Thus, the inverse Fourier series of $H(\omega_1, \omega_2)$ can be computed to obtain the sequence:

$$h(n_1, n_2) = \frac{1}{4\pi^2} \int_{-\pi}^{\pi} \int_{-\pi}^{\pi} H(\omega_1, \omega_2) e^{j(\omega_1 n_1 + \omega_2 n_2)} d\omega_1 d\omega_2. \tag{10.66}$$

The 2-D Fourier Transform was defined as

$$H(\omega_1, \omega_2) = \sum_{k_1=-\infty}^{\infty} \sum_{k_2=-\infty}^{\infty} h(k_1, k_2) e^{-j(\omega_1 k_1 + \omega_2 k_2)}. \tag{10.67}$$

This definition can be substituted for $H(\omega_1, \omega_2)$ in Eq. (10.66) to obtain

$$
\begin{aligned}
h(n_1, n_2) = \frac{1}{4\pi^2} \int_{-\pi}^{\pi} \int_{-\pi}^{\pi} & \left[\sum_{k_1=-\infty}^{\infty} \sum_{k_2=-\infty}^{\infty} h(k_1, k_2) e^{-j(\omega_1 k_1 + \omega_2 k_2)} \right] \\
& \times e^{j(\omega_1 n_1 + \omega_2 n_1)} d\omega_1 d\omega_2.
\end{aligned} \tag{10.68}
$$

$H(\omega_1, \omega_2)$ must be continuous and finite if $h(n_1, n_2)$ is a stable sequence. Thus, if $h(n_1, n_2)$ is assumed to be a stable sequence, then the order of summation and integration in Eq. (10.68) can be reversed to obtain

$$h(n_1, n_2) = \frac{1}{4\pi^2} \sum_{k_1=-\infty}^{\infty} \sum_{k_2=-\infty}^{\infty} h(k_1, k_2) \int_{-\pi}^{\pi} e^{-j\omega_1(k_1-n_1)} d\omega_1$$
$$\times \int_{-\pi}^{\pi} e^{-j\omega_2(k_2-n_2)} d\omega_2. \tag{10.69}$$

For convenience, let

$$P = \int_{-\pi}^{\pi} e^{-j\omega_1(k_1-n_1)} d\omega_1. \tag{10.70}$$

If $n_1 \neq k_1$, then

$$P = \int_{-\pi}^{\pi} e^{-j\omega_1 m_1} d\omega_1, \quad m_1 = k_1 - n_1. \tag{10.71}$$

$$P = \frac{2}{jm_1} \sin(\pi m_1) = 0 \quad \forall\, m_1. \tag{10.72}$$

If $k_1 = n_1$, then

$$P = \int_{-\pi}^{\pi} d\omega_1 = 2\pi. \tag{10.73}$$

In a similar manner,

$$\int_{-\pi}^{\pi} e^{-j\omega_2(k_2-n_2)} d\omega_2 = 0 \quad \forall k_2 \neq n_2,$$
$$= 2\pi \quad \forall k_2 = n_2. \tag{10.74}$$

It follows that

$$h(n_1, n_2) = \frac{1}{4\pi^2} [(2\pi)(2\pi)h(n_1, n_2)] = h(n_1, n_2). \tag{10.75}$$

Thus, the inverse Fourier series of $H(\omega_1, \omega_2)$ is the 2-D sequence $[h(n_1, n_2)]$.

In a manner similar to the above development, it be can show that the frequency distribution of the output for the convolution summation of the sequence $[h(n_1, n_2)]$ and the sequence $[x(n_1, n_2)]$ is given by [33]

$$Y(\omega_1, \omega_2) = H(\omega_1, \omega_2)X(\omega_1, \omega_2). \tag{10.76}$$

10.10 THE 2-D FOURIER TRANSFORM

The properties of the 2-D and M-D Fourier Transform are essentially the same as those for the 1-D Fourier Transform [33]. Recall that the 1-D DFT pair was developed as the Fourier series representation of a periodic discrete time sequence. The 1-D DFT pair is given by

$$x(n) = \frac{1}{N}\sum_{n=0}^{N-1} X(k)W_N^{-kn}, \quad 0 \le n \le N-1 \tag{10.77}$$

where $W_N = e^{-j(2\pi/N)}$ and

$$X(k) = \sum_{n=0}^{N-1} x(n)W_N^{kn}, \quad 0 \le k \le N-1. \tag{10.78}$$

The periodic sequence was formed by extending $x(n)$ as follows:

$$x(n) = f(n+kN), \quad -\infty \le k \le \infty \tag{10.79}$$

where $f(n)$ was the original sequence defined over $0 \le n \le N$.

The 2-D DFT can be developed as the Fourier series representation of a two-dimensional discrete time periodic sequence in a similar manner. In this development, the sequence is assumed to be periodic with period N_1 in the n_1 direction and N_2 in the n_2 direction. The Fourier series coefficients for such a sequence are the DFT samples and are given by:

$$X(k_1, k_2) = \sum_{n_1=0}^{N_1-1}\sum_{n_2=0}^{N_2-1} x(n_1,n_2)e^{-(j2\pi n_1 k_1/N_1 + j2\pi n_2 k_2/N_2)}$$
$$0 \le k_1 \le N_1-1, \quad 0 \le k_2 \le N_2-1. \tag{10.80}$$

Actually, the 2-D DFT is a sampled representation of the transform of the sequence given by

$$X(\omega_1,\omega_2) = \sum_{n_1=0}^{N_1-1}\sum_{n_2=0}^{N_2-1} x(n_1,n_2)e^{-(j\omega_1 n_1 + j\omega_2 n_2)}. \tag{10.81}$$

The properties of the 2-D DFT are the same as those for the 1-D DFT. The proofs of these properties are similar to the 1-D proofs except for the two variables involved.

10.11 2-D CONVOLUTION

Convolution is one of the applications of the 2-D DFT. However, the circular convolution property of the DFT which caused problems for 1-D convolution is also a problem that carries over to the 2-D DFT. The review of the solution to this problem for the 1-D case can lead to the solution for the 2-D. Recall that two 1-D sequences can be padded with zeros in order to obtain a linear convolution by computing the inverse DFT of the product of their DFTs. Thus if sequence 1 has a length of N_1 and sequence 2 has a length of N_2, then each sequence can be padded with zeros to extend its length to

$$L = N_1 + N_2 - 1. \tag{10.82}$$

The resulting output sequence has length L.

This concept can be extended to two dimensions. For example, if sequence 1 is N_1 rows by N_2 columns and sequence 2 is N_3 rows by N_4 columns, then each sequence should be padded with zeros to have

$$L_1 = N_1 + N_3 - 1 \tag{10.83}$$

rows and

$$L_2 = N_2 + N_4 - 1 \tag{10.84}$$

columns. The output has L_1 rows and L_2 columns.

In many practical cases, the size of one of the sequences may be significantly greater than the other. For example, it may be desirable to use a 25×25 kernel (FIR) filter with a 1024×1024 image. For this case, the output from the linear convolution would be 1048×1048. If an FFT algorithm is used for the computation, then data size should be a power of 2 in each dimension. Thus, the output image would be 2048×2048. This is four times the size of the original image. In such a case, it requires less online data storage to partition the image into several sub-images and perform convolutions on these rather than perform the convolution on the entire image. The overlap–save and overlap–add procedures can be used to accomplish this. These procedures are straightforward extensions of the corresponding 1-D procedures.

10.11.1 OVERLAP–ADD

The 1-D overlap–add procedure was presented in Sect. 3.9.2. The 2-D overlap–add procedure is an extension of the 1-D procedure. The concepts for partitioning the data into segments and for padding the segments with zeros are extensions of these concepts to two dimensions. A large image can be partitioned into $L_1 \times L_2$ segments such that

$$x(n_1, n_2) = \sum_{k_1=0}^{L_1-1} \sum_{k_2=0}^{L_2-1} x_{k_1 k_2}(n_1, n_2). \tag{10.85}$$

The distributive property of convolution can then be used to obtain the convolution of $x(n_1, n_2)$ with $h(n_1, n_2)$ [7]. Thus,

$$
\begin{aligned}
y(n_1, n_2) &= x(n_1, n_2) * *h(n_1, n_2) \\
&= \left[\sum_{k_1=0}^{L_1-1} \sum_{k_2=0}^{L_2-1} x_{k_1 k_2}(n_1, n_2) \right] * *h(n_1, n_2) \\
&= \sum_{k_1=0}^{L_1-1} \sum_{k_2=0}^{L_2-1} x_{k_1 k_2}(n_1, n_2) * *h(n_1, n_2). \qquad (10.86)
\end{aligned}
$$

It follows that the original convolution can be partitioned into the sum of a set of smaller convolutions.

10.11.2 OVERLAP–SAVE

The 1-D overlap–save procedure was presented in Sect. 3.9.1. The 2-D overlap–save procedure is an extension of the 1-D overlap–save procedure. The concepts for partitioning the data into segments and padding the segments with zeros are extensions of these concepts to two dimensions. Although the overlap–save procedure is comparable in efficiency to the overlap–add procedure, the overlap–add procedure is simpler, conceptually. Therefore, it is more often used.

10.11.3 COMPUTATIONAL CONSIDERATIONS

There are many methods used to measure the complexity of an FFT algorithm. It is often desirable to process a very large data set using the FFT algorithm. For example, a synthetic aperture radar (SAR) image with 8192 rows by 8192 columns may be obtained with the use of a satellite. Each frame of such an image would require 64×10^6 complex words of storage since the DFT is complex. Thus, if 8 bytes are used to represent a complex number, one gigabyte of memory would be required to store the data for the filtering operation. It is not practical to design a digital system that can randomly access the entire data set for such an image. Thus, data communication requirements are extremely important for the development of a procedure to use an FFT algorithm to process such a large image. However, it is difficult to quantify the cost of data communication requirements because the cost depends upon the architecture of the computer system.

10.11.3.1 *Direct Calculation*

The direct calculation of the 2-D DFT involves N_1 rows and N_2 columns as given in Eq. (10.80). This calculation requires $N_1^2 N_2^2$ complex multiplications and additions as shown in Table 10.1 [7,33].

10.11.3.2 *Row–Column Decomposition*

The row–column decomposition involves computing the DFT of the rows and then computing the DFT of the columns of the resulting output. The computation of the

Table 10.1 Comparison of computational complexity for implementations of the 2-D DFT

Method	Complex Multiplications	Complex Additions
Direct	$N_1^2 N_2^2$	$N_1^2 N_2^2$
Row–Column	$0.5 N_1 N_2 \log_2(N_1 N_2)$	$N_1 N_2 \log_2(N_1 N_2)$
2×2 Radix	$\frac{3N^2}{4} \log_2(N)$	$2N^2 \log_2(N)$
$(N_1 = N_2 = N)$		

DFT of the rows can be represented as

$$G(n_1, k_2) = \sum_{n_2=0}^{N_2-1} x(n_1, n_2) W_{N_2}^{n_2 k_2}. \tag{10.87}$$

This can be followed by the computation of the columns using the output from the row computations as the inputs:

$$X(k_1, k_2) = \sum_{n_1=0}^{N_1-1} G(n_1, k_2) W_{N_1}^{n_1 k_1}. \tag{10.88}$$

The 1-D DFT requires $N_1 N_2 (N_1 + N_2)$ complex multiplications and additions to compute the DFT using the row–column decomposition. The use of the radix 2, 1-D FFT requires $0.5 N_1 N_2 \log_2(N_1 N_2)$ complex multiplications and $N_1 N_2 \log_2(N_1 N_2)$ complex additions to compute the DFT using the row–column decomposition.

It is typical to use a raster-scan procedure to acquire a large image or the image is typically stored in row scan format. Thus, the typical procedure for obtaining the DFT of an image using the row–column decomposition is to

1. Obtain the 1-D FFT of each row,
2. Transpose the image and store it by columns,
3. Obtain the 1-D FFT of each column.

If a convolution is to be performed, then the same procedure must be used on the impulse response to compute its DFT. Also, both the impulse response and the image must be expanded and padded with zeros to avoid errors due to the circular convolution problem as previously discussed. Note that the row–column decomposition either requires the entire image to be accessible or a transpose operation is required if the image is read in by rows (which is the normal procedure). The additional required input/output (I/O) operations reduce the efficiency of this procedure.

10.12 THE DISCRETE COSINE TRANSFORM

The Discrete Cosine Transform (DCT) is used extensively in some image processing applications. It is the most widely used transform in a class of image coding systems

Table 10.2 A 2-D sequence as input for the 2-D DCT

1.0	2.0	3.0	4.0	5.0
2.0	3.0	4.0	5.0	6.0
3.0	4.0	5.0	6.0	7.0
4.0	5.0	6.0	7.0	8.0
5.0	6.0	7.0	8.0	9.0

Table 10.3 The extended sequence as input for the 2-D DCT

1.0	2.0	3.0	4.0	5.0	5.0	4.0	3.0	2.0	1.0
2.0	3.0	4.0	5.0	6.0	6.0	5.0	4.0	3.0	2.0
3.0	4.0	5.0	6.0	7.0	7.0	6.0	5.0	4.0	3.0
4.0	5.0	6.0	7.0	8.0	8.0	7.0	6.0	5.0	4.0
5.0	6.0	7.0	8.0	9.0	9.0	8.0	7.0	6.0	5.0
5.0	6.0	7.0	8.0	9.0	9.0	8.0	7.0	6.0	5.0
4.0	5.0	6.0	7.0	8.0	8.0	7.0	6.0	5.0	4.0
3.0	4.0	5.0	6.0	7.0	7.0	6.0	5.0	4.0	3.0
2.0	3.0	4.0	5.0	6.0	6.0	5.0	4.0	3.0	2.0
1.0	2.0	3.0	4.0	5.0	5.0	4.0	3.0	2.0	1.0

known as transform coders. The 1-DCT was discussed in Sect. 3.10. The 1-D DCT can be extended to two dimensions in a straightforward way to obtain the 2-D DCT. Example 10.3 gives an example of the extension of the 1-D DCT concept to obtain the 2-D DCT.

EXAMPLE 10.3
(A 2-D DCT)

Consider the 2-D sequence given in Table 10.2. The input is extended to a 10×10 sequence as given in Table 10.3. It is easy to see that there are no discontinuities in the periodic, extended sequence obtained by setting

$$x(n_1+10k_1, n_2+10k_2) = x(n_1, n_2), \quad -\infty \le k_1 \le \infty, \ -\infty \le k_2 \le \infty. \quad (10.89)$$

The 2-D DCT for the original sequence is the 2-D DFT of the extended sequence and is given in Table 10.4.
End of the Example

Table 10.4 The 2-D DCT for the 2-D sequence

n	0	1	2	3	4	5	6	7	8	9
$DCT(n)$	500.0	−99.5959	0.0	−8.9806	0.0	−99.5959	0.0	0.0	0.0	0.0
n	10	11	12	13	14	15	16	17	18	19
$DCT(n)$	0.0	0.0	0.0	0.0	0.0	−8.9806	0.0	0.0	0.0	0.0
n	20	21	22	23	24					
$DCT(n)$	0.0	0.0	0.0	0.0	0.0					

Appendix A: Matlab Code for Chapter 3

This Appendix provides the Matlab functions and Matlab code used in the Chap. 3 examples.

A.1 GENERATE A TEST SEQUENCE

Matlab Script A.1.

```
function y = sampdata
% y = sampdata;
%
%   This function returns a sample data sequence to be used
%       as a test input for digital signal processing problems.
%   The input data is filtered with a low pass filter to make
%       it band limited (cutoff frequency is set at 0.95*pi).
ntaps = 65;
f = [0.0 0.9 0.95 1.0];
mag = [ 1.0 1.0 0.7071 0.0];
b = fir2(ntaps, f, mag);
n1 = length(b);
len1 = 256 - n1 + 1;
%
% Generate the input sample sequence.
%
d1 = zeros(1,24);
d2 = ones(1,48);
d3 = zeros(1,48);
d4 = -1*d2;
d5 = zeros(1,23);
data = 5*[d1 d2 d3 d4 d5];
y = conv(b, data);
return
```

Digital Signal Processing. DOI: 10.1016/B978-0-12-804547-3.00019-X

Appendix B: Matlab Code for Chapter 4

B.1 HAMMING WINDOW DESIGN

The following Matlab function can be used to design a Hamming window.

Matlab Script B.1.

```
function wind = myhamming(n)
% wind = myhamming(n);
%
%  This routine returns a Hamming window of length n.
nmod = mod(n,2);
c = 2*pi/(n-1);
if (nmod == 1)
    m = fix(0.5*(n-1));
    k = -m:1:m;
    wind = 0.54 + 0.46*cos(k*c);
elseif (nmod == 0)
    m = fix(0.5*n);
    k = -m:1:(m-1);
    k = k+0.5;
    wind = 0.54+0.46*cos(k*c);
end
return
```

Digital Signal Processing. DOI: 10.1016/B978-0-12-804547-3.00020-6

Appendix C: Matlab Code for Chapter 5

C.1 IIR SECOND ORDER SECTIONS

The following Matlab script provides an example of the decomposition of a fourth order IIR filter into second order sections.

Matlab Script C.1.

```
% IIR second order section (SOS) example (iirsosex1)

x = sampdata;
wp = 0.3;
order = 4;
rp = 2.0;
rs = 45;
[b, a] = ellip(order, rp, rs, wp);
[sos, g] = tf2sos(b, a);
[n1, n2] = size(sos);

% Filter data using SOS
xin = x;
for k=1:n1
    bloc = sos(k, 1:3);
    aloc = sos(k, 4:6);
    y1 = filter(bloc, aloc, xin);
    xin = y1;
end

% Filter data using Matlab filter function
y2 = filter(b, a, x);
subplot(2,1,1)
stem(y1)
title('Output using second order sections');
xlabel('sample number');
subplot(2,1,2)
stem(y2)
title('Output for the direct form');
xlabel('sample number');
print -dps iirsosex1.ps
```

Digital Signal Processing. DOI: 10.1016/B978-0-12-804547-3.00021-8

C.2 VECTOR IMPLEMENTATION OF IIR FILTER

The following Matlab function can be used to represent IIR filters with the coefficients represented as vectors.

Matlab Script C.2.

```
function y = myfilter(b, a, x);
% y = myfilter(b, a, x);
%
%   This function implements the digital filter
% Inputs:
%   b - array of numerator filter coefficients
%   a - array of denominator filter coefficients
%   x - input sequence
% Outputs:
%   y - output sequence

n1 = length(x);
n2 = length(b);
n2m1 = n2 - 1;
n3 = length(a) - 1;
n3m1 = n3 - 1;
xvec = zeros(1, n2);
if (n3 > 0)
    yvec = zeros(1, n3);
    avec = -a(1, 2:end);
    for k=1:n1
        xvec = [x(1,k) xvec(1, 1:n2m1)];
        ynew = dot(b, xvec) + dot(avec, yvec);
        yvec = [ynew yvec(1, 1:n3m1)];
        y(1,k) = ynew;
    end
else
    for k=1:n1
        xvec = [x(1,k) xvec(1, 1:n2m1)];
        y(1,k) = dot(b, xvec);
    end
end
return
```

C.3 LINEAR PHASE IIR FILTER

The following Matlab script provides an example of the linear phase implementation of an IIR filter.

Matlab Script C.3.

```
% Linear phase IIR filtering (iir_ex2)
% This script is demonstrates linear phase IIR filtering.
```

```
% first do detailed implementation

x = sampdata;   % Obtain the input
wp = 0.3;
order = 4;
rp = 2.0;
rs = 45;
[b, a] = ellip(order, rp, rs, wp);
g1 = filter(b, a, x);
x2 = fliplr(g1); % reverse g1
g2 = filter(b, a, x2);
y1 = fliplr(g2);   % reverse g2

subplot(2,1,1)
stem(x)
title('Stem plot of the input sequence');
xlabel('sample number');
subplot(2,1,2)
stem(y1)
axis([0.0 200.0 -8.0 8.0]);
title('IIR filter output using the linear phase procedure');
xlabel('sample number');
print -dps iirex2.ps
```

Appendix D: Matlab Code for Chapter 6

D.1 SCALING OF FIR FILTERS

The Matlab function *firscale* can be used to scale FIR filters for a specified number of bits.

Matlab Script D.1.

```
function [scf, scale]  = fscale(f, bits)
% [scf, scale] = fscale(f, bits);
%
% This function can be used to scale floating point
%    signals to integers.
%
% Inputs:
%   f - floating point input sequence
%   bits - number of bits to use for the integer output
%
% Outputs:
%   scf - integer ouput sequence
%   scale - scale factor used for the conversion
%
t1 = max(max(abs(f)));
kbits = bits - 1;
scale = (2^(kbits) - 1)/t1;
scf = round(scale*f); return;
```

D.2 OUTPUT OVERFLOW SCALING OF FIR FILTERS

The Matlab *firscale2* can be used to scale the coefficients for a FIR filter based upon the criteria to avoid overflow for a given number of bits for the input and a given number of bits for the output.

Matlab Script D.2.

```
function [bsc, scfac] = firscale2(b, ibits, obits);
% [bsc, scfac] = firscale2(b, ibits, obits);
%
% This script scales the FIR filter using the
%       output overflow criteria
```

Digital Signal Processing. DOI: 10.1016/B978-0-12-804547-3.00022-X
Copyright © 2017 Elsevier Inc. All rights reserved.

```
%  Inputs:
%     b - the array of FIR filter coefficients
%     ibits - the number of bits for the samples of the input
%     obits - the number of bits for the output accumulator
%
%  Outputs
%     bsc - the array of scaled filter coefficients
%     scfac - the scale factor used for the scaling

m1 = ibits - 1;
xmax = 2^(m1) - 1;
m2 = obits - 1;
ymax = 2^(m2) - 1;
c = sum(abs(b));
scfac = ymax/(xmax*c);
bsc = round(scfac*b);
return;
```

D.3 SCALING OF IIR FILTERS

The following Matlab script can be used to scale IIR digital filters. The scale factor for the denominator coefficients is chosen to be 2^m for some integer value m to simplify implementation later. However, this constraint is not necessary for the numerator coefficients.

Matlab Script D.3.

```
function [bsc, asc, scfac] = iirscale(b, a, bits)
%[bsc, asc, scfac] = iirscale(b,a, bits);
%
%  Script to scale IIR filters.
%
%  Inputs
%  b - The floating point numerator filter coefficients
%  a - The floating point denominator filter coefficients
%  bits - the number of bits to use for the scaled coefficients
%
%  Outputs
%  bsc - the scaled numerator coefficients
%  asc - the scaled denominator coefficients
%  scfac - array of scale factors for the scaled coefficients.
%     The first number is the scale factor for the numerator
%        coefficients.
%     The second number is the scale factor for the denominator
%        coefficients.
%
t1 = max(abs(b));
t2 = max(abs(a));
mbits = bits - 1;
bscale = fix((2^(mbits) - 1)/t1);
```

```
ascale = ((2^(mbits) - 1)/t2);
% Fix scale factor for a coefficients to be a power of 2.
ka = log2(ascale);
ka = fix(ka);
ascale = 2^ka;
scfac(1,1) = bscale;
scfac(1,2) = ascale;
len1 = length(b); len2 = length(a);
len = max(len1,len2);
if (len1 < len2)
   ldif = len2 - len1;
   b = [b zeros(1,ldif)];
   end
if (len2 < len1)
    ldif = len1 - len2;
    a = [a zeros(1,ldif)];
    end
bsc = round(bscale*b);
asc = round(ascale*a); return
```

D.4 OUTPUT OVERFLOW SCALING OF IIR FILTERS

The following Matlab script can be used to scale IIR digital filters using the output overflow criteria.

Matlab Script D.4.

```
function [bsc, asc, scfac] = iirscale2(b, a, ibits, obits)
%[bsc, asc, scfac] = iirscale2(b,a, ibits, obits);
%
%  Script to scale IIR filters using the output
%       overflow criteria
%  Inputs
%  b - The floating point numerator filter coefficients
%  a - The floating point denominator filter coefficients
%  ibits - the number of bits to use for the scaled input
%  obits - the number of bits to use for the scaled output
%
%  Outputs
%  bsc - the scaled numerator coefficients
%  asc - the scaled denominator coefficients
%  scfac - array of scale factors for the scaled coeffients.
%     The first number is the scale factor for the numerator
%         coefficients.
%     The second number is the scale factor for the denominator
%         coefficients.
%     The third number is the log2 of the scale factor for the
%         denominator coefficients
%
m1 = ibits - 1;
xmax = 2^(m1) - 1;
```

```
m2 = obits - 1;
ymax = 2^(m2) - 1;
t1 = sum(abs(b));
[n1,n2] = size(a);
t2 = sum(abs(a(1,2:n2)));
t3 = t1+t2;
s2 = ymax/(xmax*t3);
m3 = fix(log2(s2));
s3 = 2^m3;
scfac = [s2 s3 m3];
len1 = length(b);
len2 = length(a);
len = max(len1,len2);
if (len1 < len2)
   ldif = len2 - len1;
   b = [b zeros(1,ldif)];
   end
if (len2 < len1)
    ldif = len1 - len2;
    a = [a zeros(1,ldif)];
    end
bsc = round(s2*b); asc = round(s3*a); return
```

D.5 SCALING THE STATE SPACE MODEL

The following Matlab script can be used to scale the matrices for the state space representation.

Matlab Script D.5.

```
function [Asc, Bsc, Csc, Dsc, s3] = ss_scale(A, B, C, D, ibits,
obits)
% [Asc, Bsc, Csc, Dsc] = ss_scale(A, B, C, D, ibits, obits);
%  This script is used to scale the filter coefficients for the
%       state space representation with coefficient matrices
%       (A, B, C, D) to obtain the scaled integer coefficient
%       matrices (Asc, Bsc, Csc, Dsc).  The state
%       variables should be divided by 2^m prior to implementing
%       the output equation.
%       s3 is the feedback scale factor for the state variables.

mat1 = [A B];
t1 = norm(mat1, inf);        % compute the infinity norm
mat2 = [C D];
t2 = max(sum(abs(transpose(mat2))));
% The norm does not work properly if
%     mat2 is a row vector.
r1 = 2^(ibits - 1) - 1;
r2 = 2^(obits -1) - 1;
m = fix(log2(r2/(r1*t1)));
s3 = 2^m;
```

```
s2 = r2/(t2*r1);
Asc = round(s3*A);
Bsc = round(s3*B);
Csc = round(s2*C);
Dsc = round(s2*D);
return
```

D.6 SCALING IIR SOS IN DIRECT FORM II

The following Matlab function can be used to partition an arbitrary order IIR filter into second order section and scale the second order sections for implementation using fixed point arithmetic.

Matlab Script D.6.

```
function [sos, gain, scsos, scfac]=iirsos2(b, a, ibits, obits)
%   [sos, gain, scsos, scfac] = iirsos2(b, a, ibits, obits);
[sos, gain] = tf2sos(b, a);
[n1, n2] = size(sos);
xmax = 2^(ibits-1) - 1;
ymax = 2^(obits-1) - 1;
m1 = obits - 1;
m2 = 2*(ibits-1);
c1 = (2^(m1) -2^(m2))/2^(ibits);
c2 = ymax/xmax;
for k1=1:n1
    bk = sos(k1, 1:3);
    ak = sos(k1, 5:6);
    tmp = c1/sum(abs(ak));
    mk = fix(log2(tmp));
    s3 = 2^(mk);
    scsos(k1, 5:6) = round(s3*ak);
    scsos(k1, 4) = s3;
    p1 = c2/sum(abs(bk));
    p2 = xmax/(max(abs(bk)));
    s2 = min(p1, p2);
    scsos(k1, 1:3) = round(s2*bk);
    scfac(k1, 1) = s2;
    scfac(k1, 2) = s3;
end
return
```

D.7 THE FIXED POINT STATE SPACE MODEL

The following Matlab function can be used to implement the scaled state space filter.

Matlab Script D.7.

```
function y = issfilt(A, B, C, D, s3, x)
% y = issfilt(A, B, C, D, x);
```

```
% This script filters the input x using the filter
%    with scaled, integer filter coefficients in the state
%     space format with coefficient matrices (A, B, C, D).
%    s3 is the feedback scale factor for the state variables.
[m1, m2] = size(A);
m3 = length(x);
Qm1 = zeros(m1, 1);
s3inv = 1/s3;
for k=1:m3
    xin = fix(x(1,k));          % ensure inputs are integers
    Q = A*Qm1 + B*xin;
    y(1,k) = C*Qm1 + D*xin;
    Qm1 = fix(s3inv*Q);
end return
```

Appendix E: Matlab Code for Chapter 7

E.1 POLYPHASE FILTER WITH DOWNSAMPLING

The following Matlab function can be used to implement an arbitrary level polyphase FIR filter with downsampling by a rate of m.

Matlab Script E.1.

```
function y = polydownfil(b, m, x)
% y = polydownfil(b, m, x);
%
%    This function implements an m level polyphase FIR filter
%         with a downsampling rate of 1:m. The first output sample,
%         with downsampling by a ratio of 1:m, is sample m.
%
%    Inputs:
%    b - FIR filter coefficients
%    m - downsampling rate (same as level of polyphase filter)
%    x - input sequence
%    Output:
%    y - output sequence

a = 1;
n1 = length(b);
k1 = mod(n1, m);
k2 = 0;
if (k1~=0)
    k2 = m-k1;
    n1 = n1 + k2;
end
b2 = [b zeros(1, k2)];

for k1 = 1:m
    bnow = b2(1, k1:m:end);
    k2 = k1 - 1;
    k3 = m - k1;
    r = [zeros(1,k2) x zeros(1, k3)];
    xnow = r(1, m:m:end);
    s(k1,:) = conv(bnow, xnow);
end
y = sum(s);
return
```

Digital Signal Processing. DOI: 10.1016/B978-0-12-804547-3.00023-1

E.2 POLYPHASE FILTER WITH UPSAMPLING

The following Matlab function can be used to implement an arbitrary level polyphase FIR filter with upsampling by a rate of m.

Matlab Script E.2.

```
function y = polyupfil(b, rate, x)
% y = polyupfil(b, rate, x);
%
%   This function implements a polyphase FIR filter with
%       upsampling by a factor of rate
%
%  Inputs:
%   b - FIR filter coefficients
%   rate - upsampling rate (same as level of polyphase filter)
%   x - input sequence
%  Output:
%   y - output sequence

n1 = length(b);
m1 = ceil(n1/rate);
m2 = length(x) + m1 - 1;
n2 = rate*m2 + rate-1;  % length of the convolutions
y = zeros(1, n2);
for k=1:rate;
    c = b(1,k:rate:end);
    g = conv(c, x);
    if (length(g) < m2)
        g = [g 0];
    end
    kend = n2 - rate + 1;
    y(1, k:rate:kend) = g;
end
return
```

Appendix F: Matlab Code for Chapter 8

F.1 MULTIPLY/ACCUMULATE CELLULAR ARRAY

The following Matlab function can be used to implement the FIR computational cell given in Fig. 8.24.

Matlab Script F.1.

```
function [rk, sk, rout] = firmac_cell(rkm1, skm1, bk, in)
%  [rk, sk, rout] = fir_mac(rkm1, skm1, bk, in);
%  Computational cell for an FIR cell using the multiplier
%      accumulator implementation using floating point
%      arithmetic
%
%  Inputs:
%      rkm1 - top left input to the cell
%      skm1 - bottom left input to the cell
%      bk - filter coefficient for this cell
%      in - 3x1 array of previous state variables
%  Outputs:
%      rout - 3x1 array of updated state variables
%      rk - upper right output from the cell
%      sk - lower right output from the cell
%
rout(1,1) = rkm1;
rout(2,1) = in(1,1);
rk = in(2,1);
sk = in(3,1);
rout(3,1) = bk*rkm1 + skm1;
return
```

F.2 MULTIPLY/ACCUMULATE FIR PIPELINE

The following Matlab function can be used to implement an arbitrary order FIR filter.

Matlab Script F.2.

```
function y = firmac_pipe(b, x);
%  y = firmac_pipe(b, x);
%
```

Digital Signal Processing. DOI: 10.1016/B978-0-12-804547-3.00024-3

```
%   This routine is used to implement a pipeline FIR filter
%       using the firmac pipeline computational cell.
%   The latency for this function is equal to the number of
%       filter coefficients.
%
%   Inputs:
%       b - FIR filter coefficients
%       x - interleaved input sequence
%
[m1, m2] = size(b);
x2 = [x zeros(1,m2)];        % Add zeros for pipeline delay
m3 = length(x2);
reg = zeros(3,2,m2);         % Array to store state variables
for k3=1:m3
    for k2=1:m2
        if(k2 == 1)
            rkm1 = x2(1,k3);
            skm1 = 0.0;
        else
            rkm1 = rk;
            skm1 = sk;
        end
        in = reg(:,2,k2);
        bk = b(1, k2);
        [rk, sk, rout] = firmac_cell(rkm1, skm1, bk, in);
        reg(:,1,k2) = rout;
    end
    y(1,k3) = sk;
    reg(:,2,:) = reg(:,1,:);     % update delays for next input
end
return
```

F.3 A SECOND INPUT SAMPLE SEQUENCE

The following Matlab function can be used to generate another input sample sequence
to use for modeling and simulation of digital filters.

Matlab Script F.3.

```
function y = sampdata2
% y = sampdata2;
%
%   This function returns a test sample data set to be used for
%       signal processing problems. The sequence is composed
%       of three sampled sinusoidal signals as follows:
%
%   x = 5*cos(0.2*pi*k + 0.5*pi) + 0.2*cos(0.75*pi*k + 0.5*pi)
%       + 0.1*cos(0.92*pi*k + 0.5*pi);
%
%   where k is the sample index.
%
```

```
%   A Hanning window is then multiplied by x to obtain y;
%
len = 256;   % Data is the same length as that from sampdata.
for k=1:len
    m = (k-1);
    t1 = 5*cos(0.1*pi*m + 0.5*pi);
    t2 = 2*cos(0.75*pi*m + 0.5*pi);
    t3 = 1*cos(0.92*pi*m + 0.5*pi);
    x(k) = t1 + t2 + t3;
    end

win = transpose(hanning(len));
y = win.*x;
return
```

F.4 MULTIPLY/ADD COMPUTATIONAL ARRAY

The following Matlab function can be used to implement the FIR computational cell given in Fig. 8.30.

Matlab Script F.4.

```
function [rk, sk, rout] = firmadd_cell(rk1, sk1, bk, in)
%  [rk, sk, rout] = firmadd_cell(rk1, sk1, bk, in);
%  Computational cell for an FIR cell using the multiply
%      add implementation and floating point arithmetic
%
%  Inputs:
%       rk1 - top left input to the cell
%       sk1 - bottom left input to the cell
%       bk - filter coefficient for this cell
%       in - 3x1 array of previous state variables
%  Outputs:
%       rout - 3x1 array of updated state variables
%       rk - upper right output from the cell
%       sk - lower right output from the cell
%
rout(1,1) = rk1;
rout(2,1) = sk1;
rout(3,1) = bk*rk1 + in(2,1);
rk = in(1,1);
sk = in(3,1);
return
```

F.5 MULTIPLY/ADD FIR PIPELINE

The following Matlab function can be used to implement an FIR filter using the *firmadd_cell* Matlab function.

Matlab Script F.5.

```
function y = firmadd_pipe(b, x);
%  y = firmadd_pipe(b, x);
%
%  This function is used to implement a pipeline FIR filter
%      using the firmac_cell computational cell.
%  The latency for this function is equal to the number of
%      filter coefficients.
%
%  Inputs:
%      b - FIR filter coefficients
%      x - interleaved input sequence
%
[m1, m2] = size(b);
x2 = [x zeros(1,m2)];        % Add zeros for pipeline delay
m3 = length(x2);
reg = zeros(3,2,m2);         % Array to store state variables
for k3=1:m3
    for k2=m2:-1:1
      if(k2 == m2)
        rk1 = x2(1,k3);
        sk1 = 0.0;
      else
        rk1 = rk;
        sk1 = sk;
      end
      in = reg(:,2,k2);
      bk = b(1, k2);
      [rk, sk, rout] = firmadd_cell(rk1, sk1, bk, in);
      reg(:,1,k2) = rout;
    end
    y(1,k3) = sk;
    reg(:,2,:) = reg(:,1,:);    % update delays for next input
end
return
```

F.6 INTERLEAVE FIR FILTER CELLULAR ARRAY

The following Matlab function can be used to implement the computational cell for an interleaved FIR filter as given in Fig. 8.37.

Matlab Script F.6.

```
function [rk1, sk, rout] = firintlv_cell(rk, sk1, bk, in)
%  [rk1, sk, rout] = firintlv_cell(rk, sk1, bk, in);
%  Regular computational cell for an interleave FIR filter
%
%  Inputs:
%      rk - top left input to the cell
%      sk1 - lower right input to the cell
```

```
%         bk - filter coefficient for this cell
%         in - 3x1 array of previous state variables
%  Outputs:
%         rout - 3x1 array of updated state variables
%         rk1 - upper right output from the cell
%         sk - lower left output from the cell
%
rout(1,1) = rk;
rout(2,1) = rk;
rout(3,1) = bk*in(2,1) + sk1;
sk = in(3,1);
rk1 = in(1,1);
return
```

F.7 INTERLEAVE FIR FILTER PIPELINE

The following Matlab *fiintlv_pipe* function can be used to implement an FIR filter using the *fiintlv_cell* Matlab function.

Matlab Script F.7.

```
function y = fiintlv_pipe(b, x);
%  y = fiintlv_pipe(b, x);
%
%  This routine is used to implement a linear
%      phase FIR filter using the FIR interleave cellular array.
%
%  Inputs:
%      b - FIR filter coefficients
%      x - input sequence
%
%  Output:
%      y - output sequence
[m1, m2] = size(b);
reg = zeros(3,2,m2);        % Array to store state variables
m3 = length(x);
for k3=1:m3
    for k2=1:m2
        if(k2 == 1)
            rk = x(1,k3);
            sk1 = reg(3,1,2);
        elseif (k2==m2)
            rk = rk1;
            sk1 = 0.0;
        else
            rk = rk1;
            sk1 = reg(3,1,k2+1);
        end
        in = reg(:,2,k2);
        bk = b(1, k2);
        [rk1, sk, rout] = fiintlv_cell(rk, sk1, bk, in);
```

```
        reg(:,1,k2) = rout;
      end
    y(1,k3) = reg(3,1,1);
    reg(:,2,:) = reg(:,1,:);      % update delays for next input
end
return
```

F.8 LINEAR PHASE FIR CELLULAR ARRAY

The following Matlab function can be used to implement the regular computational
cell for the linear phase FIR filter as given in Fig. 8.48.

Matlab Script F.8.

```
function [rk1, sk1, gk, rout] = flin_rcell(rk, sk, gk1, bk, in)
%  [rk1, sk1, gk, rout] = flin_rcell(rk, sk, gk1, bk, in);
%  Regular computational cell for a linear phase FIR filter
%
%  Inputs:
%        rk - top left input to the cell
%        sk -  middle left input to the cell
%        gk1 - lower right input to the cell
%        bk - filter coefficient for this cell
%        in - 6x1 array of previous state variables
%  Outputs:
%        rout - 6x1 array of updated state variables
%        rk1 - upper right output from the cell
%        sk1 - middle right output from the cell
%        gk - lower left output from the cell
%
rout(1,1) = rk;
rout(2,1) = in(1,1);
rout(3,1) = in(2,1);
rout(4,1) = in(1,1) + gk1;
rout(5,1) = bk*in(4,1) + sk;
rout(6,1) = gk1;
rk1 = in(3,1);
sk1 = in(5,1);
gk = in(6,1);
return
```

F.9 COMPUTATIONAL CELL FOR THE LINEAR PHASE FIR FILTER

The following Matlab function can be used to implement the end computational cell
for the linear phase FIR filter as given in Fig. 8.50.

Matlab Script F.9.

```
function [rk1, sk1, gk, rout] = flin_ecell(rk, sk, gk1, bk, in)
%   [rk1, sk1, gk, rout] = flin_ecell(rk, sk, gk1, bk, in);
%   End computational cell for a linear phase FIR filter
%
%   Inputs:
%          rk - top left input to the cell
%          sk -  middle left input to the cell
%          gk1 - lower right input to the cell
%          bk - filter coefficient for this cell
%          in - 6x1 array of previous state variables
%   Outputs:
%          rout - 6x1 array of updated state variables
%          rk1 - upper right output from the cell
%          sk1 - middle right output from the cell
%          gk - lower left output from the cell
%
rout(1,1) = rk;
rout(2,1) = in(1,1);
rout(3,1) = in(2,1);
rout(4,1) = in(1,1);
rout(5,1) = bk*in(4,1) + sk;
rout(6,1) = in(1,1) + gk1;
rk1 = in(3,1);
sk1 = in(5,1);
gk = in(6,1);
return
```

F.10 IMPLEMENTATION OF THE FIR LINEAR PHASE FILTER

The following Matlab function can be used to implement an FIR filter using the *flin_regcell* and the *flin_endcell* Matlab functions.

Matlab Script F.10.

```
function y = flin_pipe(b, x);
%   y = flin_pipe(b, x);
%
%   This function is used to implement a linear phase FIR filter
%        using the flin_rcell and flin_ecell functions
%        for a linear phase cellular array.
%
%   Inputs:
%          b - FIR filter coefficients
%          x - input sequence
%
[m1, m2] = size(b);
n1 = mod(m2,2);
```

```
if (n1 == 0)
    fpintf(1, 'The filter order must be even.\n');
    return
end
n2 = fix(0.5*m2);
n3 = n2+1;
x2 = [x zeros(1,m2)];          % Add zeros for pipeline delay
m3 = length(x2);
reg = zeros(6,2,n3);           % Array to store state variables
for k3=1:m3
    for k2=1:n2
        if(k2 == 1)
            rk = x2(1,k3);
            sk = 0.0;
        else
            rk = rk1;
            sk = sk1;
        end
        gk1 = reg(6,2,(k2+1));
        in = reg(:,2,k2);
        bk = b(1, k2);
        [rk1, sk1, gk, rout] = flin_rcell(rk, sk, gk1, bk, in);
        reg(:,1,k2) = rout;
    end
    % The end cell
    rk = rk1;
    sk = sk1;
    gk1 = 0.0;
    bk = b(1,n3);
    in = reg(:, 2, n3);
    [rk1, sk1, gk, rout] = flin_ecell(rk, sk, gk1, bk, in);
    y(1,k3) = sk1;
    reg(:,1,n3) = rout;
    reg(:,2,:) = reg(:,1,:);     % update delays for next input
end
return
```

Appendix G: Matlab Code for Chapter 9

G.1 FIR INTERLEAVE COMPUTATIONAL CELL

The following Matlab function implements a computational cell for a FIR filter using the interleaving of two inputs.

Matlab Script G.1.

```
function [sk, rkp1, qn] = fir3cell(rk, skp1, bk, qnm1)
%   [sk, uk, qn] = fir3cell(rk, vk, bk, qnm1);
%   Computational cell for FIR digital filter with
%   interleaved inputs
%
%   Inputs:
%       rk - upper left input to the cell
%       skp1 - lower right input to the cell
%       bk - filter coefficient for this cell
%       qnm1 - 3x1 array of previous state variables
%   Outputs:
%       sk - lower left output from the cell
%       rkp1 - upper right output from the cell
%       qn - 3x1 array of updated current state variables
%
qn(1,1) = rk;
qn(2,1) = bk*qnm1(1,1) + skp1;
sk = qnm1(2,1);
rkp1 = qnm1(1,1);
return
end
```

G.2 FIR INTERLEAVE ARRAY

The following Matlab function can be used to implement a computational array using the *fir3cell* function in Appendix G.1.

Matlab Script G.2.

```
function y = fir3array(b, x);
%   y = fir3array(b, x);
%
%   This function can be used to implement a pipeline FIR
%       filter using the fir3cell pipeline computational
```

Digital Signal Processing. DOI: 10.1016/B978-0-12-804547-3.00025-5

```
%       cell (systolic array cell).
%
%   Inputs:
%       b - FIR filter coefficients
%       x - interleaved input sequence
% Initialization block
[m1, m2] = size(b);
m3 = length(x);
regn = zeros(2, m2);       % Array to store state variables
regnm1 = zeros(2, m2);
r = zeros(1, m2);
s = zeros(1, m2+1);
u = zeros(1, m2);
v = zeros(1, m2);
% Computational block
for k3=1:m3
    for k2=m2:-1:1
        if(k2 == 1)        % Input/Output cell
            rk = x(1,k3);
            skp1 = regnm1(2,2);
        elseif (k2 == m2) % last cell
            k2m1 = k2 - 1;
            rk = regnm1(1, k2m1);
            skp1 = 0;
        else
            k2m1 = k2 - 1;
            k2p1 = k2 + 1;
            rk = regnm1(1, k2m1);
            skp1 = regnm1(2, k2p1);
        end
        qvar = regnm1(:, k2);
        bk = b(1, k2);
        [sk, rkp1, qout] = fir3cell(rk, skp1, bk, qvar);
        regn(:,k2) = qout;
        s(1, k2) = sk;
        u(1, k2) = rkp1;
    end
    y(1,k3) = regnm1(2,1);
    regnm1 = regn;        % update delays for next input
end
return
```

G.3 FIR INTERLEAVE ARRAY TEST

The following Matlab script can be used to test the *fir3cell* function in Appendix G.1 and the *fir3array* function in Appendix G.2.

Matlab Script G.3.

```
% testfir3
%
```

```
% This script can be used to test the fir3array
%      computational array
%
clear
close all
order = 24;
wc = 0.317;
b = fir1(order, wc);
a = 1;
xa = sampdata;
xb = sampdata2;
m1 = length(xa);
len = length(b);
n1 = 2*len - 1;            % compute length of the delay
x1 = [xa zeros(1, 2*n1)];  % add zeros for convolution
x2 = [xb zeros(1, 2*n1)];  %    and for pipeline delay
n4 = length(x1);
x3 = intleav(x1, x2);
g1 = filter(b, a, x1);          % filter
g2 = filter(b, a, x2);
m2 = fix(0.5*(len-1));
m3 = m2 + m1 - 1;
g3 = g1(1,m2:m3);
g4 = g2(1,m2:m3);
y1 = fir3array(b, x3);
[y3, y4] = deintleav(y1);
m5 = fix(0.5*(len-1)) + 1;
m6 = m5 + m1 - 1;
y5 = y3(1, m5:m6);
y6 = y4(1, m5:m6);
num = 1;
figure(num)
subplot(2,1,1); stem(g3)
title('Output from Matlab filter function');
subplot(2,1,2); stem(y5);
title('Output 1 from fir3cell pipeline function');
print -deps fir3a.eps

num = gcf + 1;
figure(num)
subplot(2,1,1); stem(g4)
title('Output from Matlab filter function');
subplot(2,1,2); stem(y6);
title('Output 2 from fir3cell pipeline function');
print -deps fir3b.eps

[err1, mea1, mea2] = ncompare(g3, y5);
print -deps fir3c.eps

[err2, mea3, mea4] = ncompare(g4, y6);
print -deps fir3d.eps
```

G.4 FIXED POINT FIR INTERLEAVE CELL

The following Matlab function can be used to implement the Interleave FIR filter computational cell using fixed point arithmetic. The inputs and outputs are assumed to be 16 bit fixed point numbers.

Matlab Script G.4.

```
function [sk, rkp1, qn] = dfir3cell(rk, skp1, bk, qnm1)
%  [sk, rkp1, qn] = dfir3cell(rk, skp1, bk, qnm1);
%  Computational cell for FIR digital filter with interleaved
%     inputs and fixed point arithmetic
%
% All inputs are 16 bit fixed point
%
%  Inputs:
%     rk - upper left input to the cell
%     skp1 - lower right input to the cell
%     bk - filter coefficient for cell
%     qnm1 - 2x1 array of previous state variables
%  Outputs:
%     sk - lower left output from the cell
%     rkp1 - upper right output from the cell
%     qn - 2x1 array of updated current state variables
%
lshift = 2^(16);
rshift = 2^(-16);
wordsize = 32;
mask = 2^(32) - 1;
qn(1,1) = rk;
tmp1 = bk*qnm1(1,1);
tmp2 = tmp1 + lshift*skp1;
qn(2,1) = fix(tmp2*rshift);
sk = qnm1(2,1);
rkp1 = qnm1(1,1);
return
end
```

G.5 FIXED POINT FIR INTERLEAVE ARRAY

The following Matlab function can be used to implement a computational array using the *dfir3cell* function in Appendix G.4.

Matlab Script G.5.

```
function y = dfir3array(b, x);
%  y = dfir3array(b, x);
%
%  This function can be used to implement a pipeline FIR
%     filter using the fir3cell pipeline computational
```

```
%       cell and fixed point arithmetic.
%
%       All inputs are 16 bit fixed point
%
%  Inputs:
%      b - FIR filter coefficients
%      x - interleaved input sequence
%
%  Output:
%      y - interleaved output (32 bit fixed point)
%
[m1, m2] = size(b);
m3 = length(x);
regn = zeros(2, m2);     % Array to store state variables
regnm1 = zeros(2, m2);
for k3=1:m3
    for k2=m2:-1:1
        if(k2 == 1)      % Input/Output cell
            rk = x(1,k3);
            vk = regnm1(2,2);
        elseif (k2 == m2) % last cell
            k2m1 = k2 - 1;
            rk = regnm1(1, k2m1);
            vk = 0;
        else
            k2m1 = k2 - 1;
            k2p1 = k2 + 1;
            rk = regnm1(1, k2m1);
            vk = regnm1(2, k2p1);
        end
        qnm1 = regnm1(:, k2);
        bk = b(1, k2);
        [sk, uk, qn] = dfir3cell(rk, vk, bk, qnm1);
        regn(:,k2) = qn;
    end
    y(1,k3) = regnm1(2,1);  % store the output
    regnm1 = regn;     % update delays for next input
end
return
```

G.6 FIXED POINT FIR INTERLEAVE TEST

The following Matlab script can be used to test the *dfir3cell* function in Appendix G.4 and the *dfir3array* function in Appendix G.5.

Matlab Script G.6.

```
% testdfir3
%
% This function is used to test the dfir3cell fixed point
%   computational array
```

```
%
clear
close all
%%  Design filter and scale filter coefficients
ibits = 16;
obits = 32;
order = 24;
wc = 0.317;
b = fir1(order, wc);
[bsc, scfac] = firscale2(b, ibits, obits);
a = 1;

%% Get input sequences and interleave them.
xa = sampdata;
[xasc, asf] = fscale(xa, ibits);
xb = sampdata2;
[xbsc, bsf] = fscale(xb, ibits);
m1 = length(xasc);
len = length(bsc);
n1 = 2*len - 1;              % compute length for delay
x1 = [xasc zeros(1, 2*n1)]; % increase length for pipeline
num = 1;
figure(num)
stem(x1)
title('Scaled input from sampdata');
xlabel('Sample Number');
print -deps dfir3a.eps
pause(4)

x2 = [xbsc zeros(1, 2*n1)];
stem(x2)
title('Scaled input from sampdata2');
xlabel('Sample Number');
print -deps dfir3b.eps
pause(4)

%% Compute Outputs using conv
n4 = length(x1);
x3 = intleav(x1, x2);
g1 = filter(b, a, x1);          % filter
g2 = filter(b, a, x2);
m2 = fix(0.5*(len-1));
m3 = m2 + m1 - 1;
g3 = g1(1,m2:m3);
g4 = g2(1,m2:m3);

%% Compute output from Computational Array
y1 = dfir3array(bsc, x3);
[y3, y4] = deintleav(y1);
m5 = fix(0.5*(len-1)) + 1;
m6 = m5 + m1 - 1;
y5 = y3(1, m5:m6);
```

```
y6 = y4(1, m5:m6);

%% Compare the outputs
num = gcf + 1;
figure(num)
subplot(2,1,1); stem(g3)
title('Output 1 from Matlab filter function');
subplot(2,1,2); stem(y5);
title('Output 1 from dfir3cell pipeline function');
print -deps dfir3c.eps
pause(4);

num = gcf + 1;
figure(num)
subplot(2,1,1); stem(g4)
title('Output 2 from Matlab filter function');
subplot(2,1,2); stem(y6);
title('Output 2 from dfir3cell pipeline function');
print -deps dfir3d.eps
pause(4);
[err1, mea1, mea2] = ncompare(g3, y5);
print -deps dfir3c.eps
pause(4)

[err2, mea3, mea4] = ncompare(g4, y6);
print -deps dfir3e.eps
```

G.7 TWO'S COMPLEMENT FIR INTERLEAVE CELL

The following Matlab function can be used to implement the Interleave FIR filter computational cell using two's complement arithmetic. The inputs and outputs are assumed to be 16 bit two's complement numbers.

Matlab Script G.7.

```
function [sk, uk, qn] = tfir3cell(rk, vk, bk, qnm1)
%  [sk, uk, qn] = tfir3cell(rk, vk, bk, qnm1);
%  Computational cell for FIR digital filter with interleaved
%       inputs using two's complement fixed point arithmetic.
%
%  All inputs and outputs are 16 bit fixed point
%
%  Inputs:
%       rk - upper left input to the cell
%       vk - lower right input to the cell
%       bk - filter coefficient for this cell
%       qnm1 - 2x1 array of previous state variables
%  Outputs:
%       sk - lower left output from the cell
%       uk - upper right output from the cell
%       qn - 2x1 array of updated current state variables
```

```
%
lshift = 16;
rshift = -16;
wordsize = 32;
mask = 2^(32) - 1;
qn(1,1) = rk;
tmp1 = mult16(bk, qnm1(1,1));
tmp2 = tmp1 + bitshift(vk, lshift, wordsize);
tmp3 = bitand(mask, tmp2);
qn(2,1) = fix(bitshift(tmp3, rshift));
sk = qnm1(2,1);
uk = qnm1(1,1);
return
end
```

G.8 TWO'S COMPLEMENT FIR INTERLEAVE ARRAY

The following Matlab function can be used to implement an arbitrary order FIR filter using the *tfir3cell* function in Appendix G.7.

Matlab Script G.8.

```
function y = tfir3array(b, x);
%  y = tfir3array(b, x);
%
%  This function is used to implement a pipeline FIR filter
%      using the tfir3cell computational cell.
%
%  All inputs and outputs are 16 bit fixed point
%
%  Inputs:
%      b - scaled FIR filter coefficients
%      x - interleaved input sequence
%
%  Output:
%      y - the output sequence

[m1, m2] = size(b);
m3 = length(x);
regn = zeros(2, m2);          % Array to store state variables
regnm1 = zeros(2, m2);
r = zeros(1, m2);
s = zeros(1, m2+1);
u = zeros(1, m2);
v = zeros(1, m2);
for k3=1:m3
    for k2=m2:-1:1
        if(k2 == 1)      % Input/Output cell
            rk = x(1,k3);
            vk = regnm1(2,2);
        elseif (k2 == m2) % last cell
```

```
            k2m1 = k2 - 1;
            rk = regnm1(1, k2m1);
            vk = 0;
        else
            k2m1 = k2 - 1;
            k2p1 = k2 + 1;
            rk = regnm1(1, k2m1);
            vk = regnm1(2, k2p1);
        end
        qnm1 = regnm1(:, k2);
        bk = b(1, k2);
        [sk, uk, qn] = tfir3cell(rk, vk, bk, qnm1);
        regn(:,k2) = qn;
        s(1, k2) = sk;
        u(1, k2) = uk;
    end
    y(1,k3) = regnm1(2,1);
    regnm1 = regn;     % update delays for next input
end
return
```

G.9 TWO'S COMPLEMENT FIR INTERLEAVE ARRAY TEST

The following Matlab script can be used to test the *tfir3cell* function in Appendix G.7 and the *tfir3array* function in Appendix G.8.

Matlab Script G.9.

```
%% testtfir3
%
% This function can be used to test the tfir3cell in
%     a computational array through a call to the
%     tfir3array function
%
clear
close all
%% Design the filter
cbits = 16;
ibits = 16;
obits = 32;
order = 24;
wc = 0.317;
b = fir1(order, wc);
[bsc, scfac] = firscale2(b, ibits, obits);
btwx = twoscomp(bsc, cbits);
a = 1;

%% Obtain the two sequences, scale them, convert them
%    to two's complement numbers and interleave them
xa = sampdata;
[xasc, asf] = fscale(xa, ibits);
```

```
xatwx = twoscomp(xasc, ibits);
num = 1;
figure(num)
stem(xatwx);
xlabel('Sample Number');
title('Twos Complement of Sequence 1');
print -deps '../../eps/tfir3a.eps';
pause (4);

xb = sampdata2;
[xbsc, bsf] = fscale(xb, ibits);
xbtwx = twoscomp(xbsc, ibits);
num = gcf + 1;
figure(num)
stem(xbtwx);
xlabel('Sample Number');
title('Twos Complement of Sequence 2');
print -deps '../../eps/tfir3b.eps';
pause (4);

m1 = length(xatwx);
len = length(bsc);
n1 = 2*len - 1;               % compute length for the delay
x1 = [xatwx zeros(1, 2*n1)];  % increase length for pipeline
x2 = [xbtwx zeros(1, 2*n1)];
n4 = length(x1);
x3 = intleav(x1, x2);
num = gcf + 1;
figure(num)
stem(x3);
xlabel('Sample Number');
title('Twos Complement of the Input Sequence');
print -deps '../../eps/tfir3c.eps';
pause (4);

%% Compute the outputs using conv
g1 = conv(b, xasc);          % filter
g2 = conv(b, xbsc);
m2 = fix(0.5*(len-1));
m3 = m2 + m1 - 1;
g3 = g1(1,m2:m3);
g4 = g2(1,m2:m3);

%% Compute the outputs using the Computational Array
y1twx = tfir3array(btwx, x3);
y1 = invtwoscomp(y1twx, ibits);
[y3, y4] = deintleav(y1);
m5 = fix(0.5*(len-1)) + 1;
m6 = m5 + m1 - 1;
y5 = y3(1, m5:m6);
y6 = y4(1, m5:m6);
num = 1;
```

```
figure(num)
subplot(2,1,1); stem(g3)
title('Output from Matlab filter function');
subplot(2,1,2); stem(y5);
title('Output 1 from fir3cell pipeline function');
print -deps '../../eps/tfir3d.eps'
pause(4);

%%  Compare the outputs
num = gcf + 1;
figure(num)
subplot(2,1,1); stem(g4)
title('Output from Matlab filter function');
subplot(2,1,2); stem(y6);
title('Output 2 from fir3cell pipeline function');
print -deps '../../eps/tfir3e.eps'
pause(4);

[err1, mea1, mea2] = ncompare(g3, y5);
print -deps '../../tfir3f.eps'
pause(4)

[err2, mea3, mea4] = ncompare(g4, y6);
print -deps dfir3d.eps
print -deps '../../tfir3g.eps'
```

G.10 TWO'S COMPLEMENT OF INTEGER NUMBERS

The following Matlab function converts the input integer array x to the corresponding two's complement integer array using m bits for the word size.

Matlab Script G.10.

```
function twx = twoscomp(x, m)
%   twx = twoscomp(x, m);
%
%   This function converts the input integer array x to
%       the corresponding two's complement number with
%       m bits.
%
bitmax = 2^m;
f = fix(x);   % In case the input is floating point
twx = f;
[m1, m2] = size(f);
for k1 = 1:m1,
    for k2=1:m2,
        if (f(k1,k2) < 0)
            twx(k1,k2) = bitmax - abs(f(k1,k2));
        end
        end
    end
return
```

G.11 INVERSE TWO'S COMPLEMENT

The following Matlab function converts the two's complement input array x with m bits per word to the corresponding signed integer array.

Matlab Script G.11.

```
function y = invtwoscomp(twx, m)
%  y = invtwoscomp(twx, m);
%
%  This function converts the two's complement input array x to
%     the corresponding signed integer array number with the
%     two's complement numbers being in m bits.
%
bitmax = 2^m;
mm1 = m - 1;
top = 2^(mm1) - 1;
f = fix(twx);   %  In case the input is actually floating point
y = f;
[m1, m2] = size(f);
for k1 = 1:m1,
    for k2=1:m2,
        if (f(k1,k2) > top)
            tmp = bitmax - abs(f(k1,k2));
            y(k1,k2) = -tmp;
        end
    end
end
return
```

G.12 TWO'S COMPLEMENT MULTIPLICATION

The following Matlab function multiplies the two inputs a and b to obtain the output c. Both a and b are assumed to be 16 bit numbers in two's complement format. The output c is returned as a 32 bit two's complement number.

Matlab Script G.12.

```
function c = mult16(b, a)
%  c = mult16(b, a);
%
%  This function multiplies the two inputs a and b
%      to obtain the output c.  Both a and b are
%      assumed to be 16 bit numbers in two's
%      complement format.  The output c is returned
%      as a 32 bit two's complement number.
ibitmax = 2^(16);
obitmax = 2^(32);
s1 = bitget(a, 16);
s2 = bitget(b, 16);
```

```
if (s1 == 1)          % a is negative
    reg1 = invtwoscomp(a, 16);
elseif (s1 == 0)
    reg1 = a;
end
if (s2 == 1)
    reg2 = invtwoscomp(b, 16);
elseif (s2 == 0)
    reg2 = b;
end
reg3 = reg1*reg2;
c = twoscomp(reg3, 32);
return
```

Bibliography

1. L. Tan, J. Jiang, Digital Singal Processing: Fundamentals and Applications, second edn., Elsevier Inc., Burlington, MA, 2013, p. 01803.
2. A.V. Oppenheim, R.W. Shafer, Discrete-Time Signal Processing, third ed., Prentice-Hall, 2010.
3. J.H. McClellan, R.W. Schafer, M.A. Yoder, Signal Processing First, Pearson Prentice Hall, Inc., Upper Saddle River, NJ, 2003.
4. J.G. Proakis, D.G. Manolakis, Digital Signal Processing: Principles, Algorithms and Applications, fourth edn., Pearson Prentice Hall, Upper Saddle River, NJ 07458, 2007.
5. J.G. Proakis, Digital Communications, fourth edn., McGraw Hill, New York, NY, 2001.
6. S. Haykin, B.V. Veen, Signals and Systems, John Wiley & Sons, Inc., New York, NY, 1999.
7. J.S. Lim, Two-Dimensional Signal and Image Processing, Prentice Hall, Englewood Cliffs, New Jersey, 1990.
8. S.K. Mitra, Digital Signal Processing: A Computer-Based Approach, third edn., McGraw Hill, New York, NY, 2006.
9. L. Tan, Digital Singal Processing: Fundamentals and Applications, Elsevier Inc., Burlington, MA 01803, 2008.
10. T.W. Parks, J.H. McClellan, Chebyshev approximation for nonrecursive digital filters with linear phase, in: IEEE Transactions on Circuit Theory, vol. CT-19, 1972, pp. 189–194.
11. J.F. Kaiser, Nonrecursive filter design using the $i_0 - \text{sinh}$ window function, in: Proceedings of the 1974 IEEE International Symposium on Circuits and Systems, vol. ASSP-25, 1974, pp. 415–422.
12. A.D. Poularikas, S. Seely, Signals and Systems, PWS Publishers, 1985.
13. K. Ogata, State Space Analysis of Control Systems, Prentice-Hall, Inc., Englewood Cliffs, NJ, 1967.
14. Microprocessor Standards Committee of the IEEE Computer Society, IEEE standard for binary floating-point arithmetic, Tech. Rep. Standard 754-1985, The Institute of Electrical and Electronic Engineers, Inc., 1985.
15. M. Vetterli, J. Kovacevic, Wavelets and Subband Coding, Prentice-Hall, 1995.
16. D. Le Gall, A. Tabatabai, Sub-band coding of digital images using symmetric short kernel filters and arithmetic coding techniques, in: International Conference on Acoustics, Speech, and Signal Processing, 1988. ICASSP-88, IEEE, 1988, pp. 761–764.
17. S.Y. Kung, VLSI Array Processors, Prentice Hall, Englewood Cliffs, New Jersey, 1988.
18. E.C. Ifeachor, B.W. Jervis, Digital Signal Processing: A Practical Approach, second edn., Prentice Hall, 2002.
19. U. Meyer-Baese, Digital Signal Processing with Field Programmable Gate Arrays, third edn., Springer, 2007.
20. National Instruments, Introduction to FPGA technology: top five benefits, Tech. rep., National Instruments, December 2008.
21. Xilinx, 7 series FPGAs overview, Tech. rep., Xilinx, May 2015.
22. W. Sung, S.K. Mitra, B. Jeren, Multiprocessor implementation of digital filtering algorithms using a parallel block processing model, IEEE Transactions on Parallel and Distributed Systems 3 (1) (1992).
23. T.P. Barnwell III, V.K. Madisetti, S.J.A. McGrath, The Georgia Tech digital signal multiprocessor, IEEE Transactions on Signal Processing 41 (7) (1993).

Digital Signal Processing. DOI: 10.1016/B978-0-12-804547-3.00026-7

24. H.T. Kung, Why systolic architectures?, Computer 15 (1) (1982) 37–46.
25. S.Y. Kung, S.C. Lo, S.N. Jean, J.N. Hwang, Wavefront array processor – concept to implementation, Computer 20 (7) (1987) 18–33.
26. D.E. Heller, Decomposition of Recursive Filters for Linear Systolic Arrays, vol. 431, SPIE, 1000 20th Street, Bellingham WA, 1983, pp. 55–59.
27. M.Y. Dabbagh, Multiprocessor Implementation of Two-Dimensional Digital Filters for Real-Time Processing, PhD thesis, North Carolina State University, 1989.
28. R.A. Roberts, C.T. Mullis, Digital Signal Processing, Addison-Wesley, 75 Arlington Street, Suite 300, Boston, MA, 1987.
29. T. Laakso, I. Hartimo, Direct form revisted: recursive filter implementation using higher-order direct form sections, in: Proceedings Internat. Symp. Circuits and Systems, 1988, pp. 791–795.
30. D.A. Schwartz, T.P. Barnwell III, A graph theoretic technique for the generation of systolic implementation for shift-invariant flow graphs, in: Proceedings of the International Conference on Acoustics, Speech and Signal Processing, 1984, pp. 8.3.1–8.3.4.
31. H. Kimura, T. Osaka, Canonical pipelining of lattice filters, IEEE Transactions on Acoustics, Speech and Signal Processing ASSP-35 (1987) 878–887.
32. Y. Tanurhan, Processors and FPGAs quo vadis, Computer 39 (November 2006) 108–110.
33. D. Dudgeon, R. Mersereau, Multidimensional Digital Signal Processing, Prentice-Hall Inc., Englewood Cliffs, NJ, 1984.
34. J.H. Lee, Y.H. Yang, Two-dimensional non-symmetric half-plane recursive doubly complementary digital filters, Signal Processing 89 (10) (2009) 2027–2035.

Index

Symbols

2-D convolution, 548, 556
2-D DCT, 559
2-D DFT, 555
2-D frequency response, 552
2-D impulse response, 554

A

Aliasing, 49
All pole lattice structure, 309
Allpass, 308
ASCS, 460

B

Basic operations, 6
BIBO stability, 95
Bilinear transformation, 242
Block IIR digital filter, 509

C

Casual, 551
Causality, 551
Causality constraint, 207
Classification of signals, 27
Cut set retiming, 468

D

DCT, 189
Decimation, 131
Decimation in time FFT, 191
Decimator, 398
Deterministic signal, 3
Difference equation, 59
Digital resonator, 171
Direct form 1 IIR implementation, 531
Direct form II IIR realization, 506
Discrete Fourier series, 178
Discrete Fourier Transform, 179
Discrete time convolution, 54
Discrete time correlation, 67
Discrete time transformations, 248
Dynamic system, 34

E

Energy signal, 32
Even signal, 28
Exponential sequence, 25

F

Filter design using analog prototypes, 244
Filtering, 6
FIR cascade form, 280
FIR direct form, 278
FIR filter design using Matlab, 226
FIR filter structures, 484
Fir interleave array implementation, 519
FIR lattice structure, 286, 287
FIR linear phase form, 280
FIR transpose direct form, 279
Fixed point, 351
Floating point, 352
FPGA, 459
Frequency domain interpolation, 124
Frequency response, 159
Frequency transformations, 246

G

General purpose computer, 456

I

IIR all pole lattice structure, 310
IIR cascade form, 300
IIR direct form, 298
IIR filter design using Matlab, 252
IIR lattice structure, 311
IIR linear phase, 317
IIR parallel form, 305
IIR transpose direct form, 299
Impulse invariant design, 240
Impulse response, 81
Initial conditions, 111
Interleave FIR filter, 490
Interleaving, 477
Interpolation, 116
Interpolation using the FFT, 400
Interpolator, 397
Inverse Z Transform, 96

Digital Signal Processing. DOI: 10.1016/B978-0-12-804547-3.00027-9

Printed in the United States
By Bookmasters